INSECT RESISTANCE MANAGEMENT

INSECT RESISTANCE MANAGEMENT

Biology, Economics, and Prediction

Second Edition

Edited by

David W. Onstad

Insect Resistance Management Science,
DuPont Agricultural Biotechnology,
Wilmington, DE, USA

AMSTERDAM • BOSTON • HEIDELBERG • LONDON
NEW YORK • OXFORD • PARIS • SAN DIEGO
SAN FRANCISCO • SINGAPORE • SYDNEY • TOKYO
Academic Press is an imprint of Elsevier

Academic Press is an imprint of Elsevier
32 Jamestown Road, London NW1 7BY, UK
225 Wyman Street, Waltham, MA 02451, USA
525 B Street, Suite 1800, San Diego, CA 92101-4495, USA

Notice
No responsibility is assumed by the publisher for any injury and/or damage to persons
or property as a matter of products liability, negligence or otherwise, or from any use
or operation of any methods, products, instructions or ideas contained in the material
herein.
Because of rapid advances in the medical sciences, in particular, independent verification
of diagnoses and drug dosages should be made.

British Library Cataloguing-in-Publication Data
A catalogue record for this book is available from the British Library

Library of Congress Cataloging-in-Publication Data
A catalog record for this book is available from the Library of Congress

ISBN: 978-0-12-396955-2

For information on all Academic Press publications
visit our website at elsevierdirect.com

Typeset by MPS Limited, Chennai, India
www.adi-mps.com

Printed and bound in United States of America

14 15 16 17   10 9 8 7 6 5 4 3 2 1

DEDICATION

To
Dawn Dockter,
my partner in life and science for 25 years,
and Nora and Emma,
who are amazing

CONTENTS

LIST OF CONTRIBUTORS

Analiza P. Alves
Insect Resistance Management Science, DuPont Pioneer, Johnston, IA

Yves Carrière
Department of Entomology, University of Arizona, Tucson, AZ

John M. Clark
Department of Veterinary and Animal Sciences, University of Massachusetts, Amherst, MA

J. Lindsey Flexner
Insect Resistance Management Science, DuPont Pioneer, Wilmington, DE

Patrick Gaffney
Department of Statistics, University of Wisconsin, Madison, WI

Aaron J. Gassmann
Department of Entomology, Iowa State University, Ames, IA

Joseph E. Huesing
Office of Agricultural Research and Policy, Bureau for Food Security, U.S. Agency for International Development, USDA ARS OIRP, Washington, DC

Sarah A. Hughson
Department of Entomology, University of Illinois, Urbana, IL

Terrance M. Hurley
Department of Applied Economics, University of Minnesota, St. Paul, MN

Lisa M. Knolhoff
Genective, AgReliant Genetics, Champaign, IL

Si Hyeock Lee
Department of Agricultural Biotechnology, Seoul National University, Seoul, South Korea

Eli Levine
Illinois Natural History Survey, Prairie Research Institute, University of Illinois, Champaign, IL

Venu Madhav Margam
Regulatory Science, DuPont Knowledge Centre, Turkapally, Shameerpet Mandal, Ranga Reddy District, Hyderabad, Andhra Pradesh, India

Paul D. Mitchell
Department of Agricultural and Applied Economics, University of Wisconsin, Madison, WI

Mark E. Nelson
Insect Resistance Management Science, DuPont Pioneer, Wilmington, DE

Brett P. Olds
Department of Animal Biology, University of Illinois, Urbana, IL

David W. Onstad
Insect Resistance Management Science, DuPont Pioneer, Wilmington, DE

Barry R. Pittendrigh
Department of Entomology, University of Illinois, Urbana, IL

Anthony M. Shelton
Department of Entomology, Cornell University, Geneva, NY

Joseph L. Spencer
Illinois Natural History Survey, Prairie Research Institute, University of Illinois, Champaign, IL

Bruce H. Stanley
Insect Resistance Management Science, DuPont Pioneer, Wilmington, DE

Laura D. Steele
Department of Entomology, University of Illinois, Urbana, IL

Lijie Sun
Synthetic Biology and Bioenergy, J. Craig Venter Institute, San Diego, CA

Kent R. Walters, Jr.
Department of Entomology, University of Illinois, Urbana, IL

FOREWORD

Resistance to pesticides has been with us for at least 100 years now, since the recognition of resistance to lime sulfur in the San Jose scale in the state of Washington in 1908[1]. At least through the early 1950s, resistance was apparently seen by many entomologists as just a curiosity, and no real threat to the emerging field of integrated control (which later developed into integrated pest management, IPM).

For example, when I arrived at Mississippi State University in 1981, I was housed in the Clay Lyle Entomology Building. Lyle was a very prominent applied entomologist, and reportedly once advised fellow entomologists that they should collect all of the houseflies they would ever need for museum specimens because DDT and other insecticides would drive them extinct. With no disrespect intended for Lyle and so many others of his time, I studied strains of houseflies in that building, by 1983, that were resistant to every major class of insecticides at the time.

Houseflies were early responders to selection, starting in the 1940s, but by 1989 there were more than 500 species of insects, mites, or ticks that had evolved resistance to at least one pesticide somewhere in the world. At least 20 of these species evolve resistance so frequently that they remain a challenge to our attempts to manage them cost-effectively. These "resistance recidivists" include *Anopheles* mosquitoes, *Helicoverpa armigera* (a pest of many food crops in addition to being a major cotton bollworm), *Plutella xylsotella* (the diamondback moth, a major pest of crucifer crops), whiteflies such as *Bemisia tabaci*, the cockroach, *Blattella germanica*, *Leptinotarsa decemlineata* (the Colorado potato beetle), stored grain pests (contributing to some 40% of grain losses in storage in developing countries), and corn rootworms (*Diabrotica* spp). Some, such as mosquitoes and *Helicoverpa*, contribute to great human illness and misery, including the transmission of malaria and other diseases (still killing more than a million people a year) and, due to crop damage and desperate financial circumstances, suicides of farmers in India.

The typical response of pesticide users to the evolution of resistance by pests is to use more pesticides or switch to more expensive pesticides, or sometimes to ones that are more environmentally risky. As the global

---

[1] A.L. Melander Can insects become resistant to sprays? J. Econ. Entomol. 7 1914 167–173

agricultural community contemplates increasing crop production by some 70–100% to meet the demands of a growing and more affluent human population, pesticide resistance continues to be a threat to both production and the hope of reducing the environmental impacts of agriculture.

Ironically, the broader significance of resistance was perhaps first recognized not by applied entomologists, but by evolutionary biologists, thanks to Theodosius Dobzhansky. In his 1937 classic, *Genetics and the Origin of Species,* arguably the most influential book of the modern Darwinian synthesis, Dobzhansky seized upon insecticide resistance to demonstrate that evolution was occurring in our lifetimes.

The significance of the evolution of resistance for both the study of evolution and to IPM has been captured in this volume by David Onstad and the team of leading experts he has orchestrated. Onstad is well placed to discuss the diversity of evolutionary responses to control tactics in arthropods due to his own curiosity, the diversity of his own career interests, and the network he has built across an even wider diversity of biologists and social scientists.

Onstad and colleagues describe and analyze the evolution of resistance to a wide range of pesticides (including the diversity of mechanisms and behavioral modification), conventionally bred and genetically modified crop varieties, crop rotation, and insect pathogens. In so doing, we are treated to a broad canvas of population dynamics, landscape processes, models, toxicology, and economics. We see the complexity and excitement of insect resistance management (IRM), some of the successes and the continuing challenges, including in regulation and compliance with IRM strategies.

Resistance has been evolving to plant defenses for millennia, but has been less well documented than for pesticides. Given that resistance has been evolving to crop plants selected by humans over at least 40 years, we can suspect the same was happening to crop varieties even in the 1930s, but escaped attention. Anyone reading this book will now likely be on the lookout for a wider range of evolutionary responses, especially wherever the "greater the effectiveness and success of arthropod pest management."

In his final chapter, Onstad advises that we should "Always consider IRM within an IPM framework." I would add that we must also "Always consider IPM within an IRM framework." Not only is IRM part of IPM, IRM is most interesting and perhaps most powerful when the best resistance-management tactics are not obvious extensions of IPM.

Good IRM is often counterintuitive and runs contrary to some of the long-held tenants of IPM, such as applying tactics only when and where needed or on thresholds. The effectiveness and relative durability of high-expression, pyramided, transgenic crops suggest the need for a rethink of this dogma, especially where such varieties have so reduced pesticide use that they have in turn opened opportunities for enhanced IPM of other pests.

As suggested above, successful IPM is more important now than ever if we are to meet global demands for food and public health while reducing the environmental impacts of agriculture and disease management. Pesticides should not be the basis of these IPM programs, but we need to maintain them and other tactics, such as resistant crop varieties, as effective tools when required.

Whether for crop varieties, crop rotation, or pesticides, this volume clearly illustrates that IPM programs can fail if we do not also implement countermeasures for evolution. This is a one-hundred-year-old story, with abundant lessons that we need to apply more vigorously if we are to meet the crop production and environmental challenges now before us.

Rick Roush
Melbourne School of Land and Environment,
University of Melbourne, Parkville, Victoria, Australia

PREFACE TO SECOND EDITION

By the time we finished writing the chapters, it was obvious that a new edition of this book with additional knowledge and case studies was needed to keep pace with the science and the evolving arthropods. Although the book's themes remain the same as before, the authors demonstrate that new challenges confront practitioners of insect resistance management.

I thank the authors of the first edition who volunteered again to address the important concepts and examples. I am also pleased and thankful that the new authors agreed to join the group to make the second edition even better than the first. Rick Roush, Dean at the University of Melbourne, was kind enough to write a forward for this edition that places insect resistance management in a broader context. Dawn Dockter edited several chapters and helped produce the index (as she did for the first edition). Bruce Stanley and I thank Stephen Irving for use of his resistance-monitoring example in Chapter 15. I also thank the Academic Press and Elsevier staff who led me through the book-production process one more time.

David Onstad

Agricultural Biotechnology, DuPont Pioneer
Department of Entomology and Wildlife Ecology, University of Delaware

The cover photograph is used with permission from Joseph Spencer (copyright 2013). It is the shadow of an adult *Diabrotica virgifera virgifera* on a corn leaf. The front cover was designed with the help of Emma Onstad.

PREFACE TO THE FIRST EDITION

I produced this book so that readers would be able to discover and gain much of the current knowledge about insect resistance in one volume. The study of resistance is a dynamic process that never ceases to surprise scholars. I hope that this book encourages readers to actively study this subject with curiosity and an open mind. As scientific editor, I asked all authors to accomplish three goals with each chapter. First, produce chapters that describe all major concepts, not just those derived from the author's own work. Second, provide important advice and conclusions for readers. And third, relate contents of a chapter to several themes expressed throughout the book. These themes are highlighted in Chapters 1 and 14. I believe that the book demonstrates our joint commitment to these goals.

Professors Al Gutowsky (Economics), Harvey Reissig (Entomology), and Christine Shoemaker (Engineering) were my mentors at C.S.U.S. and Cornell University. Some of my better ideas were developed under their guidance. Fred Gould hosted me during my sabbatical visit to North Carolina State University in 1994 and helped start my work in insect resistance management.

I thank both the authors and others who contributed to the development of this book. Andy Richford of Elsevier promoted the concept for the book at Academic Press. The following colleagues read portions or early drafts of several chapters: Dawn Dockter (Chapters 1, 9, 10, and 14), Casey Hoy (Chapters 1 and 9), Terry Hurley (Chapter 2), Jack Juvik (Chapter 9), and Ralf Nauen (Chapter 7). Lisa Knolhoff created figures for Chapters 2, 4, and 10. Christine Minihane, Cindy Minor, and Sunita Sundarajan of Elsevier guided me through the stages of production. Bruce Stanley and I thank Stephen Irving for use of his resistance-monitoring example in Chapter 13. Barry Pittendrigh and I thank Scott Charlesworth for creating Figures 6.1, 6.3, 6.4, and 6.5. The writing of Chapter 10 was facilitated by a Cooperative Agreement with USDA-ARS, "Contributions to a Framework for Managing Insect Resistance to Transgenic Crops." The ideas and conclusions may not represent those of the USDA or USEPA.

David W. Onstad
Champaign-Urbana, IL, USA

CHAPTER 1

Major Issues in Insect Resistance Management

David W. Onstad
DuPont Pioneer, Wilmington, DE

Chapter Outline

PHILOSOPHY AND HISTORY

Nature is exciting because it is dynamic; the management of nature can be equally exciting, and it is certainly challenging. This book presents a story about some of the most challenging aspects of pest management: the dynamics of society's competition and struggles with arthropods over evolutionary time. In this case, evolutionary time is not the millions of years required for macroevolution and speciation, but the tens of years required for pest populations to evolve the ability to withstand or overcome control.

Entomologists, acarologists, and practitioners of integrated pest management (IPM) know that arthropods can evolve resistance to chemicals, host-plant defenses, and cultural practices such as crop rotation. Insect resistance is a general term representing heritable traits selected by management. These traits typically permit an arthropod to overcome pest management due to changes in behavior, maturation, or biochemical

Insect Resistance Management
DOI: http://dx.doi.org/10.1016/B978-0-12-396955-2.00001-1

1

processes. Insect resistance is similar to the term *host-plant resistance*, which means that the plant has defenses against and is resistant to an arthropod. Throughout the book, the term *insect resistance* will be used, even though other arthropods, such as mites and ticks, are also frequent targets of pest management.

The greater the effectiveness and success of arthropod pest management, the greater the likelihood of the pest evolving resistance to that management tactic. This is particularly true when the goal of pest management is to reduce the pest population and maintain it at a very low level. The probability of resistance evolution will be lower when goals emphasize the prevention of damage and disease, such as the promotion of crop tolerance, which sometimes can be accomplished without harming most of the pest population. Nevertheless, if our goals or tactics involve significant pest population reduction, we likely will need to manage the evolution of resistance to the management tactics that we wish to be so effective. Insect resistance management is the scientific approach to managing pests over the long run so that resistance does not interfere with our ability to accomplish our goals.

A common attitude in the practice of pest management is to expect that effective pesticides and other tactics will always be available in the future as each current treatment fails due to resistance. This is not a sophisticated strategy, and it often is a wasteful and inefficient one. Of course, this requires the farmer and public health official to do nothing other than hope for the best. As each failure is observed, stakeholders search for a cure.

In the past, insect resistance management (IRM) has often emphasized this reactive approach to the sequential failures of insecticides. Each insecticide is used for several years until it no longer adequately controls the pest population. Population monitoring may help identify problems before regional failure occurs. Under the best circumstances of reactive IRM, a new class of toxin with a different mode of action (physiological mechanism that kills the pest; see Pittendrigh *et al.*, Chapter 3, and Nelson and Alves, Chapter 4) is introduced to manage the pest again with pesticides. This reliance on sequential use of tactics for control is the hallmark of reactive (some call it curative) IRM. This approach requires an optimistic view of science and industry's capabilities to produce new tactics and chemicals for future use in pest management.

The alternative approach is preventative IRM. In preventative IRM, resistance management plans are implemented when an IPM tactic is first introduced. In industry, this is called product stewardship. These plans alter the design and control of the management system so that the tactic (insecticide, crop rotation, host-plant resistance) can make a significant contribution to IPM for a period that otherwise would not have been possible. This approach is based on a pessimistic view of nature and industry: When we are careless, pests evolve faster than science and industry can develop new solutions. Yet, if we are careful and delay the evolution of resistance, we give our best scientists and technologists time to focus on a much wider range of management tools for the entire system. This approach does place a greater burden on practitioners and end users (ranchers, farmers, public health officials, and citizens). However, since practitioners do not "own" pest susceptibility to management tactics, they should never believe and act as though elimination of susceptibility is simply an externality of their business activities (Mitchell and Onstad, Chapter 2).

The purpose of this book is to promote scientific, predictive, and preventative IRM. The book is written for those scientists, regulators, and consultants who wish to participate in the difficult but valuable efforts to (1) incorporate IRM into IPM, (2) develop economical IRM plans, and (3) design IRM plans for local environmental and social conditions.

History and Current Status of Resistance to Pesticides

Georghiou and Lagunes-Tejeda (1991) documented the history of field observations of resistance to pesticides. They stated that the first report of resistance was published by Melander (1914), who described the resistance of orchard pests to sulfur-lime, a compound typical of the inorganic chemicals used for pest management 100 years ago. By 1989, Georghiou and Lagunes-Tejeda (1991) had counted over 500 arthropod species with strains evolving resistance in the field to toxins used against them. Within this total, 23 beneficial species were included. The resistant pests are categorized as crop pests (59%) and medical or veterinary pests (41%). By 1989, chemicals selecting for resistance included not only the modern classes of organic chemicals (cyclodiene, DDT, organophosphate, carbamate, pyrethroid) but also inorganic and elemental chemicals (e.g., arsenicals, sulfur) commonly used before 1940. Table 1.1 presents a list of the arthropod species for which resistance has been most frequently reported (Head and Savinelli, 2008).

The best source for up-to-date information about the arthropods' resistance to pesticides around the world is the Arthropods Resistant to Pesticides Database, ARPD (http://www.pesticideresistance.org), sponsored by Michigan State University, the Insecticide Resistance Action Committee

Table 1.1 Top 16 Resistant Arthropods, Based on the Number of Unique Active Ingredients for which Resistance has been Reported and the Number of Cases Reported

Species	Order: Family	Pest Type	No. Active Ingredients	Cases
Tetranychus urticae	Acari: Tetranychidae	Crop	79	325
Plutella xylostella	Lepidoptera: Plutellidae	Crop	76	278
Myzus persicae	Hemiptera: Aphididae	Crop	68	293
Leptinotarsa decemlineata	Coleoptera: Chrysomelidae	Crop	48	183
Musca domestica	Diptera: Muscidae	Urban	44	183
Boophilus microplus	Acari: Ixodidae	Livestock	43	127
Blatella germanica	Blattodea: Blattellidae	Urban	42	213
Bemisia tabaci	Hemiptera: Aleyrodidae	Crop	39	169
Panonychus ulmi	Acari: Tetranychidae	Crop	38	178
Aphis gossypii	Hemiptera: Aphididae	Crop	37	103
Culex pipiens pipiens	Diptera: Culicidae	Urban	34	119
Helicoverpa armigera	Lepidoptera: Noctuidae	Crop	33	434
Heliothis virescens	Lepidoptera: Noctuidae	Crop	33	106
Culex quinquefasciatus	Diptera: Culicidae	Urban	30	229
Spodoptera littoralis	Lepidoptera: Noctuidae	Crop	30	50
Anopheles albimanus	Diptera: Culicidae	Urban	21	72

Source: Head and Savinelli (2008) created this table in 2006 from records in the Arthropods Resistant to Pesticides Database (http://www.pesticideresistance.org).

(IRAC), and the U.S. Department of Agriculture. The database contains reports of resistance cases from 1914 to the present, including all of those reported by Georghiou and Lagunes-Tejeda (1991). Each case is defined by the time and location at which the resistance is first discovered. Whalon *et al.* (2008) provide a detailed description of the database and an analysis of its contents. As of 2008, the database contained over 7747 cases involving 553 species. Approximately 40% of resistant arthropods are medical pests, and close to 60% are agricultural pests (Whalon *et al.*, 2008). The rate of increase in species becoming resistant for the first time is slowing down; this means that only three new species were added to the list from 2002 to 2008. The public can search the database for information, and authorized experts can submit new cases.

Definitions of Insect Resistance

Evolution is the net directional change or cumulative change in the characteristics of individuals within a population over many generations.

Microevolution, as defined by population geneticists, is the change in allele frequencies in a population over generations. Clearly, in both cases, evolution is change in some state of the population or species. In IRM, we focus on the state described as resistance. As with any state or characteristic, the level of the state can increase or decrease in a population due to evolution. In an individual, resistance can be a thicker or thinner cuticle, a concentration of a molecule, or a type of protein structure. Tolerance to a toxin within an individual can increase or decrease. In reality, resistance can decline due to fitness costs and removal of the selective agent from the environment. Resistance can even decline due to evolution when a variety of selective factors in IPM favor the susceptible individuals over the resistant ones.

Thus, from a strictly biological perspective, the evolution of resistance can lead to increases or decreases in tolerance to a selective agent in an arthropod population. In biology, we would expect resistance to be measured in terms of mean population fitness. Furthermore, from a biological perspective, resistance is neither change nor solely the decrease in susceptibility. Resistance can be complete or incomplete when measured as survival (Tabashnik *et al.*, 2009). Lastly, changes in resistance can be either barely observable or extreme and obvious. One example from IPM and agriculture that matches the biological definition is host-plant resistance (Onstad and Knolhoff, Chapter 9). Host-plant resistance is defined as the state of the plant that either tolerates damage by the pest, harms the pest, or repels the pest. The effectiveness of all three possible states can decline or increase due to plant breeding.

Thus, arthropod or insect resistance is a characteristic with various levels defining the relationship between an arthropod and various selective agents, such as a toxin, a host plant, a natural enemy, crop rotation, or mating disruption. Although the characteristic can be measured in each individual arthropod, in IRM we are most interested in population-level indices of resistance such as mean population fitness due to resistance. Resistance-allele frequency is a poor index of resistance because it tells us nothing directly about fitness. Even if we could relate fitness to allele frequency through knowledge of genotypic response to a single toxin, for cases involving multiple selective agents and multiple arthropod genes, relating allele frequencies to mean population fitness or some other population-level index is difficult and not straightforward. It is particularly difficult if the various types of resistance have different measures and different maximum values (not all complete resistance). Thus, communication among stakeholders would be improved if we de-emphasized allele frequency and promoted the use of mean population fitness (Brévault *et al.*, 2013).

Not surprisingly, practitioners of IRM often broaden the semantics concerning resistance. IRM practitioners could try to discuss resistance only in terms of fitness of an insect population, but many stakeholders want to add meaning that is relevant to their goals. It is very common in IRM literature for resistance to be discussed as a dichotomous variable that is either positive or negative in terms of a stakeholder's values. In other words, a population is either resistant or it is not resistant. Even Tabashnik *et al.* (2009) commonly use the term *resistant populations* to indicate that some populations are qualitatively and discretely different from other populations. Human values are involved in defining the threshold that differentiates the negative from the positive, the resistant from the not resistant. And human values may base this threshold on factors such as probability of control failures or economic loss.

Thus, in conversations, presentations, and publications, it is common to find two meanings for the word *resistance*. The first is the strict biological meaning, with resistance defined as a state or characteristic of an arthropod that determines how fit it is in relation to a selective agent. The second meaning views resistance as a dichotomous state of an arthropod population that has either negative or positive effects on a stakeholder. Dichotomous resistance in a population depends on the stakeholder, the chosen temporal and spatial scales, and the natural and social environment. Biological resistance can be related to dichotomous resistance, but this relationship is complicated by variability in such factors as growing conditions and price for crop or livestock.

MAJOR THEMES

Insect resistance management (IRM) is often considered the management of the evolution of resistance in an arthropod species. However, this is a very narrow and restricted view of the interacting ecological and socioeconomic systems that not only are affected by resistance but determine whether resistance will evolve. Just as integrated pest management (IPM) does not simply focus on killing pests, IRM should not be limited to restraining the dynamics of genes. In this section, I introduce several major themes that are expressed throughout the book.

Integrated Pest Management

Integrated pest management was conceptualized during the 1950s when insecticide resistance, nontarget effects, and economic waste were clearly apparent (Stern *et al.*, 1959). Practitioners understood the consequences for the larger environment and the longer term, but implementation of IPM emphasized short-term economic efficiency and integration of cultural, biological, and chemical control measures. For example, by including natural enemies as biological control agents in the management of pests, IPM practitioners knew that more specific and less harmful chemicals would need to be used over the long run (Hoy, 1990; Hull *et al.*, 1997).

IRM must be considered a part of IPM (Croft, 1990; McGaughey and Whalon, 1992; Glaser and Matten, 2003). Certainly after 50 years of effort to implement rational and socially beneficial IPM, most people would agree that IRM must at least account for the consequences for IPM of managing the evolution of resistance genes. McGaughey and Whalon (1992) stated that IRM within the context of IPM is based on four factors: (1) diversification of causes of mortality so that a pest is not selected by a single mechanism, (2) reduction of selection pressure for each mortality mechanism, (3) maintenance of a refuge or immigration to promote mixing of susceptible and resistant individuals, and (4) prediction using monitoring and models.

Formal representation of long-term management within the IPM paradigm is all that is needed to bring IRM and IPM together. By combining population genetics (from IRM) with the focus on economic efficiency and environmental stewardship (from IPM) and formally considering a multiyear time period, all aspects of IPM and IRM can be combined. Chapters 6–9 provide a variety of case studies that highlight the interactions between IRM and IPM.

In essence, the linking of IRM and IPM provides a management perspective that is the same as long-term, areawide pest management (Elliott *et al.*, 2008). Areawide pest management was first promoted for the eradication of a few pests by collective efforts (Myers *et al.*, 1998; Smith, 1998; Bowman, 2006). Some researchers also recognized that if regionally coordinated IPM could occur, even annual management of a constant pest could be made more efficient (Faust and Chandler, 1998). Recently, more attempts have been made to coordinate efforts for regional IPM (Pereira, 2003; Sexson and Wyman, 2005).

Siegfried *et al.* (1998) recognized the relationship between areawide pest management and IRM when they warned proponents of areawide pest management about the increased risk of resistance evolution in areawide projects because of the reliance on uniform exposure of pest populations. They believed that some attributes of areawide management are incompatible with many conventional IRM techniques, but suggested that the use of biologically based control tactics, such as behavior-disrupting chemicals, may contribute to both areawide pest management and IRM. Siegfried *et al.* (1998) concluded that both areawide pest management and IRM require a high degree of grower compliance.

IRM requires an approach that considers not only the long-term, but also the spatial dynamics of the pest and its management over a large region. Thus, all IRM should be areawide pest management. (Note though that both IPM and IRM perspectives are generally in opposition to pest-eradication activities.) The term *integrated* in IPM can also refer to the integration of management across space and over time. Thus, integrated pest management implies that areawide and long-term approaches can be valuable. Because this type of approach is the basis for IRM, I view IRM as an important part of future IPM strategies.

Coordination

In a few cases, IRM may be strictly a private matter for one company that both produces the livestock or crop and provides the tools for managing the pest. This company would likely be interested in product stewardship if an insecticidal plant or compound were used. Nevertheless, in most cases, IRM requires the coordinated behavior of many individuals and businesses.

Coordinated behavior is necessary to provide the areawide pest management described earlier. If the actions taken by individuals are not clearly beneficial to them, especially if their individual goals differ from those who are leading or mandating the coordination, then some kind of persuasion or coercion will be necessary. Even in the case of a unique, synthetic toxin patented by a company, any product stewardship will require obtaining the cooperation of most farmers, ranchers, pet owners, or public health departments using the product.

Keiding (1986) described the coordination and cooperation involved in the management of resistance to insecticides by *Musca domestica* in livestock barns in Denmark. Keiding stated that collaboration and exchange of information must be maintained between the agrochemical industry, users of the insecticides, those who advise them (e.g., extension services or farm organizations), and research institutes. He suggested that coordination could be organized by an international agency, such as the Food and Agriculture Organization (FAO) or the World Health Organization (WHO), or by a national or state institution. In the same book, Brent (1986) also promoted the coordination of public and private sectors in managing resistance. In fact, the U.S. National Research Council's Committee on Strategies for the Management of Pesticide Resistant Pest Populations recommended that working groups involving all stakeholders should prioritize IRM efforts based on economic, environmental, and social factors (NRC, 1986, p. 275).

Forrester (1990) provides a good summary of the coordination and critical activities required to make an IRM plan succeed. He states that preventative IRM is preferable to curative IRM because curative approaches are more restrictive and have a smaller chance of long-term success. Negotiations that involve compromise and consensus among stakeholders is important, although he suggests that centrally planned and regulated IRM strategies can be very successful. In either case, compliance with the plan is a critical factor. Forrester (1990) emphasized the need to make IRM strategies match the local conditions for the pest, environment, and community.

From an economic perspective, coordination can be valuable, but only under certain conditions. Miranowski and Carlson (1986) stated that voluntary IRM cooperation among farmers will likely occur only when pests can move from farm to farm, when the benefits and costs of a farmer's participation are proportional to the level of participation, when free-riders receive minor benefits, and when coordination costs are low.

Resistance management strategies are only successful at the landscape level, which requires the coordination of all producers in a given area. How do we (1) convince producers that resistance management is essential to maintaining effective arthropod control measures and (2) devise IRM strategies that are in the economic interest of producers? Producers are likely to recognize the threat that resistance poses to pest management, but if preventing resistance becomes too burdensome from either a time or a money perspective, they will not adopt IRM techniques. Maintaining a refuge of susceptible plants (and alleles) is a common strategy, but if the pest population causes significant damage to refuge plants, producers are less likely to comply (Hurley and Mitchell, Chapter 13).

Pest Behavior

As one would expect, mortality is easier to measure than any type of arthropod behavior. As a result, pest behavior and behavioral resistance have traditionally not been investigated sufficiently during studies of population genetics and evolution. One objective of this book is to promote the study of pest behavior in order to improve IRM. Many of the cases of resistance and its management throughout the book demonstrate the importance of behavioral studies.

Toxicological resistance to pesticides has been the focus of the vast majority of IRM studies. However, evidence has been accumulating that demonstrates the importance of behavioral resistance (Lockwood et al., 1984; Gould, 1991; Hoy et al., 1998). Toxicological resistance is the evolution of a mechanism that reduces or prevents the intoxication of an individual once the toxin contacts or enters the body (Chapter 3). To avoid confusion, I do not use the term *physiological* resistance because behavior can be considered an observable consequence of physiological mechanisms (Georghiou, 1972). Behavioral resistance is the evolution of any behavioral change that permits a population to avoid or overcome management tactics. Behaviors that may be important include movement of immature stages, adult dispersal, oviposition, feeding, or any social or nonsocial interaction in a population.

Gould (1984) investigated the management of behavioral resistance using a mathematical model. Lockwood et al. (1984) reviewed early cases and described the shift in perspective that was needed to appreciate behavioral resistance. They recognized a connection between behavior and toxicological resistance and emphasized that the two may occur simultaneously. Gould (1991) related behavioral resistance to the evolutionary biology of plant–herbivore interactions. He encouraged the study of the behavioral responses of herbivorous arthropods in natural plant communities to discover clues to the evolution of resistance to pesticides. Hoy et al. (1998) emphasized the role of spatial heterogeneity in the evolution of behavioral responses to toxins, and the role of these behavioral responses in the evolution of toxicological resistance. Because natural and synthetic toxins are heterogeneously distributed in plants and across the landscapes containing plant communities, evolution of behavioral responses should be expected. Since the review of Hoy et al. (1998), dozens of publications have reported on observed or potential behavioral resistance to not only insecticides but also natural enemies, transgenic insecticidal plants, sterile-insect releases, and diatomaceous earth.

Pheromone-based mating disruption is an IPM tactic that has been implemented for management of several pests over the past 40 years (Tabata *et al.*, 2007a; Baker, 2009). Mochizuki *et al.* (2002) were the first to discover field-evolved resistance to a mating-disruption pheromone. *Adoxophyes honmai*, a major pest of tea (*Camellia sinensis*), evolved resistance after 10 years of annual mating disruption in Japan (Tabata *et al.*, 2007a). After additional selection in the laboratory, Tabata *et al.* (2007a) created a resistant strain in which male adults exhibited a broader range of behavioral responses to component blends of the pheromone. Resistance in males was inherited as a single, autosomal gene (Tabata *et al.*, 2007b). In the presence of disrupting pheromones, resistant males successfully found and mated with females (Tabata *et al.*, 2007a). Resistant females produce more of one component, which changes the ratio of a single acetate (Mochizuki *et al.*, 2008).

The potential for evolution of resistance to pheromone-based mating disruption has been demonstrated in a variety of species (Torres-Vila *et al.*, 1997; Evenden and Haynes, 2001; Shani and Clearwater, 2001; Svensson *et al.*, 2002; Spohn *et al.*, 2003). Mutations and existing genetic variation can allow female adults to produce different amounts of pheromone or new proportions of components. Mutations in male adults permit them to perceive and be attracted to new types of production by females, though not necessarily eliminating perceptions of the old pheromone used as a disruptant.

Plutella xylostella seems to be one of the best model organisms for studying behavioral resistance. It is easy to rear, has many generations per year, and can be investigated under laboratory, greenhouse, or field conditions. Therefore, the studies of behavioral responses by *P. xylostella* and their influence on toxicological resistance are summarized ahead. As you read, note the consequences of pest behavior, how behavior influences pest survival in a treated environment, and how it influences the toxin dose acquired by an insect and, therefore, the selection pressure for toxicological tolerance. These issues have traditionally not been investigated sufficiently during studies of population genetics and evolution.

Head *et al.* (1995a) investigated the genetic basis of toxicological and behavioral responses to a pyrethroid in populations of *P. xylostella* with different average levels of tolerance for the toxin. Heritabilities for behavioral avoidance of the pyrethroid were low and significant in only one population, although additive genetic variances were similar to those observed for the toxicological responses. The phenotypic and genetic correlations between the two traits varied among the populations. All correlations were negative, and significant correlations occurred in populations with relatively high levels of additive variation for both traits. Individuals with low tolerance for the toxin fed more on leaves having low concentrations of the toxin. In a subsequent study, Head *et al.* (1995b) demonstrated that two elements of behavior were affected by selection in laboratory populations: General larval activity increased with behavioral selection, and larvae displayed a greater tendency to avoid the pyrethroid. Note, however, that for *Leptinotarsa decemlineata* feeding on toxic potato plants, Hoy and Head (1995) observed a positive correlation between larval movement away from high-toxin concentrations and tolerance for the toxin. Head *et al.* (1995a,b), Hoy and Head (1995), Hoy *et al.* (1998), and Jallow and Hoy (2007) suggested that some IRM plans could take advantage of behavioral evolution by selecting for susceptibility in landscapes with heterogeneous spatial distributions of toxins.

In a series of laboratory and greenhouse experiments, Jallow and Hoy (2005, 2006, 2007) investigated the simultaneous evolution of behavioral responsiveness and toxicological resistance in *P. xylostella*. Jallow and Hoy (2005) first measured phenotypic variation in behavioral response and toxicological tolerance to permethrin in one field and one laboratory population of *P. xylostella*. In laboratory bioassays, females from both populations were less likely to oviposit on cabbage leaf disks and seedlings treated with permethrin, and this oviposition deterrence was correlated with permethrin concentration. The laboratory population was more behaviorally responsive to the insecticide and showed a greater avoidance than the field population. They measured the toxicological response of each population with feeding bioassays, and the laboratory population was more susceptible to the permethrin. Thus, there was a negative correlation between avoidance and detoxification.

Jallow and Hoy (2006) extended their study to include the genetic basis of adult behavioral response and larval toxicological tolerance to permethrin within the two populations of *P. xylostella*. The adult behavioral response was again measured as oviposition site preference. They discovered that a high proportion of phenotypic variation for adult behavioral response to permethrin was heritable genetic variation. The larval toxicological response was measured with a topical application bioassay. Significant additive genetic variances and heritabilities for toxicological tolerance to permethrin were detected in both populations. The genetic correlations between adult behavioral response and larval toxicological tolerance to permethrin were negative, but significant only in the field population (Jallow and Hoy, 2006).

In their greenhouse study of the field population of *P. xylostella*, Jallow and Hoy (2007) investigated the changes in behavioral response and toxicological tolerance of *P. xylostella* to homogeneous and heterogeneous distribution of the toxin permethrin. They utilized three selection regimes: uniform high concentration hypothesized to result in increased toxicological tolerance, heterogeneous low concentration hypothesized to result in increased susceptibility to the toxin through indirect selection on behavior, and a control with no exposure to permethrin. All life stages of the moth were exposed to the selection regimes. The insects were observed in 1 m^3 cages in a greenhouse for 33 generations. Each successive generation was started with a random selection of pupae from the previous generation. Cohorts selected with uniform high concentrations evolved high levels of resistance to permethrin by the seventeenth generation. For generations 1−20, cohorts selected with heterogeneous low concentrations were similar to the unselected control, but in generations 21−33, those selected with the heterogeneous low concentration were more susceptible than those of the control. Jallow and Hoy (2007) concluded that low heterogeneous doses could lead to increased susceptibility to permethrin by selecting indirectly on behavior.

The work of Jallow and Hoy (2006) demonstrated that female moths that are more behaviorally responsive to permethrin produce offspring that are more susceptible to the same insecticide. Jallow and Hoy (2007) concluded that selection on this behavioral response can result in greater susceptibility compared to scenarios with very high uniform concentrations or no toxin in the environment. The adult behavioral response can lower the exposure of larvae to the insecticide, reducing selection pressure for toxicological resistance in larvae. Thus, this behavioral response and associated larval survival could help preserve susceptible alleles in the population, which would contribute to the success of IRM.

Based on all of the evidence presented above, we can conclude that accounting for arthropod behavior is important for predicting the evolution and management of resistance. Behavioral resistance may evolve or behaviors may vary from environment to environment and may influence the evolution of toxicological resistance. Other evidence for the important role of behavior can be found throughout this book. This does not mean that behavioral resistance will always be observed, as the case evaluated by Hawthorne (1999) indicates. Nevertheless, the evidence does support the claim that more resources should be allocated for behavioral studies during the preparation of IRM plans.

Variability and Complexity of Management Strategies

The most common IRM strategies are briefly described ahead to provide some background from which we can draw another theme for the book. The challenge for all readers, as well as for all workers in the field of IRM, is not to rely completely on tradition when developing strategies for new pests and new pest-management systems. A full appreciation of strategies for managing insect resistance to any pest-management tactic requires an understanding of economics (Chapter 2), population genetics (Chapter 5), and other information about nature and society.

Denholm and Rowland (1992), Denholm et al. (1992), McGaughey and Whalon (1992), McKenzie (1996), Roush (1989), Roush and Tabashnik (1990), and Tabashnik (1989) provide good overviews of the strategies commonly considered when arthropods may evolve resistance to insecticides. The focus is on preventative IRM strategies for managing susceptibility before resistance genes increase in frequency in the population. Variations or even completely different plans will be needed once resistance is observable and measurable in the field (Forrester, 1990).

Kill Fewer Susceptibles

Selection pressure can be reduced by lowering the selection intensity of each treatment or by decreasing the number of treatments applied against a pest species over time. When treatments are reduced by avoiding treatments experienced by the vulnerable life stages of the pest in certain generations, the evolution of resistance can be delayed. However, resistance-allele frequency will continue to rise over many generations unless the resistant individuals have a lower relative fitness than the susceptibles (fitness cost) in the absence of treatments.

Another approach provides a spatial refuge that allows susceptible individuals a place to escape selection by the treatment. Refuges are deployed so that adequate mixing of the subpopulations occurs (Pan et al., 2011). Susceptible individuals can then mate with any resistant individuals, lowering the proportion of homozygous resistant genotypes in the population.

The treatment effect can be reduced to lower selection pressure by decreasing the concentration of the insecticide or other treatment. This approach is effective if it allows more susceptible individuals to escape mortality, thus increasing the relative fitness of the susceptibles in the population. McGaughey and Whalon (1992) noted that low-dose strategies for IRM would only work when they became a significant part of an IPM program. In that case, the reduced concentration of insecticide would promote the efficacy of natural enemies of the targeted pest. All of the tactics, particularly the insecticide, would need to maintain the pest density below the economic threshold; otherwise farmers and similar stakeholders would not accept the greater damage by the pest.

Without the support of an effective IPM program, attempts to reduce the selection on susceptibles might simply lead to more damage by the pest, with subsequent reduction in compliance by stakeholders. Gray (2000) has suggested that transgenic insecticidal crops be planted only with permission from an independent agent, similar to the need for prescriptions from physicians for medicines.

Kill all the Heterozygotes

If there are very few resistant homozygotes in the population, then an effective strategy may be to increase the concentration or efficacy of the treatment so that all heterozygotes are killed. This lowers the fitness of the resistant individuals relative to susceptibles. More recently, with the use of transgenic insecticidal crops, this has been called the high-dose strategy (Onstad and Knolhoff, Chapter 9). A refuge for susceptibles is often included to prevent evolution of resistance by promoting the mating of homozygous susceptibles with any rare homozygous resistant individuals; heterozygote offspring in treated areas will all die in the next generation.

When the concentration of the toxin is either decreased or increased, the effect on evolution of resistance depends on the population dynamics and environment of the targeted pest and its natural enemies. Thus, information on interactions and complexities in the system should be gathered before predicting the long-term effectiveness of a strategy.

Use Two Treatments

When two or more treatments have different effects on the arthropods (e.g., different modes of actions by toxins without cross-resistance), then it may be possible to use them either in mixtures or rotations to delay the evolution of resistance. A mixture is the simultaneous application of two treatments to the same individuals in a population. Both parts of the mixture must remain effective for the same period of time over the same region of the landscape. A refuge may be needed, as described previously, to provide a source of susceptible individuals that can mate with any rare homozygous resistant individuals. With mixtures we expect each treatment to kill any individuals resistant to the other treatment. When multiple genes for pest control are incorporated together in a crop, this mixture is called a pyramid. Roush (1994, 1998) explained some of the advantages and limitations of insecticide mixtures and transgenic insecticidal crops with pyramided traits. Gould *et al.* (2006) demonstrated that fitness costs due to resistance can also be important in IRM with pyramided crops or mixtures. Difficulties encountered when implementing an IRM strategy with mixtures include ensuring the equal persistence of both treatments and the possibility that resistance genes will interact through epistasis in ways that reduce the effectiveness of the mixture (Onstad and Gassmann, Chapter 5).

A core assumption of the pyramid strategy is that insects resistant to one toxin will be killed by the other toxin, which is called redundant killing (Gould, 1986; Roush, 1998). The redundant killing is apparent when homozygotes susceptible to both toxins suffer a very low survival based on the multiplication of the values for each toxin (e.g., 0.1 for toxin A and 0.01 for toxin B, $0.1 \times 0.01 = 0.001$ survival on pyramid). As plants age, toxin concentrations often decline, and resistance to a single toxin could significantly enhance survival on two-toxin Bt crops, if the concentration of the other toxin is not high enough to kill resistant individuals (Carrière *et al.*, 2010). It is also possible that epistasis within a transgenic insecticidal plant interferes with the expression and toxicity of one or more of the pyramided insecticidal traits.

Records of the United States Environmental Protection Agency (2009) indicated that pyramided Bt corn (*Zea mays*) expressing Cry3Bb1 and Cry34/35Ab1 causes mortality to *Diabrotica virgifera virgifera* similar to that caused by either of the toxins alone. The mortality caused by Cry34/35Ab1 is 0.9420−0.9918, that caused by Cry3Bb1 is 0.962−0.9996, and that caused by corn expressing both traits is 0.9822−0.9997. Thus, the variability for a given trait is greater than the variability across the highest mortality rates for all three conditions. Onstad and Meinke (2010) concluded that the combination of traits appears to increase mortality very little, if at all. Therefore, they assumed in their model that insecticidal mortality is more likely equal to the minimum survival caused by either toxin rather than the multiplicative product of the survival rates observed when the pest is exposed to each insecticide. Without multiplicative or independently acting survival, there is no redundant killing of susceptible insects. Onstad and Meinke (2010) evaluated hypothetical IRM strategies for *D. v. virgifera* with several models and concluded that evolution of resistance is delayed more when the survival rates on pyramids act independently and are multiplicative compared to the case using the minimum survival caused by one of the toxins.

The assumption of redundant killing on cotton (*Gossypium hirsutum*) producing Cry1Ac and Cry2Ab was tested in *Helicoverpa zea* by Brévault et al. (2013). After selecting an *H. zea* strain for resistance to Cry1Ac, survival of the unselected and resistant strain was evaluated on field-grown cotton producing Cry1Ac and Cry2Ab. In diet bioassays, resistance to Cry2Ab did not differ significantly between the unselected and selected strain, indicating that there was little if any cross-resistance. However, survival to adulthood on pyramided cotton was significantly higher in the resistant than nonselected strain, showing that the concentration of Cry2Ab in pyramided cotton was not sufficient to kill insects resistant to Cry1Ac. Brévault et al. (2013) concluded that the concentration of Bt toxins in pyramided Bt cotton may not be sufficient to ensure redundant killing in *H. zea* during the entire growing season. They also proposed a formula for calculating the influence of less-than-full redundant killing in any system.

A rotation involves alternating the use of multiple treatments across generations of the targeted pest. In essence, treatments are applied to the same space at different times. In this approach, we assume that individuals resistant to one treatment will be killed by the next treatment in the rotation. When large fitness costs are associated with resistance, rotations may be especially effective. Curtis et al. (1993), however, reviewed experimental evidence demonstrating that rotations are not always superior to sequential treatments (reactive IRM). It is generally not recommended to alternate insecticides within a single pest generation (Roush, 1989).

A mosaic of treatments is the simultaneous application of tactics, each to a different area infested by the pest population. This is the opposite of the rotation strategy. In general, a spatial mosaic without spatial refuge should not be considered for IRM because it is the least likely to succeed; there is simultaneous selection for resistance to both toxins in the total population. Roush (1989) used a mathematical model to show that rotations are superior to mosaics. Zhao et al. (2010) experimentally evaluated two insecticide rotation strategies and a spatial mosaic of insecticide use. Greenhouse cages contained *P. xylostella* on broccoli (*Brassica oleracea*) plus three insecticides with different modes of action: indoxacarb, spinosad, and *Bacillus thuringiensis*. Two of the 14 plants per cage were refuges that were never sprayed. For the mosaic, four plants were treated with each insecticide every generation. The rotations either alternated them each generation or switched them after three generations. In both rotations, each insecticide was sprayed in three of nine generations. After nine generations, densities were lowest in annual rotation cages. Survival to insecticides generally increased over time in all treatments. Zhao *et al.* (2010) found that the population exposed to the annual rotation was least resistant with regard to spinosad compared to the mosaic and three-generation rotation. There was no clear pattern for resistance to the other two insecticides.

Scientists must always be skeptical about claims that two chemicals have such different modes of action that an insect cannot evolve resistance to both simultaneously. Certainly within a given class of chemicals, cross-resistance is a common phenomenon observed in the field when resistance to one chemical is followed by rapid, if not immediate, evolution of resistance to the second chemical used in the sequence. Unless resistant populations already exist in laboratories, cross-resistance is difficult if not impossible to evaluate. When these laboratory colonies do exist, they may not contain the rare mutants with cross-resistance genes. Thus, it is very difficult to experimentally provide evidence demonstrating lack of cross-resistance in a real population. Perhaps this means that future work should emphasize strategies that use two treatments, only one of which is a chemical. The other treatment would be cultural control, biological control, or environmental manipulation. This does not guarantee lack of cross-resistance, but broadening our scope forces stakeholders and developers to face the complexity of pest management and perhaps take advantage of it.

The Future is not the Past
The complexities and dynamics of nature and its management will likely require IRM strategies that do not fit easily into these three categories (Gardner *et al.*, 1998). Spencer *et al.* (Chapter 7) describe resistance to crop rotation: a different kind of problem with a variety of IRM solutions. Pittendrigh *et al.* (Chapter 11) explain how negative cross-resistance can be used as an effective IRM strategy. In these and other cases, scientists are focusing their attention on IRM strategies that are not simple extensions of traditional approaches.

The success of any strategy depends on coordination of treatments over time and space, particularly within a region inhabited by a pest that can disperse from one field to another. For example, mixtures require coordination to avoid simultaneous use of single components of the mixture that would lead to sequential evolution of resistance, first to the single component and then to the other component encountered in areas with mixtures. Rotations require coordination to avoid the creation of a spatial mosaic in a region.

One of the most difficult problems in IRM is the design and implementation of a strategy for multiple pests (McGaughey and Whalon, 1992; Gould, 1994; Wearing and Hokkanen, 1994). This is especially true when the simplest approach for each pest interacts with and affects the other. Both the timing of the pests and the mortality caused by a toxin may be different, and the pests may generally have behaviors that differ over time and space. Tabashnik and Croft (1982) stated, "Even when the conditions are appropriate for using a high-dose strategy to delay resistance in one pest species, this strategy may greatly accelerate the rate of resistance development of other pests in the species complex" (p. 1143). Furthermore, when the pests infest multiple crops in a landscape and are selected by multiple control tactics (Chapter 16), IRM becomes even more complicated. New ideas and much hard work will be needed to deal with these issues in the future.

ENCOURAGEMENT

As the themes we have presented indicate, IRM is certainly more than just the study of insect evolution. Both theoretical and practical IRM requires the study and appreciation of socioeconomic factors and related human behaviors that contribute to coordination, goal setting, and risk aversion (Mitchell and Onstad, Chapter 2; Hurley and Mitchell, Chapter 13). Pittendrigh *et al.* (Chapter 3) and Nelson and Alves (Chapter 4) present the background on toxins and resistance mechanisms in arthropods. A variety of case studies are described in Chapters 6–9 and 16. Onstad and Gassmann (Chapter 5), Onstad (Chapter 14), and Stanley (Chapter 15) discuss techniques and concepts concerning population genetics, modeling, and monitoring. Several chapters provide more details about the roles that abiotic and biotic environmental factors and landscape design play in IRM (Chapters 10–12). The concluding chapter summarizes the major themes and expresses these and other important issues as a set of rules for IRM practitioners.

The purpose of this book is to present the complexity and excitement of IRM. Practical solutions may not be readily available in the following chapters, but important concepts and some techniques will be discussed. Case studies will give the reader an understanding of how each arthropod species and its environment must be carefully considered before developing an IRM strategy. Throughout the book, the major themes—IPM, co-ordination and human behavior, and pest behavior—should be apparent. I hope the reader maintains an openness to the great diversity and complexity of populations and individual behavior that is necessary to develop the skills needed to confront, if not prevent, the evolution of resistance in arthropod pests. Populations and their environments are dynamic, and those of us investigating and managing them must be dynamic, too.

REFERENCES

Baker, T.C., 2009. Use of pheromones in IPM. In: Radcliffe, E.B., Hutchison, W.D., Cancelado, R.E. (Eds.), Integrated Pest Management: Concepts, Tactics, Strategies and Case Studies. Cambridge University Press, Cambridge.

Bowman, D.D., 2006. Successful and currently ongoing parasite eradication programs. Vet. Parasitol. 139, 293−307.

Brent, K.J., 1986. Detection and monitoring of resistant forms: an overview. In: National Research Council, Pesticide Resistance: Strategies and Tactics for Management. National Academy Press, Washington, DC, pp. 298−312.

Brévault, T., Heuberger, S., Zhang, M., Ellers-Kirk, C., Ni, X., Masson, L., et al., 2013. Potential shortfall of pyramided Bt cotton for resistance management. Proc. Natl. Acad. Sci. USA. 110, 5806−5811.

Carrière, Y., Crowder, D.W., Tabashnik, B.E., 2010. Evolutionary ecology of adaptation to Bt crops. Evol. Appl. 3, 561−573.

Croft, B.A., 1990. Developing a philosophy and program of pesticide resistance management. In: Roush, R.T., Tabashnik, B.E. (Eds.), Pesticide Resistance in Arthropods. Chapman and Hall, New York (Chapter 11).

Curtis, C.F., Hill, N., Kasim, S.H., 1993. Are there effective resistance management strategies for vectors of human disease? Biol. J. Linn. Soc. 48, 3−18.

Denholm, I., Rowland, M.W., 1992. Tactics for managing pesticide resistance in arthropods: theory and practice. Annu. Rev. Entomol. 37, 91−112.

Denholm, I., Devonshire, A.L., Hollomon, D.W., 1992. Resistance 91: Achievements and Developments in Combatting Pesticide Resistance. Elsevier Science Publishers, Ltd., Essex, England, p. 367.

Elliott, N.C., Onstad, D.W., Brewer, M.J., 2008. History and ecological basis for areawide pest management. In: Koul, O., Cuperus, G.W., Elliott, N. (Eds.), Area-wide Pest Management: Theory to Implementation. CABI Publishing, Wallingford, UK, 590pp (Chapter 2, pp. 15−33).

Evenden, M.L., Haynes, K.F., 2001. Potential for the evolution of resistance to pheromone-based mating disruption tested using two pheromone strains of the cabbage looper, *Tricoplusia ni*. Entomol. Exp. Appl. 100, 131−134.

Faust, R.M., Chandler, L.D., 1998. Future programs in areawide pest management. J. Agric. Entomol. 15, 371—376.

Forrester, N.W., 1990. Designing, implementing and servicing an insecticide resistance management strategy. Pestic. Sci. 28, 167—179.

Gardner, S.N., Gressel, J., Mangel, M., 1998. A revolving dose strategy to delay the evolution of both quantitative vs major monogene resistances to pesticides and drugs. Int. J. Pest. Manag. 44, 161—180.

Georghiou, G.P., 1972. The evolution of resistance to pesticides. Annu. Rev. Ecol. Syst. 3, 133—168.

Georghiou, G.P., Lagunes-Tejeda, A., 1991. The Occurrence of Resistance to Pesticides in Arthropods. FAO UN, Rome, p. 318.

Glaser, J.A., Matten, S.R., 2003. Sustainability of insect resistance management strategies for transgenic Bt corn. Biotech. Adv. 22, 45—69.

Gould, F., 1984. Role of behavior in the evolution of insect adaptation to insecticides and resistant host plants. Bull. Entomol. Soc. Am. 30, 34—41.

Gould, F., 1986. Simulation models for predicting durability of insect-resistant germ plasm: a deterministic diploid, two-locus model. Environ. Entomol. 15, 1—10.

Gould, F., 1991. Arthropod behavior and the efficacy of plant protectants. Annu. Rev. Entomol. 36, 305—330.

Gould, F., 1994. Potential and problems with high-dose strategies for pesticidal engineered crops. Biocontrol Sci. Tech. 4, 451—461.

Gould, F., Cohen, M.B., Bentur, J.S., Kennedy, G.G., Van Duyn, J., 2006. Impact of small fitness costs on pest adaptation to crop varieties with multiple toxins: a heuristic model. J. Econ. Entomol. 99, 2091—2099.

Gray, M.E., 2000. Prescriptive use of transgenic hybrids for corn rootworms: an ominous cloud on the horizon? Proceedings of the Crop Protection Technology Conference. University of Illinois, Urbana-Champaign, pp. 97—103.

Hawthorne, D.J., 1999. Physiological not behavioral adaptations of leafminers to a resistant host plant: a natural selection experiment. Environ. Entomol. 28, 696—702.

Head, G., Savinelli, C., 2008. Adapting insect resistance management programs to local needs. In: Onstad, D.W. (Ed.), Insect Resistance Management: Biology, Economics and Prediction, first ed. Academic Press, Burlington, MA (Chapter 5).

Head, G., Hoy, C.W., Hall, F.R., 1995a. Quantitative genetics of behavioral and physiological response to permethrin in diamondback moth (Lepidoptera: Plutellidae). J. Econ. Entomol. 88, 447—453.

Head, G., Hoy, C.W., Hall, F.R., 1995b. Direct and indirect selection on behavioral response to permethrin in larval diamondback moths (Lepidoptera: Plutellidae). J. Econ. Entomol. 88, 461—469.

Hoy, C.W., Head, G., 1995. Correlation between behavioral and physiological responses to transgenic potatoes containing *Bacillus thuringiensis* δ-endotoxin in *Leptinotarsa decemlineata* (Coleoptera: Chrysomelidae). J. Econ. Entomol. 88, 480—486.

Hoy, C.W., Head, G., Hall, F.R., 1998. Spatial heterogeneity and insect adaptation to toxins. Annu. Rev. Entomol. 43, 571—594.

Hoy, M.A., 1990. Pesticide resistance in arthropod natural enemies: variability and selection responses. In: Roush, R.T., Tabashnik, B.E. (Eds.), Pesticide Resistance in Arthropods. Chapman and Hall, New York, pp. 203—236.

Hull, L.A., McPheron, B.A., Lake, A.M., 1997. Insecticide resistance management and integrated mite management in orchards: can they coexist? Pestic. Sci. 51, 359—366.

Jallow, M.F.A., Hoy, C.W., 2005. Phenotypic variation in adult behavioral response and offspring fitness in *Plutella xylostella* (Lepidoptera: Plutellidae) in response to permethrin. J. Econ. Entomol. 98, 2195—2202.

Jallow, M.F.A., Hoy, C.W., 2006. Quantitative genetics of adult behavioral response and larval physiological tolerance to permethrin in diamondback moth (Lepidoptera: Plutellidae). J. Econ. Entomol. 99, 1388–1395.

Jallow, M.F.A., Hoy, C.W., 2007. Indirect selection for increased susceptibility to permethrin in the diamondback moth (Lepidoptera: Plutellidae). J. Econ. Entomol. 100, 526–533.

Keiding, J., 1986. Prediction or resistance risk assessment. In: National Research Council, Pesticide Resistance: Strategies and Tactics for Management. National Academy Press, Washington, DC, pp. 279–297.

Lockwood, J.A., Sparks, T.C., Story, R., 1984. Evolution of resistance to insecticides: a reevaluation of the roles of physiology and behavior. Bull. Entomol. Soc. Am. 30, 41–51.

McGaughey, W.H., Whalon, M.E., 1992. Managing insect resistance to *Bacillus thuringiensis* toxins. Science. 258, 1451–1455.

McKenzie, J.A., 1996. Ecological and Evolutionary Aspects of Insecticide Resistance. Academic Press, Austin, TX.

Melander, A.L., 1914. Can insects become resistant to sprays? J. Econ. Entomol. 7, 167–173.

Miranowski, J.A., Carlson, G.A., 1986. Economic issues in public and private approaches to preserving pest susceptibility. In: National Research Council, Pesticide Resistance: Strategies and Tactics for Management. National Academy Press, Washington, DC, pp. 436–448.

Mochizuki, F., Fukumoto, T., Noguchi, H., Sugie, H., Morimoto, T., Ohtani, K., 2002. Resistance to a mating disruption composed of (Z)-11-tetradecenyl acetate in the smaller tea tortrix, *Adoxophyes honmai* (Yasuda) (Lepidoptera: Tortrididae). Appl. Entomol. Zool. 37, 299–304.

Mochizuki, F., Noguchi, H., Sugie, H., Tabata, J., Kainoh, Y., 2008. Sex pheromone communication from a population resistant to mating disruptant of the smaller tea tortrix, *Adoxophyes honmai* Yasuda (Lepidoptera: Tortricidae). Appl. Entomol. Zool. 43, 293–298.

Myers, J.H., Savoie, A., van Randen, E., 1998. Eradication and pest management. Annu. Rev. Entomol. 43, 471–491.

National Research Council, 1986. Pesticide Resistance: Strategies and Tactics for Management. National Academy Press, Washington, DC.

Onstad, D.W., Meinke, L.J., 2010. Modeling evolution of *Diabrotica virgifera virgifera* (Coleoptera: Chrysomelidae) to transgenic corn with two insecticidal traits. J. Econ. Entomol. 103, 849–860.

Pan, Z., Onstad, D.W., Nowatzki, T.M., Stanley, B.H., Meinke, L.J., Flexner., J.L., 2011. Western corn rootworm (Coleoptera: Chrysomelidae) dispersal and adaptation to single-toxin transgenic corn. Environ. Entomol. 40, 964–978.

Pereira, R.M., 2003. Areawide suppression of fire ant populations in pastures: project update. J. Agric. Urban Entomol. 20, 123–130.

Roush, R.T., 1989. Designing resistance management programs: how can you choose? Pestic. Sci. 26, 423–441.

Roush, R.T., 1994. Managing pests and their resistance to *Bacillus thuringiensis*: can transgenic crops be better than sprays?. Biocontrol Sci. Tech. 4, 501–516.

Roush, R.T., 1998. Two-toxin strategies for management of insecticidal transgenic crops: can pyramiding succeed where pesticide mixtures have not? Phil. Trans. R. Soc. Lond. B. 353, 1777–1786.

Roush, R.T., Tabashnik, B.E., 1990. Pesticide Resistance in Arthropods. Chapman and Hall, New York.

Sexson, D.L., Wyman, J.A., 2005. Effect of crop rotation distance on populations of Colorado potato beetle (Coleoptera: Chrysomelidae): development of areawide Colorado potato beetle pest management strategies. J. Econ. Entomol. 98, 716−724.

Shani, A., Clearwater, J., 2001. Evasion of mating disruption in *Ephestia cautella* (Walker) by increased pheromone production relative to that of undisrupted populations. J. Stored Prod. Res. 37, 237−252.

Siegfried, B.D., Meinke, L.J., Scharf, M.E., 1998. Resistance management concerns for areawide management programs. J. Agric. Entomol. 15, 359−369.

Smith, J.W., 1998. Boll weevil eradication: area-wide pest management. Ann. Entomol. Soc. Am. 91, 239−247.

Spohn, B.G., Zhu, J., Chastain, B.B., Haynes, K.F., 2003. Influence of mating disruptants on the mating success of two strains of cabbage loopers, *Trichoplusia ni* (Hübner) (Lepidoptera: Noctuidae). Environ. Entomol. 32, 736−741.

Stern, V.M., Smith, R.F., van den Bosch, R., Hagen., K.S., 1959. The integrated control concept. Hilgardia. 29, 81−101.

Svensson, G.P., Ryne, C., Löfstedt, C., 2002. Heritable variation of sex pheromone and the potential for evolution of resistance to pheromone-based control of the Indian meal moth, *Plodia interpunctella*. J. Chem. Ecol. 28, 1447−1461.

Tabashnik, B.E., 1989. Managing resistance with multiple pesticide tactics: theory, evidence, and recommendations. J. Econ. Entomol. 82, 1263−1269.

Tabashnik, B.E., Croft, B.A., 1982. Managing pesticide resistance in crop-arthropod complexes: Interactions between biological and operational factors. Environ. Entomol. 11, 1137−1144.

Tabashnik, B.E., Van Rensburg, J.B.J., Carrière, Y., 2009. Field-evolved insect resistance to Bt crops: definition, theory, and data. J. Econ. Entomol. 102, 2011−2025.

Tabata, J., Noguchi, H., Kainoh, Y., Mochizuki, F., Sugie, H., 2007a. Sex pheromone production and perception in the mating disruption-resistant strain of the smaller tea leafroller moth, *Adoxophyes honmai. Entomol. Exp. Appl.* 122, 145−153.

Tabata, J., Noguch, H., Kainoh, Y., Mochizuki, F., Sugie, H., 2007b. Behavioral response to sex pheromone-component blends in the mating disruption-resistant strain of the smaller tea tortrix, *Adoxophyes honmai* Yasuda (Lepidoptera : Tortricidae), and its mode of inheritance. Appl. Entomol. Zool. 42, 675−683.

Torres-Vila, L.M., Stockel, J., Lecharpentier, P., Rodríguez-Molina, M.C., 1997. Artificial selection in pheromone permeated air increases mating ability of the European grape vine moth *Lobesia botrana* (Lep., Tortricidae). J. Appl. Entomol. 121, 189−194.

United States Environmental Protection Agency, 2009. Pesticide fact sheet. <www.epa.gov/oppbppd1/biopesticides/pips/smartstax-factsheet.pdf>, 29 July 2009, (accessed 12.10.09.).

Wearing, C.H., Hokkanen, H.M.T., 1994. Pest resistance to *Bacillus thuringiensis*: case studies of ecological crop assessment for Bt gene incorporation and strategies of management. Biocontrol Sci. Tech. 4, 573−590.

Whalon, M.E., Mota-Sanchez, D., Hollingworth, R.M., 2008. Global Pesticide Resistance in Arthropods. CABI Publishing, Wallingford, UK.

Zhao, J.-Z., Collins, H.L., Shelton, A.M., 2010. Testing insecticide resistance management strategies: mosaic versus rotations. Pest. Manag. Sci. 66, 1101−1105.

CHAPTER 2

Valuing Pest Susceptibility to Control

Paul. D. Mitchell[1] and David. W. Onstad[2]

[1]Agricultural and Applied Economics, University of Wisconsin-Madison, Madison, WI
[2]DuPont Pioneer, Wilmington, DE

Chapter Outline

Resistance management involves understanding both the evolution of arthropods and the value of the evolved pest. Most chapters in this book describe prediction of the evolution of resistance as well as its management. This chapter, however, focuses entirely on the issues of preference and value, of which most biologists have only a vague understanding. Understanding the valuation of natural resources such as pest susceptibility, particularly from the perspective of economics, is an important foundation for the management of insect resistance.

Management implies that decision makers have goals and that resources and labor will be allocated to achieve these goals. Hence, as noted in the first chapter, insect resistance management (IRM) must be based on the goals of the decision makers. These goals require focus not only on particular resources and their values, but also on the time horizons over which these goals will be achieved and how to address the uncertainty of knowledge. For example, if stakeholders place a high value on low frequencies of resistance alleles, then one goal could be to minimize the expected resistance-allele frequency after a certain number of years, within the constraints of the decision maker. We often take our values for granted and express them implicitly when we are stating our goals. Thus, IRM goals are usually the starting point for studies of resource values and economics, with the values implicit in the stated goals.

This chapter separates the discussion into five sections. First, we provide a general overview of the classification of goods from an economic perspective, focusing on those types that pertain to IRM and integrated pest management (IPM). Second, we discuss the valuation of pest density and of pest susceptibility at a single point in time. These attributes of pest quantity and quality are the primary factors in most discussions of IPM and IRM, as well as the critical variables in most economic models. We realize that a complete evaluation of the costs and benefits of IRM and IPM should consider impacts beyond pest quantity and quality to address factors such as environmental quality and human health, but this additional evaluation would extend our efforts beyond the intended scope of the book. Third, we consider time preferences by discussing the use of discounting of future values. Given that IRM requires management over multiple years, we must quantitatively compare values from different times to evaluate different strategies. Fourth, we consider risk preferences by discussing methods commonly used to incorporate uncertainty into the decision-making process. Fifth, we develop simple illustrations of how these methods have been applied, with a brief overview of economic IRM models. At the end of these five sections, we draw some conclusions and suggest future work.

GOODS AND VALUES
Goods

To better understand the management of natural resources and environmental goods, economists classify goods based on the properties of rivalry and excludability. Rivalry describes how one person's consumption of the good changes the availability of the good for others, while excludability describes the extent to which others can be prevented from consuming the good. A private good is excludable and rival—one person's use excludes all others, and when consumed, the good no longer has value. A simple example is an apple purchased from a vendor—the buyer owns it and decides its use, and once consumed, the apple is gone. At the other extreme is a pure public good, which is nonexcludable and nonrival—all people consume the good (none are excluded), and each person's consumption does not reduce what is available for others. The air we breathe is a simple but not quite perfect example (one person's use of air can lessen or pollute its availability for another's use). However, the definition is a useful theoretical construct, though it is difficult to identify real-world goods that are absolutely nonrival and nonexcludable. Most public goods are, in some sense, not completely nonrival and nonexcludable.

Several types of impure public goods exist (OECD, 2001a; Cornes and Sandler, 1996). In terms of managing pests and resistance, open access and common property resources may occur. An open-access resource is nonexcludable and rival—anyone who wants to can obtain the good and once consumed it is gone, which leads to the classical "tragedy of the commons" problem (Hardin, 1968). Fishing stocks in the open ocean are probably the most well-known example. Common property resources are also rival goods but are excludable to outsiders, though with open access to those in the commons. Typical examples are aquifers and commonly held pastures (Bromley and Cernea, 1989). Before providing examples of these different types of private and public goods resulting from pest and resistance management, we discuss economic valuation.

Values

Deciding how to manage pests and their resistance to control requires placing a value on the pests, on the damage they cause, and on the parts and processes in the ecosystem affected by their management. For this discussion of IRM, we focus primarily on pest control and pest susceptibility as goods to be managed, noting that these goods are commonly measured with the pest population density and the frequency of susceptibility (or resistance) alleles among this population. These goods do not capture all the values that IRM can consider, but they serve as a convenient and important subset to illustrate the methods and issues. Incorporating other values would not change the general methodology illustrated here, but would extend the discussion beyond our intended scope. Let it suffice to say that we are not forgetting or ignoring such values, but rather are not explicitly including them here for convenience.

Economists define two types of value for goods—use value and non-use value (Figure 2.1). As the name implies, use value is the value a good possesses because it can be used or consumed by a person. Use values for a good or resource can include its direct use value for consumption or experience and/or its indirect use (or functional) value as a supplier of ecosystem services (Barbier, 1991, 2000; Young, 1992; Heal, 2000). For example, honeybees provide both honey (direct use value) and pollination (indirect use value), so that the use value of a honeybee population includes the sum of the value of the honey and pollination it produces. In addition, use value includes the value of the option to potentially use the good or resource in the future, which can be either a direct or an indirect use value. Such option values are often used to argue for the preservation of species biodiversity or ecosystems, since some species will likely be directly and/or indirectly useful for future problems or needs (Pearce and Moran, 1994; OECD, 2001b). Finally, a good can also have nonuse values that arise from the value attached to the existence of the good (existence value) and the possibility of maintaining the good for future generations to use (bequest value) (OECD, 2001a).

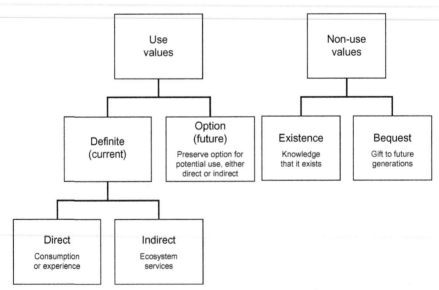

Figure 2.1 Diagram of economic values of goods. Analysis of insect resistance management focuses on use values.

The primary source of value identified in this book will be of the use type. For example, the value of pest density is related to the damage that the pest causes either to (a) some other good such as a crop, which can be consumed directly by humans, or (b) an ecosystem service, which benefits humans indirectly. Furthermore, some stakeholders likely place a positive option value on pest susceptibility that might be taken advantage of in the future, while others attach existence and bequest values for pest susceptibility as well. However, with the exception of Wesseler's (2003) option-value approach, the predominant value of pest susceptibility emphasized in discussions is the direct-use value of susceptibility for managing pest damage, now and in the future (e.g., Hueth and Regev, 1974; Regev et al., 1983; Plant et al.,1985; Mangel and Plant, 1983; Onstad and Guse, 1999; Onstad et al., 2003; Hurley et al., 2001, 2002; Hurley, 2005; Livingston et al., 2004; Mitchell and Onstad, 2005). Nevertheless, we note that, because pest populations and their genetic composition have bequest and existence values to individuals and societies, as well as indirect use values, pest control and susceptibility also have some public-good qualities. Empirical estimates of these values vary widely and consensus has yet to emerge, so that such values have not been incorporated into a quantitative IRM analysis.

Pest mobility and ecology together with social institutions and the environment determine the level of excludability in pest control and susceptibility. Control of a highly localized and genetically isolated pest population can be treated as a private good, with the benefits of control and resistance management largely captured by the local landowner. Scale insects (Hemiptera: Coccoidea) are an example of such pests, due to the low mobility many of the species exhibit in all life stages and their preference for long-lived hosts (Banks and Denno, 1994).

As a pest species becomes more mobile and has greater genetic exchange, pest density and susceptibility are more like common property resources than open-access resources. The distinction between common property and open-access pest populations often revolves around whether or not the "owners" of the pest population can potentially organize. If organized management is possible, then the population is a common-property resource; if it is not possible, then it is an open-access resource. Pest ecology, the natural environment, and social institutions must converge for successful common-property management of a pest (Knipling, 1979; Kogan, 1998; Gray, 1995; Olmstead and Rhode, 2008).

In some locations, pest mobility and natural barriers together create pest populations that are common-property resources (Regev *et al.*, 1976). For example, the surrounding desert and mountains in parts of the western United States create barriers that reduce entry of some pests, so that residents of the area can commonly manage pest control and susceptibility, as, for example, proposed by Carrière *et al.* (2006) for *Lygus hesperus* in Arizona. Social institutions can also create such barriers for common property management of some pests, as the successful eradication programs for boll weevil (*Anthonomous grandis*) and screwworm (*Cochliomyia hominivorax*) illustrate (Myers *et al.*, 1998; Olmstead and Rhode, 2008).

For some pests, mobility and/or the lack of barriers do not permit effective areawide management, which seems to be the case for *Diabrotica virgifera virgifera* in the United States (Gray, 1995; Spencer, Hughson, and Levine, Chapter 7 in this volume). As a result, the density and susceptibility of *D. virgifera virgifera* are managed as open-access resources, which has contributed to (but is not the sole cause of) the pest developing resistance to various insecticides, crop rotation, and recently Bt corn (Levine and Oloumi-Sadeghi, 1996; Metcalf, 1983; Meinke *et al.*, 1998; Wright *et al.*, 2000; Levine *et al.*, 2002, Spencer, Hughson, and Levine, Chapter 7; Gassman *et al.*, 2011). Clark and Carlson (1990) also find empirical support, based on analysis of the demand for insecticides, that farmers manage insect pests as common-property resources without group coordination. Numerous insect pests have developed resistance to control (CAST, 2004), providing additional evidence that pest susceptibility is used as an open-access resource without coordinated management.

For some pests, however, even uncoordinated, open-access use of pest control can generate areawide pest management benefits, even though the population is mobile and no geographic barriers exist. For example, Hutchison *et al.* (2010) demonstrate that voluntary, widespread farmer adoption of Bt corn has suppressed *Ostrinia nubilalis* in the U.S. Corn Belt, substantially benefiting both Bt corn and non-Bt corn acres. Similarly, Lu *et al.* (2012) demonstrate that widespread farmer adoption of Bt cotton in China has suppressed several major insect pests, thus reducing use of insecticide sprays and increasing biocontrol services from predatory insect species in cotton and other crops. The effects of areawide suppression on the evolution of insect resistance remain unclear.

VALUATION OF PESTS
Valuation of Pest Population Densities and Damage

In IPM, entomologists have traditionally focused on the economic damage that a population causes at a particular place and time. The concept of the economic threshold (Stern *et al.*, 1959; Onstad, 1987) shows that pest density must be considered in the economic context in which it occurs. For instance, a low pest density is acceptable as long as the damage that it could cause in the future does not exceed the cost of preventing that damage. At the farm/field level, most economic analyses of pest management focus on the effect of a pest population on the value of a crop. Usually, crop loss from a pest is some combination of a reduction in usable crop biomass (yield) and/or in crop quality, which imply a decrease in the market value of the harvested crop. Typically, this crop loss is conceptualized as some function of the pest population density (a damage function), while pest control reduces this population density or the damage it causes (a control function) (e.g., Mitchell *et al.*, 2004; Dun *et al.*, 2010; Tinsley *et al.*, 2012). The net value of pest control to the owner/manager is the value of the prevented crop loss from pest damage, minus the cost of pest control.

Two types of data are generally available for economic analysis of the value of pest damage and management: either *observational data* of actual (or aggregate) farm use of insecticides or *experimental data* consisting of measures of pest population density, crop damage or yield loss, and the efficacy of different control methods. Observational data on farmer behavior suffer what economists call an endogeneity problem and so should not be used to directly estimate pest damage or a pest-control production function as in the early analyses of Headley (1968) and Carlson (1977). Rather, because both the input (pesticide) and output (crop yield) are endogenous to the farmer's decision, the decision-making process must be explicitly modeled, with most analyses assuming profit-maximization or cost-minimization behavior by farmers. Lichtenberg and Zilberman's (1986) seminal paper effectively argued that for damage control inputs such as insecticides, standard econometric methods must explicitly specify the damage function (technically they express the model in terms of damage prevented, but the implication is the same). Their paper generated several responses and evaluations of their methodology (e.g., Babcock, Lichtenberg, and Zilberman, 1992; Carrasco-Tauber and

Moffit, 1992; Blackwell and Pagoulatos, 1992; Saha *et al.*, 1997; Carpentier and Weaver, 1997; Fox and Weersink, 1995; Hennessy, 1998), but in general, their econometric method has been accepted as the proper method for using observational data in estimating the productivity of pest-control inputs.

Experimental data avoids the endogenity problem because pest management is controlled independent of yield or pest pressure, and so the damage function can be directly estimated using more traditional regression techniques. Several empirical applications extend the literature to account for pest population dynamics (Shoemaker, 1973; Talpaz and Borosh, 1974; Talpaz *et al.*, 1978), uncertainty (Feder, 1979; Moffit *et al.*, 1984), interactions with secondary pests and other inputs (Harper and Zilberman, 1989), global concavity of the production function (Hennessy, 1998), management of insects as virus vectors (Marsh *et al.*, 2000), and separate identification of experimental errors and damage variability (Mitchell *et al.*, 2004; Dun *et al.*, 2010; Tinsley *et al.*, 2012).

Such analyses usually ignore any positive or negative externalities that farm-/field-level pest control generates, such as the value that pest control on one farm has for the pest population and crop losses on other farms or the cost of human health impacts and environmental damages from pest control. Exceptions to this generalization exist. Harper and Zilberman (1989) examine on-farm externalities from input interactions and secondary pests, and Regev *et al.* (1976) incorporate population effects on other farms. Theoretically, incorporating the effect of these and similar externalities in pest-control decisions is straightforward using taxes or subsidies, command and control policies, or other policy instruments so that on-farm decisions account for the actual human health and environmental costs of their pest control decisions (Baumol and Oates, 1988; Cornes and Sandler, 1996). However, practical application is problematic because individual farm contributions are difficult to measure and to value (Knight and Norton, 1989).

The difficulty measuring farm-specific contributions to environmental pollution implies that pest control inputs usually become nonpoint source pollution, which has remained notoriously difficult to regulate using policy tools based on economic theory (Ribaudo *et al.*, 1999). Even if accurate measurement of individual contributions to environmental pollution and human exposure were available, valuing the cost of these contributions is not clear. Tremendous advances in market and nonmarket valuation methods for estimating such costs have occurred (Champ *et al.*,

2003; Freeman, 2003; Maler and Vincent, 2005; Bazerman *et al.*, 1997; Willis and Corkindale, 1995; Vatn and Bromley, 1995), but practical application to pest-control externalities remains to be established. On a more positive note, some farmers earn price premiums for their products owing to some consumers' willingness to pay more for ecolabeled products (e.g., organic, pesticide-free, or IPM), which would seem to be a method for farmers to internalize the full cost/benefit of their pest-control decisions (see Wessells *et al.*, 2001 for a review of the economics of ecolabeling). However, it remains to be established whether these price premiums compensate farmers an amount equal to the actual value of the environmental damage they do not cause (Dosi and Moretto, 1998). Nor would these premiums necessarily lead to optimal supply of nonuse values from pest populations due to the public nature of these goods (Cornes and Sandler, 1996).

Valuation of Pest Susceptibility

The various stakeholders in pest susceptibility likely have different values for pest susceptibility. Those concerned with environmental damages and indirect use-value goods often want to protect pest susceptibility separately from concern for pest density. Companies selling toxins and transgenic insecticidal plants want to maintain susceptibility to valuable products to maintain their sales at least until patents expire. Insecticide users are generally less concerned about a particular product and tend to relate the value of pest susceptibility to pest control and damage. The U. S. Environmental Protection Agency (EPA), which regulates commercialization of insecticides, has not made maintenance of pest susceptibility a requirement for registration of insecticides, with the exception of transgenic insecticidal crops.

Defining and valuing pest susceptibility may depend on the control tactic. Should the susceptibility of a pest population to a synthetic toxin developed by a private company be considered a private good of the patent holder, or is pest susceptibility an open-access resource that should be regulated for the public good? If the number of molecular or behavioral mechanisms for resistance are limited and already exist in the pest population, and if cross-resistance to several toxins or IPM tactics is a real possibility, then should pest susceptibility be considered a common property resource of those developing new products and tactics? Society may hold a high option value for a lack of cross-resistance among pest populations when a corporation commercializes a new and unique toxin—should it regulate the toxin differently than other commercialized toxins? If a pest develops resistance to an unpatented IPM tactic, whose resource was consumed?

From a legal perspective, at this time in the United States, property rights for ownership of pest susceptibility remain incompletely enforced. For synthetic toxins, the patent holder must register the product for commercialization under the Federal Insecticide, Fungicide, and Rodenticide Act (FIFRA), which the EPA enforces. Thus far, the EPA has not chosen to impose resistance-management requirements on those registering synthetic pest-control products under FIFRA, with the exception of transgenic insecticidal crops (Bt corn, Bt sweet corn, Bt cotton). Hence, ownership of pest susceptibility to synthetic toxins (other than transgenic insecticidal toxins) rests in some legal sense with the product registrants, in the sense that companies manage resistance to their registered compounds as they see fit. However, this property right to susceptibility is not completely enforceable. For example, suppose two companies patent insecticides at the same time with similar modes of action; one company registers and markets its insecticide immediately, while the second company waits. Resistance develops to the first insecticide, including cross-resistance to the second insecticide, so that the second company's product is worthless. Under current U. S. legal precedent, the second company cannot successfully sue the first company for damages because the first company made the second company's insecticide worthless. Hence, companies do not completely own pest susceptibility.

Ownership of pest susceptibility to other control methods also remains undefined. For example, who owns pest susceptibility to crop rotation? As Spencer, Hughson, and Levine (Chapter 7) indicate, crop rotation was an effective pest control method for *Diabrotica virgifera virgifera* and *Diabrotica barberi* for many years, but both species evolved resistance by changing egg-laying behavior and extending egg diapauses, respectively (Levine and Oloumi-Sadeghi, 1996; Krysan *et al.*, 1986; Levine *et al.*, 1992). If a pest develops resistance to biological or cultural control, whose resource was consumed? Generally, it seems that pest susceptibility to IPM is an open-access or common property resource with an undefined legal status, though this need not be the case. Ownership of other goods of this sort has been legally defined. For example, the federal government has auctioned the right to use different radio frequencies to private companies (Ahrens, 2006; MacAfee and MacMillan, 1996).

Without any legal definition of property rights, susceptibility remains an open-access resource without institutional barriers to its access. Economic theory and historical experience for other open-access resources suggest that pest susceptibility is then subject to the tragedy of the commons problem of overexploitation and depletion. The only notable exception to this generalization is pest susceptibility to transgenic insecticidal crops. The EPA chose to require resistance-management plans for Bt crops because the insecticidal proteins were found to be "in the pubic interest," with the EPA's goal being to prevent "unreasonable adverse effects on the environment" as mandated by FIFRA (Berwald et al., 2006, pp. 23–24). Indeed, Berwald et al. (2006, p. 33) go on to explain: "EPA considers pest susceptibility to Bt a common property resource, where a policy goal is to avoid depletion of this resource." Hurley (2005) reaches a similar conclusion concerning EPA policy goals for IRM.

The EPA could use the same arguments to justify requiring resistance management for all registered pesticides, not just those expressed in transgenic insecticidal plants. Why the EPA has not done so is not clear, though conjectures are possible. Perhaps the public benefits of transgenic insecticidal crops were considered greater than for conventionally delivered synthetic insecticides, or perhaps the threat of resistance evolving rapidly was perceived as greater due to the expected (and realized) rapid adoption of transgenic insecticidal crops. Another possibility is that setting a regulatory precedent for a radically new class of insecticides was politically easier than trying to change regulatory policy for the numerous conventional pesticides already registered. Regardless of EPA motives, given the long history of the evolution of resistance to many products and the potential for cross-resistance and other interactions, how can regulations logically omit a large class of chemicals? Perhaps the new regulatory precedent set for transgenic insecticidal crops will lead to IRM requirements for other pesticides.

Conceptual and empirical issues remain for economists analyzing IRM. Whose objective should be modeled, that of farmers, companies, or regulators? What values should be incorporated into the analysis: direct-use values, indirect-use values, and/or nonuse values? A quick examination of the research literature indicates that most economic analyses of IRM focus on the use value of the insecticide for controlling damaging pest populations (e.g., Hueth and Regev, 1974; Regev et al., 1983; Plant et al., 1985; Mangel and Plant, 1983; Onstad and Guse, 1999;

Onstad *et al.*, 2003; Hurley *et al.*, 2001, 2002; Hurley, 2005; Livingston *et al.*, 2004; Mitchell and Onstad, 2005). The use value to farmers is often the focus, but some analyses focus on the use value to others as well. For example, Hueth and Regev (1974) explain how their results would change if farmers managed the pest for the common good instead of their individual good. Regev *et al.* (1983) examine the difference between decentralized decision making by farmers and centralized decision making by a planner maximizing the social benefit (the sum of consumer and producer surplus). Hurley *et al.* (2002) include changes in economically optimal insecticide use in an IRM model for insect resistance to transgenic insecticidal corn. The analyses of Morel *et al.* (2003) and Wesseler (2003) both include the social benefit and indirect-use values in a conceptual (nonempirical) analysis of IRM for transgenic insecticidal crops to illustrate general effects and principles.

Alix-Garcia and Zilberman (2005) examine the effect of the pesticide market structure on the evolution of resistance. A standard theoretical and common empirical finding is that unregulated monopolists raise prices to restrict the supply of their goods and increase their profits, which reduces social welfare. In the context of pesticides, this implies that a patent-holding monopolist will sell less pesticide than is socially optimal. However, because the problem for pest susceptibility as an open-access resource is overexploitation, the monopolist's restriction of pesticide supply offsets this overexploitation. The issue then, as Alix-Garcia and Zilberman (2005) point out, is whether the welfare loss due to restriction of pesticide supply exceeds the welfare gain from slower consumption of pest susceptibility. They show for reasonable parameterizations of their model that, indeed, it is possible for the monopolist to delay the evolution of resistance more than is socially optimal (considering only the direct use values of pesticides for agricultural production). The main policy implication for IRM is that the open-access nature of pest susceptibility is not necessarily a reason to impose resistance management on pesticides, because the structure of the pesticide market also matters and can even offset distortions due to the open-access problem.

Besides providing an excellent review of the pertinent economic and public health literature on resistance management, Goeschl and Swanson (2001) extend the standard economic analysis of IRM by modeling a coevolutionary process in which a pest population evolves resistance and a research and development market creates new technologies to sell to farmers to control a pest population. They explicitly model the research and development process and incorporate economic optimality into farmer pest-control decisions (i.e., farmers only treat when it is economically beneficial). Their results are too rich to fully summarize here, but a key insight they offer is that, conceptually, evolving pests are like a competitor costlessly developing new products that erode the market share of companies developing new products, which can cause research and development to collapse.

DISCOUNTING AND VALUING THE FUTURE

The previous discussion concerned the problem of valuing resources such as a pest population or pest susceptibility in the present time. However, IRM usually requires valuing these goods over long periods of time, often years or decades, which leads to the problem of how to value a resource in the future. Valuation of future benefits and costs typically uses the concept of (time) discounting, which is a method to convert the value of a future cost or benefit to its value in another time period, most commonly the present time. Discounting assumes that the value of a good in the future is different from the value of the same good in the current time, and discounting provides a method for comparing these values by converting between them. For example, the current value of possessing $100 today is not the same to most people as the current value of possessing $100 ten years from now, and a discount factor converts the future $100 into its present value.

The justification for time discounting arises from the common practice in financial markets and from human behavior. Financial markets use discount rates to determine the price for trading assets with future value, and these discount rates define the interest charged for loans or paid for deposits. Studies of human behavior demonstrate the consistent devaluation (discounting) of future costs and benefits; people have a strong preference for immediate gratification over delayed gratification (e.g., Soman *et al.*, 2005). Commonly, the psychological discounting consistent with human behavior has been implemented for valuing goods intertemporally using the same methods as financial markets (i.e., use of interest or discount rates) (Frederick *et al.*, 2002).

Discounting for resource management also requires a time horizon—how far into the future do we measure the value of the resource? The time horizon is the final point for the time-discounted economic analysis.

It can also be thought of as the endpoint defining the period during which a stakeholder will evaluate resource-management decisions. All resource values after the time horizon are ignored as either too small (discounted too much) or irrelevant (e.g., because the farmer will have retired), so that they do not need to be considered. Alternatively, these values can be captured by the resource's "salvage value"—the value of the resource after the time horizon.

Salvage value, a concept borrowed from financial analysis, is the value of a capital investment at the end of its useful life for an investor. In resource economics, the salvage value of a resource is its value after the time horizon into the infinite future in its best possible uses (including nonuse values). For example, Secchi *et al.* (2006) in an IRM analysis use a salvage value derived from the annualized net present value of agricultural production after the time horizon, assuming the introduction of new pest-control technologies to replace those that become obsolete due to the evolution of resistance.

The discount rate, which is comparable to the interest rate on a loan or a deposit, is used to derive the discount factor. Mathematically, the per-period discount rate d determines the discount factor δ that converts a future value into an equivalent present value with the following formula: $\delta(t) = [1/(1+d)]^t$, where t is the number of time periods until the time horizon (Figure 2.2). In continuous time, the discount factor $\delta(t) = \exp(-dt)$, where

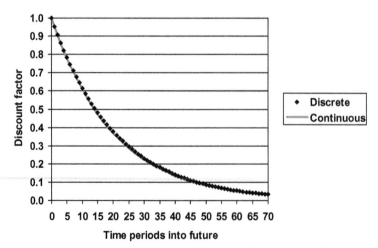

Figure 2.2 Discount factor declines over time. As the factor approaches zero, so too will the present (perceived) economic value of a good produced in the future.

t is now the length of time between the present and the time horizon. With a 5% annual discount rate, the discount factor for a ten-year time horizon is $\delta = [1/(1+0.05)]^{10} = 0.4632$ with discrete time and $\delta = \exp(-0.05*10) = 0.4493$ with continuous time, so that the present value of $100 paid ten years in the future is $46.32 or $44.93 today, depending on which discount formula is used. Note how the discount factors are insignificant (less than 0.03) by year 70 with a 5% discount rate (Figure 2.2). The discount factor would decline even faster with a higher discount rate. Economists and economic models place little value on goods produced or maintained beyond the time at which the discount factor approaches zero.

Often, an asset or activity generates a stream of income, that is, a series of net returns over several time periods. In such cases, the entire income stream is discounted back to its present value, which is termed its net present value (NPV). For example, suppose a crop field generates a net return of π_t each year t, where π_t varies depending on the crop planted in the rotation. The NPV of the stream of returns generated by the crops from this field over a ten-year time horizon is

$$\text{NPV} = \sum_{t=1}^{10} \pi_t/(1+d)^t = \sum_{t=1}^{10} \delta(t)\pi_t \tag{2.1}$$

In some cases, the NPV of an asset or activity is converted into an annuity—the constant return each period that generates the same NPV as the varying payment. An annuity value (sometimes called an annualized NPV) is calculated as NPV/k, where k, the present-value annuity factor, equals the sum of the discount factors $\delta(t)$ from the current period to the time horizon. Continuing the crop-return example, the present-value annuity factor is

$$k = \sum_{t=1}^{10} \delta(t) = \sum_{t=1}^{10} 1/(1+d)^t \tag{2.2}$$

so that the annuity equivalent NPV is

$$\text{NPV}_{ae} = \sum_{t=1}^{10} \delta(t)\pi_t / \sum_{t=1}^{10} \delta(t) = \sum_{t=1}^{10} \delta(t)\pi_t/k = \frac{1}{k}\sum_{t=1}^{10} \delta(t)\pi_t \tag{2.3}$$

For example, suppose π_t is $100/ha for corn and $75/ha for soybeans in a rotation; the NPV of this income over ten years with a discount rate of 5% is $678/ha (beginning with corn in year 1), the present-value annuity factor is 7.7217, and the annuity equivalent NPV is $87.80/ha per year.

The debate concerning the use of discounting for valuing natural resources concerns many issues, most of which are beyond the scope of this chapter. For the purposes of this chapter, we highlight two areas. First, deriving a technical form for the discount function so that it is more consistent with human behavior is an active area of research. Frederick et al. (2002) review many of these areas, such as hyperbolic discounting (a discount rate r that varies with the time period t) and loss aversion (a higher discount rate for a future gain than for a future loss) (also see Groom et al., 2005 and Gollier, 2001). Second, much debate exists concerning the appropriate discount rate to use for environmental goods. This research finds empirical support for use of lower discount rates for environmental goods and shows that different discount rates should be used for the different (use and nonuse) values of the same resource (Weitzman, 1998; Henderson and Bateman, 1995; Luckert and Adamowicz, 1993; Weikard and Zhu, 2005). For example, Luckert and Adamowicz (1993) analyze survey data to show that, empirically, people express lower discount rates for publicly managed and environmental goods relative to private goods and financial assets. In terms of IRM, this research implies that lower discount rates than are typical for private goods and hyperbolic discounting are more appropriate, both of which would increase the value of maintaining pest susceptibility.

RISK

Thus far, our discussion of pest control and IRM has assumed no uncertainty in the information used to make the management decision, though typically few decisions are made under such conditions. Understanding and modeling human decision making under uncertainty is a large and active area of research beyond the scope of this chapter. The goal of this section is to explain the intuition of methods commonly used in economic models of IRM to account for uncertainty.

The traditional economic approach to incorporating uncertainty into decision making is first to convert all uncertainty into monetary outcomes with associated probabilities, which converts the uncertainty into risk—known events with known probabilities that can be expressed as a cumulative distribution function or probability density function. Next, the preferences of the decision maker in terms of monetary risk are specified. Three general types of risk preferences are recognized—risk neutral, risk averse, and risk loving, which are easiest to explain in terms of how a person responds to an uncertain outcome relative to the case with no uncertainty and the same expected (mean) outcome. Risk-neutral persons are neutral to uncertainty in the sense that they are indifferent between a certain outcome and an uncertain outcome with a mean value equal to the certain outcome. A risk-averse person prefers the certain outcome to the uncertain outcome with the same mean, while a risk-loving person prefers the uncertain outcome to the certain outcome with the same mean (Chavas, 2004, pp. 31–51; Eeckhoudt *et al.*, 2005, pp. 3–23). For example, in a game with a 10% chance of gaining $1000 and a 90% chance of losing $111.11, the mean gain is essentially zero dollars. The risk lover would play the game, the risk-averse person would not play the game, and the risk-neutral person could choose either behavior.

The empirical evidence indicates that most people exhibit risk-averse behavior for most decisions, so that they value uncertain outcomes at some level less than the mean. Hence, the issue for conceptual and empirical analyses is how to incorporate into the economic analysis a cost or reduction in benefits due to uncertainty. The standard method is to assume some form of utility function to transform monetary outcomes into utility. Utility is a theoretical construct that measures the satisfaction a good gives to a consumer or user. Another term used to describe utility is preferences. Risk aversion implies that individual preferences exhibit diminishing marginal utility with respect to monetary outcomes; that is, the more money a person receives, the smaller the increase in utility. In the case of pest management and IRM, pest control is one of the goods (Figure 2.3). By using an efficacy function and/or a damage function, pest control is converted into monetary outcomes and then into utility. Given constant cost per unit of control, as more and more pest control measures are applied, total utility increases with additional pest control, but each new increment of control provides a decreasing amount of marginal utility (Figure 2.3). Thus, risk aversion implies diminishing marginal utility in pest control.

Economic analyses usually focus on the expected value of utility, not the expected monetary value. For decisions under uncertainty, a utility

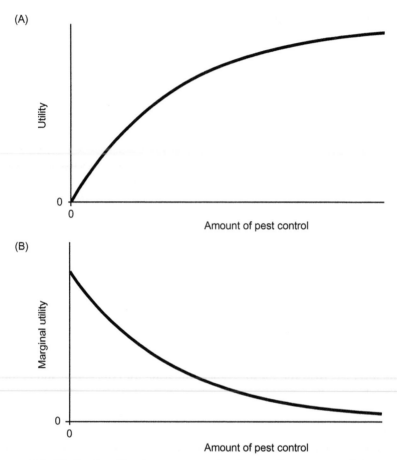

Figure 2.3 (A) Total utility increases as the level of pest control increases; (B) Marginal (incremental) utility to a farmer declines for each additional amount of pest control.

function imposes a cost to risk in much the same way that a discount function imposes a cost on a benefit not realized until the future. Combining a utility function with time discounting to model the simultaneous management of risk and intertemporal substitution significantly complicates the optimization process. Such problems are well studied in finance and macroeconomics (see Gollier, 2001, for an overview) and, to some extent, in resource and agricultural economics (e.g., Peltola and Knapp, 2001; Knapp and Olson, 1996; Lence, 2000), but are beyond the scope of this book. Specific to insect management, Mitchell and Hutchison (2008) provide a concise and readable overview of methods for incorporating risk into IPM, with examples to illustrate applications and differences.

This description of uncertainty and risk from an economic perspective is far from complete; the economics of risk is a large literature, even in the context of just pest management. For example, we did not discuss generalized expected utility, safety-first preferences, safe minimum standards, or stochastic dominance. Chavas (2004) provides a readable introduction with several empirical examples, but unfortunately none concerning insects or IRM. Gollier (2001) also provides a useful overview and summary of the economics of risk management.

OVERVIEW OF ECONOMIC MODELS

For this section, we first define notation and then illustrate the concepts presented in previous sections of this chapter and overview recent papers. To capture the essence of the IRM problem, assume that a population genetics model generates two key outputs that are used for economic analysis: the frequency of a single resistance allele (r_t) and population density of a single pest (n_t) in a landscape, where the subscript t indicates the time period (or generation). This notation assumes a discrete time period; equivalent notation for a continuous time model is $r(t)$ and $n(t)$. This fairly simplified assumption abstracts from important issues to capture the essence of population genetics models as used by economic analyses of IRM. The manager chooses ϕ_t, the proportion of the landscape to treat for the pest, or equivalently, the proportion of the pest population to treat in period t. In some manner defined by the population genetics model (but unnecessary to explain here), both the frequency of the resistance allele and the pest population density depend on the manager's choice of ϕ_t and the previous level of resistance and the pest density, which we denote as $r_t(\phi_t, r_{t-1}, n_t)$ and $n_t(\phi_t, n_{t-1}, r_t)$.

The manager derives different types of value from goods or services provided by the pest population and its level of resistance, such as the various direct- and indirect-use and nonuse values as described in previous sections. In some manner defined by the economic model (but unnecessary to explain here), the monetary value of each of these goods depends on the pest population and its level of resistance, which we denote as $V_{tj}(r_t(\phi_t, r_{t-1}, n_t), \ n_t(\phi_t, n_{t-1}, r_t))$ for value type j in period t.

For example, $j = 1$ may denote all direct use values, $j = 2$ may denote all indirect use values, and $j = 3$ may denote all nonuse values, but more values are possible, since each of these types of values can be further separated into subtypes.

Based on this simplified abstract model, the manager's IRM problem can be expressed as

$$\max_{\phi_t, \forall t} \sum_{t=1}^{T} \sum_{j=1}^{J} \delta_{tj} V_{tj}(r_t(\cdot), n_t(\cdot)) \quad \text{with} \quad \delta_{tj} = [1/(1+d_{tj})]^t \qquad (2.4)$$

where δ_{tj} is the discount factor for time period t for value type j, d_{tj} is the discount rate for period t for value type j, T is the time horizon, and J is the number of value types. The population genetics model determines how $r_t(\phi_t, r_{t-1}, n_t)$ and $n_t(\phi_t, n_{t-1}, r_t)$ change over time and are affected by the manager's choice of ϕ_t. Equation 2.4 represents a scenario in which the manager's IRM problem is to choose the proportion of the pest population to treat in each time period to maximize the net present value of the discounted stream of the different types of value derived from the pest population and its level of resistance, where the underlying population-genetics model describes pest population dynamics and the evolution of resistance. Note that this general specification (Equation 2.4) uses d_{tj} to allow time-specific discount rates (such as implied by hyperbolic discounting) and value-specific discount rates (so the private- and public-good aspects of a pest population can be discounted at different rates).

Equation 2.4 is the most popular specification for economic analysis of IRM. Hurley et al. (2001) (based on Hurley et al.'s 1997 working paper) and Onstad and Guse (1999) were among the first to analyze the economics of IRM for managing Ostrinia nubilalis resistance to transgenic insecticidal corn. Their analyses differ in terms of the details of the population–genetics and economic models, but they both use one type of value (farmer returns) ($J = 1$), a single discount rate ($d_{tj} = d$ for all t and j), and the manager's choice variable (the proportion of conventional refuge corn to plant) does not vary across time ($\phi_t = \phi$ for all t). Other studies extend these initial analyses in different ways by relaxing key assumptions.

Secchi et al. (2006) also examine IRM for Ostrinia nubilalis and transgenic insecticidal corn. Their analysis uses the same general assumptions ($J = 1$, $d_{tj} = d$ for all t and j), but compares results for a static (time invariant) refuge and a dynamic (time varying) refuge. In addition, they add a salvage value to farmer returns derived from the annualized net present value of agricultural production, which captures the effect of allowing the introduction of new technologies to replace control methods that become obsolete due to the evolution of resistance. Livingston *et al.* (2004) examine IRM for two pests (*Heliothis verescens* and *Helicoverpa zea*) and two control methods (transgenic insecticidal cotton and conventional insecticide) using refuge either treated or untreated with an insecticide. Their analysis also uses the same general assumptions ($J = 1$, $d_{tj} = d$ for all t and j) and compares results with static and dynamic refuge. Onstad *et al.* (2003) use the same general assumptions to examine the economics of different strategies to manage *D. virgifera virgifera* resistance to crop rotation (Levine *et al.*, 2002). Specifically, they use a single value (farmer income) and a single discount rate, and they assume time-invariant implementation of each management practice. Laxminarayan and Simpson (2002) develop and solve the same basic model ($J = 1$, $d_{tj} = d$ for all t and j, and $\phi_t = \phi$ for all t) in a continuous time framework. In a primarily conceptual analysis, Brock and Xepapadeas (2003) develop a continuous time model for the economic management of genetic diversity and pest resistance using the same basic assumptions ($J = 1$, $d_{tj} = d$ for all t and j, and $\phi_t = \phi$ for all t). Finally, Mitchell and Onstad (2005) also use the same basic assumptions ($J = 1$, $d_{tj} = d$ for all t and j, and $\phi_t = \phi$ for all t) when exploring the impact of extended diapause on the development of resistance to Bt corn among *Diabrotica barberi* populations.

Incorporating uncertainty into the IRM problem is the next model extension. Uncertainty can arise for a variety of reasons, such as weather variability or lack of knowledge concerning biological parameters. Regardless of the source, such factors imply that the pest population density $n_t(\phi_t, n_{t-1}, r_t)$ and/or the level of resistance $r_t(\phi_t, r_{t-1}, n_t)$ are random. For this abstract model, the manager's IRM problem can be expressed as

$$\max_{\phi_t, \forall t} E_{r,n} \left[\sum_{t=1}^{T} \sum_{j=1}^{J} \delta_{tj} V_{tj}(r_t(\cdot), n_t(\cdot)) \right] \qquad (2.5)$$

where $E_{r,n}[\cdot]$ denotes the expected value over the random variables r and n. Because the order of integration and summation are interchangeable here, the problem can also be expressed as

$$\max_{\phi_t, \forall t} \sum_{t=1}^{T} \sum_{j=1}^{J} \delta_{tj} E_{r,n}[V_{tj}(r_t(\cdot), n_t(\cdot))] \qquad (2.6)$$

as long as the discount factor δ_{tj} does not depend on n_t or r_t. This problem is the same as for the deterministic case (Equation 2.4), except that the manager chooses the proportion of the pest population to treat in each time period to maximize the *expected* net present value of the discounted stream of the different types of value derived from the pest population and its level of resistance. In other words, the stochastic problem is the same as the previous deterministic problem, except that the value function $V_{tj}(\cdot)$ is replaced with its expected value $E_{r,n}[V_{tj}(\cdot)]$. As for Equation 2.4, the underlying population-genetics model describes pest-population dynamics and the evolution of resistance.

Conceptually, the replacement of the value function with its expected value is simple, but empirical implementation is difficult because the population-genetics model outputs $n_t(\cdot)$ and $r_t(\cdot)$ and/or the value function $V_{tj}(\cdot)$ are typically highly nonlinear functions of the random variable(s), so that closed-form solutions for the expected value $E_{r,n}[V_{tj}(\cdot)]$ cannot be derived analytically. As a result, empirical analysis requires use of numerical procedures such as Monte Carlo integration or quadrature (Press *et al.*, 1992), so that empirical applications to IRM for insecticides or transgenic insecticidal crops are less numerous.

Hurley *et al.* (2002) developed an IRM model for *Ostrinia nubilalis* and transgenic insecticidal corn with a random annual pest population density, as well as random parameters to capture uncertainty about specific genetic factors (i.e., initial frequency of a resistance allele, heterozygote survival rate). The analysis uses two values ($J = 2$): the value of agricultural production for farmers and revenue collected by the company selling the transgenic insecticidal corn technology, which are both direct use values. However, the same discount rate is used for both values ($d_{tj} = d$ for all t and j), and the decision maker is a social planner who chooses a time-invariant refuge ($\phi_t = \phi$ for all t) to maximize the expected net present value of the sum of farmer returns and company revenue. Monte Carlo integration is used to solve for expected values. The required simulations also allow calculation of the risk of resistance—the probability that the resistance-allele frequency exceeds a set threshold after a set number of time periods or generations (they use a threshold of 50% and 30 generations = 15 years). Hurley (2005) uses a similar model to examine the effects of partial adoption of transgenic insecticidal corn and partial compliance with refuge requirements on the evolution of resistance. Both adoption and compliance depend on the difference between expected farmer returns with transgenic insecticidal corn and with conventional corn in each period (which depends on the current level of resistance), with equations calibrated to fit the general characteristics of observed transgenic insecticidal corn adoption and compliance data.

Incorporation of the manager's risk preferences into the IRM problem is the next key extension. However, as noted in the previous section, combining a manager's utility function with time discounting to model the simultaneous management of risk and intertemporal substitution significantly complicates the optimization process. We found no applications that analyzed IRM with such models, though Secchi and Babcock (2001, 2002) analyze the economics of managing bacterial resistance to antibiotics, combining a simple utility function with time discounting in a model. It is not clear how much the analysis is improved by the increased complexity and difficulty in solving such models in the context of IRM. A practical approach used in some analyses is to use measures, such as the risk of resistance, to provide some quantification of risk in stochastic IRM models, maximizing the expected net present value of agricultural productivity (e.g., Hurley et al., 2002; Hurley, 2005).

CONCLUSIONS

We hope that we have demonstrated the importance of economics for the management of pests and pest resistance. Unfortunately, institutions that significantly influence those working in this field do not always facilitate the use of economics. For example, Onstad and Knolhoff (2009) surveyed several important journals representing the discipline of economic entomology and discovered that less than 1% of research papers published from 1972 to 2008 include economic evaluations of pest management tactics. They also found that at least 85% of these analyses were performed by entomologists, not economists. Onstad and Knolhoff (2009) concluded that academia rarely supports training in economics for entomologists, and agencies that provide grants often fail to encourage economic analyses in funded projects.

Both IPM and IRM share a foundation in economics. Some may argue that we cannot place a monetary value on many goods and services affected by pest management and IRM. This may be true, but rational decisions still depend on the relative valuation of these goods and services in some manner. The risks of resistance evolution are not just ecological changes but also the potential losses to the health and livelihoods of millions of people who benefit from pest management. Thus, IRM must consider more than biology to design and implement practical, feasible, and effective strategies, a point emphasized by Onstad *et al.* (2011). Furthermore, in this book, Onstad (Chapters 1 and 16) and Hurley and Mitchell (Chapter 13) indicate that social, regulatory, and educational factors must be considered if coordination or cooperation of stakeholders is necessary to implement successful IRM.

Valuing a resource in the present is usually feasible, but valuing future resources is often difficult. For instance, how long should susceptibility be preserved and at what cost to individuals and society? To make predictions and assess risks, we must decide how far into the future we need to place values on resources. The choice of time horizon is based not only on a concern for the future but also on practical issues. Can our institutions make plans and maintain efforts over the long term? Can politicians focus on time periods beyond the next election?

Much of this chapter has been about philosophy as well as technique. Who owns susceptibility to an insecticide manufactured by a corporation? Who benefits from and controls susceptibility to crop rotation? Who should own and control these goods? How do we balance and account for the variety of values each stakeholder group places on pest density and pest susceptibility? How do we determine whether the social value of pest susceptibility justifies governmental regulation of IRM? None of these questions have easy answers, but they must be discussed and debated in academia and society.

REFERENCES

Ahrens, F., 2006. FCC Wireless Auction Could Open Up Airwaves. Washington Post, September 19, p. D1.

Alix-Garcia, J., Zilberman, D., 2005. The Effect of Market Structure on Pest Resistance Buildup. Working Paper, Department of Economics, University of Montana.

Babcock, B.A., Lichtenberg, E., Zilberman, D., 1992. Impact of damage control and quality of output: estimating pest control effectiveness. Amer. J. Agric. Econ. 74, 163–172.

Banks, L.M., Denno, R.F., 1994. Local adaptation in the armored scale insect *Pseudaulacaspis pentagona*—Homoptera: Diaspididae. Ecology. 75 (8), 2301–2310.

Barbier, E.B., 1991. Environmental degradation in the Third World. In: Pearce, D. (Ed.), Blueprint 2: Greening the World Economy. Earthscan, London.

Barbier, E.B., 2000. Valuing the environment as input: review of applications to mangrove-fishery linkages. Ecol. Econ. 35, 47–61.

Baumol, W.J., Oates, W.A., 1988. The Theory of Environmental Policy. Cambridge University Press, Cambridge.

Bazerman, M.H., Messick, D.M., Tenbrunsel, A.E., Wade-Benzoni, K.A., 1997. Environment, ethics, and behavior: the psychology of environmental valuation and degradation. The New Lexington Press management series and the New Lexington Press social and behavioral science series, 393 pp.

Berwald, D., Matten, S., Widawsky, D., 2006. Economic analysis and regulating pesticide biotechnology at the U. S. environmental protection agency. In: Just, R.E., Alston, J.M., Zilberman, D. (Eds.), Regulating Agricultural Biotechnology: Economics and Policy. Springer, New York.

Blackwell, M., Pagoulatos, A., 1992. The econometrics of damage control: comment. Amer. J. Agric. Econ. 74, 1040−1044.

Brock, W.A., Xepapadeas, A., 2003. Valuing biodiversity from an economic perspective: a unified economic, ecological, and genetic approach. Amer. Econ. Rev. 93, 1597−1614.

Bromley, D.W., Cernea, M.M., 1989. The Management of Common Property Natural Resources, World Bank Discussion Papers, No. 57. World Bank, Washington, DC.

Carlson, G.A., 1977. Long-run productivity of insecticides. Am. J. Agric. Econ. 59, 543−548.

Carpentier, A., Weaver, R.D., 1997. Damage control productivity: why econometrics matters. Amer. J. Agric. Econ. 79, 47−61.

Carrasco-Tauber, C., Moffit, L.J., 1992. Damage control econometrics: functional specification and pesticide productivity. Amer. J. Agric. Econ. 74, 158−162.

Carrière, Y., Ellsworth, P.C., Dutilleul, P., Ellers-Kirk, C., Barkley, V., Antilla, L., 2006. A GIS-based approach for areawide pest management: the scales of *Lygus Hesperus* movements to cotton from alfalfa, weeds, and cotton. Entomol. Exp. Appl. 118, 203−210.

Champ, P.A., Boyle, K.J., Brown, T.C., 2003. A Primer on Nonmarket Valuation. Kluwer Academic Publishers, Dordrecht, The Netherlands.

Chavas, J.P., 2004. Risk Analysis in Theory and Practice.. Elsevier Academic Press, San Diego, CA.

Clark, J.S., Carlson, G.A., 1990. Testing for common versus private property: the case of pesticide resistance. J. Environ. Econ. Manage. 19, 45−60.

Cornes, R., Sandler, T., 1996. The Theory of Externalities, Public Goods and Club Goods, second ed. Cambridge University Press.

Council for Agricultural Science and Technology (CAST), 2004. Management of pest resistance: strategies using crop management, biotechnology, and pesticides. Special Publication 24, Ames, IA, CAST.

Dosi, C., Moretto, M., 1998. Is Ecolabeling a Reliable Environmental Policy Measure? Working Paper 1999.9. Fondazione Eni Enrico Mattei, Milan, Italy.

Dun, Z., Mitchell, P.D., Agosti, M., 2010. Estimating *Diabrotica virgifera virgifera* damage functions with field data: applying an unbalanced nested error component model. J. Appl. Entomol. 134, 409−419.

Eeckhoudt, L., Gollier, C., Schlesinger, H., 2005. Economic and Financial Decisions under Risk. Princeton University Press, Princeton, NJ.

Feder, G., 1979. Pesticides, information, and pest management under uncertainty. Amer. J. Agric. Econ. 61, 97−103.

Fox, G., Weersink, A., 1995. Damage control and increasing returns. Amer. J. Agric. Econ. 77, 33−39.

Frederick, S., Loewenstein, G., O'Donoghue, T., 2002. Time discounting and time preference: a critical review. J. Econ. Lit. 40, 351−401.

Freeman, M.A., 2003. The Measurements of Environmental and Resource Values: Theory and Methods. Resources for the Future, Washington, DC.

Gassman, A.J., Petzold-Maxwell, J.L., Keweshan, R.S., Dunbar, M.W., 2011. Field-evolved resistance to Bt Maize by Western corn rootworm. PLoS ONE. 6 (7), e22629. Available from: http://dx.doi.org/10.1371/journal.pone.0022629.

Goeschl, T., Swanson, T., 2001. On the economic limits to technological potential: will industry resolve the resistance problem? In: Swanson, T. (Ed.), The Economics of Managing Biotechnologies. Kluwer, Netherlands, pp. 99−128.

Gollier, C., 2001. The Economics of Risk and Time. MIT Press, Cambridge.

Gray, M.E., 1995. Areawide pest management for corn rootworms: fantasy or realistic expectations? Proceedings of the Illinois Agricultural Pesticides Conference. Illinois Cooperative Extension Service, Urbana-Champaign, pp. 101−106.

Groom, B., Hepburn, C., Koundouri, P., Pearce, D., 2005. Declining discount rates: the long and short of It. Environ. Resour. Econ. 32, 445−493.

Hardin, G., 1968. The tragedy of the commons. Science. 162, 1243−1248.

Harper, C.R., Zilberman, D., 1989. Externalities from agricultural inputs. Amer. J. Agric. Econ. 71, 692−702.

Headley, J.C., 1968. Estimating the productivity of agricultural pesticides. Am. J. Agric. Econ. 50, 13−23.

Heal, G., 2000. Nature and the marketplace: capturing the value of ecosystem services. Island Press, Washington, DC.

Henderson, N., Bateman, I., 1995. Empirical and public choice evidence for hyperbolic social discount rates and the implications for intergenerational discounting. Environ. Resour. Econ. 5, 413−423.

Hennessy, D., 1998. Damage control and increasing returns: Further results. Am. J. Agric. Econ. 79, 786−791.

Hueth, D., Regev, U., 1974. Optimal agricultural pest management with increasing pest resistance. Am. J. Agric. Econ. 56, 543−552.

Hurley, T.M., 2005. Bt resistance management: experiences from the U.S. In: Wessler J. (Ed.), Environmental Costs and Benefits of Transgenic Crops in Europe. Wageningen UR Frontis Series, vol. 7. Springer, Dordrecht, pp. 81−93.

Hurley, T.M., Babcock, B.A. Hellmich, R.L., 1997. Biotechnology and pest resistance: an economic assessment of refuges. Working Paper 97-WP 1839, Center for Agricultural and Rural Development, Iowa State University, Ames, IA.

Hurley, T.M., Babcock, B.A., Hellmich, R.L., 2001. Bt Corn and insect resistance: an economic assessment of refuges. J. Agric. Resour. Econ. 26, 176−194.

Hurley, T.M., Secchi, S., Babcock, B., Hellmich, R., 2002. Managing the risk of European corn borer resistance to Bt Corn. Environ. Resour. Econ. 22, 537−558.

Hutchison, W.D., Burkness, E.C., Mitchell, P.D., Moon, R.D., Leslie, T.W., Fleischer, S. J., et al., 2010. Areawide suppression of European corn borer with Bt maize reaps savings to non-Bt maize growers. Science. 330, 222−225.

Knapp, K.C., Olson, L.J., 1996. The economics of conjunctive groundwater management with stochastic surface supplies. J. Environ. Econ. Manag. 78, 1004−1014.

Knight, A.L., Norton, G.W., 1989. Economics of agricultural pesticide resistance in arthropods. Ann. Rev. Entomol. 34, 293−313.

Knipling, E.F., 1979. The Basic Principles of Insect Population Suppression and Management. Washington, DC, USDA, Agric. Handbook 512, 659 pp.

Kogan, M., 1998. Integrated pest management: historical perspectives and contemporary developments. Annu. Rev. Entomol. 43, 243−270.

Krysan, J.L., Foster, D.E., Branson, T.F., Ostlie, K.R., Cranshaw, W.S., 1986. Two years before the hatch: rootworms adapt to crop rotations. Bull. Entomol. Soc. Am. 32, 250−253.

Laxminarayan, R., Simpson, R.D., 2002. Refuge strategies for managing pest resistance in transgenic agriculture. Environ. Resour. Econ. 22, 521−536.

Lence, S.H., 2000. Using consumption and asset return data to estimate farmers' time preferences and risk attitudes. Amer. J. Agric. Econ. 82, 934−947.

Levine, E., Oloumi-Sadeghi, H., 1996. Western corn rootworm (Coleoptera: Chrysomelidae) larval injury to corn grown for seed production following soybeans grown for seed production. J. Econ. Entomol. 89, 1010–1016.

Levine, E., Oloumi-Sadeghi, H., Fisher, J.R., 1992. Discovery of multiyear diapause in Illinois and South Dakota northern corn rootworm (Coleoptera: Chrysomelidae) eggs and incidence of the prolonged diapause trait in Illinois. J. Econ. Entomol. 85, 262–267.

Levine, E., Spencer, J.L., Isard, S.A., Onstad, D.W., Gray, M.E., 2002. Adaptation of the Western corn rootworm to crop rotation: evolution of a new strain in response to a management practice. Am. Entomol. 48, 64–107.

Lichtenberg, E., Zilberman, D., 1986. The econometrics of damage control: why specification matters. Am. J. Agric. Econ. 68, 261–273.

Livingston, M.J., Carlson, G.A., Fackler, P.L., 2004. Managing resistance evolution in two pests to two toxins with refugia. Am. J. Agric. Econ. 86, 1–13.

Lu, Y., Wu, K., Jiang, Y., Guo, Y., Desneux, N., 2012. Widespread adoption of Bt cotton and insecticide decrease promotes biocontrol services. Nature. 487, 362–365.

Luckert, M., Adamowicz, W., 1993. Empirical measures of factors affecting social rates of discount. Environ. Res. Econ. 3, 1–22.

MacAfee, R.P., MacMillan, J., 1996. Analyzing the airwaves auction. J. Econ. Perspect. 10, 159–175.

Maler, K.-G., Vincent, R.J., 2005. Valuing Environmental Changes: Handbook of Environmental Economics, vol. 2. Elsevier, Amsterdam, The Netherlands.

Mangel, M., Plant, R.E., 1983. Multiseasonal management of an agricultural pest I: development of the theory. Ecol. Modell. 20, 1–19.

Marsh, T.L., Huffaker, R.G., Long, G.E., 2000. Optimal control of vector-virus-plant interactions: the case of potato leafroll virus net necrosis. Amer. J. Agric. Econ. 82, 556–569.

Meinke, L.J., Siegfried, B.D., Wright, R.J., Chandler, L.D., 1998. Adult susceptibility of Nebraska western corn rootworm populations to selected insecticides. J. Econ. Entomol. 91, 594–600.

Metcalf, R.L., 1983. Implications and prognosis of resistance to insecticides. In: Georghiou, G.P., Saito, T. (Eds.), Pest Resistance to Pesticides. Plenum, New York, pp. 703–769.

Mitchell, P., Hurley, T., Babcock, B., Hellmich, R., 2002. Insuring the Stewardship of Bt Corn: a carrot versus a stick. J. Agric. Resour. Econ. 27, 390–405.

Mitchell, P.D., Hutchison, W.D., 2008. Decision making and economic risk in IPM. In: Radcliffe, E.B., Hutchison, W.D., Cancelado, R.E. (Eds.), Integrated Pest Management. Cambridge University Press, Cambridge, pp. 33–50.

Mitchell, P.D., Onstad, D.W., 2005. Effect of extended diapause on the evolution of resistance to transgenic *Bacillus thuringiensis* Corn by Northern corn rootworm (Coleoptera: Chrysomelidae). J. Econ. Entomol. 98, 2220–2234.

Mitchell, P.D., Gray, M., Steffey, K., 2004. Composed error model for estimating pest-damage functions and the impact of the Western Corn rootworm soybean variant in Illinois. Am. J. Agric. Econ. 86, 332–344.

Moffit, L.J., Hall, D.C., Osteen, C.D., 1984. Economic thresholds under uncertainty with application to corn nematode management. Am. J. Agric. Econ. 16 (5), 151–157.

Morel, B., Farrow, R.S., Wu, F., Casman, E.A., 2003. Pesticide resistance, the precautionary principle and the regulation of Bt Corn. In: Laxminarayan, R. (Ed.), Battling Resistance to Antibiotics and Pesticides: An Economic Approach. Resources for the Future, Washington, DC, pp. 184–213.

Myers, J.H., Savoie, A., van Randen, E., 1998. Eradication and pest management. Annu. Rev. Entomol. 43, 471–491.

Olmstead, A.L., Rhode, P.W., 2008. Cotton and its enemies. In: Olmstead, A.L., Rhode, P.W. (Eds.), Creating Abundance: Biological Innovation and American Agricultural Development. Cambridge University Press, Cambridge.

Onstad, D.W., 1987. Calculation of economic-injury levels and economic thresholds for pest management. J. Econ. Entomol. 80, 297–303.

Onstad, D.W., Guse, C.A., 1999. Economic analysis of transgenic maize and nontransgenic refuges for managing European Corn Borer (Lepidoptera: Pyralidae). J. Econ. Entomol. 92, 1256–1265.

Onstad, D.W., Knolhoff, L.M., 2009. Finding the economics in economic entomology. J. Econ. Entomol. 102, 1–7.

Onstad, D.W., Crowder, D.W., Mitchell, P.D., Guse, C.A., Spencer, J.L., Levine, E., et al., 2003. Economics versus alleles: balancing IPM and IRM for rotation-resistant western corn rootworm (Coleoptera: Chrysomelidae). J. Econ. Entomol. 96, 1872–1885.

Onstad, D.W., Mitchell, P.D., Hurley, T.M., Lundgren, J.G., Porter, R.P., Krupke, C.H., et al., 2011. Seeds of change: corn seed mixtures for resistance management and IPM. J. Econ. Entomol. 104, 343–352.

Organisation for Economic Co-Operation and Development, 2001a. Multifunctionality: towards an analytical framework. Organisation for Economic Co-Operation and Development, Paris.

Organisation for Economic Co-Operation and Development, 2001b. Valuation of biodiversity benefits: selected studies. Organisation for Economic Co-Operation and Development, Paris.

Pearce, D., Moran, D., 1994. The Economic Value of Biodiversity. World Conservation Union, Earthscan Publications, London.

Peltola, J., Knapp, K.C., 2001. Recursive preferences in forest management. Forest Sci. 47, 455–465.

Plant, R.E., Mangel, M., Flynn, L.E., 1985. Multiseasonal management of an agricultural pest II: the economic optimization problem. J. Environ. Econ. Manage. 12, 45–61.

Press, W.H., Teukolsky, S.A., Vetterling, W.T., Flannery, B.P., 1992. Numerical Recipes in C++: The Art of Scientific Computing, second ed. Cambridge University Press, Cambridge.

Regev, U., Gutierrez, A.P., Feder, G., 1976. Pests as a common property resource: a case study of Alfalfa Weevil control. Am. J. Agric. Econ. 58 (2), 186–197.

Regev, U., Shalit, H., Gutierrez, A.P., 1983. On the optimal allocation of pesticides with increasing resistance: the case of the alfalfa weevil. J. Environ. Econ. Manage. 10, 86–100.

Ribaudo, M.O., Horan, R.D., Smith, M.E., 1999. Economics of water quality protection from nonpoint sources: theory and practice. Agricultural Economic Report No. (AER782). USDA-ERS, Washington, DC.

Saha, A., Shumway, C.R., Havenar, A., 1997. The economics and econometrics of damage control. Amer. J. Agric. Econ. 79, 773–785.

Secchi, S., Babcock, B.A., 2001. Optimal antibiotic use with resistance and endogenous technological change. Working Paper 01-WP 269, Center for Agricultural and Rural Development, Iowa State University, Ames, IA.

Secchi, S., Babcock, B.A., 2002. Pearls before swine? potential tradeoffs between the human and animal use of antibiotics. Amer. J. Agric. Econ. 84, 1279–1286.

Secchi, S., Hurley, T.M., Babcock, B.A., Hellmich, R.L., 2006. Managing European corn borer resistance to Bt corn with dynamic refuges. In: Just, R.E., Alston, J.M., Zilberman, D. (Eds.), Regulating Agricultural Biotechnology: Economics and Policy. Springer, New York.

Shoemaker, C., 1973. Optimization of agricultural pest management. Math. Biosci. 18, 1–22.

Soman, D., Ainslie, G., Frederick, S., Li, X., Lunch, J., Moreau, P., et al., 2005. The psychology of intertemporal discounting: why are distant events valued differently from proximal ones? Mark. Lett. 16, 347–360.

Stern, V.M., Smith, R.F., van den Bosch, R., Hagen, K.S., 1959. The integrated control concept. Hilgardia. 29, 81–101.

Talpaz, H., Borosh, I., 1974. Strategy for pesticide use: frequency and applications. Am. J. Agric. Econ. 56, 769–775.

Talpaz, H., Curry, G.L., Sharpe, P.J., DeMichele, D.W., Frisbie, R., 1978. Optimal pesticide applications for controlling boll weevil on cotton. Am. J. Agric. Econ. 60, 470–475.

Tinsley, N.A., Estes, R.E., Gray, M.E., 2012. Validation of a nested error component model to estimate damage caused by corn rootworm larvae. J. Appl. Entomol. (early view). Available from: http://dx.doi.org/10.1111/j.1439-0418.2012.01736.x.

Vatn, A., Bromley, D.W., 1995. Choices without prices without apologies. In: Bromley, D.W. (Ed.), The Handbook of Environmental Economics. Blackwell Publishers, Cambridge, USA, 705 pp (Chapter 1).

Weikard, H.P., Zhu, X., 2005. Discounting and environmental quality: when should dual rates be used? Econ. Modell. 22, 868–878.

Weitzman, M.L., 1998. Why the far-distant future should be discounted at its lowest possible rate. J. Environ. Econ. Manage. 36, 201–208.

Wesseler, J.H.H., 2003. Resistance economics of transgenic crops under uncertainty: a real option approach. In: Laxminarayan., R. (Ed.), Battling Resistance to Antibiotics and Pesticides. An Economic Approach. RFF press, Washington D.C, pp. 214–237.

Wessells, C.R., Cochrane, K., Deere, C., Wallis, P., Willmann, R., 2001. Product Certification and Ecolabelling for Fisheries Sustainability. FAO Fisheries Technical Paper 422. United Nations Food and Agriculture Organization, Rome, Italy.

Willis, K.G., Corkindale, J.T., 1995. Environmental Valuation: New Perspectives. CAB International, Wallingford, UK, 249 pp.

Wright, R.J., Scharf, M.E., Meinke, L.J., Zhou, X., Siegfried, B.D., Chandler, L.D., 2000. Larval susceptibility of an insecticide-resistant Western Corn Rootworm (Coleoptera: Chrysomelidae) population to soil insecticides: laboratory bioassays, assays of detoxification enzymes, and field performance. J. Econ. Entomol. 93, 7–13.

Wu, K.M., Lu, Y.H., Feng, H.Q., Jiang, Y.Y., Zhao, J.Z., 2008. Suppression of cotton Bollworm in multiple crops in China in areas with Bt toxin–containing cotton. Science. 321, 1676–1678.

Young, M.D., 1992. Sustainable investment and resource use: equity, environmental integrity and economic efficiency, UNESCO, Paris, 176 pp.

CHAPTER 3

Understanding Resistance and Induced Responses of Insects to Xenobiotics and Insecticides in the Age of "Omics" and Systems Biology

Barry Robert Pittendrigh[1], Venu Madhav Margam[2], Kent R. Walters, Jr.[1], Laura D. Steele[1], Brett P. Olds[3], Lijie Sun[4], Joseph Huesing[5], Si Hyeock Lee[6] and John M. Clark[7]
[1]Department of Entomology, University of Illinois Urbana Champaign, Urbana, IL, USA
[2]Regulatory Science, DuPont Knowledge Centre, Turkapally, Shameerpet Mandal, Ranga Reddy District, Hyderabad, Andhra Pradesh, India
[3]Department of Animal Biology, University of Illinois Urbana Champaign, Urbana, IL, USA
[4]Synthetic Biology and Bioenergy, J. Craig Venter Institute, San Diego, CA
[5]USDA-ARS/USAID-BFS, Washington, DC, USA
[6]Department of Agricultural Biotechnology, Seoul National University, Seoul, Korea
[7]Department of Veterinary and Animal Sciences, University of Massachusetts, Amherst, MA, USA

Chapter Outline

Insect Resistance Management
DOI: http://dx.doi.org/10.1016/B978-0-12-396955-2.00003-5

INTRODUCTION

We tend to think of the word *resistance* in terms of evolutionary changes in an insect population that occur in response to repetitive exposures to insecticides or other xenobiotics used to manage insect pests in crops, homes, gardens, or on livestock or humans (including disease vectors). Resistance can also be defined in broader terms since insects are "resistant" to many naturally occurring abiotic and biotic factors they encounter in their environment. In this chapter, we will outline the concepts associated with insecticide resistance, as well as provide examples of some of the known mechanisms associated with resistance. In addition, we will discuss the broader context of how we can use emergent *omics* tools, such as genomics, proteomics, and metabolomics, to better understand and discover resistance mechanisms. Finally, we will discuss how we can use this information to develop strategies that minimize the impact of insects on human health, food, and property.

The advent of a staggering array of *omics* technologies coupled with next-generation sequencing allows us to develop comprehensive descriptions of virtually all components and processes in an organism (Joyce and Palsson, 2006). Indeed, there is such a wealth of information that the challenge becomes how to integrate and extract usable information from these amassed datasets. When viewed through an appropriate filter, these techniques may permit us to address "resistance" in insect populations in a completely different manner than we have in the past. One such filter is evolutionary time.

If we view insect resistance within a broad evolutionary context, the first major evolutionary event we might consider is the divergence of the common ancestor of insects and vertebrates some 540 MYA (Grimaldi and Engel, 2005) (Figure 3.1). At this time, the earliest known animals with a brain, the flatworm and acorn worm, are thought to have evolved. Some consider the acorn worm to be the evolutionary link between

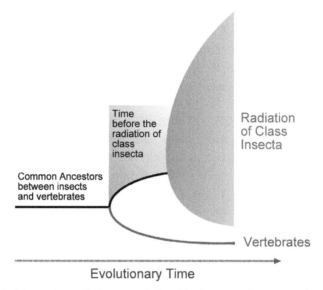

Figure 3.1 Schematic evolutionary relationship between insects and vertebrates. Common ancestors of insects and vertebrates branched into early arthropods and vertebrates approximately 540 MYA (Grimaldi and Engel, 2005). Over the following approximately several hundred million years the ancestor(s) of modern insects evolved, followed by the radiation of class Insecta.

vertebrates and invertebrates. Thus, many of the basic physiological functions were established at this time and remain conserved across these taxa. Over the following several hundred million years, the ancestors of modern insects evolved. The radiation of class Insecta was thought to have begun approximately 310–390 MYA at about the same time that plants evolved seeds. In evolutionary terms, only very recently has humankind developed and exploited inorganic and organic chemistries for the control of insects. Each of these evolutionary events (i.e., the divergence of invertebrates and vertebrates, adaptive radiation of insects, and selection pressure by humankind) has contributed unique, and thus exploitable, traits that enable insects to resist the variety of abiotic and biotic challenges they experience. In addition, the coevolutionary arms race between insect herbivores and plants has added to this trait expansion or specialization and species adaptive divergence (Fischer *et al.*, 2008; Toju, 2009; Whiteman *et al.*, 2011).

The aforementioned evolutionary events have been well studied over the past several decades. Many scientists studying *Drosophila melanogaster* (hereafter referred to as *Drosophila*) have focused on the conserved evolutionary processes common to insects and mammals. This process has given us important insights into the nervous system (Wang *et al.*, 2000; Li *et al.*, 2004a; Savare *et al.*, 2005), developmental biology (Fristrom, 1970; Lawrence and Morata, 1992; Bejsovec *et al.*, 2004), and human diseases (Pan *et al.*, 2004a; Bilen and Bonini, 2005; Wolf *et al.*, 2006). Less well studied are those traits that evolved in insects during the last 310–390 million years. It is these traits and specialized adaptations to their food and environment that may provide the best opportunities for human interventions for pest control. For example, one of the best studied plant–insect interactions is the relationship between butterflies in the family Papilionidae and their host plants, specifically the detoxification enzymes that these insects use to defend themselves against the plant's secondary compounds that they experience in their diets (Berenbaum, 1995; Berenbaum *et al.*, 1996; Hung *et al.*, 1996; Berenbaum and Zangerl, 1998; Li *et al.*, 2004b; Pan *et al.*, 2004b; Wen *et al.*, 2006a,b). Although the use of insecticides and the associated evolution of insecticide resistance represent a mere "blip" in the evolution of insects, the topic has received a great deal of attention by molecular entomologists because of the consequences of insecticide resistance for humans in terms of food production and insects that vector diseases.

The *omics* approach has the tremendous potential to allow us to understand insect-specific responses to the insect's environment at the molecular, cellular, and organismal levels, and to do so in an integrated manner. This level of "systems understanding" will enable us to develop novel pest control strategies that will be of immense practical value since the development of insecticides that interact with insect-specific target sites, for example, the peritrophic matrix or exoskeleton, are likely to have minimal or no direct impact on mammals.

Recently, as proof of concept for using the *omics* approach at an organismal level, the combined transcriptome, proteome, and metabolome of *Drosophila*, along with systems-scale analyses of these combined datasets, were monitored in response to the dietary administration of a toxic compound, methamphetamine (Sun *et al.*, 2011). Understanding what these *omics* responses mean, in terms of the biology of the organism's response to a given xenobiotic, requires that we have an in-depth knowledge of the biological effects that would be expected from that given xenobiotic at the organismal level. Methamphetamine, albeit not a pesticide, was the xenobiotic of choice because of its well-known behavioral and toxicological effects in mammalian systems (Ferguson *et al.*, 1993; Schmued and Bowyer, 1997; Bowyer *et al.*, 2007, 2008; Ito *et al.*, 2007; Páleniček *et al.*, 2011). This approach revealed that many of the pathways involved in methamphetamine toxicity appear to be evolutionarily conserved from *Drosophila* to humans. For example, increased oxidative stress, disruption of mitochondrial function, altered muscle homeostasis, and disruption of spermatogenesis were observed in *Drosophila* and in other models of methamphetamine toxicity. There exists considerable opportunities for researchers to begin to apply these same tools to understanding how insects respond to insecticides, both in terms of induction of pathways at subtoxic doses in susceptible strains and the constitutive expression and induction in resistant strains that have been selected for over multiple generations.

In the last century, our options for managing insects using chemicals were restricted mostly due to our limited knowledge of the unique biochemical and molecular aspects of insects. Of major concern was that some of the insecticides also affected higher vertebrates, because these compounds targeted evolutionarily conserved biological processes (e.g., the impact of organophosphates on acetylcholinesterases). Obviously, there has been and continues to be the long-term need for the development and use of pest control strategies that are specific to the target organisms, with minimal impact on nontarget organisms, especially vertebrates.

In this chapter, we briefly review what is known about resistance to select classes of insecticides and then discuss some exciting new possibilities that systems-scale analyses using *omics* may provide. We also discuss the potential importance of understanding the molecular mechanisms by which insects resist environmental challenges that they experience in their life histories. Inhibition of these other resistance mechanisms, where the resistance mechanism is critical for the survival of the insect (we term these *Achilles' heel traits*), holds the possibility for the development of novel insect control methods. We define an Achilles' heel trait as a target molecule that, when inhibited (or negatively impacted), reduces the ability of an organism or a population of organisms to persist under specific conditions.

First, we will explore resistance mechanisms to traditional insecticides (Figures 3.2−3.8). Insecticides are often classified according to their modes of action on arthropod physiological processes (Table 3.1). The Insecticide Resistance Action Committee maintains a list of insecticides and their modes of action on its Internet website. Resistance can be broadly classified into the following categories: (i) reduced penetration,

Toxin   Inactive toxin   Target site   Modified
target site

Figure 3.2 Symbols used in subsequent figures 3.3 to 3.8.

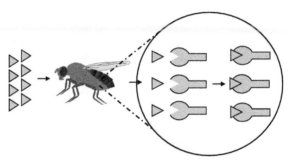

Figure 3.3 A given pesticide interacts with a target site in a pesticide susceptible insect causing death of the insect.

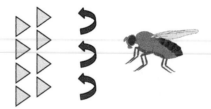

Figure 3.4 Reduced penetration – The insect population evolves a heritable mechanism/mechanisms to reduce (or prevent) the entry of the toxins into the insect's body.

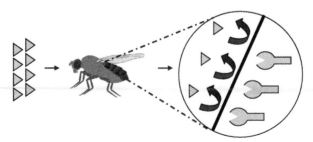

Figure 3.5 Sequestration–After entry of the pesticide into the insect's body, enzymes or proteins bind to the toxin and transfer them away from the target site to various organelles such as fat body and hemolymph for safe storage.

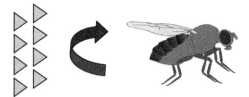

Figure 3.6 Behavioral Resistance — A given insect population evolves a heritable mechanism/mechanisms to avoid the toxin molecules by changing their behavior.

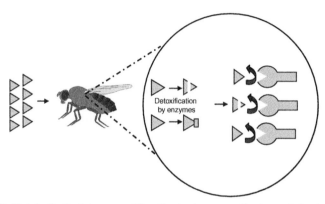

Figure 3.7 Metabolic Resistance — After the toxin enters the insect's body, enzymes in the insect alter the toxin such that it no longer binds to its intended target site, thereby allowing the insect to survive the given dose of the toxin.

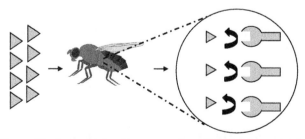

Figure 3.8 Target Site Insensitivity—Due to conformational changes in the target site of the toxin, the pesticide becomes less toxic to the insect. Such changes in the target site may reduce the ability of the toxin to bind to the target site or it may change the target site's response to the toxin.

Table 3.1 Mode of Action Classification of Insecticides

Group	Primary Site of Action	Examples
1	Acetylcholine esterase inhibitors	Carbamates, organophosphates
2	GABA-gated chloride channel antagonists	Organochlorines, fiproles
3	Sodium channel modulators	Pyrethroids, DDT
4, 5	Nicotinic acetylcholine receptor agonists	Neonicotinoids, spinosyns
6	Chloride channel activators	Avermectins
7	Juvenile hormone mimics	Juvenile hormone analogues
10	Mite growth inhibitors	Clofentezine
11	Microbial disruptors of insect midguts	*Bacillus* crystalline proteins
12	Inhibitors of ATP synthase	Diafenthiuron, propargite
13	Uncouplers of oxidative phosphorylation	Chlorfenapyr
15, 16	Inhibitors of chitin biosynthesis	Benzoylureas, buprofezin
17, 18	Molting disruptors	Cyromazine, azadirachtin
19	Octopaminergic agonists	Amitraz
20, 21, 24, 25	Mitochondrial electron transport inhibitors	Hydramethylnon, rotenone
22	Voltage-dependent sodium channel blockers	Indoxacarb
23	Inhibitors of lipid synthesis	Tetronic acid derivatives
28	Ryanodine receptor modulators	Diamides

Derived from Insecticide Resistance Action Committee's classification scheme Version 7.2, which represents current expert consensus as of April 2012. Not all known mode of action groupings are shown. A full list can be found at http://www.irac-online.org/teams/mode-of-action/

(ii) increased sequestration and/or excretion, (iii) behavioral resistance, (iv) metabolic resistance, and (v) target site insensitivity. In the following paragraphs, we will define, and provide examples of each of these forms of resistance.

GENERAL MECHANISMS OF RESISTANCE
Reduced Penetration

Resistance by reduced penetration occurs when insects develop a heritable mechanism that reduces or prevents the entry or penetration of the toxin or insecticide into the insect's body (Figure 3.4). It has been hypothesized that decreased penetration can give detoxifying enzymes more time to metabolize the insecticide or to efflux and excrete it before it reaches its target (Plapp and Hoyer, 1968). Plapp and Hoyer (1968) observed a form of decreased penetration of dichloro-diphenyl-trichloroethane (DDT) and dieldrin in resistant *Musca domestica*. Farnham (1971, 1973) later demonstrated that the reduced penetration was due to a gene termed *pen* (penetration) located on chromosome III. A similar mechanism and perhaps gene was also observed in permethrin-resistant flies (DeVries and Georghiou, 1981). In fact, resistance due to decreased penetration is often seen in combination with other resistance mechanisms. For example, in pyrethroid-resistant (Learn-Pyr; LPR) *M. domestica*, resistance is due to *knockdown resistance*, overexpression of cytochrome P450 CYP6D1, and decreased penetration (Seifert and Scott, 2002). The term *knockdown resistance* (*kdr*) describes cases in which insects have become resistant to DDT or pyrethroids through a reduced sensitivity in the nervous system (Soderlund and Knipple, 2003).

In combination with other resistance traits, reduced penetration, at least in the mosquito species *Culex pipiens*, appears to have a multiplicative effect on resistance (Raymond et al., 1989 2001). Thus, insects that carry the reduced penetration trait, coupled with other resistance mechanisms, may be much more resistant to insecticides than if other resistance mechanisms are combined. For example, target site insensitivity combined with increased enzymatic detoxification is merely additive in this model. If this model can be shown to apply to other pest systems, then from a resistance management perspective, reduced penetration as a resistance mechanism could have important economic and societal impacts. Unfortunately, we do not have the degree of understanding of the molecular mechanisms of reduced penetration that we have for other resistance mechanisms, notably target site insensitivity and metabolic resistance.

Beyond the important issue of insecticide resistance, understanding the molecular mechanisms by which insects reduce penetration of harmful compounds through their exoskeletons and digestive systems may also provide new opportunities to develop novel strategies to compromise that ability. For example, the peritrophic matrix (peritrophic membrane) of insects is a complex digestive system tissue composed of proteins, glycosaminoglycans, and chitin. The peritrophic matrix serves as an integral part of the controlled enzymatic degradation and absorption of food and also provides an exclusionary barrier to bacteria, viruses, and damaging mechanical elements (Lehane, 1997; Tellam *et al.*, 1999; Hegedus *et al.*, 2009; Kuraishi *et al.*, 2011). In this role, the peritrophic matrix minimizes the impact of negative biotic and abiotic factors in the insect diet. Inhibition of the production of some or all of the components of the peritrophic matrix would potentially be a highly effective and selective way to control a pest insect.

Increased Sequestration or Excretion

Increased sequestration occurs when enzymes or proteins in an insect's body bind to insecticide molecules and subsequently transfer them away from the target site to various organelles such as the fat body and hemolymph for safe storage (Lee and Clark, 1998; Nicholson *et al.*, 2006; Opitz and Müller, 2009) (Figure 3.5). Sequestering of toxins may have arisen early in the evolution of insects and was perhaps influenced by their interactions with flowering plants. A well-studied example of this phenomenon is the relationship of *Danaus plexippus* (Lepidoptera: Danainae) with otherwise poisonous milkweed plants (*Asclepias spp.*). Milkweed plants produce noxious cardenolide (cardiac glycosides) molecules. *D. plexippus* larvae feeding on milkweed plants sequester these molecules in their bodies, which in turn makes the insect unpalatable. In this case, an insect herbivore developed a mechanism to first sequester a poison and then to use that poison as a defense against predation (Nishida, 2002). The ability to sequester plant toxins seems to be particularly prevalent in the Lepidoptera. This is one area where systems–scale *omics* analyses might shed light on the common features of the sequestering mechanisms involved.

Many resistant insects sequester insecticides, and esterases frequently mediate this process. Esterase-based resistance can be classified into two types: (1) increased levels of insecticide sequestration, which involves a rapid binding of the insecticide resulting in broad spectrum resistance, and (2) changes in substrate specificity due to point mutations, wherein a group of insecticides with a common ester bond are metabolized into less toxic forms, which typically confers narrow spectrum resistance.

Examples of esterase-mediated sequestration include two aphid species, *Myzus persicae* and *Myzus nicotianae*, and a mosquito species, *C. pipiens*. Overexpression of carboxylesterases has been associated with binding to, but not necessarily the metabolism of, insecticides (e.g., organophosphates and carbamates) (Field *et al.*, 1988, 1994; Raymond *et al.*, 1998). Other examples where esterases play a role in sequestration include the plant hopper *Nilaparvata lugens*, the beetle *Leptinotarsa decemlineata*, and several mosquito species, including *Culex quinquefasciatus, Culex tarsalis, and Culex tritaeniorhynchus* (Lee *et al.*, 1998; Small and Hemingway, 2000; Karunaratne and Hemingway, 2000). In the case of *C. tritaeniorhynchus*, the specific carboxylesterase gene involved in the sequestration of organophosphates, *CtrEstbeta1*, has been elucidated (Karunaratne and Hemingway, 2000). In *Tenebrio molitor*, glutathione S-transferases (GSTs) may be involved in sequestering pyrethroids (Kostaropoulos *et al.*, 2001).

Similarly, the upregulation of cellular efflux mechanisms (phase III metabolism), such as those producing multiple drug resistance [e.g., ATP-binding cassette transporter (ABCTs) like P-glycoproteins], decreases drug concentrations at receptor sites, resulting in tolerance and resistance, particularly to anthelmintics macrocyclic lactones, such as ivermectin and the related spinosyns (Prichard and Roulet, 2007). As ivermectin is itself a negatively charged sugar conjugate, it is likely a direct substrate for ABCTs (e.g., ABCC1). There is accumulating evidence that ivermectin: (i) selects for certain alleles of P-glycoproteins (ABCB-type) and other ABCT genes (ABCC-type), (ii) selects for constitutive overexpression of some of these gene products, and (iii) induces overexpression of some P-glycoproteins in nematodes (Prichard and Roulet, 2007). In addition, dogs with a 4-bp deletion in the multiple drug resistance *MDR1* gene (Macdonald and Gledhill, 2007) and mice with a deficient amount of P-glycoprotein in their blood-brain barriers (Schinkel *et al.*, 1994) are hypersensitive to ivermectin treatments. Taken together, these data are highly suggestive that ABCTs are involved in ivermectin efflux (ABCB1), excretion (ABCC1 and ABCG2), and possibly resistance (Prichard and Roulet, 2007; Wolstenholme *et al.*, 2004; Lespine *et al.*, 2006; Ardelli and Prichard, 2007; Tribble *et al.*, 2007; Lespine *et al.*, 2007; James and Davey, 2009; Buss and Callaghan, 2008). Like ivermectin, spinosad is itself a sugar conjugate, and ABCTs are likely also to be involved in its efflux (ABCB1), in the excretion of its glutathione conjugates (ABCC1), and perhaps in its resistance. Specifically, Phase III xenobiotic metabolism appears to be the first line of defense against toxins in the environment in that all organisms, from bacteria to humans, have ABC transporters. All organisms also maintain many ABCT genes in a variety of subfamilies, indicating their importance in either effluxing toxins out of a cell or excreting toxins from a multicellular organism. Based on this assumption, it is expected that induced Phase III metabolism, particularly by ABC transporters, can serve as an early and basic tolerance mechanism. Given the importance of ABC transporters in multidrug resistance and drug–drug interactions, it is surprising that so little research has been done on the importance of phase III metabolism and insecticide resistance. Very little work has been done on insecticide excretion and even less on the excretion carried out by ABC transporters.

Behavioral Resistance

Any avoidance behavior that results in an increased chance of survival for an insect or its offspring can be defined as behavioral resistance (Figure 3.6). For example, behavioral changes in oviposition have been observed as an avoidance behavior against insecticides in the diamondback moth, *Plutella xylostella*. Sarfraz *et al.* (2005) observed that when laboratory-raised *P. xylostella* were given a choice to lay eggs on insecticide-treated host plants, the moths preferentially laid more eggs closer to the soil rather than on the stem and foliage.

Aversion behavior has been observed in the German cockroach, *Blatella germanica*, to food ingredients contained in gel baits (Wang *et al.*, 2004). Glucose, fructose, maltose, and sucrose are typically phagostimulants to nonaverse (susceptible) laboratory strains of German cockroaches. However, the Cincy strain of *B. germanica* avoided all of these compounds when they were incorporated into an agar diet substrate. The trait appears to be weakly sex linked, with females showing a higher degree of the aversion trait (Wang *et al.*, 2006). However, there also appeared to be a cost to resistance, as the Cincy strain produced fewer progeny than their nonaverse counterparts (Wang *et al.*, 2004).

Metabolic Resistance

Metabolic resistance refers to a situation in which organisms increase the levels of metabolism of a given insecticide or toxin. This process can occur (i) by increasing the levels of given enzymes that "break down" or alter the insecticide to a less toxic form or (ii) by a structural change in an enzyme that allows it to more efficiently metabolize the insecticide substrate (Figure 3.7).

Metabolic resistance to insecticides is a common mechanism by which a diversity of insects, including lepidopteran, coleopteran, and dipteran species, evolve resistance (Hemingway *et al.*, 1991; Ottea *et al.*, 1995; Rose *et al.*, 1997; Chandre *et al.*, 1998; Kasai *et al.*, 1998; Stuart *et al.*, 1998; Feyereisen,

1999; Kasai *et al.*, 2000; Li *et al.*, 2000; Scharf *et al.*, 2000; Alizadeh *et al.*, 2011). The literature on metabolic resistance is vast, and so it is not possible to cover all of the different examples in this chapter; please see Ishaaya (1993), Feyereisen (1995), Keseru (1998), Scott (1999), and Li *et al.* (2007) for reviews on the topic of metabolic resistance. We will focus instead on a few select examples to illustrate some generalities associated with metabolic resistance. Most studies to date have elucidated the role of cytochrome P450 monooxygenases (cytrochrome P450s), glutathione S transferases (GSTs), or esterases in metabolic resistance.

Cytochrome P450s are a class of oxidative enzymes found across a diverse range of organisms, including bacteria, plants, fungi, insects, and mammals. They comprise a superfamily of heme-thiolate proteins, which act on endogenous compounds, such as steroid hormones, as well as toxic compounds that insects encounter in their environment. P450s metabolize insecticides by N-, O-, and S-alkyl hydroxylation, aromatic hydroxylation, aliphatic hydroxylation and epoxidation, ester oxidation, as well as thioether and nitrogen oxidation (Schuler, 2011).

As mentioned earlier, GSTs are a family of enzymes that play a variety of biological roles in the cell, including the detoxification of xenobiotics such as insecticides, carcinogens, and drugs. All eukaryotes have cytosolic and membrane-associated GSTs. In some cases, the expression levels of given GSTs are directly related to the tolerance to toxic chemicals (Hayes and Pulford, 1995). GSTs are involved in the resistance of insects to the organophosphate, organochlorine, DDT, and pyrethroid insecticides (Ranson and Hemingway, 2005; Li *et al.*, 2007).

An esterase, for example, carboxylesterase, is a hydrolytic enzyme that cleaves ester bonds in insecticides to yield an acid and an alcohol. There are many kinds of esterases that differ in their substrate specificity, protein structure, and biological function. Esterases have been associated with resistance of insects to organophosphate, carbamate, and pyrethroid insecticides (Li *et al.*, 2007; Yan *et al.*, 2009).

Metabolic insecticide resistance in insects is typically polygenic and is often associated with overtranscription of detoxification enzymes, such as cytochrome P450s and GSTs (Houpt *et al.*, 1988; Heckel *et al.*, 1998; Maitra *et al.*, 2000; Tang *et al.*, 2000; Kranthi *et al.*, 2001; Rajurkar *et al.*, 2003). The cloning of numerous cytochrome P450 and GST genes that are overtranscribed in resistant insects but that fail to map back to a major resistance locus has led workers in the field to hypothesize that there is a *trans*-regulatory gene that controls the expression of these detoxification enzymes (Grant and Hammock, 1992; Liu and Scott, 1997; Dombrowski *et al.*, 1998); in susceptible insects a repressor acts to reduce the expression levels of detoxification enzymes associated with resistance. It has also been hypothesized that a mutation in this regulatory gene no longer allows it to suppress expression of these genes, which results in overtranscription of cytochrome P450s, and in turn results in resistance (Grant and Hammock, 1992; Carino *et al.*, 1994; Maitra *et al.*, 1996; Liu and Scott, 1997; Dombrowski *et al.*, 1998; Kasai *et al.*, 1998). To date, no such trans-acting repressor has been identified in insects. Additionally, in bacterial systems this repressor-mechanism hypothesis has been challenged (Shaw *et al.*, 1998). Selective induction of cytochrome P450s in response to insecticides was more apparent in a cypermethrin-resistant strain of *P. xylostella* than in a susceptible strain, suggesting that it may play a more crucial role in metabolic resistance of *P. xylostella* rather than constitutive overexpression of cytochrome P450s (Baek *et al.*, 2010). Because constitutive overexpression of cytochrome P450s could generate excess reactive oxygen species, leading to oxidative stress, tissue damage, and fitness disadvantage (Zangar *et al.*, 2004), the selective P450 induction strategy as a resistance mechanism would allow insects to minimize the fitness cost particularly when they are not under the insecticide selection pressure.

There is considerable evolutionary sequence divergence among cytochrome P450s and GSTs in class Insecta, even among relatively closely related species such as *Drosophila* and *Anopheles gambiae* (Ranson *et al.*, 2001). However, it remains to be determined if the regulation of P450s and GSTs across insect taxa are more conserved than the actual genes themselves (Handschin *et al.*, 2004).

Although the role of P450s in insecticide resistance is often associated with overtranscription of these enzymes, structural changes in the P450 can also lead to insecticide resistance. For example, Amichot *et al.* (2004) observed that three mutations in *Cyp6a2*, from *Drosophila*, increased the mutant CYP6A2's ability to metabolize DDT. Their results are consistent with observations in humans, where P450 polymorphisms have been shown to be associated with drug and insecticide metabolism (Guengerich *et al.*, 1999; Eaton, 2000). Such structural changes in P450s, leading to changes in insecticide resistance, have also been observed in fungi (Lamb *et al.*, 1997; Delye *et al.*, 1998; Ma *et al.*, 2006).

Although much of the research on metabolic insecticide resistance has focused on P450s, GSTs, and esterases, it remains to be determined if other genes or pathways are critical for insecticide resistance. For example, changes in glucose utilization have been associated with DDT exposure and metabolic resistance in a variety of organisms, including marine microorganisms, insects such as *Drosophila*, and mammals (Ela *et al.*, 1970; Maltseva *et al.*, 1982; Plapp, 1970; Bauer and Capone, 1985; Ahuja *et al.*, 2001; Ahuja and Kumar, 2003; Okazaki and Katayama, 2003; Pedra *et al.*, 2004, 2005). Additionally, a genome-wide comparison of two metabolically DDT-resistant strains of *Drosophila* (*Wisconsin* and *91-R*), as compared to one susceptible strain (*Canton-S*), revealed dozens of putatively differentially overtranscribed genes in the resistant strains. These putatively overtranscribed genes included: P450s, GSTs, oxidoreductases, as well as UPD-glucuronosyltransferases (UGTs), diazepam binding inhibitor, other lipid metabolism genes, peptidases, immunity/defense proteins, as well as other gene categories (Pedra *et al.*, 2004). It has been demonstrated in rats that dietary DDT increases enzymatic activity of certain hepatic UGTs (Okazaki and Katayama, 2003). However, for many of the differentially expressed genes observed by Pedra *et al.* (2004) it is still not clear what role, if any, they may actually play in metabolic resistance.

It is important to remember that differential expression does not mean that the gene and its resultant protein product are actually conferring resistance. First, any given differentially expressed gene could be regulated under a common mechanism with another gene whose protein product is actually involved in resistance. Furthermore, overtranscription of the gene does not mean that there is necessarily more protein formed to perform a given detoxification process. Thus, caution is warranted in equating over-transcription of a specific gene with a role in metabolic insecticide resistance (Pedra *et al.*, 2004).

Target Site Insensitivity

Target site insensitivity involves an alteration of the target molecule(s) that directly interacts with the insecticide, which results in the toxin being less toxic to the target pest (Figure 3.8). Target site insensitivity has been observed in a variety of insect species in response to a diversity of insecticides. Some of these are outlined below.

Cross-Resistance

In 1979, Chapman and Penman (1979) defined cross-resistance as the result of being resistant to multiple compounds due to the ability of one compound to confer resistance to additional compounds of the same group. The term *cross-resistance* is often used to describe resistance conferred to compounds with different structural groups from the original compound from which the insect population was selected for resistance (Chapman and Penman, 1979; Boyer *et al.*, 2012). Cross-resistance has been documented in a number of different insects and across multiple insecticide types. *Ceratitis capitata* populations resistant to malathion were observed to have differing levels of cross-resistance to a range of organophosphates, a carbamate, a pyrethroid, and a benzoylphenylurea (Couso-Ferrer *et al.*, 2011). A population of human lice (*Pediculus humanus*) resistant to permethrin were identified as having cross-resistance to both pyrethrum and DDT (Yoon *et al.*, 2004). In *Aedes aegypti*, cross-resistance of pyrethroid and DDT was shown to occur as a result of mutations within the voltage-gated sodium channel gene (Brengues *et al.*, 2003). Identification of the different types of cross-resistance across insecticides is an ongoing issue that public and private sector researchers have to deal with as it relates to the deployment and use of divergent insecticides in the changing insecticide marketplace.

RESISTANCE TO CLASSES OF INSECTICIDES
Resistance to DDT and the Pyrethroids

Resistance to DDT was first reported in 1947, only a few years after the introduction of this compound into the marketplace (Brown, 1986). Crow (1954) demonstrated that resistance in *Drosophila* to DDT was polygenic. Subsequent mapping studies demonstrated that several major loci contribute to metabolic DDT resistance in *Drosophila*, the best studied being the *Rst(2)DDT* locus (loci) on the second chromosome (Dapkus and Merrell, 1977; Dapkus, 1992). Work by Daborn *et al.* (2002) and Brandt *et al.* (2002) suggests that the *Rst(2)DDT* locus (or closely linked loci) may be due to overtranscription of one (*Cyp6g1*) or two (*Cyp6g1* and *Cyp12d1*) P450 genes. Overtranscription of *Cyp6g1* appears to be commonly found across diverse DDT-resistant strains of *Drosophila* (Daborn *et al.*, 2002), and *Cyp12d1* appears to be inducible (in some *Drosophila* strains) in the presence of DDT (Brandt *et al.*, 2002; Festucci-Buselli *et al.*, 2005). Metabolic resistance to DDT in *Drosophila* is associated with one of the major metabolites having an OH group added to the DDT molecule. In mosquitoes, DDT resistance has also been associated with elevated GST levels. Target site insensitivity in the neuronal voltage-gated sodium channel confers resistance to both pyrethroids and DDT (Narahashi, 1992).

For pyrethroids and DDT, the major target site is thought to be the α-subunit of the voltage-sensitive sodium channel (VSSC) (also known as the voltage-gated sodium channel), and in *Drosophila* VSSC is encoded by the *para* gene (Williamson *et al.*, 1996; Pittendrigh *et al.*, 1997). Pyrethroids and DDT are thought to cause prolonged opening of the VSSC by stabilizing the open configuration of the channel and prolonging its open state. Amino acid changes in the VSSC have been shown to confer pyrethroid resistance and DDT resistance in a variety of insect species (Table 3.1).

A leucine to phenylalanine amino acid substitution in the hydrophobic IIS6 transmembrane segment of a *Musca domestica* VSSC, termed *kdr*, resulted in moderate increases in resistance to DDT and certain pyrethroids. This mutation coupled with a second methionine to threonine substitution in the intracellular S4-S5 linker domain II (intracellular IIS4-S5 loop) conferred high levels of resistance that was termed *super-kdr* (Williamson *et al.*, 1996). Subsequent work by Pittendrigh *et al.* (1997) showed amino acid changes in IIIS6 in the temperature-sensitive *para^74* *Drosophila* strain conferred moderate DDT resistance. The *para* temperature-sensitive lines *para^{t1}/para^{ts2}*, and *para^{DN7}* had amino acid changes, respectively, in intracellular IS4-S5 and IIIS4-S5 loops; all three strains were DDT resistant. Heterozygous *para^74/para^{DN7}* flies, carrying *kdr-like* and *super-kdr-like* alleles *in trans*, showed elevated levels of DDT resistance.

The most prevalent resistance-associated mutation in *kdr* insects results from a leucine-to-phenylalanine substitution in the S6 hydrophobic segment of VSSC domain II (Williamson *et al.*, 1996; Dong, 1997, 1998; Jamroz *et al.*, 1998; Martinez-Torres *et al.*, 1998, 1999b). Two alternative substitutions at this position also confer resistance to DDT and/or pyrethroids: A leucine–histidine substitution is associated with pyrethroid resistance in *Heliothis virescens* (Fabricius) (Park and Taylor, 1997), and a leucine–serine substitution confers DDT resistance and low levels of permethrin resistance in a strain of *C. pipiens* from China (Martinez-Torres *et al.*, 1999a). An additional methionine–threonine replacement is found in strains of housefly and horn flies showing very high levels of pyrethroid resistance (*super-kdr* phenotype) (Williamson *et al.*, 1996; Jamroz *et al.*,

1998). A list of sodium channel mutations, across multiple insect species, conferring resistance to pyrethroids is summarized in Table 3.2.

Resistance to Organophosphates and Carbamates

The function of acetylcholinesterase (AChE) is to degrade the neurotransmitter acetylcholine in the cholinergic synapses of animals, including insects. Mutations in the AChE-encoding locus, known as *Ace* in *Drosophila*, have been shown to confer target site insensitivity to organophosphate and carbamate insecticides, which primarily target AChE (Table 3.1). A range of other amino acid substitutions in *M. domestica* and *Drosophila* AChE confer insecticide resistance, and these mutations typically reside near to or within the active site of the enzyme (Feyereisen, 1995). Such AChE mutations, associated with insecticide resistance, have also been observed in other species, including *L. decemlineata* (Zhu and Clark, 1997), *Bactrocera oleae*, (Vontas *et al.*, 2002), *Aedes aegypti* (Vaughan *et al.*, 1998), *Aphis gossypii* (Li and Han, 2004), *Helicoverpa armigera* (Ren *et al.*, 2002), *C. quinquefasciatus* (Liu *et al.*, 2005a), *Cydia pomonella* (Cassanelli *et al.*, 2006), *Bactrocera dorsalis* (Hsu *et al.*, 2006), and *C. pipiens* (Alout *et al.*, 2007). Additionally, Mazzarri and Georghiou (1995) observed that oxidase and nonspecific esterase enzymes were also involved in organophosphate and carbamate resistance in *Aedes aegypti* populations from Venezuela.

Resistance to Dieldrin

In *Drosophila*, the *resistance to dieldrin* (*Rdl*) gene encodes the γ-aminobutyric acid (GABA) receptor subunit RDL (ffrench-Constant *et al.*, 1998). The *Rdl* gene was cloned from a mutant line of *Drosophila* that was resistant to both picrotoxin (PTX) and cyclodiene insecticides (ffrench-Constant *et al.*, 1991). Picrotoxins were previously known to be vertebrate GABA$_A$ receptor antagonists. Dieldrin-resistant populations of *Drosophila*, collected from a variety of locations around the world, all shared the same alanine to serine substitution (A302S) (ffrench-Constant *et al.*, 1993a). This amino acid change results in the RDL subunit becoming insensitive to both dieldrin and picrotoxin. In *Drosophila simulans* (ffrench-Constant *et al.*, 1993b) and the aphid *M. persicae* (Anthony *et al.*, 1998), there is a serine to glycine substitution in the resistant insects. More recently, Le Goff *et al.* (2005) have observed two amino acid substitutions, namely, an alanine to glycine (A301G) and a

Table 3.2 Species with Voltage-Sensitive Sodium Channel Mutations Associated with Pyrethroids and DDT Resistance

Species	Amino Acid Change[‡]	Reference
kdr and kdr-like		
Musca domestica	L to F	Williamson *et al.*. 1996, Miyazaki *et al.*, 1996
Blattella germanica	L to F	Miyazaki *et al.*, 1996; Dong 1997, Dong *et al.*, 1998
Plutella xylostella	L to F	Schuler *et al.*, 1998; Endersby *et al.*, 2011
Myzus persicae	L to F	Martinez-Torres *et al.*, 1999b
Anopheles gambiae	L to F	Martinez-Torres *et al.*, 1998; Ranson *et al.*, 2000; Choi *et al.*, 2010
Anopheles stephensi	L to S	Singh *et al.*, 2011
Culex pipiens	L to F	Martinez-Torres *et al.*, 1999a
Culex quinquefasciatus	L to F	Xu *et al.*, 2005; Wondji *et al.*, 2008; Sarkar *et al.*, 2009
Haematobia irritans	L to F	Guerrero *et al.*, 1997
Leptinotarsa decemlineata	L to F	Lee *et al.*, 1999
Frankliniella occidentalis	L to F	Forcioli *et al.*, 2002
Cydia pomonella	L to F	Brun-Barale *et al.*, 2005
Ctenocephalides felis	L to F	Bass *et al.*, 2004
Cx. pipiens	L to S	Martinez-Torres *et al.*, 1999a
A. gambiae	L to S	Ranson *et al.*, 2000; Choi *et al.*, 2010
A. stephensi	L to S	Singh *et al.*, 2011
Heliothis virescens	L to H	Park and Taylor 1997
Super-kdr (and Super-kdr-like)		
Musca domestica	M to T & L to F	Williamson *et al.*, 1996; Miyazaki *et al.*, 1996
Haematobia irritans	M to T & L to F	Guerrero *et al.*, 1997
Myzus persicae	M to T & L to F	Eleftherianos *et al.*, 2008
Plutella xylostella	L to F & T to I	Endersby *et al.*, 2011
Sitophilus zeamais	T to I	Araújo *et al.*, 2011
Drosophila melanogaster	V to M	Zhao *et al.*, 2000
	—	*Pittendrigh *et al.*, 1997

[*]Temperature sensitive strains, with VSSC mutations, that also showed resistance to pyrethroids and DDT (para^{DN7}, para^{ts1}/para^{ts2}, para^{74} para^{DTS2}/para^{DN43})

[‡]*L- leucine, F- phenyl alanine, S-serine, H- histidine, M- methionine, T-threonine, V- valine ; I - isoleucine*

threonine to methionine (T350M), in the RDL GABA receptor, which conferred around 20,000-fold resistance to the insecticide fipronil in the resistant *D. simulans* line (Table 3.1). In *P. xylostella*, an A302S amino acid change in the GABA receptor (*PxRdl*) has also been associated with the fipronil resistance phenotype (Li *et al.*, 2006).

Resistance to Imidacloprid

Imidacloprid is a member of the neonicotinoid class of insecticides (chlor-onicotinyls) (Nauen *et al.*, 2002) and is a known nicotinic acetylcholine receptor (nAChR) agonist (Table 3.1). Resistance mechanisms to imida-cloprid have been observed across multiple insect species, including *Nilaparvata lugens* (Liu *et al.*, 2005b), *Ctenocephalides felis* (Rust, 2005), *Bemisia tabaci* (Prabhaker *et al.*, 2005; El Kady *et al.*, 2003), and *L. decemli-neata* (Alyokhin *et al.*, 2007), with different forms of resistance evolving in these different species.

B. *tabaci* that are imidacloprid susceptible typically do not metabolize ^{14}C-imidacloprid into P450-mediated metabolites (Rauch and Nauen, 2003). The imidacloprid/neonicotinoid resistance of the Q- and B-type *B. tabaci* strains does not appear to be based on target site insensitivity (Rauch and Nauen, 2003). The resistance appears to be associated with monooxygenase-mediated activity, with 5-hydroxy-imidacloprid being the only resultant metabolite after topical application of imidacloprid (Rauch and Nauen, 2003).

Imidacloprid binds to nAChR with high affinity in *B. tabaci* and *M. domestica*, whereas the mono-hydroxy metabolite exhibits a much lower affinity (Nauen *et al.*, 1998; Rauch and Nauen, 2003). *M. domestica* produces significant amounts of the mono-hydroxy and olefin derivatives of imidacloprid, and it is likely that detoxification of imidacloprid by *M. domestica* cytochrome P450s may account for the lower toxicity of the insecticide toward this insect as compared with the insecticide susceptible strains of *B. tabaci* (Byrne *et al.*, 2003; Nishiwaki *et al.*, 2004).

Cytochrome P450-mediated resistance to imidacloprid is not limited to insects. A study of the enzymatic basis of imidacloprid metabolism in humans showed that the human cytochrome P450, CYP3A4, oxidizes and reduces imidacloprid at the imidazolidine and nitroimine moieties, respectively (Schulz-Jander and Casida, 2002).

Another mechanism for resistance is amino acid modification of the target site (nAChRs). To date, target-site insensitivity to imidacloprid has only been observed in laboratory-selected *N. lugens* (Liu *et al.*, 2005b). Resistance was conferred by a single-point mutation at Tyrosine151 to Serine (Y151S) in the alpha subunit of nAChR, and a correlation was observed between the frequency of the point mutation and imidacloprid resistance (Liu *et al.*, 2005b). Additionally, Wang *et al.* (2009) and Li *et al.* (2012) each hypothesize that target-site modification could play a role in resistance seen in *B. tabaci* and *M. domestica*, respectively. Recent studies by Puinean *et al.* (2010a,b), however, argue that the target site mechanism for resistance has not been seen in field populations of either *N. lugens* or *M. persicae* and that overexpression of certain P450 genes may play a significant role in resistance.

Resistance to Spinosad

Spinosad, an insecticide derived from a soil fungus, is also thought to target nAChR in insects (Table 3.1) (Narahashi, 2002). Resistance to spinosad has been documented in several insect species, both in laboratory-selected and field populations (Sparks *et al.*, 2012). In *Drosophila*, a knock-out mutation of *Dalpha6*, a gene coding for a nAChR subunit, resulted in a 1181-fold increase in resistance to spinosad (Perry *et al.*, 2007). In *M. domestica*, spinosad resistance is recessive and has been mapped to autosome 1. Recently, it has been suggested that certain P450s may play a role in *M. domestica* resistance to spinosad in female flies collected in Denmark (Markassen and Kristensen, 2011). Widespread resistance to spinosad has been noted in *Plutella xylostella* in Hawaii and Thailand (Zhao *et al.*, 2002). Insects taken from the fields in Hawaii and further selected in the laboratory displayed incompletely recessive resistance.

Resistance to Indoxacarb

Indoxacarb (DPX-JW062) is an oxadiazine insecticide useful in killing a wide variety of insect pests. Insects use an esterase/amidase to decarbomethoxylate indoxacarb to N–decarbomethoxyllate JW062 (DCJW). Both indoxacarb and DCJW are VSSC blockers (Table 3.1) (Shono *et al.*, 2004). Some *M. domestica* strains appear to be partially resistant due to an increased P450 activity (Shono *et al.*, 2004). Differential sensitivity to indoxacarb in cockroach VSSC is due to amino acid changes that influence voltage dependence of slow and fast inactivation, as well as channel sensitivity to DCJW (Song *et al.*, 2006).

Resistance to RNAi

In recent years, RNA interference (RNAi) technology has been developed as a means of controlling crop pest arthropods, especially insects (Whyard et al., 2009). Several examples of the feasibility of this approach have been published covering insects in the major pest orders of the Coleoptera, Hemiptera, and Lepidoptera (Andersen et al., 2009; Huvenne and Smagghe, 2010; Zha et al., 2011). Similar to human medicine, RNAi shows great promise in curing pathogen-induced diseases in important beneficial agricultural insects such as the honey bee (Hunter et al., 2010; Beelogics, 2012)

RNAi has the potential to target insect pests and pathogens with a high degree of specificity because it takes advantage of the ability of eukaryotic cells to recognize double-stranded RNA (dsRNA) and to degrade only homologous RNA sequences, resulting in sequence-specific gene silencing (Hannon, 2002). This fact makes the technology particularly attractive since the high degree of specificity greatly reduces the potential harm to nontarget organisms (Huesing et al., 2009, 2010).

In terms of crop protection, the conventional spray approaches and transgenic plants that express arthropod-specific dsRNAs have both been shown to be useful in controlling insects when presented via an oral route of exposure (Whyard et al., 2009). When RNAi-expressing plants are consumed by pests, the dsRNA is delivered into the insect's digestive tract, eliminating the need to spray crops, while simultaneously reducing the likelihood of exposure of nontarget species. Though not all insect species are susceptible to the approach, a number of pest arthropods represent logical targets (Huvenne and Smagghe 2010). For example, when Diabrotica virgifera virgifera larvae fed on transgenic corn producing vacuolar H^+ ATPase dsRNA, a drastic reduction of the corresponding mRNA was observed in the larvae's guts (leading to decreased growth and survivorship of the larvae, resulting in reduced damage of the infested transgenic corn roots (Baum et al., 2007; Andersen et al., 2009). This finding has been extended to include adult D. virgifera virgifera where greater than 95% adult mortality was observed within two weeks following feeding of dsRNA-treated artificial diet (Rangasamy and Siegfried, 2012). It appears likely that this RNAi technology will be registered for the control of D. virgifera virgifera within the coming decade (Monsanto, 2012).

Given the nascent nature of RNAi technology for use in insect control, the development of resistance to RNAi in arthropod pests has not been documented. However, RNAi resistance in mammalian cell cultures is known to occur, and many RNA viruses escape RNAi-mediated suppression through mutation of the targeted region, by use of viral suppressors, or by cellular factors such as ADAR1 (adenosine deaminases acting on RNA), which is responsible for editing-mediated RNAi resistance (Zheng et al., 2005). Of course, it is also possible that a combination of any of the resistance mechanisms described in this chapter may occur as well.

At least two pathways for dsRNA uptake in insects have been described. The transmembrane channel-mediated uptake mechanism based on the *Caenorhabditis elegans'* SID-1 protein and an endocytosis-mediated uptake mechanism (Huvenne and Smagghe, 2010). In addition, in some cases RNAi can amplify and spread systemically from the initial site of dsRNA delivery, producing interference phenotypes throughout the treated animal (Hunter et al., 2010). Based on the known modes of action of RNAi some speculation about potential resistance in insects can be made.

In one model organism, *Caenorhabditis elegans*, at least 15 different mutations can lead to resistance to dietary RNAi (Whangbo and Hunter, 2008). Furthermore, insensitive congeners appear to be naturally insensitive to dietary dsRNA, with no apparent cost to fitness, because dsRNA is apparently not internalized by the cells of the gut. For example, in a screen of eight congeners of *C. elegans*, only one additional species was found to respond to dietary dsRNA (Winston et al., 2007). Examples do not yet exist demonstrating resistance to RNAi either at the cellular uptake level or point of systemic spread, but these are likely routes of resistance as well.

Insect resistance management (IRM) is a fundamental issue addressed in all insect protected GM crops (Siegfried et al., 2007). Industry best practice and oversight by regulatory agencies ensure that insect protected crops, long term, be deployed with multiple mechanisms (sometimes termed modes) of resistance (U.S. EPA, 1998; ILSI, 1998). This practice is encouraged for RNAi-based GM crops as well (ILSI, 2011). As with other technologies, for example, *Bt*-based crops, if RNAi technology is employed as a stand-alone technology under very high selection without adequate provision for refugia, the potential for rapid development of resistance is highly likely.

Other Insecticides

As we cannot cover every class of insecticide, we instead recommend the following key publications and reviews of other insecticides; please see: (i) Clark *et al.* (1995) for avermectins and milbemycins insecticides; (ii) Mordue and Blackwell (1993) for azadirachtin/neem; (iii) Arena (1963) for rotenone and ryania; (iv) Sattellite *et al.* (1985) for nereistoxin analogues; and Ashok *et al.* (1998), Dhadialla *et al.* (1998), and Mohandass *et al.* (2006) for juvenile hormone mimics. Additionally, please see Table 3.1, where we summarize the mode of action of additional insecticides beyond those previously discussed. In Chapter 4 Alves and Nelson review the mechanisms of resistance to several plant-incorporated toxins, particularly toxins derived from *Bacillus thurigiensis*.

EMERGING *OMICS* TECHNOLOGIES

Use of Genomics and Proteomics to Understand Insecticide-Resistance Genes

Genomics holds tremendous opportunities for us to understand how insects become "resistant" to human-made or natural poisons and how insects become "resistant" to biotic and abiotic challenges they experience in their environments.

First, we can think of the recent genomics and proteomics revolution in terms of "Henry Ford and mass production meet molecular biology." Prior to the development of many current genomics techniques, most researchers investigated a single or a limited number of genes and their potential role in a given biological process. With genomics and proteomics, large-scale genome- or proteome-wide comparisons are routinely performed between susceptible and resistant organisms or challenged and unchallenged organisms (Pedra *et al.*, 2004, 2005). For example, the expression levels of thousands or tens of thousands of genes can be determined in a single experiment through the use of microarray or RNAseq methodology (Vontas *et al.*, 2005; Wang *et al.*, 2011; Mamidala *et al.*, 2012). The differential expression of proteins can also be determined in a given treatment or tissue. These technologies enable researchers to rapidly discover genes and their associated proteins that play a critical role in an organism's response to challenges in their environment.

Specifically, this approach has and will continue to allow researchers to investigate differences between susceptible and resistant insects without the need for *a priori* knowledge of the potential genes involved in resistance.

Using whole-genome oligo-array gene chips, investigators determined the specific genes that are differentially expressed between DDT susceptible and metabolically resistant strains of fruit flies (Pedra *et al.*, 2004). Some of these genes were previously determined to be associated with metabolic insecticide resistance, such as cytochrome P450s (e.g., *Cyp12d1/Cyp12d2*, *Cyp6g1*, *Cyp6a2*, and *Cyp6a8*), GSTs, and oxidoreductases (Daborn *et al.*, 2001, 2002; Pedra *et al.*, 2004). However, genes previously not known to be associated with resistance were also observed (e.g., diazepam binding inhibitor). Nevertheless, care needs to be exercised not to equate differential expression with actual direct, or even indirect, involvement in resistance. Genes can be differentially expressed because of several factors: (i) genetic hitchhiking, (ii) the genes may also be under the control of the same regulatory process as the genes that code for the proteins that actually confer resistance, and hence are also up- or downregulated, or (iii) the genes may be differentially expressed as a response to cellular or organismal changes that occur due to differential changes in the expression of the resistance genes. Nonetheless, these techniques have the potential to reveal previously unknown components of resistance mechanisms.

It is important to note that oligoarrays and cDNA spotted arrays are used to detect changes in transcription (mRNA expression); however, changes in transcription do not necessarily mean there are changes in translation (protein expression). Since the protein is typically the critical molecule involved in resistance, a determination must be made of any real and meaningful differences in protein levels. This is especially true where resistance is thought to be associated with differential expression of metabolic enzymes.

To screen the proteome for differences, high-throughput proteomics techniques can be used to identify proteins that may not have previously been known to be associated with resistance. Most current proteomic studies utilize multiple techniques to elucidate proteome differences (Festucci-Buselli *et al.*, 2005; Pedra *et al.*, 2005; Teese *et al.*, 2010; Hall *et al.*, 2011; Jurat-Fuents *et al.*, 2011; Whitehill *et al.*, 2011). Each of these studies began with a method of gel electrophoresis; difference gel electrophoresis (DIGE; Whitehill *et al.*, 2011), two-dimensional gel electrophoresis (2-DGE; Pedra *et al.*, 2005; Jurat-Fuentes *et al.*, 2011), or sodium dodecyl sulfate polyacrylamide gel electrophoresis (SDS-PAGE; Festucci-Buselli *et al.*, 2005; Teese *et al.*, 2010; Hall *et al.*, 2011). Festucci-Buselli *et al.* (2005) further used western blots to demonstrate differences in protein expression (CYP6G1 and CYP12D1) between DDT-susceptible and -resistant insects. Pedra *et al.* (2005) used 2-DGE and matrix-assisted laser desorption ionization time of flight mass spectrometry (MALDI-TOF) to investigate differences between strains, but found neither CYP6G1 nor CYP12D1 significant. However, metabolism-associated proteins were more highly expressed in the DDT-resistant strains. Mass spectrometry has become widespread and invaluable in identifying all protein products present between samples and is often used as the second step in many proteomic studies (Pedra *et al.*, 2005; Teese *et al.*, 2010; Hall *et al.*, 2011; Whitehill *et al.*, 2011). Recently, quantitative real-time PCR (qRT-PCR) has begun to be used as an additional verification step that the identified proteins are in fact differentially expressed between samples (Hall *et al.*, 2011; Jurat-Fuentes *et al.*, 2011).

Additionally, transgenic insects (Daborn *et al.*, 2002) and RNAi (Bellés, 2010; Huvenne and Smagghe, 2010; Boerjan *et al.*, 2012) have been used to verify the roles that a given gene and its resultant protein(s) may play in a specific biological process.

The noninvasive induction assay in conjunction with transcriptional profiling of detoxification genes can be employed to identify putative metabolic resistance factors (Yoon *et al.*, 2011). Ivermectin-induced cytochrome P450 (*CYP6CJ1, CYP9AG1, CYP9AG2*) and ABCT (*PhABCC4*) genes from body lice were identified by quantitative PCR analyses using the noninvasive induction assay. The genes that were most significantly induced showed high basal expression levels and were most closely related to genes from other organisms that metabolized ivermectin, suggesting their potential involvement in metabolic resistance. Once identified, these inducible detoxification genes may be used in proactive resistance monitoring schemes and in the construction of metabolic maps using a variety of insecticides to establish cross-expression and negative cross-expression patterns during the acquisition of tolerance following induction.

Genomics and Proteomics for Discovery of Resistance Mechanisms for Abiotic and Biotic Challenges

Genomic and proteomic techniques have provided researchers with the tools to more rapidly discover how insects evolved resistance to a variety of abiotic and biotic factors. In the case of pest insects, such resistance mechanisms may have the potential to be used as target sites for the development of novel pest control agents. The genes or proteins that confer the resistance mechanism to a given stressor can now become the target site for compounds that alter (e.g., inhibit) the protein's function.

These targets may include plant defensive compounds contained in the insects' diet, oxidative stress, temperature, desiccation, or other stressors experienced in the insect's life history.

An example of the kind of environmental challenge we wish to emphasize is an environmental stress that nearly all eukaryotic organisms experience: oxidative stress. Oxidative stress occurs when the cellular antioxidant machinery, both enzymatic (e.g., catalase and superoxide dismutase) and nonenzymatic (e.g., reduced glutathione) antioxidants, cannot keep pace with the formation of reactive oxygen species (ROS), and, to a lesser extent, reactive nitrogen species. As a result, cellular damage occurs when ROS react with cellular components, including proteins, lipids, and DNA, and cause irreversible structural changes. ROS are by-products of normal aerobic metabolism, and accordingly, the mitochondria are the most significant sources of ROS under normal conditions. Oxidative stress can result from either a lack of antioxidants in the cell or an excess of ROS.

Oxidative stress can be due to exogenous effects or endogenous reactions. For example, energetic radiation, such as ultraviolet (UV) rays, can lead to hydroxyl radicals, resulting in an increase in ROS in the cell. Additionally, exogenous oxidants (e.g., peroxides), redox recycling agents (e.g., quinone compounds), hormones, and endotoxins all can lead to increased intracellular ROS production. Compounds such as hydrogen peroxide can lead to ROS by uncoupling cytochrome P450 reactions. In mammals, physiological signaling, including immune system responses, contributes to intracellular ROS production.

Conversely, low levels of intracellular antioxidants can also lead to the accumulation of ROS. For example, (i) if glutathione production is reduced, (ii) if there are fewer antioxidant vitamins in the cell, or (iii) if ROS-scavenging enzymes are inhibited, (iv) or a combination of these factors exist, then the levels of ROS will increase in the cell. It is critical that cells are able to neutralize ROS, since oxidative stress can prolong cell cycles causing arrested development of the overall organism (Wiese et al., 1995). Cells use multiple systems to protect themselves against ROS, including glutathione, which acts as an intracellular antioxidant buffer system in the cell. The thiol-containing moiety on the cysteine residue of glutathione has reducing power (supplies electrons) that nullifies the oxidative potential of the ROS. Glutathione homeostasis occurs by the balance between glutathione (GSH) and glutathione disulfide (oxidized glutathione; GSSG); GSH is oxidized by glutathione peroxidase to generate GSSG, which in turn can be reduced back to GSH by glutathione reductase. However, the rate-limiting step in the production of GSH is the enzyme γ-Glutamylcystein synthase, which converts N-acetylecysteine into GSH. The ratio of GSH to GSSG in the cell is typically 100:1, which means that the oxidation of GSH can dramatically influence the redox status in the cell. Fluorochrome probes (e.g., 2', 7'-dichlorofluorescein) can be used to assay this change in the oxidative status of the cell. Cells also employ proteins and vitamins to reduce oxidative stress, including superoxide dismutase, catalase, quinone reductase (detoxifies quinone compounds), metallothionein (traps heavy metal cations), and vitamins such as E and C that trap free radicals.

Oxidative stress influences the regulation of gene expression, causing both induction of some genes and repression of others. For example, ROS are known to induce the expression of antioxidant proteins as well as the enzymes that the cell uses to regenerate these proteins (e.g., Trx and glutathione reductases). Conversely, ROS at the same time repress such genes as α-actin, troponin I, some cytochrome P450s, as well as genes that code for proteins in mammals associated with sugar regulation (e.g., insulin) and the immune system (IL-2) (Beiqing et al., 1996; Matsuoka et al., 1997; Barker et al., 1994). These aforementioned genes represent only a subset of the total genes differentially expressed due to oxidative stress. For example, it has also been demonstrated that UV-B radiation strongly inhibits mitochondrial transcription, which results in a repression of mitochondrial function; the mitochondria, which is a major generator of ROS, is very susceptible to oxidative stress (Vogt et al., 1997).

Work in Drosophila and Spodoptera littoralis has shown that there is evolutionary conservation between mammals and insects in some of the mechanisms by which both groups of organisms deal with oxidative stress (e.g., superoxide dismutase, catalase, ascorbate peroxidase, and glutathione S-transferase peroxidase) (Krishnan and Kodrík, 2006; Magwere et al., 2006). Krishnan and Kodrík (2006) demonstrated that, in S. littoralis, these aforementioned enzymes are associated with the digestive system, suggesting that they potentially play a role in dealing with oxidative radicals associated with their food.

In addition to the conserved mechanisms for combating oxidative stress, there is growing evidence of insect-unique systems for neutralizing oxidative stress. Dubuisson et al. (2004) recently observed that luciferin (which is involved in bioluminescence) is a scavenger for the oxidant peroxynitrite. Their observations are consistent with hypotheses proposed for marine organisms, suggesting that bioluminescence may have initially evolved as an antioxidant mechanism and secondarily as a light-producing system. These findings hold out the possibility that if insect-unique anti-oxidant systems occur in other insect species, then synthetic inhibitors targeting these insect-specific antioxidant systems may be used to selectively interfere with the ability of insects to protect themselves from the effects of oxidative stress in their environment.

Understanding oxidative stress certainly has more immediate implications for issues concerning insecticide resistance. For example, *An. gambiae* mosquitoes that are DDT resistant, via GST activity, appear to also be more responsive/resistant to oxidative stress (Enayati *et al.*, 2005; Ranson and Hemingway, 2005). Vontas *et al.* (2001) also demonstrated that pyrethroids induced oxidative stress responses in *Nilaparvata lugens*. Thus, it is possible that insect strains that live in environments where they experience more oxidative stress may be predisposed to being more resistant to insecticides (a hypothesis that remains to be tested). Alternatively, since P450s are typically downregulated during times of oxidative stress, it would be logical that overexpression of P450s may, in some circumstances, be a means to inhibiting an insect's ability to mitigate the effects of oxidative stress in their environment. Recently, it was determined that DDT resistance in four strains of *Drosophila* was inversely correlated with their ability to survive hydrogen peroxide (Sun *et al.*, 2011). Thus, the more insecticide-resistant strains, presumably due to increased P450 expression/activity, showed increased susceptibility to oxidative stress.

It remains to be determined whether inhibition of systems that allow insects to respond to biotic and abiotic stressors may prove useful in practical insect control in the future (Figure 3.9). Existing and emerging *omics* tools can be used to gain insights into the diversity of insect responses to stress, and in turn further our understanding of the molecular basis for these responses. Ultimately, with this understanding, we may be able to manipulate the environments of pests to maximize the costs of resistance alleles.

CONCLUSIONS

In this chapter we have reviewed the current status of the mechanisms of action of select insecticidal agents and known forms of resistance

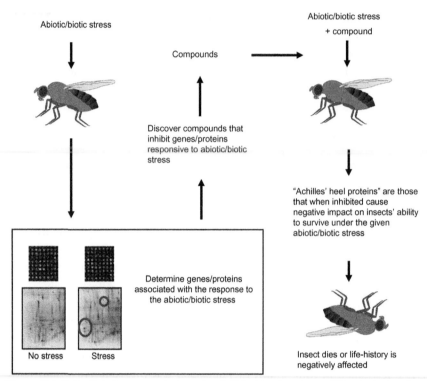

Figure 3.9 Potential discovery strategy for Achilles' heel proteins. 2D gel electrophoresis and microarray technologies were chosen to illustrate proteomic and transcriptomic technologies, respectively, due to the visual nature of the data. However, newer generation proteomic and transcriptomic technologies, such as GC/MS peptide fingerprinting and RNA seq, respectively, are more likely to be used at the time of writing of this chapter.

to them. We have used examples to illustrate that insects use evolutionarily conserved resistance mechanisms common to all animals (e.g., some aspects of oxidative stress) as well as more conserved mechanisms restricted to the insects (e.g., the peritrophic membrane in the digestive system of insects). Finally, we have illustrated how the "omics" revolution is just beginning to reveal more in-depth knowledge of the bases of these mechanisms (e.g., metabolic insecticide resistance). Perhaps not surprising, but nonetheless exciting, are the emerging examples of the involvement of hitherto unidentified genes and mechanisms involved in resistance. These findings should allow us to identify novel and safe insecticides as well as better design resistance management strategies to ensure their long-term utility.

REFERENCES

Ahuja, R., Kumar, A., 2003. Metabolism of DDT [1,1,1-Trichloro-2,2-bis(4-chlorophenyl) ethane] by *Alcaligenes denitrificans* ITRC-4 under aerobic and anaerobic conditions. Curr. Microbiol. 46 (1), 65−69.

Ahuja, R., Awasthi, N., Manickam, N., Kumar, A., 2001. Metabolism of 1, 1-dichloro-2, 2-bis (4-chlorophenyl) ethylene by *Alcaligenes denitrificans*. Biotechnol. Lett. 23, 423−426.

Alizadeh, A., Talebi, K., Hosseininaveh, V., Ghadamyari, M., 2011. Metabolic resistance mechanisms to phosalone in the common pistachio psyllid, *Agonoscena pistaciae* (Hem.: Psyllidae). Pestic. Biochem. Phys. 101, 59−64.

Alout, H., Berthomieu, A., Hadjivassilis, A., Weill, M., 2007. A new amino-acid substitution in acetylcholinesterase 1 confers insecticide resistance to *Culex pipiens* mosquitoes from Cyprus. Insect. Biochem. Mol. Biol. 37, 41−47.

Alyokhin, A., Dively, G., Patterson, M., Castaldo, C., Rogers, D., Mahoney, M., et al., 2007. Resistance and cross-resistance to imidacloprid and thiamethoxam in the Colorado potato beetle *Leptinotarsa decemlineata*. Pest. Manag. Sci. 63, 32−41.

Amichot, M., Tares, S., Brun-Barale, A., Arthaud, L., Bride, J.M., Berge, J.B., 2004. Point mutations associated with insecticide resistance in the *Drosophila* cytochrome P450 Cyp6a2 enable DDT metabolism. Eur. J. Biochem. 271 (7), 1250−1257.

Andersen, S., Hicks, G., Huesing, J., Romano, C., Vetsch, C., 2009. U.S. Patent. 7,612,194 B2. Nucleic acid sequences from *Diabrotica virgifera virgifera* LeConte and uses thereof.

Anthony, N., Unruh, T., Ganser, D., ffrench-Constant, R., 1998. Duplication of the Rdl GABA receptor subunit gene in an insecticide-resistant aphid, *Myzus persicae*. Mol. Gen. Genet. 260 (2-3), 165−175.

Araújo, R.A., Williamson, M.S., Bass, C., Field, L.M., Duce, I.R., 2011. Pyrethroid resistance in *Sitophilus zeamais* is associated with a mutation (T929I) in the voltage-gated sodium channel. Insect. Mol. Biol. 20, 437−445.

Ardelli, B.F., Prichard, R.K., 2007. Reduced genetic variation of an *Onchocera volvulus* ABC transporter gene following treatment with ivermectin. R. Soc. Trop. Med. Hyg. 101, 1223−1232.

Arena, J.M., 1963. Report from the Duke University poison control center. Rotenone and Ryania. N. C. Med. J. 24, 388−389.

Ashok, M., Turner, C., Wilson, T.G., 1998. Insect juvenile hormone resistance gene homology with the bHLH-PAS family of transcriptional regulators. Proc. Natl. Acad. Sci. USA. 95 (6), 2761−2766.

Baek, J.H., Clark, J.M., Lee, S.H., 2010. Cross-strain comparison of cypermethrin-induced cytochrome P450 transcription under different induction conditions in diamondback moth. Pestic. Biochem. Physiol. 96, 43−50.

Barker, C.W., Fagan, J.B., Pasco, D.S., 1994. Down-regulation of P4501A1 and P4501A2 mRNA expression in isolated hepatocytes by oxidative stress. J. Biol. Chem. 269, 3985−3990.

Bass, C., Schroeder, I., Turberg, A., Field, L.M., Williamson, M.S., 2004. Identification of mutations associated with pyrethroid resistance in the *para*-type sodium channel of the cat flea, *Ctenocephalides felis*. Insect. Biochem. Mol. Biol. 34, 1305−1313.

Bauer, J.E., Capone, D.G., 1985. Effects of four aromatic organic pollutants on microbial glucose metabolism and thymidine incorporation in marine sediments. Appl. Environ. Microbiol. 49, 828−835.

Baum, J.A., Bogaert, T., Clinton, W., Heck, G.R., Feldmann, P., Ilagan, O., et al., 2007. Control of coleopteran insect pests through RNA interference. Nat. Biotechnol. 25, 1322−1326.

Beeologics, 2012. 2011 Clinical Trials FAQ. <http://www.beeologics.com/2011_clinical_trials. asp?subject = 2011%20Clinical%20Trials>.

Beiqing, L., Chen, M., Whisler, R.L., 1996. Sublethal levels of oxidative stress stimulate transcriptional activation of c-jun and suppress IL-2 promoter activation in Jurkat T cells. J. Immunol. 157, 160–169.

Bejsovec, A., Lecuit, T., Modolell, J., 2004. The fly Olympics: faster, higher and stronger answers to developmental questions. Conference on the Molecular and Developmental Biology of Drosophila. EMBO Rep 5: pp. 1037–1040.

Bellés, X., 2010. Beyond Drosophila: RNAi In Vivo and Functional Genomics in Insects. Annu. Rev. Entomol. 55, 111–128.

Berenbaum, M.R., 1995. The chemistry of defense: theory and practice. Proc. Natl. Acad. Sci. USA. 92, 2–8.

Berenbaum, M.R., Zangerl, A.R., 1998. Chemical phenotype matching between a plant and its insect herbivore. Proc. Natl. Acad. Sci. USA. 95, 13743–13748.

Berenbaum, M.R., Favret, C., Schuler, M.A., 1996. On defining "key innovations" in an adaptive radiation: cytochrome P450s and Papilionidae. Am. Nat. 148, S139–S155.

Bilen, J., Bonini, N.M., 2005. Drosophila as a model for human neurodegenerative disease. Annu. Rev. Genet. 39, 153–171.

Boerjan, B., Tobback, J., Vandersmissen, H.P., Huybrechts, R., Schoofs, L., 2012. Fruitless RNAi knockdown in the desert locust, Schistocerca gregaria, influences male fertility. J. Insect. Physiol. 58, 265–269.

Bowyer, J.F., Pogge, A.R., Delongchamp, R.R., O'Callaghan, J.P., Patel, K.M., Vrana, K.E., et al., 2007. A threshold neurotoxic amphetamine exposure inhibits parietal cortex expression of synaptic plasticity-related genes. Neuroscience. 144 (1), 66–76.

Bowyer, J.F., Thomas, M., Schmued, L.C., Ali, S.F., 2008. Brain region-specific neurodegenerative profiles showing the relative importance of amphetamine dose, hyperthermia, seizures, and the blood-brain barrier. Ann. NY Acad. Sci. 1139, 127–139.

Boyer, S., Zhang, H., Lempérière, G., 2012. A review of control methods and resistance mechanisms in stored-product insects. Bull. Entomol. Res. 102 (2), 213–229.

Brandt, A., Scharf, M., Pedra, J.H.F., Holmes, G., Dean, A., Kreitman, M., et al., 2002. Differential expression and induction of two Drosophila cytochrome P450 genes near the Rst (2) DDT locus. Insect. Mol. Biol. 11, 337–341.

Brengues, C., Hawkes, N.J., Chandre, F., McCarroll, L., Duchon, S., Guillet, P., et al., 2003. Pyrethroid and DDT cross-resistance in Aedes aegypti is correlated with novel mutations in the voltage-gated sodium channel gene. Med. Vet. Entomol. 17, 87–94.

Brown, A.W.A., 1986. Insecticide resistance in mosquitoes: a pragmatic review. J. Am. Mosq. Control Assoc. 2, 123–140.

Brun-Barale, A., Bouvier, J.C., Pauron, D., Berge, J.B., Sauphanor, B., 2005. Involvement of a sodium channel mutation in pyrethroid resistance in Cydia pomonella L, and development of a diagnostic test. Pest. Manag. Sci. 61, 549–554.

Buss, D.S., Callaghan, A., 2008. Interaction of pesticides with p-glycoprotein and other ABC proteins: a survey of the possible importance to insecticide, herbicide and fungicide resistance. Pestic. Biochem. Physiol. 90, 141–153.

Byrne, F.J., Castle, S., Prabhaker, N., Toscano, N.C., 2003. Biochemical study of resistance to imidacloprid in B biotype Bemisia tabaci from Guatemala. Pest. Manag. Sci. 59, 347–352.

Carino, F.A., Koener, J.F., Plapp Jr., F.W., Feyereisen, R., 1994. Constitutive overexpression of the cytochrome P450 gene CYP6A1 in a house fly strain with metabolic resistance to insecticides. Insect. Biochem. Mol. Biol. 24, 411–418.

Cassanelli, S., Reyes, M., Rault, M., Carlo Manicardi, G., Sauphanor, B., 2006. Acetylcholinesterase mutation in an insecticide-resistant population of the codling moth Cydia pomonella (L.). Insect. Biochem. Mol. Biol. 36, 642–653.

Chandre, F., Darriet, F., Darder, M., Cuany, A., Doannio, J.M., Pasteur, N., et al., 1998. Pyrethroid resistance in *Culex quinquefasciatus* from West Africa. Med. Vet. Entomol. 12, 359−366.

Chapman, R.B., Penman, D.R., 1979. Negatively correlated cross-resistance to a synthetic pyrethroid in organophosphorus-resistant *Tetranychus urticae*. Nature. 281, 298−299.

Choi, K.S., Spillings, B.L., Coetzee, M., Hunt, R.H., Koekemoer, L.L., 2010. A comparison of DNA sequencing and the hydrolysis probe analysis (TaqMan assay) for knockdown resistance (kdr) mutations in *Anopheles gambiae* from the Republic of the Congo. Malar. J. 9, 278−284.

Clark, J.M., Scott, J.G., Campos, F., Bloomquist, J.R., 1995. Resistance to avermectins: extent, mechanisms, and management implications. Annu. Rev. Entomol. 40, 1−30.

Couso-Ferrer, F., Arouri, R., Beroiz, B., Perera, N., Cervera, A., Navarro-Llopis, V., et al., 2011. Cross-Resistance to Insecticides in a Malathion-Resistant Strain of *Ceratitis capitata* (Diptera: Tephritidae). J. Econ. Entomol. 104 (4), 1349−1356.

Crow, J.F., 1954. Analysis of a DDT-resistant strain of *Drosophila*. J. Econ. Entomol. 47, 393−398.

Daborn, P., Boundy, S., Yen, J., Pittendrigh, B.R., ffrench-Constant, R., 2001. DDT resistance in *Drosophila* correlates with *Cyp6g1* over-expression and confers cross-resistance to the neonicotinoid imidacloprid. Mol. Genet. Genomics. 266 (4), 556−563.

Daborn, P.J., Yen, J.L., Bogwitz, M.R., Le Goff, G., Feil, E., Jeffers, S., et al., 2002. A single P450 allele associated with insecticide resistance in *Drosophila*. Science. 297, 2253−2256.

Dapkus, D., 1992. Genetic localization of DDT resistance in *Drosophila melanogaster* (Diptera: Drosophilidae). J. Econ. Entomol. 85, 340−347.

Dapkus, D., Merrell, D.J., 1977. Chromosomal analysis of DDT-resistance in a long-term selected population of *Drosophila melanogaster*. Genetics. 87, 685−697.

DeVries, D, Georghiou, G., 1981. Decreased nerve sensitivity and decreased cuticular penetration as a mechanism of resistance to pyrethroids in a (IR)-trans-permethrin-selected strain of the house fly. Pestic. Biochem. Physiol. 15, 242−252.

Delye, C., Bousset, L., Corio-Costet, M.F., 1998. PCR cloning and detection of point mutations in the eburicol 14alpha-demethylase (CYP51) gene from *Erysiphe graminis f. sp. hordei*, a "recalcitrant" fungus. Curr. Genet. 34, 399−403.

Dhadialla, T.S., Carlson, G.R., Le, D.P., 1998. New insecticides with ecdysteroidal and juvenile hormone activity. Ann. Rev. Entomol. 43, 545−569.

Dombrowski, S.M., Krishnan, R., Witte, M., Maitra, S., Diesing, C., Waters, L.C., et al., 1998. Constitutive and barbital-induced expression of the Cyp6a2 allele of a high producer strain of CYP6A2 in the genetic background of a low producer strain. Gene. 221, 69−77.

Dong, K., 1997. A single amino acid change in the para sodium channel protein is associated with *knockdown-resistance* (*kdr*) to pyrethroid insecticides in German cockroach. Insect. Biochem. Mol. Biol. 27, 93−100.

Dong, K., Valles, S.M., Scharf, M.E., Zeichner, B., Bennett, G.W., 1998. The knockdown resistance (*kdr*) mutation in pyrethroid-resistant German cockroaches. Pestic. Biochem. Physiol. 60, 195−204.

Dubuisson, M., Marchand, C., Rees, J.F., 2004. Fire fly luciferin as antioxidant and light emitter: the evolution of insect bioluminescence. Luminescence. 19, 339−344.

Eaton, D.L., 2000. Biotransformation enzyme polymorphism and pesticide susceptibility. Neurotoxicology. 21, 101−111.

El Kady, H., Devine, G.J., 2003. Insecticide resistance in Egyptian populations of the cotton whitefly, *Bemisia tabaci* (Hemiptera: Aleyrodidae). Pest. Manag. Sci. 59, 865−871.

Ela, R., Chefurka, W., Robinson, J.R., 1970. In vivo glucose metabolism in the normal and poisoned cockroach, *Periplaneta americana*. J. Insect. Physiol. 16, 2137−2156.

Eleftherianos, I., Foster, S.P., Williamson, M.S., Denholm, I., 2008. Inheritance of L1014F and M918T sodium channel mutations associated with pyrethroid resistance in *Myzus persicae*. Biol. Lett. 4 (5), 545−548.

Enayati, A.A., Ranson, H., Hemingway, J., 2005. Insect glutathione transferases and insecticide resistance. Insect. Mol. Biol. 14 (1), 3−8.

Endersby, N.M., Viduka, K., Baxter, S.W., Saw, J., Heckel, D.G., McKechnie, S.W., 2011. Widespread pyrethroid resistance in Australiam diamondback moth, *Plutella xylostella* (L.), is related to multiple mutations in the *para* sodium channel gene. Bull. Ent. Res. 101 (4), 393−405.

Farnham, A.W., 1971. Changes in cross-resistance patterns of house flies selected with natural pyrethrins or resmethrin (5-benzyl-3-furylmethyl-cis-transchrysanthemate). Pest Sci. 2, 138−143.

Farnham, A.W., 1973. Genetics of resistance of pyrethroid selected house flies, *Musca domestica* L. Pest Sci. 4, 513−520.

Ferguson, S.A., Racey, F.D., Paule, M.G., Holson, R.R., 1993. Behavioral effects of methylazoxymethanol-induced micrencephaly. Behav. Neurosci. 107 (6), 1067−1076.

Festucci-Buselli, R.A., Carvalho-Dias, A.S., de Oliveira-Andrade, M., Caixeta-Nunes, C., Li, H.M., Stuart, J.J., et al., 2005. Expression of Cyp6g1 and Cyp12d1 in DDT resistant and susceptible strains of *Drosophila melanogaster*. Insect. Mol. Biol. 14, 69−77.

Feyereisen, R., 1995. Molecular biology of insecticide resistance. Toxicol. Lett. 82-83, 83−90.

Feyereisen, R., 1999. Insect P450 enzymes. Annu. Rev. Entomol. 44, 507−533.

Ffrench-Constant, R.H., Mortlock, D.P., Shaffer, C.D., MacIntyre, R.J., Roush, R.T., 1991. Molecular cloning and transformation of cyclodiene resistance in *Drosophila*: an invertebrate gamma-aminobutyric acid subtype A receptor locus. Proc. Natl. Acad. Sci. USA. 88, 7209−7213.

Ffrench-Constant, R.H., Rocheleau, T.A., Steichen, J.C., Chalmers, A.E., 1993a. A point mutation in a *Drosophila* GABA receptor confers insecticide resistance. Nature. 363, 449−451.

Ffrench-Constant, R.H., Steichen, J.C., Rocheleau, T.A., Aronstein, K., Roush, R.T., 1993b. A single-amino acid substitution in a gamma-aminobutyric acid subtype A receptor locus is associated with cyclodiene insecticide resistance in *Drosophila* populations. Proc. Natl. Acad. Sci. USA. 90, 1957−1961.

Ffrench-Constant, R.H., Pittendrigh, B., Vaughan, A., Anthony, N., 1998. Why are there so few resistance-associated mutations in insecticide target genes?. Philos. Trans. R. Soc. Lond. B. Biol. Sci. 353, 1685−1693.

Field, L.M., Devonshire, A.L., Forde, B.G., 1988. Molecular evidence that insecticide resistance in peach-potato aphids (*Myzus persicae* Sulz.) results from amplification of an esterase gene. Biochem. J. 251, 309−312.

Field, L.M., Javed, N., Stribley, M.F., Devonshire, A.L., 1994. The peach-potato aphid *Myzus persicae* and the tobacco aphid *Myzus nicotianae* have the same esterase-based mechanisms of insecticide resistance. Insect. Mol. Biol. 3, 143−148.

Fischer, H.M., Wheat, C.W., Heckel, D.G., Vogel, H., 2008. Evolutionary origins of a novel host plant detoxification gene in butterflies. Mol. Biol. Evol. 25 (5), 809−820.

Forcioli, D., Frey, B., Frey, J.E., 2002. High nucleotide diversity in the *para*-like voltage-sensitive sodium channel gene sequence in the western flower thrips (Thysanoptera: Thripidae). J. Econ. Entomol. 95, 838−848.

Fristrom, J.W., 1970. The developmental biology of *Drosophila*. Annu. Rev. Genet. 4, 325−346.

Grant, D.F., Hammock, B.D., 1992. Genetic and molecular evidence for a trans-acting regulatory locus controlling glutathione S-transferase-2 expression in *Aedes aegypti*. Mol. Gen. Genet. 234, 169–176.

Grimaldi, D., Engel, M.S., 2005. Evolution of the Insects. Cambridge U. Press, Cambridge.

Guengerich, F.P., Parikh, A., Turesky, R.J., Josephy, P.D., 1999. Inter-individual differences in the metabolism of environmental toxicants: cytochrome P450 1A2 as a prototype. Mutat. Res. 428, 115–124.

Guerrero, F.D., Jamroz, R.C., Kammlah, D., Kunz, S.E., 1997. Toxicological and molecular characterization of pyrethroid-resistant horn flies, *Haematobia irritans*: identification of *kdr* and *super-kdr* point mutations. Insect. Biochem. Mol. Biol. 27, 745–755.

Hall, D.E., Robert, J.A., Keeling, C.I., Domanski, D., Quesada, A.L., Jancsik, S., et al., 2011. An integrated genomic, proteomic and biochemical analysis of (+)-3-carene biosynthesis in Sitka spruce (*Picea sitchensis*) genotypes that are resistant or susceptible to white pine weevil. Plant J. 65, 936–948.

Handschin, C., Blattler, S., Roth, A., Looser, R., Oscarson, M., Kaufmann, M.R., et al., 2004. The evolution of drug-activated nuclear receptors: one ancestral gene diverged into two xenosensor genes in mammals. Nucl. Recept. 2, 7. Available from: http://dx.doi.org/10.1186/1478-1336-2-7.

Hannon, G.J., 2002. RNA interference. Nature. 418, 244–251.

Hayes, J.D., Pulford, D.J., 1995. The glutathione S-transferase supergene family: regulation of GST and the contribution of the isoenzymes to cancer chemoprotection and drug resistance. Crit. Rev. Biochem. Mol. Biol. 30, 445–600.

Heckel, D.G., Gahan, L.J., Daly, J.C., Trowell, S., 1998. A genomic approach to understanding *Heliothis* and *Heliocoverpa* resistance to chemical and biological insecticides. Philos. Trans. R. Soc. 353, 1492–1502.

Hegedus, D., Erlandson, M., Gillott, C., Toprak, U., 2009. New insights into peritrophic matrix synthesis, architecture, and function. Annu. Rev. Entomol. 54, 285–302.

Hemingway, J., Miyamoto, J., Herath, P.R.J., 1991. A possible novel link between organophosphorus and DDT insecticide resistance genes and in *Anopheles-* supporting evidence from fenitrothion metabolism studies. Pestic. Biochem. Physiol. 39, 45–56.

Houpt, D.R., Pursey, J.C., Morton, R.A., 1988. Genes controlling malathion resistance in a laboratory-selected population of *Drosophila melanogaster*. Genome. 30, 844–853.

Hsu, J.C., Haymer, D.S., Wu, W.J., Feng, H.T., 2006. Mutations in the acetylcholinesterase gene of *Bactrocera dorsalis* associated with resistance to organophosphorus insecticides. Insect. Biochem. Mol. Biol. 36, 396–402.

Huesing, J., Lloyd, F., Levine, S. Vaughn, T., 2009. Approaches to tier-based NTO testing of RNAi pest control traits. Regulation of transgenic crops: The state of the science. Annual Meeting of the Entomology Society of America. Indianapolis, IN. <http://esa.confex.com/esa/2009/webprogram/Paper40251.html>.

Huesing, J., Lloyd, F., Levine, S. Vaughn, T., Moar, W., Heck, G., 2010. Approaches to tier-based NTO testing of RNAi pest control traits. Regulation of transgenic crops: The state of the science. Annual Meeting of the American Chemical Society. San Francisco, CA. <http://abstracts.acs.org/chem/239nm/program/view.php?obj_id = 26350&terms>.

Hung, C.F., Holzmacher, R., Connolly, E., Berenbaum, M.R., Schuler, M.A., 1996. Conserved promoter elements in the CYP6B gene family suggest common ancestry for cytochrome P450 monooxygenases mediating furanocoumarin detoxification. Proc. Natl. Acad. Sci. USA. 93, 12200–12205.

Hunter, W., Ellis, J., vanEngelsdorp, D., Hayes, J., Westervelt, D., Glick, E., et al., 2010. Large-scale field application of RNAi technology reducing Israeli acute paralysis virus disease in honey bees (*Apis mellifera*, Hymenoptera: Apidae). PLoS Pathog. 6 (12), e1001160. Available from: http://dx.doi.org/10.1371/journal.ppat.1001160.

Huvenne, H., Smagghe, G., 2010. Mechanisms of dsRNA uptake in insects and potential of RNAi for pest control: a review. J. Insect. Physiol. 56, 227–235.

ILSI [International Life Sciences Institute], 1998. An evaluation of insect resistance management in Bt field corn: A science-based framework for risk assessment and risk management. ILSI Press, Washington, D.C, 78 pp.

ILSI [International Life Sciences Institute], 2011. Problem formulation for the environmental risk assessment of RNAi plants. ILSI Press, Washington, D.C, 48 pp.

Ishaaya, I., 1993. Insect detoxifying enzymes: their importance in pesticide synergism and resistance. Arch. Insect. Biochem. Physiol. 22, 263–276.

Ito, S., Mori, T., Kanazawa, H., Sawaguchi, T., 2007. Differential effects of the ascorbyl and tocopheryl derivative on the methamphetamine-induced toxic behavior and toxicity. Toxicology. 240 (1-2), 96–110.

James, C.E., Davey, M.W., 2009. Increased expression of ABC transport proteins is associated with ivermectin resistance in the model nematode *Caenorhabditis elegans*. I J. Parsitol. 39, 213–220.

Jamroz, R.C., Guerrero, F.D., Kammlah, D.M., Kunz, S.E., 1998. Role of the *kdr* and *super-kdr* sodium channel mutations in pyrethroid resistance: correlation of allelic frequency to resistance level in wild and laboratory populations of horn flies (*Haematobia irritans*). Insect. Biochem. Mol. Biol. 28, 1031–1037.

Joyce, A., Palsson, B., 2006. The model organism as a system: integrating 'omics' data sets. Nat. Rev. Mol. Cell. Biol. 7 (3), 198–210.

Jurat-Fuentes, J.L., Karumbaiah, L., Jakka, S.R.K., Ning, C., Liu, C., Wu, K., et al., 2011. Reduced levels of membrane-bound alkaline phosphatase are common to Lepidopteran strains resistant to Cry toxins from *Bacillus thuringiensis*. PLoS One. 6 (3), e17606. Available from: < http://dx.doi.org/10.1371/journal.pone.0017606 >.

Karunaratne, S.H., Hemingway, J., 2000. Insecticide resistance spectra and resistance mechanisms in populations of Japanese encephalitis vector mosquitoes, *Culex tritaeniorhynchus* and *Cx. gelidus*, in Sri Lanka. Med. Vet. Entomol. 14, 430–436.

Kasai, S., Shono, T., Yamakawa, M., 1998. Molecular cloning and nucleotide sequence of a cytochrome P450 cDNA from a pyrethroid-resistant mosquito, *Culex quinquefasciatus* say. Insect. Mol. Biol. 7, 185–190.

Kasai, S., Weerashinghe, I.S., Shono, T., Yamakawa, M., 2000. Molecular cloning, nucleotide sequence and gene expression of a cytochrome P450 (CYP6F1) from the pyrethroid-resistant mosquito, *Culex quinquefasciatus* Say. Insect. Biochem. Mol. Biol. 30, 163–171.

Keseru, G.M., 1998. Cytochrome P450 catalysed insecticide metabolism: chemical and theoretical models. Sci. Prog. 81 (3), 245–272.

Kostaropoulos, I., Papadopoulos, A.I., Metaxakis, A., Boukouvala, E., Papadopoulou-Mourkidou, E., 2001. Glutathione S-transferase in the defence against pyrethroids in insects. Insect. Biochem. Mol. Biol. 31, 313–319.

Kranthi, K.R., Jadhav, D., Wanjari, R., Kranthi, S., Russell, D., 2001. Pyrethroid resistance and mechanisms of resistance in field strains of *Helicoverpa armigera* (Lepidoptera: Noctuidae). J. Econ. Entomol. 94, 253–263.

Krishnan, N., Kodrik, D., 2006. Antioxidant enzymes in *Spodoptera littoralis* (Boisduval): are they enhanced to protect gut tissues during oxidative stress? J. Insect. Physiol. 52, 11–20.

Kuraishi, T., Binggeli, O., Opota, O., Buchon, N., Lemaitre, B., 2011. Genetic evidence for a protective role of the peritrophic matrix against intestinal bacterial infection in *Drosophila melanogaster*. PNAS. 108 (38), 15966–15971.

Lamb, D.C., Kelly, D.E., Schunck, W.H., Shyadehi, A.Z., Akhtar, M., Lowe, D.J., et al., 1997. The mutation T315A in *Candida albicans* sterol 14alpha-demethylase causes reduced enzyme activity and fluconazole resistance through reduced affinity. J. Biol. Chem. 272, 5682–5688.

Lawrence, P.A., Morata, G., 1992. Developmental biology. Lighting up *Drosophila*. Nature. 356, 107−108.

Le Goff, G., Hamon, A., Berge, J.B., Amichot, M., 2005. Resistance to fipronil in *Drosophila simulans*: influence of two point mutations in the RDL GABA receptor subunit. J. Neurochem. 92, 1295−1305.

Lee, S.H., Clark, J.M., 1998. Permethrin Carboxylesterase functions as nonspecific sequestration proteins in the hemolymph of colorado potato beetle. Pestic. Biochem. Phys. 62, 51−63.

Lee, S.H., Dunn, J.B., Clark, J.M., Soderlund, D.M., 1999. Molecular analysis of *kdr*-like resistance in a permethrin-resistant strain of Colorado potato beetle. Pestic. Biochem. Physiol. 63, 63−75.

Lehane, M., 1997. Peritrophic Matrix structure and function. Annu. Rev. Entomol. 42, 525−550.

Lespine, A., Dupuy, J., Orlowski, S., Nagy, T., Glavinas, H., Krajcsi, P., et al., 2006. Interaction of ivermectin with multidrug resistance proteins (MRP1, 2 and 3). Chem-Biol. Interact. 159, 169−179.

Lespine, A., Martin, S., Dupuy, J., Roulet, A., Pineau, T., Orlowski, S., et al., 2007. Interaction of macrocyclic lactones with P-glycoprotein: structure−affinity relationship. Eur. J. Pharm. Sci. 30, 84−94.

Li, A., Yang, Y., Wu, S., Li, C., Wu, Y., 2006. Investigation of resistance mechanisms to fipronil in diamondback moth (Lepidoptera: Plutellidae). J. Econ. Entomol. 99, 914−919.

Li, F., Han, Z., 2004. Mutations in acetylcholinesterase associated with insecticide resistance in the cotton aphid, *Aphis gossypii* Glover. Insect. Biochem. Mol. Biol. 34, 397−405.

Li, J., Wang, Q., Zhang, L., Gao, X., 2012. Characterization of imidacloprid resistance in the housefly *Musca domestica* (Diptera: Muscidae). Pestic. Biochem. Phys. 102 (2), 109−114.

Li, X., Zangerl, A.R., Schuler, M.A., Berenbaum, M.R., 2000. Cross-resistance to alpha-cypermethrin after xanthotoxin ingestion in *Helicoverpa zea* (Lepidoptera: Noctuidae). J. Econ. Entomol. 93, 18−25.

Li, X., Baudry, J., Berenbaum, M.R., Schuler, M.A., 2004b. Structural and functional divergence of insect CYP6B proteins: from specialist to generalist cytochrome P450. Proc. Natl. Acad. Sci. USA. 101, 2939−2944.

Li, X., Schuler, M.A., Berenbaum, M.R., 2007. Molecular mechanisms of metabolic resistance to synthetic and natural xenobiotics. Annu. Rev. Entomol. 52, 231−253.

Li, Y., Liu, T., Peng, Y., Yuan, C., Guo, A., 2004a. Specific functions of *Drosophila* amyloid precursor-like protein in the development of nervous system and nonneural tissues. J. Neurobiol. 61, 343−358.

Liu, H., Xu, Q., Zhang, L., Liu, N., 2005a. Chlorpyrifos resistance in mosquito *Culex quinquefasciatus*. J. Med. Entomol. 42, 815−820.

Liu, N., Scott, J.G., 1997. Phenobarbital induction of CYP6D1 is due to a *trans* acting factor on autosome 2 in house flies, *Musca domestica*. Insect. Mol. Biol. 6, 77−81.

Liu, Z., Williamson, M.S., Lansdell, S.J., Denholm, I., Han, Z., Millar, N.S., 2005b. A nicotinic acetylcholine receptor mutation conferring target-site resistance to imidacloprid in *Nilaparvata lugens* (brown planthopper). P. Natl. Acad. Sci. USA. 102 (24), 8420−8425.

Ma, Z., Proffer, T.J., Jacobs, J.L., Sundin, G.W., 2006. Overexpression of the 14α-Demethylase target gene (*CYP51*) mediates fungicide resistance in *Blumeriella jaapii*. Appl. Environ. Microbiol. 72, 2581−2585.

Macdonald, M., Gledhill, A., 2007. Potential impact of ABCB1 (p-glycoprotein) polymorphisms on avermectin toxicity in humans. Arch. Toxicol. 81, 553−563.

Magwere, T., West, M., Riyahi, K., Murphy, M.P., Smith, R.A., Partridge, L., 2006. The effects of exogenous antioxidants on lifespan and oxidative stress resistance in *Drosophila melanogaster*. Mech. Ageing. Dev. 127, 356–370.

Maitra, S., Dombrowski, S.M., Waters, L.C., Ganguly, R., 1996. Three second chromosome-linked clustered Cyp6 genes show differential constitutive and barbital-induced expression in DDT-resistant and susceptible strains of *Drosophila melanogaster*. Gene. 180, 165–171.

Maitra, S., Dombrowski, S.M., Basu, M., Raustol, O., Waters, L.C., Ganguly, R., 2000. Factors on the third chromosome affect the level of cyp6a2 and cyp6a8 expression in *Drosophila melanogaster*. Gene. 248, 147–156.

Maltseva, O.V., Golovleva, L.A., 1982. [Central metabolic characteristics of a *Pseudomonas aeruginosa* culture degrading DDT]. Mikrobiologiia. 51, 5–11.

Mamidala, P., Wijeratne, A.J., Wijeratne, S., Kornacker, K., Sudhamalla, B., Rivera-Vega, L.J., et al., 2012. RNA-Seq and molecular docking reveal multi-level pesticide resistance in the bed bug. BMC Genomics. 13, 6. Available from: http://dx.doi.org/10.1186/1471-2164-13-6.

Markussen, M.D., Kristensen, M., 2011. Spinosad resistance in female *Musica domestica* L. from a field-derived population. Pest. Manag. Sci. 68, 75–82.

Martinez-Torres, D., Chandre, F., Williamson, M.S., Darriet, F., Berge, J.B., Devonshire, A.L., et al., 1998. Molecular characterization of pyrethroid knockdown resistance (kdr) in the major malaria vector *Anopheles gambiae* s.s. Insect. Mol. Biol. 7, 179–184.

Martinez-Torres, D., Chevillon, C., Brun-Barle, A., Berge, J.B., Pasteur, N., Pauron, D., 1999a. Voltage-dependent Na + channels in pyrethroid resistant *Culex pipiens* L mosquitoes. Pest Sci. 55, 1012–1020.

Martinez-Torres, D., Foster, S.P., Field, L.M., Devonshire, A.L., Williamson, M.S., 1999b. A sodium channel point mutation is associated with resistance to DDT and pyrethroid insecticides in the peach-potato aphid, *Myzus persicae* (Sulzer) (Hemiptera: Aphididae). Insect. Mol. Biol. 8, 339–346.

Matsuoka, T., Kajimoto, Y., Watada, H., Kaneto, H., Kishimoto, M., Umayahara, Y., et al., 1997. Glycation-dependent, reactive oxygen species–mediated suppression of the insulin gene promoter activity in HIT cells. J. Clin. Invest. 99 (1), 144–150.

Mazzarri, M.B., Georgihiou, G.P., 1995. Characterization of resistance to organophosphate, carbamate and pyrethroid in field populations of *Aedes aegypti* from Venezuela. J. Amer. Mosq. Con. Assoc. 11, 315–322.

Miyazaki, M., Ohyama, K., Dunlap, D.Y., Matsumura, F., 1996. Cloning and sequencing of the *para*-type sodium channel gene from susceptible and *kdr*-resistant German cockroaches (*Blattella germanica*) and house fly (*Musca domestica*). Mol. Gen. Genet. 252, 61–68.

Mohandass, S.M., Arthur, F.H., Zhu, K.Y., Throne, J.E., 2006. Hydroprene: Mode of action, current status in stored-product pest management, insect resistance, and future prospects. Crop Prot. 25, 902–909.

Monsanto, 2012. Corn Rootworm III (*Advanced to Phase 3*). Corn Rootworm III would offer increased control and durability against the corn rootworm by providing two distinct modes of actions. <http://www.monsanto.com/products/Pages/corn-pipeline.aspx>.

Mordue, A.J., Blackwell, A., 1993. Azadirachtin: an update. J. Insect Phys. 39, 903–924.

Narahashi, T., 1992. Nerve membrane Na + channels as targets of insecticides. Trends. Pharmacol. Sci. 13, 236–241.

Narahashi, T., 2002. Nerve membrane ion channels as the target site of insecticides. Mini. Rev. Med. Chem. 2 (4), 419–432.

Nauen, R., Tietjen, K., Wagner, K., Elbert, A., 1998. Efficacy of plant metabolites of imidacloprid against *Myzus persicae* and *Aphis gossypii* (Homoptera: Aphididae). Pestic Sci. 52, 53–57.

Nauen, R., Stumpf, N., Elbert, A., 2002. Toxicological and mechanistic studies on neonicotinoid cross resistance in Q-type *Bemisia tabaci* (Hemiptera: Aleyrodidae). Pest. Manag. Sci. 58, 868–875.

Nicholson, R.A., Botham, R.P., Collins, C., 2006. The use of [³H] permethrin to investigate the mechanisms underlying its differential toxicity to adult and larval stages of the sheen blowfly *Lucilia sericata*. Pest Sci. 14 (1), 57–63.

Nishida, R., 2002. Sequestration of defensive substances from plants by Lepidoptera. Ann. Rev. Entomol. 47, 57–92.

Nishiwaki, H., Sato, K., Nakagawa, Y., Miyashita, M., Miyagawa, H., 2004. Metabolism of Imidacloprid in houseflies. J. Pest Sci. 29, 110–116.

Okazaki, Y., Katayama, T., 2003. Effects of dietary carbohydrate and myo-inositol on metabolic changes in rats fed 1,1,1-trichloro-2,2-bis (p-chlorophenyl) ethane (DDT). J. Nutr. Biochem. 14, 81–89.

Opitz, S.E., Müller, C., 2009. Plant chemistry and insect sequestration. Chemoecology. 19, 117–154.

Ottea, J.A., Ibrahim, S., Youins, A.M., Young, R.J., Leonard, B.R., McCaffery, A.R., 1995. Biochemical and physiological mechanisms of pyrethroid resistance in *Heliothis virescens* (F.). Pestic. Biochem. Physiol. 51, 117–128.

Páleníček, T., Balíková, M., Rohanová, M., Novák, T., Horáček, J., Fujáková, M., et al., 2011. Behavioral, hyperthermic and pharmacokinetic profile of para-methoxymethamphetamine (PMMA) in rats. Pharmacol. Biochem. Be. 98 (1), 130–139.

Pan, D., Dong, J., Zhang, Y., Gao, X., 2004a. Tuberous sclerosis complex: from *Drosophila* to human disease. Trends. Cell. Biol. 14, 78–85.

Pan, L., Wen, Z., Baudry, J., Berenbaum, M.R., Schuler, M.A., 2004b. Identification of variable amino acids in the SRS1 region of CYP6B1 modulating furanocoumarin metabolism. Arch. Biochem. Biophys. 422, 31–41.

Park, Y., Taylor, M.F., 1997. A novel mutation L1029H in sodium channel gene hscp associated with pyrethroid resistance for *Heliothis virescens* (Lepidoptera:Noctuidae). Insect. Biochem. Mol. Biol. 27, 9–13.

Pedra, J.H., McIntyre, L.M., Scharf, M.E., Pittendrigh, B.R., 2004. Genome-wide transcription profile of field- and laboratory-selected dichlorodiphenyltrichloroethane (DDT)-resistant. *Drosophila*. Proc. Natl. Acad. Sci. USA. 101, 7034–7039.

Pedra, J.H., Festucci-Buselli, R.A., Sun, W., Muir, W.M., Scharf, M.E., Pittendrigh, B.R., 2005. Profiling of abundant proteins associated with dichlorodiphenyltrichloroethane resistance in *Drosophila melanogaster*. Proteomics. 5, 258–269.

Perry, T., McKenzie, J.A., Batterham, P., 2007. A Dalpha6 knockout strain of Drosophila melanogaster confers a high level of resistance to spinosad. Insect Biochem. 37 (2), 184–188.

Pittendrigh, B., Reenan, R., ffrench-Constant, R., Ganetsky, B., 1997. Point mutations in the *Drosophila para* voltage-gated sodium channel gene confer resistance to DDT and pyrethroid insecticides. Mol. Gen. Genet. 256, 602–610.

Plapp Jr., F.W., 1970. Changes in glucose metabolism associated with resistance to DDT and dieldrin in the house fly. J. Econ. Entomol. 63, 1768–1772.

Plapp, F.W., Hoyer, R.F., 1968. Insecticide resistance in the house fly: decreased rate of absorption as the mechanism of action of a gene that acts as an intensifier of resistance. J. Econ. Entomol. 61, 1298–1303.

Prabhaker, N., Castle, S., Henneberry, T.J., Toscano, N.C., 2005. Assessment of cross-resistance potential to neonicotinoid insecticides in *Bemisia tabaci* (Hemiptera: Aleyrodidae). Bull. Entomol. Res. 95, 535–543.

Prichard, R.K., Roulet, A., 2007. ABC transporters and β-tubulin in macrocyclic lactone resistance: prospects for marker development. Parasitology. 134, 1123–1132.

Puinean, A.M., Denholm, I., Millar, N.S., Nauen, R., Williamson, M.S., 2010a. Characterization of imidacloprid resistance mechanisms in the brown planthopper, Nilaparvata lugens Stal (Hemiptera: Delphacidae). Pestic. Biochem. Phys. 97 (2), 129–132.

Puinean, A.M., Foster, S.P., Oliphant, L., Denholm, I., Field, L.M., Millar, N.S., et al., 2010b. Amplification of a cytochrome P450 gene is associated with resistance to neonicotinoid insecticides in the aphid Myzus persicae. PLoS Genet. 6 (6), 1–10, e1000999.

Rajurkar, R.B., Khan, Z.H., Gujar, G.T., 2003. Studies on levels of glutathione S-transferase, its isolation and purification from Helicoverpa armigera. Curr. Sci. Ind. 85, 1355–1360.

Rangasamy, M., Siegfried, B., 2012. Validation of RNA interference in western corn rootworm Diabrotica virgifera virgifera LeConte (Coleoptera: Chrysomelidae) adults. Pest. Manag. Sci. 68 (4), 587–591.

Ranson, H., Hemingway, J., 2005. Mosquito glutathione transferases. Methods Enzymol. 401, 226–241.

Ranson, H., Jensen, B., Vulule, J.M., Wang, X., Hemingway, J., Collins, F.H., 2000. Identification of a point mutation in the voltage-gated sodium channel gene of Kenyan Anopheles gambiae associated with resistance to DDT and pyrethroids. Insect. Mol. Biol. 9, 491–497.

Ranson, H., Rossiter, L., Ortelli, F., Jensen, B., Wang, X., Roth, C.W., et al., 2001. Identification of a novel class of insect glutathione S-transferases involved in resistance to DDT in the malaria vector Anopheles gambiae. Biochem. J. 359, 295–304.

Rauch, N., Nauen, R., 2003. Identification of biochemical markers linked to neonicotinoid cross resistance in Bemisia tabaci (Hemiptera: Aleyrodidae). Arch. Insect. Biochem. Physiol. 54, 165–176.

Raymond, M., Heckel, D.G., Scott, J.G., 1989. Interactions between pesticide genes: model and experiment. Genetics. 123, 543–551.

Raymond, M., Chevillon, C., Guillemaud, T., Lenormand, T., Pasteur, N., 1998. An overview of the evolution of overproduced esterases in the mosquito Culex pipiens. Philos. Trans. R. Soc. Lond. B. Biol. Sci. 353, 1707–1711.

Raymond, M., Berticat, C., Weill, M., Pasteur, N., Chevillion, C., 2001. Insecticide resistance in the mosquito Culex pipens: what have we learned about adaptation. Genetica. 112-113, 287–296.

Ren, X., Han, Z., Wang, Y., 2002. Mechanisms of monocrotophos resistance in cotton bollworm, Helicoverpa armigera (Hubner). Arch. Insect. Biochem. Physiol. 51, 103–110.

Rose, R.L., Goh, D., Thompson, D.M., Verma, K.D., Heckel, D.G., Gahan, L.J., et al., 1997. Cytochrome P450 (CYP)9A1 in Heliothis virescens: the first member of a new CYP family. Insect. Biochem. Mol. Biol. 27, 605–615.

Rust, M.K., 2005. Advances in the control of Ctenocephalides felis (cat flea) on cats and dogs. Trends. Parasitol. 21, 232–236.

Sarfraz, M., Dosdall, L.M., Keddie, B.A., 2005. Evidence for behavioural resistance by the diamondback moth, Plutella xylostella (L.). J. Appl. Entomol. 129, 340–341.

Sarkar, M., Borkotoki, A., Baruah, I., Bhattacharyya, I.K., Srivastava, R.B., 2009. Molecular analysis of knock down resistance (kdr) mutation and distribution of kdr genotypes in a wild population of Culex quinquefasciatus from India. Trop. Med. Int. Health. 14, 1097–1104.

Satellite, D.B., Harrow, I.D., David, J.A., Pelhate, M., Callec, J.J., 1985. Nereistoxin: Actions on a CNS acetylcholine receptor/ion channel in the cockroach Periplanata americana. J. Exp. Biol. 118, 37–52.

Savare, J., Bonneaud, N., Girard, F., 2005. SUMO represses transcriptional activity of the Drosophila SoxNeuro and human Sox3 central nervous system-specific transcription factors. Mol. Biol. Cell. 16, 2660—2669.

Scharf, M.E., Siegfried, B.D., Meinke, L.J., Chandler, L.D., 2000. Fipronil metabolism, oxidative sulfone formation and toxicity among organophosphate and carbamate-resistant and susceptible western corn rootworm populations. Pest. Manag. Sci. 56, 757—766.

Schinkel, A.H., Smit, J.J., van Tellingen, O., Beijnen, J.H., Wagenaar, E., van Deemter, L., et al., 1994. Disruption of the mouse mdr1a P-glycoprotein gene leads to a deficiency in the blood-brain barrier and to increased sensitivity to drugs. Cell. 77, 491—502.

Schmued, L.C., Bowyer, J.F., 1997. Methamphetamine exposure can produce neuronal degeneration in mouse hippocampal remnants. Brain. Res. 759 (1), 135—140.

Schuler, M.A., 2011. P450s in plant-insect interactions. Biochim. Biophys. Acta. 1814, 36—45.

Schuler, T.H., Martinez-Torres, D., Thompson, A.J., Denholm, I., Devonshire, A.L., Duce, I.R., et al., 1998. Toxicological, electrophysiological and molecular characterization of knockdown resistance to pyrethroid insecticides in the diamond-back moth, Plutella xylostella (L.). Pestic. Biochem. Physiol. 59, 169—182.

Schulz-Jander, D.A., Casida, J.E., 2002. Imidacloprid insecticide metabolism: human cytochrome P450 isozymes differ in selectivity for imidazolidine oxidation versus nitroimine reduction. Toxicol. Lett. 132, 65—70.

Scott, J.G., 1999. Cytochromes P450 and insecticide resistance. Insect. Biochem. Mol. Biol. 29, 757—777.

Seifert, J., Scott, J.G., 2002. The CYP6D1v1 allele is associated with pyrethroid resistance in the house fly Musca domestica. Pestic. Biochem. Physiol. 72, 40—44.

Shaw, G.C., Sung, C.C., Liu, C.H., Lin, C.H., 1998. Evidence against the Bm1P1 protein as a positive transcription factor for barbiturate-mediated induction of cytochrome P450BM-1 in Bacillus megaterium. J. Biol. Chem. 273, 7996—8002.

Shono, T., Zhang, L., Scott, J.G., 2004. Indoxacarb resistance in the house fly, Musca domestica. Pestic. Biochem. Phys. 80, 106—112.

Siegfried, B., Spencer, T., Crespo, A., Storer, N., Head, G., Owens, E., et al., 2007. Ten years of Bt resistance monitoring in the European corn borer: what we know, what we don't know, and what we can do better. Amer. Entomol. 53, 208—214.

Singh, O.P., Dykes, C.L., Lather, M., Agrawal, O.P., Adak, T., 2011. Knockdown resistance (kdr)-like mutations in the voltage-gated sodium channel of a malaria vector Anopheles stephensi and PCR assays for their detection. Malar. J. 10, 59—66.

Small, G.J., Hemingway, J., 2000. Molecular characterization of the amplified carboxylesterase gene associated with organophosphorus insecticide resistance in the brown planthopper, Nilaparvata lugens. Insect. Mol. Biol. 9, 647—653.

Soderlund, D.M., Knipple, D.C., 2003. The molecular biology of knockdown resistance to pyrethroid insecticides. Insect. Biochem. Mol. Biol. 33 (6), 563—577.

Song, W., Liu, Z., Ke, D., 2006. Molecular basis of differential sensitivity of insect sodium channels to DCJW, a bioactive metabolite of the oxadiazine insecticide indoxacarb. Neurotoxicology. 27 (2), 237—244.

Sparks, T.C., Dripps, J.E., Watson, G.B., Paroonagian, D., 2012. Resistance and cross-resistance to the spinosyns—A review and analysis. Pestic. Biochem. Phys. 102, 1—10.

Stuart, J.J., Ray, S., Harrington, B.J., Neal, J.J., Beeman, R.W., 1998. Genetic mapping of a major locus controlling pyrethroid resistance in Tribolium castaneum (Coleoptera: Tenebrionidae). J. Econ. Entomol. 91, 1232—1238.

Sun, L., Schemerhorn, B., Jannasch, A., Walters Jr., K.R., Adamec, J., Muir, W.M., et al., 2011. Differential transcription of cytochrome P450s and glutathione S transferase in DDT-susceptible and—resistant Drosophila melanogaster strains in response to DDT and oxidative stress. Pestic. Biochem. Phys. 100 (1), 7—15.

Tang, F., Yue, Y.D., Hua, R.M., 2000. The relationships among MFO, glutathione S-transferases, and phoxim resistance in *Helicoverpa armigera*. Pestic. Biochem. Physiol. 68, 96—101.

Teese, M.G., Campbell, P.M., Scott, C., Gordon, K.H.J., Southon, A., Hovan, D., et al., 2010. Gene identification amd proteomic analysis of the esterases of the cotton bollworm, *Helicoverpa armigera*. Insect. Mol. Biol. 40, 909—925.

Tellam, R., Wijffels, G., Willadsen, P., 1999. Peritrophic matrix proteins. Insect. Biochem. Mol. Biol. 29 (2), 87—101.

Toju, H., 2009. Natural selection drives the fine-scale divergence of a coevolutionary arms race involving a long-mouthed weevil and its obligate host plant. BMC. Evol. Biol. 9, 273—293.

Tribble, N.D., Burka, J.F., Kibenge, F.S.B., 2007. Evidence for changes in the transcription levels of two putative P-glycoprotein genes in sea lice (*Lepeophtheirus salmonis*) in response to emamectin benzoate exposure. Mol. Biochem. Parasitol. 153, 59—65.

U.S. Environmental Protection Agency, 1998. Scientific Advisory Panel, Subpanel on *Bacillus thuringiens* (*Bt*) Plant-Pesticides and Resistance Management, February 9-10, 1998 (Docket Number: OPP 00231).

Vaughan, A., Chadee, D.D., ffrench-Constant, R., 1998. Biochemical monitoring of organophosphorus and carbamate insecticide resistance in *Aedes aegypti* mosquitoes from Trinidad. Med. Vet. Entomol. 12, 318—321.

Vogt, T.M., Welsh, J., Stolz, W., Kullmann, F., Jung, B., Landthaler, M., et al., 1997. RNA fingerprinting displays UVB-specific disruption of transcriptional control in human melanocytes. Cancer. Res. 57, 3554—3561.

Vontas, J., Blass, C., Koutsos, A.C., David, J.P., Kafatos, F.C., Louis, C., et al., 2005. Gene expression in insecticide resistant and susceptible *Anopholes gambiae* strains constitutively or after insecticide exposure. Insect. Mol. Biol. 14 (5), 509—521.

Vontas, J.G., Small, G.J., Hemingway, J., 2001. Glutathione S-transferases as antioxidant defence agents confer pyrethroid resistance in *Nilaparvata lugens*. Biochem. J. 357 (1), 65—72.

Vontas, J.G., Hejazi, M.J., Hawkes, N.J., Cosmidis, N., Loukas, M., Janes, R.W., et al., 2002. Resistance-associated point mutations of organophosphate insensitive acetylcholinesterase, in the olive fruit fly *Bactrocera oleae*. Insect. Mol. Biol. 11, 329—336.

Wang, C., Scharf, M.E., Bennett, G.W., 2004. Behavioral and physiological resistance of the German cockroach to gel baits (Blattodea: Blattellidae). J. Econ. Entomol. 97, 2067—2072.

Wang, C., Scharf, M.E., Bennett, G.W., 2006. Genetic basis for resistance to gel baits, fipronil, and sugar-based attractants in German cockroaches (Dictyoptera: blattellidae). J. Econ. Entomol. 99, 1761—1767.

Wang, W., Lo, P., Frasch, M., Lufkin, T., 2000. Hmx: an evolutionary conserved homeobox gene family expressed in the developing nervous system in mice and *Drosophila*. Mech. Dev. 99, 123—137.

Wang, Y., Zhang, H., Li, H., Miao, X., 2011. Second-generation sequencing supply an effective way to screen RNAi targets in large for potential application in pest insect control. PLoS ONE. 6 (4), e18644. Available from: http://dx.doi.org/10.1371/journal.pone.0018644.

Wang, Z., Yao, M., Wu, Y., 2009. Cross-resistance, inheritance and biochemical mechanisms of imidacloprid resistance in B-biotype *Bemisia tabaci*. Pest. Manag. Sci. 65, 1189—1194.

Wen, Z., Berenbaum, M.R., Schuler, M.A., 2006a. Inhibition of CYP6B1-mediated detoxification of xanthotoxin by plant allelochemicals in the black swallowtail (*Papilio polyxenes*). J. Chem. Ecol. 32, 507—522.

Wen, Z., Rupasinghe, S., Niu, G., Berenbaum, M.R., Schuler, M.A., 2006b. CYP6B1 and CYP6B3 of the black swallowtail (*Papilio polyxenes*): adaptive evolution through subfunctionalization. Mol. Biol. Evol. 23, 2434−2443.

Whangbo, J.S., Hunter, C.P., 2008. Environmental RNA interference. Trends. Genet. 24, 297−305.

Whitehill, J.G.A., Popova-Butler, A., Green-Church, K.B., Koch, J.L., Herms, D.A., Bonello, P., 2011. Interspecific proteomic comparisons reveal ash phloem genes potentially involved in constitutive resistance to the Emerald Ash Borer. PLoS One. 6 (9), e24863. Available from: http://dx.doi.org/10.1371/journal.pone.0024863.

Whiteman, N.K., Groen, S.C., Chevasco, D., Bear, A., Beckwith, N., Gregory, T.R., et al., 2011. Mining the plant-herbivore interface with a leafmining *Drosophila* of *Arabidopsis*. Mol. Ecol. 20 (5), 995−1014.

Whyard, S., Singh, A.D., Wong, S., 2009. Ingested double-stranded RNAs can act as species-specific insecticides. Insect. Biochem. Mol. Biol. 39 (11), 824−832.

Wiese, A.G., Pacifici, R.E., Davies, K.J., 1995. Transient adaptation of oxidative stress in mammalian cells. Arch. Biochem. Biophys. 318 (1), 231−240.

Williamson, M.S., Martinez-Torres, D., Hick, C.A., Devonshire, A.L., 1996. Identification of mutations in the housefly *para*-type sodium channel gene associated with *knockdown resistance* (*kdr*) to pyrethroid insecticides. Mol. Gen. Genet. 252, 51−60.

Winston, W.M., Sutherlin, M., Wright, A.J., Feinberg, E.H., Hunter, C.P., 2007. Caenorhabditis elegans SID-2 is required for environmental RNA interference. Proc. Natl. Acad. Sci. USA. 104 (25), 10565−10570.

Wolf, M.J., Amrein, H., Izatt, J.A., Choma, M.A., Reedy, M.C., Rockman, H.A., 2006. *Drosophila* as a model for the identification of genes causing adult human heart disease. Proc. Natl. Acad. Sci. USA. 103, 1394−1399.

Wolstenholme, A.J., Fairweather, I., Prichard, R., von Samson-Himmelstjerna, G., Sangster, N.C., 2004. Drug resistance in veterinary helminthes. Trends. Parasitol. 20, 471−476.

Wondji, C.S., Priyanka De Silva, W.A.P., Hemingway, J., Ranson, H., Parakrama Karunaratne, S.H.P., 2008. Characterization of knockdown resistance in DDT- and pyrethroid-resistant *Culex quinquefasciatus* populations from Sri Lanka. Trop. Med. Int. Health. 13, 548−555.

Xu, Q., Liu, H., Zhang, L., Liu, N., 2005. Resistance in the mosquito, *Culex quinquefasciatus*, and possible mechanisms for resistance. Pest. Manag. Sci. 61, 1096−1102.

Yan, S., Cui, F., Qiao, C., 2009. Structure, Function and Applications of Carboxylesterases from Insects for Insecticide Resistance. Protein Peptide Lett. 16, 1181−1188.

Yoon, K.S., Gao, J.-R., Lee, S.H., Coles, G.C., Meinking, T.L., Taplin, D., et al., 2004. Resistance and cross-resistance to insecticides in human head lice from Florida and California. Pestic. Biochem. Physiol. 80, 192−201.

Yoon, K.S., Strycharz, J.P., Baek, J.H., Sun, W., Kim, J.H., Kang, J.S., et al., 2011. Brief exposures of human body lice to sublethal amounts of ivermectin over-transcribes detoxification genes involved in tolerance. Insect. Mol. Biol. 20, 687−699.

Zangar, R.C., Davydov, D.R., Verma, S., 2004. Mechanisms that regulate production of reactive oxygen species by cytochrome P450. Toxicol. Appl. Pharmacol. 199, 316−331.

Zha, W., Peng, X., Chen, R., Du, B., Zhu, L., He, G., 2011. Knockdown of midgut genes by dsRNA-transgenic plant-mediated RNA interference in the hemipteran insect *Nilaparvata lugens*. PLoS One 2011. 6 (5), e20504 (Epub 2011 May 31).

Zhao, J.Z., Li, Y.X., Collins, H.L., Gusukuma-Minuto, L., Mau, R.F., Thompson, G.D., et al., 2002. Monitoring and characterization of diamondback moth (Lepidoptera: Plutellidae) resistance to spinosad. J. Econ. Entomol. 95 (2), 430−436.

Zhao, Y., Park, Y., Adams, M.E., 2000. Functional and evolutionary consequences of pyrethroid resistance mutations in S6 transmembrane segments of a voltage-gated sodium channel. Biochem. Biophys. Res. Commun. 278, 516–521.

Zheng, Z., Tang, S., Tao, M., 2005. Development of resistance to RNAi in mammalian cells. Ann. NY. Acad. Sci. 1058, 105–118. Available from: http://dx.doi.org/10.1196/annals.1359.019.

Zhu, K.Y., Clark, J.M., 1997. Validation of a point mutation of acetylcholinesterase in Colorado potato beetle by polymerase chain reaction coupled to enzyme inhibition assay. Pestic. Biochem. Physiol. 57, 28–35.

CHAPTER 4

Plant Incorporated Protectants and Insect Resistance

Mark E. Nelson[1] and Analiza P. Alves[2]

[1]DuPont Pioneer, Wilmington, DE
[2]DuPont Pioneer, Johnston, IA

Chapter Outline

INTRODUCTION

Plant incorporated protectants (PIPs) render the plant unsuitable for pest insects. Obviously, this does not require insect lethality and could be achieved through expression of a suitable repellent. From a practical

Insect Resistance Management
DOI: http://dx.doi.org/10.1016/B978-0-12-396955-2.00004-7

context, however, effective protectants typically act by compromising insect viability by disrupting normal physiology. Insecticidal chemistries have achieved this mostly by targeting the insect nervous system, which can be accomplished through both oral delivery and/or absorption following surface contact. PIPs, however, must be ingested, overcome both digestive and physical barriers, and then reach the target site where they act. Systems for overcoming similar challenges in achieving activity in target hosts by oral delivery exist in nature and can be thought of as models for developing PIP technologies. For example, bacteria and other microorganisms that utilize foreign hosts (such as insects) to promote self-propagation have evolved methods for invading, immobilizing, and often killing the host when the main route of access is through the digestive system.

Entomopathogenic bacteria, such as *Bacillus thuringiensis* (Bt) and its relatives, have served as sources for insect-control agents for over a century. These insect pathogens selectively target their host through the digestive system. The digestive tract represents a readily accessible tissue that in reality is a continuous tube running through the body cavity, with the lumen exposed indirectly to the external environment. The lining of the digestive system therefore represents a critical barrier mediating nutrient absorption while providing protection against pathogens. To pathogenic bacteria, the digestive tract represents an accessible host vulnerability. These microbes have evolved mechanisms to compromise host viability by directly interfering with digestive system functioning or by utilizing molecular targets within the digestive system as a means to penetrate it. Targeting the insect digestive system also contributes to insect species selectivity because the gut environment can impact the mode of action of many protein actives (see the section on Bt mode of action later in this chapter) and therefore is a determinant of toxicity and susceptibility. Differences between the gut environments of insects and mammals also have contributed to the safe use of entomopathogenic bacteria for crop protection for decades. For these reasons, disruption of gut function has been a common theme in the discovery and development of new leads for transgenic insect control applications. Targeting of the nervous system or other tissues beyond the digestive system also has potential. However, because insects and mammals have many nervous system similarities, this would require selectivity for targeting the insect site of action in addition to a specific mechanism for penetrating the digestive system to reach the target tissue.

While insecticidal proteins from Bt are particularly suited for use as PIPs, many other proteins also have insecticidal activity and have been explored as potential PIPs. These include other bacterial proteins, plant-produced proteins that act as deterrents of feeding (e.g., snowdrop lectin (*Galanthus nivalis* agglutinin), various trypsin inhibitors, as well as α-amylase), and proteins that interfere with the uptake of essential nutrients (e.g., avidin). Although none of these proteins have thus far proven to be as useful as Bt-derived PIPs, they illustrate the variety of mechanisms through which plant protection might be achieved.

INSECTICIDAL PROTEINS
Bacillus thuringiensis

The *Bacillus* family of microorganisms was first described in the 1870s as aerobic, rod-shaped bacteria that undergo endospore formation. The insecticidal properties of *Bacillus thuringiensis* were first described for silkworm (*Bombyx mori*) in Japan in 1902 (see Milner, 1994). Slightly more than a decade later, another strain was described after being isolated from dead larvae of moths that had infested stored grain (see Milner, 1994). Bt are ubiquitous gram positive bacteria commonly found in soil throughout the world. Considerable research has been devoted to characterizing the wide variety of subspecies that exist, the spectrum of insects targeted by each, and the identification and characterization of the proteins that confer target insect specificity (van Frankenhuyzen, 2009).

Agriculture has utilized Bt for over 70 years, first as foliar spray formulations and in the last 20 years as a source for insecticidal proteins to be expressed in transgenic plants. Because of its efficacy and excellent safety profile (Glare and O'Callaghan, 2000), Bt offers an appealing alternative to use of synthetic chemistries, not only because of the environmental benefits, but also because of target pest selectivity and the utility of *in planta* delivery. The successful development and widespread adoption of Bt crops make them the model for broadening insect control options through biotechnological advances. Numerous excellent reviews have been published that detail various aspects of Bt and their insecticidal proteins (Bravo *et al.*, 2007; Bravo and Soberón, 2008; de Maagd *et al.*, 2001; Gill *et al.*, 1992; Knowles, 1994; Pigott and Ellar, 2007; van Frankenhuyzen, 2009).

The Bt life cycle includes a vegetative stage for growth and division when conditions are favorable and a dormant sporulation phase that occurs when nutrients are limited. Leading into its sporulation phase, Bt produces high amounts of the insecticidal proteins that form crystal inclusions along with its spore. Once ingested, the Bt crystal proteins (known as δ-endotoxins or Cry proteins) provide a mechanism through which the bacterium can immobilize and kill the host. The host cadaver then serves as a source of nutrients to support the germination of the spores, leading to propagation of the next generation. Bt strains are subclassified microbiologically into "serovars" that are differentiated by the presence of antigenic determinants expressed in their flagella (De Barjac and Frachon, 1990; Lecadet et al., 1999). By varying the type of δ-endotoxin proteins that are expressed in the crystal inclusions, different strains of Bt target different insect orders or species as hosts (Bravo et al., 2013; de Maagd et al., 2001; Schnepf et al., 1998). The type of proteins expressed by different strains also results in different crystal lattices or shapes. The correlation between crystal appearance and insect selectivity led to the early classification of Bt proteins based on these parameters so that four general classes existed: those that were active on Lepidoptera (I), those that were active on Lepidoptera and Diptera (II), those that were active on Coleoptera (III), and those that had activity on Diptera (IV) (Höfte and Whiteley, 1989). This system of nomenclature quickly ran into challenges. As more toxins were identified through the introduction of cloning technologies, increasing exceptions were found, and the system began to fail. Some strains produced crystals composed of multiple toxins, different strains were found to express the same toxins, strains that produced highly homologous toxins had differing spectra of activity, and some strains produced insecticidal proteins independent of crystal formation. Additionally, new Bt proteins were found to have unexpected inter order activities (e.g., Cry1B activity on Lepidoptera and Coleoptera) that were not covered in the four established categories. These limitations led to the proposal for a more systematic approach that relied on gene sequence homology (Höfte and Whiteley, 1989). This system was later refined into the current rules of nomenclature to impose explicit guidelines to be used when new Bt insecticidal protein sequences are discovered, especially to accommodate highly homologous variants that could not be differentiated by earlier systems (Crickmore et al., 1998).

To date, over 70 classes of Cry proteins have been identified and assigned a primary rank based solely on primary sequence homology guidelines (Crickmore *et al.*, 2013). However, relatively few possess the necessary spectrum of activity and potency to be utilized in commercial agriculture (James, 2009; van Frankenhuyzen, 2009). Nevertheless, the insect pests that damage crops in major agricultural regions that can be controlled by Bt proteins are members of the orders Lepidoptera or Coleoptera. Currently, commercialized transgenic insect control crops utilize very few out of the hundreds of different naturally occurring Bt

Table 4.1 Bt Proteins Commercialized for Transgenic Insect Control

Active	Commercial Trait Name[a]	Event Name	Year Active First Deregulated in U.S.A.[b]
Corn			
Cry1Ab	YieldGard Corn Borer®	MON810	1996
	AgriSure Corn Borer®	BT11	
Cry1Fa	Herculex I®	TC1507	2001
Cry1A.105	YieldGard VT Pro®	MON89034	2008
Cry2Ab	YieldGard VT Pro®	MON89034	2008
Cry3Bb	YieldGard Rootworm®	MON863	2003
mCry3A	AgriSure Rootworm®	MIR604	2006
Cry34/Cry35	Herculex Rootworm®	DAS59122	2005
Cry9C	StarLink®	CBH351	1998
Vip3Aa20	Viptera®	MIR162	2008
Cotton			
Cry1Ac	Bollgard ®	MON531	1995
	Widestrike®	DAS21023	
Cry1Ab	VipCot®	COT67B	2008
	Twinlink®	T304-40	
Cry1F	Widestrike®	DAS24236	2004
Cry2Ab	Bollgard II®	MON15985	2002
Cry2Ae	Twinlink®	GHB119	2012
Vip3Aa19	VipCot®	COT102	2008
Potato			
Cry3Aa	Newleaf®	BT6 (and others)	1994
Soybean			
Cry1Ac	Intacta™	MON87701	2010[c]

[a]Trait names are provided solely for the purpose of cross-reference.
[b]Identified in http://www.epa.gov/pesticides/biopesticides/pips/pip_list.htm (last update May 12, 2012).
[c]Not fully deregulated in the United States.

Cry proteins (or slight variants of each), including Cry1Ab, Cry1Ac, Cry1F, Cry2Ab, and Cry2Ae for lepidopteran control, and Cry3Bb, Cry3Aa, and Cry34/Cry35 for coleopteran control (see Table 4.1).

Genetic engineering of what are essentially chimeras of some of these proteins have also been commercialized, including the lepidopteran–active Cry1A.105 (consisting of sequences derived from Cry1Ab, Cry1F, and Cry1Ac) (Bogdanova et al., 2007) and the coleopteran–active eCry3.1Ab (consisting of sequences derived from a slightly modified Cry3Aa and Cry1Ab)(Walters et al., 2010). Discovery of new classes of Bt insecticidal proteins or improved understanding of what limits their spectrum of activity could allow their broader application in biotechnology, but this remains to be seen and likely will take significant investment in research and development.

Bt Vegetative Insecticidal Proteins (Vips)

Although the crystalline proteins produced by Bt during sporulation have been the source of many PIPs, proteins that are expressed and secreted during the vegetative stage (Vegetative Insecticidal Proteins or Vips) also have proven to be useful. Over 50 different Vip proteins have been identified to date and these have been divided among four classes according to sequence homology, using a nomenclature system similar to that applied to the Cry family of proteins (Crickmore et al., 2013). The Vip1 and Vip2 classes, acting together as binary toxins, have been reported to have activity against certain Coleoptera, while the Vip3 class exhibits broad spectrum lepidopteran activity. For the binary Vips, the Vip1 component is considered the membrane binding component that mediates the entry of the Vip2 component across the midgut membrane. Once inside breached cells, Vip2 achieves toxicity by disrupting normal cell function. In the case of Vip2Ac, it modifies cytoplasmic actin by ADP-ribosyltransferase activity, causing disruption of cellular architecture leading to cell death (Han et al., 1999).

Vip3A proteins, on the other hand, cause toxicity by pore formation similar to Cry proteins (see later section; Lee et al., 2003). However, Vip3A proteins have been shown to act by binding to different midgut receptors than those utilized by Cry1A or Cry2A proteins and therefore represent a different mode of action than these PIPs, having little likelihood of significant cross-resistance to them (Lee et al., 2003, 2006). Vip3Aa is the only Vip to date to be commercialized for transgenic insect control (Kurtz, 2010; see also Table 4.1).

Other Bacterial Proteins

Another class of insecticidal proteins that have been explored for transgenic insect control are the toxin complex (Tc) proteins that are expressed by *Photorhabdus* and *Xenorhabdus* bacteria (Forst *et al.*, 1997). These are gram negative bacteria that exist in a symbiotic state in the gut of soil-dwelling nematodes. The nematode serves as a vector providing a means for the bacteria to penetrate the host insect hemocoel. There, the bacteria are regurgitated into the hemolymph and enter a pathogenic stage producing an array of endo- and exotoxins that kill the infected insects. Similar to Bt, this process allows the bacteria and the nematode to utilize the insect cadaver to support self-propagation (ffrench-Constant *et al.*, 2003; Forst *et al.*, 1997). Although during infection the bacteria are naturally delivered to the hemocoel by their nematode vectors, it has been demonstrated that some of their insecticidal proteins are orally active and do not require delivery to the hemocoel.

Tc proteins are produced by a large family of related genes and have been classified into three categories (A, B, and C) according to sequence homologies (ffrench-Constant and Waterfield, 2006; ffrench-Constant *et al.*, 2007). The oral insecticidal properties of the Tc proteins were first reported after purification from supernatants of *Photorhabdus luminescens* cultures and shown to kill *Manduca sexta* (Bowen *et al.*, 1998; Bowen and Ensign, 1998). Histological characterization of *M. sexta* midguts following feeding with purified TcA toxin complex revealed severely affected columnar epithelial cells that exhibited blebbing (membrane bulges) similar to what has been reported to occur in the midguts of insects intoxicated with Bt proteins (Blackburn *et al.*, 1998). The Tc complex from *P. luminescens* consists of several different proteins ranging in size from 30 to 200 kDa that assemble together into a structure of over 1,000 kDa (Bowen and Ensign, 1998). The TcA-like components are generally thought to mediate binding to cell membranes as tetrameric oligomers (Lee *et al.*, 2007), while the TcB-like components are thought to serve as links between TcA and TcC components (Waterfield *et al.*, 2005). The TcC-like proteins were reported to cause cell death by inducing actin clustering through ADP-ribosyltransferase activity (Lang *et al.*, 2010). A more in-depth characterization of Tc proteins from *P. luminescens* revealed that the A component alone exhibited insecticidal activity, even causing mortality to first instar *M. sexta* with transgenic expression in *Arabidopsis* (Liu *et al.*, 2003; Waterfield *et al.*, 2001). However, only when the A component was combined with the B and C components in artificial diet assays was the activity of the recombinant proteins similar to that of the native *P. luminescens* strain (Waterfield *et al.*, 2005). This suggested that the native activity was mediated by a hetero-oligomeric complex of the components.

Recent detailed studies on the mode of action of the Tc proteins from *Xenorhabdus nematophilus* provided evidence that the TcA-like proteins form a tetrameric structure capable of generating pores in artificial membranes (Sheets *et al.*, 2011). Additionally, this study demonstrated the inter-compatibility of B- and C-like proteins from *P. luminescens* with the A-like protein from *X. nematophilus* in forming a heteroligomeric complex having high insecticidal activity consistent with a conserved mode of action among Tc protein-expressing bacteria.

Enzyme Inhibitors

The digestive systems of herbivorous insects have evolved to break apart plant tissue for optimal absorption of vital nutrients, including proteins, carbohydrates, and lipids. The breakdown of proteins is achieved by expression of a wide array of proteases that will liberate amino acids for subsequent uptake by specific transporters in the midgut (Terra and Ferreira, 1994). Similarly, an array of glycosidases break down complex carbohydrates into simple sugars to enable uptake, while lipases carry out breakdown of lipids into free fatty acids and diacylglycerol prior to uptake and lipophorin-mediated transport (Canavoso *et al.*, 2001; Terra and Ferreira, 1994). Partly as a defense mechanism against insect attack, plants express inhibitors of these enzymes to help deter feeding by disrupting the digestive process (Macedo and Freire, 2011; Ryan, 1990). Inhibitors can be found in various plant parts, but are particularly enriched in storage tissues such as seeds (Birk, 1996) and tubers (as high as 10% of total protein; Pouvreau *et al.*, 2001) where preservation of nutrient stores is paramount to future propagation. Inhibitor expression also can be evoked as part of the plant wounding response (Ryan, 1990). Transgenic enzyme inhibitors have been explored as a PIP technology, and some have been demonstrated to provide protection against insect damage (Habib and Fazili, 2007; Lawrence and Koundal, 2002). With the exception of α-amylase inhibitors, development of transgenic enzyme inhibitors has been most extensive for protease inhibitors, which will be the focus of the current discussion.

Proteases are classified according to the characteristics of their catalytic center which determines their sequence specificity in cleaving the peptide bond that links amino acids together in proteins. Many protease inhibitors act by binding at the catalytic center of the enzyme, which therefore serves as a selectivity determinant for the inhibitors. Major proteases found in the insect midgut can be grouped into three main categories: serine proteases, cysteine proteases, and aspartic/metallo-proteases. Serine proteases are the most widely characterized class and include trypsins that cleave proteins on the C-terminal side of arginine or lysine, chymotrypsins that cleave on the C-terminal side of large hydrophobic amino acids such as phenylalanine, tyrosine, or leucine, and elastases that cleave after small neutral amino acids such as alanine and glycine (Ryan, 1990). The cysteine proteases encompass the broad papain family, including various endopeptidases, dipeptidases, and some cathepsins (B, H, and L subtypes). The aspartic/metallo-proteases include some cathepsins (such as cathepsin D), aminopeptidases (including aminopeptidase-*N*), and carboxypeptidases (A- and B-like). In the midgut of Lepidoptera, serine proteases have been estimated to account for as much as 95% of the proteolytic activity, while in Coleoptera, cysteine proteases are more prominent and are co-expressed with aspartic and metallo-proteases (Murdock *et al.*, 1987; Srinivasan *et al.*, 2006; Terra and Ferreira, 1994, 2012). The difference in the expression profiles of proteases between these orders reflects the differences in their respective diets that require different conditions for efficient digestion. For example, the prevalence of cysteine proteases in certain species of Coleoptera has been postulated to reflect an adaptation to allow feeding on plant parts such as legume seeds that can be enriched in serine protease inhibitors (Terra and Ferreira, 1994). In *Diabrotica virgifera virgifera*, the cathepsin class of cysteine proteases has been reported to be the major contributor to protein digestion (Bown *et al.*, 2004).

The cowpea (*Vigna unguiculata*) trypsin inhibitor was the first protease inhibitor to be expressed transgenically for insect resistance (Hilder *et al.*, 1987). Since that time, various protease inhibitors have been expressed heterologously in plants and shown to have activity against certain pest insects. These have included inhibitors of serine proteases derived from a range of plant sources including soybean (*Glycine max*) (Bowman-Birk and Kunitz type inhibitors), potato (*Solanum tuberosum*) (inhibitors I and II), barley (*Hordeum vulgare*), squash (*Cucurbita moschata*), and cabbage (*Brassica oleracea*), as well as the cystatins or cysteine protease inhibitors from soybean, rice (*Oryza sativa*), and corn (*Zea mays*) (Lawrence and Koundal, 2002; Macedo and Freire, 2011; Mosolov and Valueva, 2008).

Although some success has been reported with transgenic protease inhibitor expression, the efficacy achieved in most cases is much lower than what has been observed for transgenic Bt. In some instances, the low efficacy might be attributed to low levels of expression, but even when strong promoters were used to boost expression, relatively low efficacy was observed. For example, the cowpea trypsin inhibitor when expressed in tobacco (*Nicotiana tabacum*) at levels reported to be ~0.05% (or 3–5 ppm) caused mainly growth inhibition of *Spodoptera littura* (Sane et al., 1997). The barley trypsin inhibitor expressed in wheat (*Triticum aestivum*) at ~1.1% of total protein provided no leaf feeding protection against *Stitotroga cerealella*, and only early instars were killed when feeding on seed where inhibitor levels are much higher (Altpeter et al., 1999). Very high expression of rice cystatin at ~2% of total protein in rapeseed (*Brassica napus*) resulted in efficacy against *Myzus persicae* (green peach aphid; Rahbe et al., 2003), and transgenic tobacco expressing high levels of Kunitz-type soybean trypsin inhibitor exhibited protection against damage from *Heliothis virescens* (Sharma et al., 2000).

The low efficacy of protease inhibitors is likely related to the insect response to inhibitor ingestion and the wide array of proteases expressed within the midgut. Many midgut proteases exhibit overlapping substrate specificities, and therefore some level of digestive redundancy exists. Coevolution over thousands of generations has allowed many herbivorous insects to adapt to the presence of plant secondary chemicals (Gassmann et al., 2009). Additionally, insects can alter expression of targeted proteases relatively rapidly even in direct response to inhibitor exposure (Bolter and Jongsma, 1997; De Leo et al., 1998). This creates a significant challenge for reducing proteolysis sufficiently with a single protease inhibitor, so chimeric protease inhibitors that target multiple classes of proteases have also been explored (Mosolov and Valueva, 2008). Recently, a naturally occurring multi-domain protease inhibitor known as NaPI identified in the ornamental tobacco plant (*Nicotinia alata*) was shown to confer significant insect protection in transgenic cotton (*Gossypium hirsutum*) (Dunse et al., 2010b). NaPI is a member of the potato type II class of inhibitors and consists of six inhibitor centers, four that are trypsin-like and two that are chymotrypsin-like (Heath et al., 1995). But NaPI with its multi-inhibitor centers also was prone to the same limitations of other protease inhibitors where efficacy was reduced against insects that express enzyme isoforms that are insensitive. In this case, the chymotrypsin targeting domains were undermined by a protective amino acid loop around the catalytic center that prevented docking of the inhibitor on the enzyme of *Helicoverpa punctigera* (Dunse *et al.*, 2010a). Ultimately, protease inhibitors alone might fail to provide sufficient protection to be used as PIPs, but might complement other types of insect protectants such as Bt proteins.

Proteases

Similar to the potential of plant protease inhibitors as PIPs, proteases themselves from various organisms including plants could also be useful. Instead of inhibiting activity that supports nutrient absorption or host defense, proteases could target critical nutrients, metabolic enzymes, or protective barriers compromising viability. A few examples that have exhibited *in planta* efficacy will be considered here. Mir1-CP is produced by inbred maize lines as part of a defense response to insect feeding. Mir1-CP is a papain-like cysteine protease that degrades the insect peritrophic membrane that protects the epithelial lining of the gut, rendering it vulnerable to ingested pathogens (Pechan *et al.*, 2000). The potency of purified recombinant Mir1-CP was reported to be similar to that of Cry2Ab against various corn pests including *Helicoverpa zea*, *Spodoptera frugiperda*, and *Diatraea grandiosella*. Furthermore, the combination of Mir1-CP with Cry2Ab exhibited some level of synergism (Mohan *et al.*, 2008).

Enhancin, a protein expressed by some baculoviruses during infection is a metalloprotease that also targets the insect peritrophic membrane (Derksen and Granados, 1988). When expressed in transgenic tobacco, enhancin has been shown to cause larval growth retardation (Cao *et al.*, 2002) and, similar to Mir1-CP, boosted the activity of Cry proteins (Granados *et al.*, 2001). V-CATH, another baculovirus-expressed protease, targets the insect cuticle during viral infection, playing a key role in host tissue liquefaction. V-CATH is a cathepsin-L protease that caused growth inhibition in *Helicoverpa armigera* when expressed in transgenic tobacco (Zhang *et al.*, 2008). Because access to the cuticle would be minimal without viral delivery, the bioactivity of plant-expressed V-CATH is thought to be due to degradation of a midgut protein that could be part of the peritrophic membrane, but this remains unclear. As demonstrated by the examples presented here, transgenic expression of proteases as PIPs can provide protection against insect damage, but similar to protease inhibitors, they might lack the potency necessary for a stand-alone active ingredient. However, when co-expressed with another PIP, proteases could augment the activity of other insecticidal actives such as Cry proteins (Mohan et al., 2008).

Lectins

A final class of proteins to be considered that has potential as PIPs is the lectins. The term *lectin* was derived from the Latin word *legere*, which means to choose or pick out. They were originally identified in plant extracts in the late nineteenth century and have been used in medicine to assign blood types for donor/recipient transfusion compatibilities (Sharon and Lis, 2004). Lectins have the ability to bind various carbohydrate groups that are present on the surface of cell membranes and play an important role in cell—cell recognition and cell—ligand interactions. Their use in blood-type assignments is due to their ability to cause agglutination (clumping) of red blood cells when they bind to surface carbohydrate groups. One of the most widely used tools in glycobiology is the lectin concanavalin A (isolated from Jack bean, *Canavalia ensiformis*) that allows identification of proteins bearing certain glycosylation modifications. Perhaps the most notorious lectin is the highly toxic protein ricin from the seeds of the castor tree (*Ricinus communis*). Similar to plant protease inhibitors, plant lectins are enriched in the storage compartments of reproductive tissues presumably to serve as protectants against pathogens and insects. Also, their expression has been reported to be induced by insect damage (Van Damme et al., 1998; Vandenborre et al., 2011). One of the first studies on the insecticidal properties of lectins showed that black bean (*Phaseolus vulgaris*) lectin killed *Callosobruchus maculatus* larvae (Janzen et al., 1976). Since that time, a wide range of lectins have been reported to have toxic effects on a variety of insects, including Lepidoptera, Coleoptera, Diptera, and Hemiptera. Transgenic plants expressing lectins similar to the snow drop lectin, *Galanthus nivalis* agglutinin (GNA), exhibit resistance to insect damage. For example, maize expressing GNA had resistance to aphids (Rhopalosiphum maidis; Wang et al., 2005). Rice expressing garlic leaf lectin (*Allium sativum* agglutinin) exhibited resistance to various planthoppers (Yarasi et al., 2008). A leek lectin (*Allium porrum* agglutinin) expressed in cotton protected against damage by *Spodoptera littoralis* (Sadeghi et al., 2009) and lectin-expressing sugarcane (*Sacharum officinarum*) exhibited protection against stalk borer damage (Setamou et al., 2002).

MODE OF ACTION OF BT PROTEINS

By far the most successful PIP technology has been derived from Bt, which has led to increasing development and adoption of transgenic Bt crops in commercial agriculture. This success has in turn driven extensive research into Bt mode of action resulting in significant progress over the last 20 years, but many details remain to be clarified (Vachon *et al.*, 2012). Numerous excellent reviews have been published that cover various aspects of Bt and their insecticidal proteins (Bravo *et al.*, 2007; de Maagd *et al.*, 2001; Gill *et al.*, 1992; Knowles, 1994; Pigott and Ellar, 2007; Schnepf *et al.*, 1998; van Frankenhuyzen, 2009). Here, we will provide a brief overview of Bt mode of action.

The selectivity of certain Bt Cry proteins against lepidopteran and coleopteran pests has been appreciated for some time, and the general mechanism of pore formation as the primary cause of toxicity has been well established (Crickmore *et al.*, 1998; Knowles, 1994; Soberon *et al.*, 2009). Yet, details concerning the determinants of selectivity remain poorly understood, and so they continue to receive much attention. The issue of selectivity has two components: (1) the regions on the Bt proteins themselves that control the targeting of midgut receptors, and (2) the factors within the insect gut, including those receptors that impact selectivity. Establishing the activity determinants is critical not only to having a complete understanding of Bt mode of action, but also to assessing how resistance may develop, which could aid the design of effective strategies for insect resistance management (IRM). Complexities on both sides of the selectivity question raise significant challenges for the discovery and development of novel Cry proteins for insect control.

The crystals produced during Bt sporulation were the focus of early work on Bt mode of action as the correlation between insect susceptibility and the presence of crystals became apparent (Angus, 1954). Studies conducted in the 1950s revealed that the cessation of feeding that followed ingestion of Bt crystal proteins correlated with changes in midgut cell morphology. These studies identified the "brush border" epithelium as the cells targeted by these toxins (Gill *et al.*, 1992; Heimpel and Angus, 1959; Knowles, 1994). Subsequent studies with various Lepidoptera determined a general time course of the physiological and morphological changes that occur in Bt-poisoned larvae and are outlined in Figure 4.1 (Knowles, 1994).

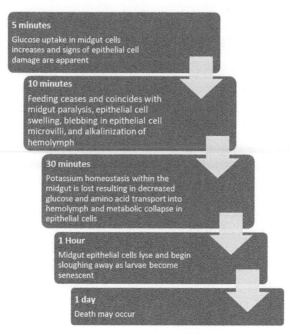

Figure 4.1 Time course of symptomology and physiological changes occurring during Bt intoxication.

Of course the exact timing and extent of these changes will depend on the particular toxin, its dose, and the developmental stage of the insect, but they reflect the events that impact insect physiology leading to death. In terms of a molecular mechanism that might explain the changes observed in the midgut, some of the early theories that were considered included direct inhibition of active transport by ATPases, modulation of endogenous potassium channels, and *de novo* pore formation after insertion into the epithelial cell apical membrane. Considering the data available at the time, the most unifying theory was that of pore formation (Gill *et al.*, 1992), which remains the most widely accepted mechanism to date (Bravo *et al.*, 2013).

As evidence for pore formation as the mechanism of Bt toxicity was established, the mode of action of Cry proteins began to be described as a sequential progression of events leading up to pore insertion into the membrane (Bravo *et al.*, 2007; de Maagd *et al.*, 2001; Knowles, 1994; Soberon *et al.*, 2010). Once ingested, Cry proteins in their native protoxin form (the precursor form of the protein) are solubilized in the lumen of the midgut where they undergo processing to their toxic or "activated"

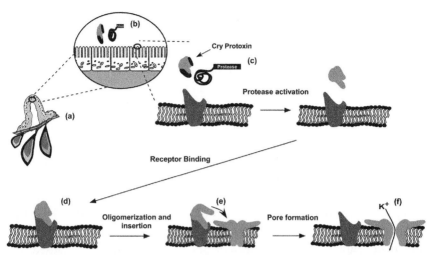

Figure 4.2 Key steps involved in the mode of action of Bt proteins. Following ingestion of transgenic material (a), PIPs are solubilized in the midgut where proteases
process (b and c) them into active form. Fingerlike projections on the surface of the
epithelial cells (b) that line the lumen of the midgut contain receptor proteins localized to the membrane bilayer that bind to activated PIPs (d). Interaction with these
receptors causes conformational changes in the activated PIP (e) that promotes oligomerization and insertion into the bilayer membrane. There they form pores allowing flow of potassium and other cations back into midgut cells, (f) thus disrupting
the ability of the cells to maintain the electrochemical gradient important to nutrient
absorption and other gut functions. Illustration inspired by de Maagd *et al.*, (2001).

state by gut digestive enzymes. At this point, or perhaps simultaneously,
the activated Cry proteins bind to specific midgut proteins that serve as
receptors to coordinate association and/or insertion into the epithelial-
cell apical membrane. Here, the fully assembled "oligomer" of multiple
Cry protein molecules forms a pore that results in the uncontrolled ionic
flux and collapse of normal cellular function (see Figure 4.2).

The toxicity following pore formation stems from the fact that it disrupts ionic gradients that are critical to normal functioning in all living
cells. Ionic gradients represent a readily usable form of energy that in the
case of the digestive system is utilized to absorb nutrients (Giordana *et al.*,
1982). The high end of the gradient is in the midgut lumen where the
concentrated ions drive co-transport of amino acids, sugars, and other
nutrients across the apical membrane to be utilized in various metabolic
pathways. The ionic gradient also contributes to the osmotic and pH
homeostasis of midgut cells (Dow and Harvey, 1988; Moffett and

Cummings, 1994). Active transport by ATP-driven pumps within subcellular compartments indirectly generates the gradient by concentrating protons (H^+) that are exchanged for cations (insects typically utilize potassium, K^+) that are released into the midgut lumen, which results in large electrical potentials. The metabolic demand to maintain these ionic gradients can be demonstrated by the rapid decline in the electrical potential across the tissue that is caused by anaerobic conditions or when metabolic inhibitors are introduced (Moffett and Cummings, 1994; Zeiske et al., 2002).

Although strong experimental evidence supporting pore formation as the primary mechanism of Bt toxicity has been generated, an alternative theory has also been proposed (Bravo et al., 2007; Ibrahim et al., 2010; Soberon et al., 2010; Zhang et al., 2006). This model was based on the observation that when the cadherin-like protein was cloned from M. sexta and then expressed in transfected Sf9 cells, cytotoxicity caused by Cry1Ab exposure was magnesium dependent and followed increased adenylyl cyclase activity along with activation of protein kinase A (Zhang et al., 2005, 2006). Thus, it was proposed that the Cry protein mode of toxicity was related to the adenylyl cyclase/protein kinase A pathway (Ibrahim et al., 2010). However, since that study was published, it has been shown that the binding of Cry1Ab to the cadherin receptor is not essential for toxicity (Soberon et al., 2007; Tabashnik et al., 2011), not to mention that other classes of Bt toxins do not utilize cadherin proteins in their mode of action (Pigott and Ellar, 2007). However, it is interesting to note that an area of recent focus in Bt toxicology has been on the nonpore-forming activities of Cry proteins (e.g., on cell signaling pathways), especially at sublethal concentrations. These pathways could be affected by Bt toxin exposure and could ultimately have long-term impact on Bt susceptibility, but current evidence suggests that they are not the primary mechanism of toxicity.

Determinants of Bt Activity

As mentioned in an earlier section, Cry proteins can be classified according their sequence homologies and target pest selectivity (Bravo et al., 2013; Crickmore et al., 1998; de Maagd et al., 2001, 2003; Schnepf et al., 1998), but the determinants of activity against a particular target insect remain an area of intense research. Three-dimensional structures have been determined by X-ray crystallography for several Cry toxins, including the lepidopteran-actives Cry1Aa (Grochulski et al., 1995), Cry1Ac

(Derbyshire *et al.*, 2001), and Cry2Aa (Morse *et al.*, 2001), as well as the coleopteran-actives Cry3Aa (Li *et al.*, 1991), Cry3Bb (Galitsky *et al.*, 2001) and Cry8Ea (Guo *et al.*, 2009). The tertiary organizations of these Cry proteins are all very similar and consist of three distinct domains: in their N-terminal region, ***domain I*** is a bundled array of seven α-helices; ***domain II*** consists of three antiparallel β-sheets that are connected by hairpin loops; and at their C-terminal region, ***domain III*** consists of two antiparallel β -sheets (see Figure 4.3).

A role for each domain in the Cry protein mode of action has been established. Domain I serves as the pore-forming region, while domains II and III play major roles in receptor interactions during binding (Bravo *et al.*, 2007; de Maagd *et al.*, 2001; Li *et al.*, 2001; Schnepf *et al.*, 1998). The hairpin loops in domain II are especially important determinants of receptor interactions (Griffitts and Aroian, 2005; Pigott *et al.*, 2008; Saraswathy and Kumar, 2004). An illustration of their role in target selectivity was demonstrated by an elegant study where a lepidopteran-selective toxin, Cry1A, was transformed to become toxic to mosquitoes

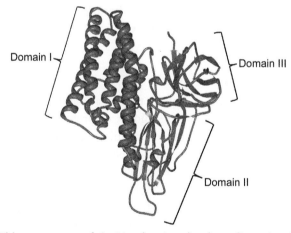

Figure 4.3 Ribbon structure of Cry1Aa showing the three-dimensional organization of the carbon backbone of the amino acid sequence into three distinct domains. All three-domain Cry proteins have a similar structural organization: ***domain I*** is a bundled array of seven α-helices connected by short loop sequences; ***domain II*** consists of three antiparallel β-sheets in a prism-like orientation that are connected by hairpin loops (projecting downward) involved in receptor interactions; ***domain III*** consists of two antiparallel β-sheets forming a sandwich-like structure the surface of which also contributes to receptor interactions. The Cry1Aa crystallization and structural determination was reported in Grochulski *et al.* (2001); Research Collaboratory for Structural Bioinformatics Protein Data Bank ID:1ciy.

by altering its three hairpin loops in domain II to mimic the corresponding loops of Cry4Ba, a mosquito-selective toxin (Liu and Dean, 2006). An important role for domain III in dictating activity/selectivity against certain species has been illustrated by a number of studies where entire domain III sequences were exchanged between toxins by molecular cloning to show that the insecticidal activity of the resulting chimera could be predicted by the identity of the domain III donor parent (Bosch et al., 1994; de Maagd et al., 1996; Naimov et al., 2001). However, exchange of domain III did not alter the apparent binding site of the recipient protein, suggesting a more complex role of domain III than simply receptor recognition in Bt mode of action.

The conserved three-dimensional organization of Bt Cry proteins supports the contention that even with diverse amino acid sequences, they act through a common mechanism of pore formation (Bravo et al., 2007; de Maagd et al., 2003; Li et al., 1991; Soberon et al., 2009). Therefore, recent work has focused on the determinants of activity against various insect species. Bt bacteria have achieved this over the course of evolution by varying amino acid sequence to capitalize on important differences in the steps leading up to pore formation. The impact of each of these steps on toxicity of Bt will be considered now.

Solubilization

The solubility of Cry proteins appears to play a role in target pest susceptibility. Gut pH can be a major determinant of protein solubility due to its effect on surface charges contributed by amino acid side chains (most Cry proteins exhibit better solubility under alkaline conditions). The pH of the typical Lepidoptera midgut is very alkaline, while in many Coleoptera, including D.v. virgifera, the midgut is slightly acidic. Interestingly, Bt proteins that are active on Lepidoptera generally have preference for the amino acid arginine in their sequences compared to proteins that have activity on Coleoptera, which typically exhibit a preference for lysine residues instead (de Maagd et al., 2001; Grochulski et al., 1995). Both arginine and lysine have basic side chains, but the pKa (pH where half of the amino acids would exist in the charged state) of arginine (pKa = 12.5) is much more alkaline than that of lysine (pKa = 10.5), consistent with target selectivity evolving around solubility. Also consistent with this reasoning, it has been reported that changes in Cry3Aa (a coleopteran-active) amino acid sequence that improved protein solubility also improved bioactivity against *D. v. virgifera* (Walters *et al.*, 2008).

Proteolytic Activation

As Bt proteins are solubilized, midgut proteases begin to process them into "activated" or toxic form. Appropriate activation is critical for maintaining protein integrity, conformation, and solubility as well as promoting other steps in Bt mode of action. Analysis of toxin processing and the enzymes that are typically found in insect midguts has revealed generalities about what enzymes are involved in the activation process. In lepidopteran insects, these enzymes are serine endopeptidases such as trypsin, chymotrypsin, and elastase, while in coleopteran insects, the primary players are cysteine proteases, including cathepsins, carboxypeptidases, and metalloproteases (Terra and Ferreira, 1994, 2012).

Cry1 protoxins are processed from ~130−145 kilodaltons (kDa) by gut enzymes to proteolycally stable core structures ranging ~55−68 kDa, depending on the subclass. Close to half of Cry1 protoxin sequence is part of a carboxy terminal tail (referred to as the crystal domain) that contributes to crystal formation during bacterial sporulation, and a short segment of the amino-terminal region is also processed *en route* to pore formation.

Cry2 protoxins, in contrast, are approximately 70 kDa, and are processed by midgut enzymes to a slightly smaller form of approximately 68 kDa and then further to a significantly smaller form between 50 and 55 kDa. Similar to Cry2, Cry3 proteins are also produced in a naturally truncated form of approximately 70 kDa that undergo processing predominately in their N-terminal region *en route* to toxicity. The Cry3 proteins have been shown to undergo additional processing deeper into the N-terminus within the pore-forming domain I region at a site that has been suggested to be critical for maximal activity against *D. v. virgifera* (Carroll *et al.*, 1997). Consistent with this theory, native Cry3Aa has activity against certain insects, such as *Diabrotica undecimpunctata* and *Tenebrio molitor*, but only after modifying this region with mutations to improve proteolytic processing was activity sufficient against *Diabrotica* spp. for commercial deployment (Walters *et al.*, 2008). Interestingly, Cry1B proteins also undergo processing at an analogous site and have been shown to have activity on some Coleoptera as well as Lepidoptera (Martins *et al.*, 2010; Tailor *et al.*, 1992). However, the significance of this additional processing in domain I in the mode of action or selectivity of Bt toxins requires further investigation.

Non-Cry Bt proteins, including the lepidopteran-active Vip3Aa and the coleopteran-active binary proteins Cry34 and Cry35, also undergo some level of proteolysis within the insect gut. The Vip3Aa protein is processed in the N-terminal region, and it has been reported that this processing is necessary for insecticidal activity (Yu et al., 1997). Cry34 and Cry35 are processed by serine proteases in vitro and by midgut extracts from D. v. virgifera, but other than affecting solubility of the proteins, it remains unclear whether the processing is necessary for insecticidal activity. Both unprocessed and processed forms of each component of the binary protein were reported to be able to undergo pore formation when present in combination (Masson et al., 2004).

Receptor Binding

After solubilization and initial processing by midgut proteases, Bt proteins must interact with what are referred to as "receptors" to induce conformational changes that are necessary for further processing leading to their insertion into the membrane for pore formation. Unlike many ligand—receptor interactions such as those involved in the mode of action of insecticidal chemistries, acute toxicity of Bt proteins does not result from changes in the functionality of the receptor caused by binding of the Bt. The binding step in Bt mode of action localizes the toxin molecules at the membrane and causes critical conformational changes in the three-dimensional structure of the protein. The exact nature of these conformational changes has not been clearly elucidated, but likely involves exposure of proteolytic sites for additional processing, exposure of contact domains to facilitate oligomerization or additional receptor interactions, or exposure of regions involved with membrane insertion. As will be discussed later, some studies on resistant insects have revealed evidence for altered receptors that reduce toxicity, but do not eliminate the ability of resisted toxins to bind (González-Cabrera et al., 2003; Tabashnik et al., 1997a). This observation illustrates the complexity of Bt mode of action and indicates that multiple receptor interactions could be involved, that a critical receptor is shared with multiple toxins, or simply that binding determinants overlap but are incompletely shared.

Interactions of Bt proteins with receptors is commonly studied using a midgut tissue preparation referred to as brush border membrane vesicles

(BBMV). BBMVs are formed by homogenizing midgut tissue using a method that isolates the surface membrane of the cells that line the midgut interior (illustrated in Fig. 4.2 part b) and therefore contains the membrane-associated receptors for Bt proteins (Wolfersberger et al., 1987). This preparation provides a convenient system to assess the specific interaction of Cry proteins with their target receptors as well as assessing whether two Cry proteins have receptors in common, which could indicate potential for cross-resistance. However, in interpreting binding studies using BBMVs, it has been reported that two types of interactions occur, reversible and irreversible, and that the correlation between binding and toxicity is greater for irreversible binding (Garczynski et al., 1991; Liang et al., 1995). Although the conclusion about the correlation to toxicity is reasonable and relevant mechanistically, the reference to "irreversible binding" is confusing in the context of Bt mode of action since the irreversible nature likely reflects steps that follow receptor binding (i.e., membrane insertion), and not the binding step at all. To accurately assess the kinetics of binding of Bt proteins with their receptors, including their reversibility (off-rates), the characterization should be performed in the absence of membranes or lipids. Insect BBMVs serve as an extremely useful model system to investigate the binding site relationships among different Bt proteins, but measures of apparent affinity are qualitative and should only be used for relative comparisons.

Early attempts to identify putative receptors for Cry1 proteins revealed interactions with the midgut protease aminopeptidase-N (APN). The APN receptors that are expressed on the surface of midgut epithelial cells are part of a large family of metalloprotease (zinc binding) enzymes (Pigott and Ellar, 2007). Their role in midgut function is to remove amino-terminal amino acids from peptides during digestion. The "N-class" of aminopeptidases removes neutral amino acids, while other classes are selective for other types of amino acids. Important features of APN receptors include the presence of glycosylation sites and anchoring in the apical membrane of midgut epithelial cells by the addition of a glycophosphatidylinositol group (a modified lipid). Although interactions of Cry1 proteins with APN receptors have been appreciated for some time, intensive work over the last decade has revealed that multiple subclasses of APNs (at least five) are expressed by a given insect species and that multiple forms are expressed within the midgut (Pigott and Ellar, 2007; Tiewsiri and Wang, 2011; Yang et al., 2010). Expression of various APN subtypes can vary during insect development and in response to Bt exposure. Extensive studies on the binding determinants of Bt proteins with various APN subtypes from various insect species show that although several Cry1 proteins can bind to several APN receptors, some Bt proteins rely on specific APN interactions (see Pigott and Ellar, 2007). Results from studies with *M. sexta* APN revealed complex protein/APN interactions that involve multiple contact sites within domain II as well as a pocket in domain III that bound to a glycosylation site on APN. Comparing the binding of Cry1Aa, Cry1Ab, and Cry1Ac in the presence and absence of the sugar *N*-acetylgalactose revealed important differences in the role of the glycosylation site as a binding determinant among these Cry proteins (Valaitis *et al.*, 1997). An elegant study validated a critical receptor role for APN by transgenic expression of *M. sexta* APN in *Drosophila melanogaster* to confer Cry1Ac susceptibility to an otherwise insensitive insect (Gill and Ellar, 2002). Finally, reducing APN expression by RNA interference techniques in *S. litura* reduced the toxicity of Cry1Ac, validating APN as a viable receptor (Rajagopal *et al.*, 2002).

The cadherin class of receptors represents one of the more thoroughly characterized receptors of Bt proteins. Cadherin-like proteins were iden-tified as Cry1Ac and Cry1Ab receptors using ligand blotting approaches in parallel with various biochemical and proteomic approaches. Cloning and heterologous expression of the cadherins from *M. sexta* and *Ostrinia nubilalis* revealed that Sf9 or Sf21 cells that were normally insensitive to Cry1Ab exhibited cytotoxicity upon Cry1Ab exposure when they expressed the cadherin protein (Flannagan *et al.*, 2005; Keeton and Bulla, 1997; Vadlamudi *et al.*, 1995).

The insect cadherin receptors are related to the cadherin superfamily of integral membrane proteins defined by the presence of conserved repeating extracellular domains that contain calcium-binding motifs. The name cad-herin was adopted to reflect the idea that proteins within this superfamily serve as calcium-dependent adhesion molecules. The cadherin superfamily is extremely large and has evolved in higher organisms to play a variety of functional and structural roles including involvement in cell adhesion and cell morphology. They are especially important during tissue growth and development in early embryogenesis (Hulpiau and van Roy, 2009). A definitive role for cadherin proteins in the insect midgut physiology has not been established, so they are technically "cadherin-like." However, the presence of cadherin in midgut epithelial cells is consistent with the role played by cadherins in other organisms (e.g., cell-to-cell adhesion, extra-cellular matrix structure, or cell signaling). Bt binding to cadherin receptors has been shown to be important in the mode of action of several (but not all) naturally occurring lepidopteran-active Cry toxins, especially for the Cry1A class. Binding to cadherin induces a conformational change in the Cry protein that allows more extensive proteolytic processing within domain I that promotes oligomerization of Cry1A toxins (Bravo *et al.*, 2007). It remains unclear if similar receptor-driven domain I processing is necessary for all 3-domain proteins, but it is interesting to note that cleavage in domain I is a normal process for Cry1B, Cry1I, Cry2, and Cry3 proteins. It is tempting to speculate that domain I processing is necessary for oligomerization of all 3-domain proteins and that the interaction with a cadherin or similar receptor ensures proper exposure of the cleavage site. This binding might also ensure release of the cleaved portion of the protein should it remain associated due to noncovalent attraction following breakage of the peptide bond as has been reported for Cry3 proteins (Carroll *et al.*, 1997). More studies around these questions will help to clarify the need for certain proteins to bind to cadherin receptors.

Alkaline phosphatase proteins represent another class of midgut enzymes that have been found to bind to Bt proteins. They exist in both soluble and membrane-associated forms where they are responsible for dephosphorylating ingested dietary substances including proteins, nucleic acids, and alkaloids. The membrane-associated form has been shown to be involved in the binding of Bt proteins in a number of insects including *M. sexta* (Arenas *et al.*, 2010; McNall and Adang, 2003), *H. virescens* (Jurat-Fuentes *et al.*, 2004), *Aedes aegypti* (Fernandez *et al.*, 2006), and *H. armigera* (Ning *et al.*, 2010; Upadhyay and Singh, 2011).

The glycoconjugate class of proteins is not as thoroughly characterized as the other Bt receptors. The term *glycoconjugate* could refer to any midgut protein with a sugar modification including both APN and alkaline phosphatase receptors, but in the context of Bt mode of action, glycoconjugate refers to the interaction of Bt proteins with high-molecular-weight molecules on ligand blots (Pandian *et al.*, 2008; Valaitis *et al.*, 2001). When isolated and subjected to further analysis, one of these molecules known as BtR-270 was found to consist of a high percentage of sugars (>70% of its mass) and a relatively high number of negatively charged amino acids. This molecule was able to bind Cry1A toxins with high affinity, but no functional role for this receptor has been established (Valaitis *et al.*, 2001).

Finally, it is important to consider that although receptor binding is absolutely necessary for toxicity, it is not sufficient. Furthermore, affinity of protein binding *in vitro* is not a reliable predictor of toxicity *in vivo*. The reason for the lack of correlation between binding affinity and toxicity is not clear, but it is likely related to the fact that the binding step in Bt mode of action is a transient state leading to pore formation, the toxic state (Pacheco *et al.*, 2009). This is not to say that binding of Bt proteins to receptors does not evoke some cellular response, but simply that the cellular response evoked by binding alone is less of a threat to insect survival than pore formation.

RNA INTERFERENCE

A promising emerging technology for the development of genetically engineered crops is RNA interference (RNAi). Also known as post-transcriptional gene silencing, RNAi is commonly found in eukaryotic organisms, including insects, for sequence-specific gene silencing that is triggered by the presence of double-stranded RNA (dsRNA) (Burand and Hunter, 2013; Hannon, 2002; Whangbo and Hunter, 2008). In plants, some RNAi pathways are endogenous and serve to control an organism's own gene expression (Lelandais-Briere *et al.*, 2010; Vazquez *et al.*, 2010). Other pathways are induced by invasive nucleic acids, such as viruses or transposons, and serve to defend the host plant against such invaders (Alvarado and Scholthof, 2009). After molecular pathways responsible for this mechanism were discovered in the late 1990s, RNAi methods were quickly adopted by the research community to investigate gene function (Bosher and Labouesse, 2000; Dykxhoorn *et al.*, 2003). Briefly, the first step in the RNAi pathway involves cleavage of dsRNAs by an enzyme called dicer into sequence-specific effector molecules, called small- or short-interfering RNAs (siRNAs), which target homologous RNAs for destruction (Dykxhoorn *et al.*, 2003; Hannon, 2002). These siRNA are then incorporated into a large protein complex called RISC (for RNA-inducing silencing complex) (Nykanen *et al.*, 2001). Subsequently, the duplex siRNA is unwound, leaving the antisense strand to guide RISC to its target mRNA for endonucleolytic cleavage (Elbashir *et al.*, 2001), requiring perfect siRNA-mRNA sequence complementarity (Zamore, 2002). The target mRNA is cleaved at a single site in the center of the duplex region between the guide siRNA and the target mRNA, resulting in gene silencing (Elbashir *et al.*, 2001; Hannon, 2003). A short description of the RNAi machinery using siRNA- (inducible RNAi pathways) or miRNA- (micro-RNA; endogenous RNAi pathways) mediated silencing, their use in strategies for plant biotechnology as well as potential limitations, are described in ILSI (2011). Additionally, more discussion of RNAi can be found in Chapter 3 of this edition.

Not surprisingly, the first studies involving RNAi in insect species were completed using model organisms such as *Drosophila melanogaster* (Kennerdell and Carthew, 1998) and *T. castaneum* (Brown *et al.*, 1999). In subsequent years, a cascade of studies have been published on a number of genes, currently representing more than 30 species of insects in a variety or orders (see Belles, 2010). In spite of uncertainties regarding the feasibility of RNAi to protect plants against herbivore insects (see Price and Gatehouse, 2008), recent results have shown that meaningful levels of insect resistance can be achieved by producing dsRNA in plants (Baum *et al.*, 2007; Kumar *et al.*, 2012; Mao *et al.*, 2007; Zhu *et al.*, 2012). Furthermore, the application of RNAi to produce genetically engineered crops with improved agronomic, nutritional, industrial, and food-processing traits is becoming increasingly common (see Frizzi and Huang, 2010 for review). The key elements in successfully using RNAi to produce insect-resistant plants are (i) identification of a suitable insect molecular target, (ii) dsRNA delivery, and (iii) specificity of the target dsRNA (Price and Gatehouse, 2008).

The use of RNAi in plant biotechnology to protect crops against insect damage has great potential, especially because of the wide array of genes that can be exploited for silencing. However, limited data on the use of RNAi against *D. v. virgifera* suggest that the efficacy of RNAi-expressing plants is inferior to that observed in transgenic maize expressing Cry3Bb insecticidal proteins (Baum *et al.*, 2007; Vaughn *et al.*, 2005). Importantly, it has been shown that dsRNA uptake, presence of dsRNA systemic distribution of dsRNA within the target organism, dsRNA sequence and length, and persistence of gene silencing can impact the efficiency of RNAi in insect diet bioassays (Baum *et al.*, 2007; Bolognesi *et al.*, 2012; Huvenne and Smagghe, 2010; Whyard *et al.*, 2009). Additionally, research suggests that long dsRNAs have greater potential for higher efficacy against *D. v. virgifera* because of the larger number of siRNAs (with a high degree of mRNAs sequence match) that can be produced (Baum *et al.*, 2007). It is believed that enhancing the effectiveness of RNAi PIP strategies might be achieved by utilizing multiple targets, which could be accomplished by (i) pyramiding multiple RNAi targets, and (ii) combining it with existing crop protection strategies (i.e., Bt and/or non-Bt proteins).

RESISTANCE TO BT PROTEINS

The sequence of steps in the mode of action of Bt proteins that are necessary for toxicity, beginning with solubilization and culminating in pore formation, serves as a general outline for the various mechanisms through which resistance might occur. The activation and receptor binding steps are particular vulnerabilities, and inactivation or detoxification through overprocessing by midgut proteases might compromise activities as well. Evidence for each of these mechanisms in resistance or tolerance has been reported for various Cry proteins and with various insects under either field or laboratory selection. Although the focus of the present discussion is resistance to Bt proteins, these mechanisms could be just as applicable to Bt alternatives since similar responses to exposure could compromise the efficacy of other PIPs as well. However, it is important to note that in terms of cross-resistance potential, insecticidal proteins that are more divergent in primary sequence and structure and in mode of action will be impacted differently by each of these parameters and therefore are less likely to exhibit cross-resistance due to any single mechanism.

As mentioned previously, solubility can impact the bioactivity of Bt proteins regardless of whether the protein is plant-expressed or is from a bacterial crystal in a spray formulation. Changes in the gut environment that alter solubility could easily result in reduced susceptibility. High midgut pH reduces the impact of plant tannins that tend to reduce protein solubility during digestion, and it has been suggested that the alkaline midgut environment represents a dietary adaptive response in Lepidoptera (Berenbaum, 1980). Additionally, Cry protein solubilities can change significantly as they are processed by midgut enzymes to their active form. Insect midguts also have reducing capacity that promotes protein solubility by eliminating disulfide bonds that maintain higher order protein structure (Appel and Martin, 1990; Barbehenn et al., 2001). Considering the implications of protein solubility for toxicity, changes in pH, redox potential, and abundance of surfactants could all contribute to reduced toxicity.

Because Bt proteins are processed or "activated" by midgut proteases for toxicity, altered protease activity represents a potential mechanism for resistance. In fact, altered protein processing has been reported for certain cases of resistance to Bt and could explain reduced Bt susceptibility in these insects (Forcada *et al.*, 1996; Li *et al.*, 2004; Oppert *et al.*, 1997). Altered proteolytic activity has been reported in *Plodia interpunctella* where reduced processing of Cry1Ac was shown to correlate with reduced susceptibility (Oppert *et al.*, 1997). Reduced protoxin activation also was reported for resistance to sprayable Bt in a laboratory-selected population of *O. nubilalis* (Li *et al.*, 2005; Li *et al.*, 2004). Both reduced activation of protoxin and increased degradation of mature toxin by midgut proteases were described for a Cry1Ab-resistant colony of *H. virescens* (CP73-3; Forcada *et al.*, 1996). More recently, reduced formation of activated Cry1Ac was described for resistant *H. armigera* (Rajagopal *et al.*, 2009).

Identifying altered protease activity in mechanisms of resistance can be a challenge. Differential susceptibility to protoxin versus *in vitro* "activated" toxin has been used as an indicator for protease-mediated resistance. However, this conclusion has not always been supported. An example of resistance in *P. xylostella* illustrates this point where protoxin was less potent than activated toxin, yet normal protoxin processing by gut enzymes was observed (SERD5; Sayyed *et al.*, 2005). This indicated the presence of a more complex resistance scenario than simply reduced activation.

Interestingly, additional processing of Cry1A "activated" toxins after the first α-helix in domain I appears to be important for toxicity. In *M. sexta*, it has been shown that removal of the first α-helix normally occurs after binding to the cadherin receptor and that the removal of this α-helix promotes oligomerization of the toxin *in vitro*, highlighting the significance of this step leading to pore formation (Gomez *et al.*, 2002; Pardo-Lopez *et al.*, 2006; Soberon *et al.*, 2007). The importance of this step was further demonstrated by the finding that Cry1A proteins engineered to lack the first α-helix were able to kill insects resistant to the native full-length toxin (Tabashnik *et al.*, 2011).

Although resistance to Bt proteins has been attributed to altered protease activity in some cases, the degree of reduced susceptibility is typically lower than what has been reported for resistance caused by binding site modifications (Ferré and Van Rie, 2002). Because midgut enzymes are responsible for processing nutrients, their modification could affect insect digestion and therefore could lead to significant fitness costs. This might be a self-limiting mechanism keeping some types of metabolic resistance from taking hold, but this possibility remains unclear (Bauer, 1995).

Receptor interactions are key determinants in differentiating the mode of action of Bt proteins. Failed receptor binding prevents proper processing, oligomerization, and insertion into the membrane for pore formation. Because of this, modifications in receptor expression or sequence typically result in very high levels of resistance. Early characterizations of receptor-mediated resistance in insects such as *P. xylostella* were used to define "Mode I" resistance, which was defined as showing high levels of resistance to Cry1A, but not Cry1C, accompanied by lost or reduced binding of the Cry1A protein (Tabashnik *et al.*, 1998). Such profiles have been reported for several insects including *H. virescens* (Gahan *et al.*, 2001), *Pectinophora gossypiella* (Morin *et al.*, 2003), and *H. armigera* (Xu *et al.*, 2005), and appear to show consistent patterns of cross-resistance across species. However, cross-resistance between specific proteins in specific insect species should be determined empirically.

Receptor-mediated resistance has been demonstrated by the reduced binding of Bt proteins to BBMVs *in vitro* and changes in specific midgut receptors have been confirmed in several cases. However, correlations between resistance and changes in midgut protein expression have been reported without direct evidence that the altered protein is the actual receptor. Such reports can lead to the erroneous impression that the altered protein is the receptor for that toxin and is causal in the resistance mechanism. Altered binding defines receptor-mediated resistance, but causality for a particular receptor protein requires direct evidence for altered interaction between the Bt protein and the putative receptor. When evaluating midgut transcriptomes or proteomes in resistant insects, the results are often very complex, with expression of several different transcripts or proteins decreasing or increasing when compared to susceptible insects (Candas *et al.*, 2003; Tetreau *et al.*, 2012). Furthermore, in some instances insects exhibit cross-resistance to several homologous proteins, but not all proteins were affected in binding (Estada and Ferre, 1994; Lee *et al.*, 1995; Tabashnik *et al.*, 1997b). Therefore, conclusions about causality and resistance should be made with care. Minimal evidence for establishing receptor alterations might include ligand blotting if the appropriate probes are available to validate interaction with a particular receptor protein. If lost binding is accompanied by reduced protein expression in resistant insect tissue, then mimicking the resistance by *in vivo* knockdown techniques or with a receptor-binding antagonist would strongly support the conclusion of a receptor role in resistance. Finally, binding to a receptor is necessary, but not sufficient for toxicity, and what appears to be normal binding might remain despite receptor-mediated resistance (Griffitts and Aroian, 2005). This is because subtle changes in binding that do not significantly alter affinity could interfere with conformational changes in the Cry protein that are critical for downstream events.

Although other classes of proteins have been identified as receptors for different Bt proteins, cadherin-like proteins, aminopeptidases, and alkaline phosphatases have received the most attention largely due to their interaction with lepidopteran-active Cry1A proteins and the importance of Cry1As in agriculture. The widespread use of Cry1As for insect control also has resulted in the first cases of field resistance and has served to justify the development of laboratory-generated resistant strains that serve as model systems of what might occur under field use. Evidence that various binding site modifications play a role in resistance to certain proteins will be reviewed briefly.

Cadherin Receptors

Several instances of Cry1 resistance in Lepidoptera have been attributed to changes in cadherin protein expression or sequence. For example, in *H. virescens*, a retrotransposon-mediated insertion of a stop codon in the cadherin gene was identified in resistance to Cry1Ac (Gahan *et al.*, 2001). In *P. gossypiella*, deletions in the cadherin gene resulted in high levels of resistance to Cry1Ac (Morin *et al.*, 2003). Similarly, disruption of the cadherin gene was associated with Cry1Ac resistance in *H. armigera* (Xu *et al.*, 2005). It should also be noted that these cadherin mutations typically result in cross-resistance to both Cry1Ab and Cry1Aa to varying degrees. These studies led to the idea for a central role for cadherin proteins in Bt mode of action.

The link between changes in cadherin receptors and resistance to Cry1A proteins spawned the development of two technologies to augment toxicity and to counter cadherin-mediated resistance. The first concept was based on the belief that if binding to cadherin protein was necessary for toxicity, then delivering a peptide corresponding to the toxin binding region of the receptor would promote the steps in the mode of action that followed binding to cadherin and might be useful to augment the activity of some toxins (Chen *et al.*, 2007). Several studies reported enhancement of toxin activity with this approach, but the degree of enhancement seems to vary depending on toxin and insect target (Chen *et al.*, 2007; Park *et al.*, 2009; Rahman *et al.*, 2012; Rodriguez-Almazan *et al.*, 2012). The other technology also was developed around the idea that binding to cadherin was a necessary initial step in toxicity. In this case, the focus was on additional domain I proteolysis at the N-terminus of Cry1A proteins that followed cadherin binding. When it was demonstrated that this processing promoted oligomerization of Cry1A proteins, the idea emerged that bypassing this step by engineering the protein into a preprocessed form (Cry1AMod) would eliminate the need to bind cadherin for toxicity. Importantly, several different species of Lepidoptera that exhibited reduced response to Cry1A proteins due to altered cadherin were found to be highly susceptible to Cry1AMod proteins, including insects where cadherin was suppressed by RNAi-induced silencing (Soberon *et al.*, 2007). It also was noted that the activity of the modified proteins was slightly lower than the unmodified proteins on susceptible strains of these insects (Soberon *et al.*, 2007; Tabashnik *et al.*, 2011), which could reflect an advantage in reaching the toxic step through natural controlled activation coordinated by cadherin binding.

Aminopeptidase Receptors

Several different cases of resistance to Bt proteins have been attributed to alterations in aminopeptidase-N (APN) receptors. *S. exigua* selected for resistance to Cry1C exhibited a large reduction in binding of Cry1C to midgut tissue(Moar *et al.*, 1995) that was shown later to be accompanied by lost expression of APN-1 (Herrero *et al.*, 2005). A role for reduced APN-1 in Cry1C resistance was supported by a study that showed experimental depression of APN-1 by RNAi in *S. litura* resulted in tolerance to Cry1C (Rajagopal *et al.*, 2002). Interestingly, high levels of resistance to Cry1Ac in *H. armigera* were also attributed to a deletion in the APN-1 gene (Zhang *et al.*, 2009). Furthermore, changes in expression of two APNs were associated with resistance to Cry1Ac in *Trichoplusia ni* where expression of the APN-1 gene was drastically reduced and the APN-6 gene was significantly elevated (Tiewsiri and Wang, 2011). Because reduced APN-1 expression was accompanied by lost Cry1Ac binding (Wang *et al.*, 2007), it would appear that APN-1 is a critical receptor for Cry1Ac in *T. ni*. However, this resistance was described as recessive and on a single autosomal locus (Wang *et al.*, 2007), and a later study showed that genetic linkage was not to any APN gene (all are on a single locus), but to an ATP binding cassette (ABC) transporter gene (Baxter *et al.*, 2011; see below). Such apparently conflicting results indicate that much remains to be understood about Bt mode of action, midgut gene regulation, and resistance mechanisms.

Finally, aminopeptidase-P (APP) was implicated in Cry1Ab-resistant *O. nubilalis*. Resistance to Cry1Ab was correlated to mutations in APP that resulted in an amino acid substitution. The role for APP in Cry1Ab toxicity was demonstrated by RNAi silencing techniques to mimic the reduced binding caused by the mutation (Khajuria *et al.*, 2011). Separate work revealed reduced Cry1Ab binding to midgut tissue from these insects supporting a receptor role for APP in *O. nubilalis* (Crespo *et al.*, 2011), but it should be noted that direct demonstration of Cry1Ab binding to APP was not reported (Khajuria *et al.*, 2011).

Alkaline Phosphatase Receptors
Reduced expression of alkaline phosphatase (ALP) has been found in Cry1Ab-resistant *H. virescens* (Jurat-Fuentes *et al.*, 2004) and more recently in Cry1F-resistant *S. frugiperda* (Jurat-Fuentes *et al.*, 2011). Reduced ALP enzymatic activity was reported in Cry1Ac-resistant *H. armigera* (Ning *et al.*, 2010), where reduced APN activity was also observed (Jurat-Fuentes *et al.*, 2011). The frequency of reduced ALP expression in resistant insects has led to the proposal that reduced ALP could serve as a generalized indicator of resistance to Bt proteins even in the absence of direct involvement of ALP receptor in the resistance mechanism (Jurat-Fuentes *et al.*, 2011). A generic marker for the development of resistance would be highly useful for monitoring purposes, but it remains to be seen how predictive ALP expression is for the evolution of resistance to Cry proteins in the field.

ABC Transporter Receptors
Recent studies with Bt-resistant insects have revealed linkage to an ABC transporter that was previously unassociated with Bt mode of action. ABC transporters are typified by highly conserved nucleotide binding domains and are known in other organisms to play a range of roles. These include a role in development and in cellular defense (by extruding toxic substances)(Dean and Annilo, 2005). In insects, ABC transporters likely play similar roles, including detoxification and defense responses that might be responsible for xenobiotic resistance (Labbe *et al.*, 2011).

Although an exact role for ABC transporters in insect midgut has not been established, high levels of Cry1A resistance in *H. virescens* that was accompanied by reduced Cry1Ab and Cry1Ac binding were linked to a mutation in a gene that encodes the ABC-C2 transporter (Gahan *et al.*, 2010). Previous studies on parental lines of the resistant strain revealed lower levels of resistance that were correlated with altered cadherin protein and lost Cry1Aa binding, but normal Cry1Ab or Cry1Ac binding remained (Gahan *et al.*, 2001). Only after continued selection for higher levels of resistance were Cry1Ab and Cry1Ac binding reduced as well (Jurat-Fuentes *et al.*, 2004). Since ABC transporters were not suspected previously in Bt mode of action, the discovery of their linkage to resistance was largely enabled by recent advancements in DNA sequencing technologies and the high degree of synteny to *B. mori*, for which a fine genetic sequence map was available. These developments allowed map-based sequence analysis to identify the mutation that correlated with the resistance phenotype. Utilizing Cry1Ac-susceptibility screening with a series of back-crosses between genotyped insects, linkage of the resistance to a region on the *B. mori* chromosome that encodes several ABC- transporters was made. Sequence analysis of clones from susceptible and resistant *H. virescens* revealed a major deletion in the ABC-C2 transporter gene. This along with qualitative binding assays that revealed reduced Cry1Ac binding suggested that the mechanism of resistance was due to lost binding of Cry1Ac to the ABC-C2 transporter.

Another study linked Cry1Ac resistance to orthologous genes in both *T. ni* and *P. xylostella* bolstering the idea that ABC transporters are involved in Cry1A toxicity (Baxter *et al.*, 2011). However, Cry1Ac resistance in the *T. ni* strain was reported previously to be correlated with reduced expression of APN-1 (Tiewsiri and Wang, 2011), and its gene is located on a completely different chromosome than the *T. ni* ABC-C transporter gene. In the resistant *P. xylostella* strain where reduced binding was previously established (NO-QAGE; Tabashnik *et al.*, 1994; Tabashnik *et al.*, 2000), a 30 base pair (10 amino acid) deletion in the ABC-C2 transporter gene was identified, but no evidence for Cry1Ac binding to an ABC transporter has been reported (Baxter *et al.*, 2011). Follow-up studies will be necessary to clarify the role of the ABC transporter in Cry1Ac resistance in these insects.

Finally, a separate group confirmed an ABC transporter defect in Cry1A-resistant *B. mori* through positional cloning and rescue of resistant insects by germline transformation with a nonmutant form of the ABC-transporter gene (Atsumi *et al.*, 2012). A role for altered ABC transporters in Bt resistance seems likely, but much remains to be understood about their role in Cry protein toxicity.

Detoxification

Reduced toxicity of Bt proteins also could result from detoxification through overprocessing by midgut proteases or binding by midgut scavenger proteins leading to toxin sequestering or aggregation. An example of overprocessing was reported to explain the reduced susceptibility of later instars of *S. litteralis* to Cry1C (Keller *et al.*, 1996). *In vitro* incubation of Cry1C toxin with gut enzyme extracts from late instars resulted in complete degradation of toxin, while extracts from earlier instars resulted in less degradation consistent with increasing survival to Cry1C intoxication with age and revealing a mechanism through which resistance could evolve. In another example, laboratory Cry3A-selected *Leptinotarsa decemlineata* exhibited increased expression of midgut proteases, which could explain, in part, their reduced susceptibility to Cry3A, although enhanced degradation of Cry3A by midgut protease extracts was not found (Loseva *et al.*, 2002).

The scavenger mechanism for reducing toxicity was postulated to explain tolerance to Cry proteins in some insects such as *H. armigera* where elevated levels of lipid particles were found in the gut lumen (Ma *et al.*, 2005). Glycolipids present in these lipid particles were reported to bind Cry1Aa and Cry2Ab, thus preventing interaction of the toxins with receptors on midgut cells (Ma *et al.*, 2005). Sequestration of Cry1Ac in another tolerant *H. armigera* strain also has been attributed to trapping by midgut esterases (Gunning *et al.*, 2005).

Finally, elevated immune responses were reported to cause tolerance to Cry proteins in *Ephestia kuehniella* and resistance in *H. armigera*. In these insects, the rate of melanization reactions was increased, resulting in tolerance to Cry protein intoxication (Ma *et al.*, 2005; Rahman *et al.*, 2004). In *S. exigua*, a role for elevated immune response in tolerance was suggested based on differential gene expression in resistant and susceptible populations (Hernandez-Martinez *et al.*, 2010). These data indicate that innate immune responses in the gut could contribute to resistance to *Bt* proteins.

Receptor Shedding

Some of the Bt receptor proteins described above (ALP and APN) are anchored in the midgut membrane by glycosylphosphatidylinositol (GPI) modifications instead of existing as integral membrane proteins. The GPI anchor likely plays a role in the targeting and movement of these proteins in normal physiological conditions, but they also provide a means through which the proteins can be quickly released by phospholipase activity under conditions of stress as part of the host defense response. Insects that have fed on Bt have been shown to exhibit shedding of GPI-anchored proteins (Valaitis, 2008). Whether GPI shedding occurs in all insects or with all toxins remains to be investigated, but the shedding process itself could result in reduced efficacy of toxins that bind to these receptors.

Resistance to RNAi

Because use of the RNAi technology as PIPs is so novel, literature related to potential mechanisms of resistance to this technology and their implications for resistance management is limited (resistance development through laboratory selection has not been documented). Recently, Bolognesi *et al.* (2012) showed that ingestion of *DvSnf7* dsRNA, a gene involved in intracellular trafficking in *D. v. virgifera*, followed by its uptake, target mRNA and protein suppression, and systemic spreading, lead to larval growth inhibition and mortality. This suggests that these are the main events comprising the mechanism of action of dsRNA activity through oral RNAi in this species. Much like what we believe could be possible with Bt proteins, one could hypothesize that resistance could arise in any step along the RNAi pathway (RNAi machinery shut down), or by a change in the nucleotide sequence of the target mRNA (target sequence specific). The implications of the former could be devastating for the use of this technology for insect management, especially if it confers cross-resistance against all other RNAi events. The implications of the latter would certainly be concerning, but limited to the target gene. Importantly, much like the insecticidal pathway of Bt, it is expected that the mechanism of action (mode of action) of all dsRNA used in RNAi biotechnology crops will be the same, but the site of action (i.e., midgut receptor sites for Bt and mRNAs for RNAi) could differ across events. At the same time, while one could argue that all Bts kill insects by pore formation, even though they may take different routes (receptors) getting there, RNAi PIPs theoretically could kill by a very large number of downstream mechanisms (disrupting/targeting any critical insect biochemical pathway).

PIP DOSE AND IRM

Protein expression and insect–plant interactions determine exposure and dose, which are important factors affecting the selection for resistance in field populations. Especially important for PIPs, dose and exposure can be affected by temporal and tissue-specific expression, as well as the homogeneity of expression. Variation of protein expression across tissue types represents a challenge in resistance management, especially for insect species that feed on reproductive tissues (e.g., corn ears, cotton blossoms, soybean seed and pollen), which often exhibit lower protein expression than vegetative tissues. Most currently registered insecticidal corn and cotton events express insecticidal proteins constitutively throughout the whole life cycle of the plant, but with significant variability in different tissues, as well as during different phenological stages (temporal expression).

The interaction among protein expression, pest susceptibility, and dominance of insect resistance certainly plays a major role in resistance evolution. Insecticidal proteins impose selection pressure on the pest population. Susceptibility to different Cry proteins can vary across species even within the same genus. For example, LC_{50} estimates (lethal concentration that kills 50% of the susceptible population) for *Spodoptera cosmioides, S. frugiperda*, and *S. eridania* varied from 0.37, 0.88, and 62 µg/ml for Cry1Ab, and from 24, 1.9, and 1.0 µg/ml for Cry2Aa (Dos Santos *et al.*, 2009). This highlights the importance of understanding the target species exposure and susceptibility to a given PIP. Recessive expression of resistance is key to the success of the high-dose/refuge IRM strategy. Fortunately, most laboratory-selected populations thus far characterized for resistance inheritance have shown recessive or partially recessive expression (reviewed in Huang *et al.*, 2011). Additionally, several studies have shown that effective or functional dominance of resistance (based on the phenotypic response of heterozygotes measured at specific concentrations) is inversely related to protein concentration (i.e., recessive at higher concentrations and dominant at lower concentrations) (Alves *et al.*, 2006; Crespo *et al.*, 2009; Pereira *et al.*, 2008; Wu *et al.*, 2009). The assessment of effective dominance of heterozygotes upon direct exposure to transgenic plants is critical. If the dose expressed in a transgenic event is not high enough to render heterozygous individuals susceptible (recessive expression of resistance), one of the major assumptions of the high–dose/refuge strategy is unmet. In such a scenario, the risk of resistance evolution in those insect populations is increased.

CONCLUSIONS

The rapid and widespread adoption of PIPs worldwide reflects the true benefits provided by this technology. Simplified season-long delivery of insect protection in addition to reduced environmental footprint make this technology extremely attractive to growers. However, the use of PIPs alone for insect management imposes risks to the durability of the technology, especially in areas of high pest pressure (e.g., tropical geographies), high adoption rate, extended use of the same PIP, and when pest susceptibility is low (i.e., in case of non–high-dose traits). Although new toxins are needed to replace or support current PIPs, the identification of novel modes of action (Bt or non-Bt) with sufficient insecticidal activity against key target pest species is extremely challenging. Therefore, a greater appreciation by growers for the need to employ responsible IPM tactics associated with deployment of PIPs is crucial to long-term sustainability. Options that should be considered to augment the use of PIPs include appropriate refuge deployment, crop rotation, planting time, adapted germplasm; scouting, and application of synthetic insecticides with appropriate damage thresholds. As the discussions about pest behavior in Chapters 1 and 16 indicate, any combination of PIPs and other IPM tactics should be evaluated with regard to affects on and influences by pest behavior.

A common IRM tactic is the use of two insecticidal proteins either as a rotation, in which proteins are alternated every other growing season, or as a pyramid, in which the PIPs are expressed simultaneously in the crop. Pyramids of toxins are usually valuable for IRM, but care should be taken when deploying pyramided PIPs so that the modes of actions are appropriate (i.e., different from each other) for the pest insects within the geography in question. Cross-resistance between traits can certainly negate the value of a pyramid and should be assessed for key target pest species. All potential approaches using PIPs must also consider potential variability in susceptibility among target species infesting the same insecticidal crop.

Overall, the first generation of transgenic insect control crops has been a major success but also has revealed the need for continued research and development for improvements in PIP deployment and stewardship. Cases of field-derived resistance to Bt crops have been reported for *S. frugiperda* in Puerto Rico (Matten *et al.*, 2008; Storer *et al.*, 2010), *Busseola fusca* in South Africa (Van Rensburg, 2007), *P. gossypiella* in India (Dennehy *et al.*, 2011; Dhurua and Gujar, 2011; Tabashnik and Carrière, 2010), and *D.v. virgifera* in the United States (Gassmann *et al.*, 2011). In all of these instances, failure to meet high-dose and/or improper utilization of refuge and other IPM practices likely contributed to the development of resistance, which highlights the impact these factors can have on PIP durability. Moreover, most of the cases described above have one or more of the following commonalities: extremely high pest pressure, favorable environment, and ability of pest to evolve resistance to other control tactics (e.g., chemistries and crop rotation in the case of *D. v. virgifera*). This emphasizes the importance of understanding and devising IRM and IPM approaches tailored to specific geographies, as factors affecting resistance evolution as well as regulatory requirements vary in what is a global market for transgenic insect control.

REFERENCES

Altpeter, F., Diaz, I., McAuslane, H., Gaddour, K., Carbonero, P., Vasil, I.K., 1999. Increased insect resistance in transgenic wheat stably expressing trypsin inhibitor CMe. Mol. Breed. 5, 53–63.

Alvarado, V., Scholthof, H.B., 2009. Plant responses against invasive nucleic acids: RNA silencing and its suppression by plant viral pathogens. Semin. Cell Dev. Biol. 20, 1032–1040.

Alves, A.P., Spencer, T.A., Tabashnik, B.E., Siegfried, B.D., 2006. Inheritance of resistance to the Cry1Ab *Bacillus thuringiensis* toxin in *Ostrinia nubilalis* (Lepidoptera: Crambidae). J. Econ. Entomol. 99, 494–501.

Angus, T.A., 1954. A bacterial toxin paralysing silkworm larvae. Nature. 173, 545–546.

Appel, H.M., Martin, M.M., 1990. Gut redox conditions in herbivorous lepidopteran larvae. J. Chem. Ecol. 16, 3277–3290.

Arenas, I., Bravo, A., Soberon, M., Gomez, I., 2010. Role of alkaline phosphatase from *Manduca sexta* in the mechanism of action of *Bacillus thuringiensis* Cry1Ab toxin. J. Biol. Chem. 285, 12497–12503.

Atsumi, S., Miyamoto, K., Yamamoto, K., Narukawa, J., Kawai, S., Sezutsu, H., et al., 2012. Single amino acid mutation in an ATP-binding cassette transporter gene causes resistance to Bt toxin Cry1Ab in the silkworm, *Bombyx mori*. Proc. Natl. Acad. Sci. USA. 109, E1591–E1598.

Barbehenn, R.V., Bumgarner, S.L., Roosen, E.F., Martin, M.M., 2001. Antioxidant defenses in caterpillars: role of the ascorbate-recycling system in the midgut lumen. J. Insect Physiol. 47, 349–357.

Bauer, L.S., 1995. Resistance: a threat to the insecticidal crystal proteins of *Bacillus thuringiensis*. Fla. Entolmol. 78, 414–443.

Baum, J.A., Bogaert, T., Clinton, W., Heck, G.R., Feldmann, P., Ilagan, O., et al., 2007. Control of coleopteran insect pests through RNA interference. Nat. Biotechnol. 25, 1322–1326.

Baxter, S.W., Badenes-Perez, F.R., Morrison, A., Vogel, H., Crickmore, N., Kain, W., et al., 2011. Parallel evolution of *Bacillus thuringiensis* toxin resistance in Lepidoptera. Genetics. 189, 675–679.

Belles, X., 2010. Beyond Drosophila: RNAi in vivo and functional genomics in insects. Annu. Rev. Entomol. 55, 111–128.

Berenbaum, M., 1980. Adaptive significance of midgut pH in larval Lepidoptera. Am. Nat. 115, 138–146.

Birk, Y., 1996. Protein proteinase inhibitors in legume seeds–overview. Arch. Latinoam. Nutr. 44, 26S–30S.

Blackburn, M., Golubeva, E., Bowen, D., ffrench-Constant, R.H., 1998. A novel insecticidal toxin from *Photorhabdus luminescens*, toxin complex a (Tca), and its histopathological effects on the midgut of *Manduca sexta*. Appl. Environ. Microbiol. 64, 3036–3041.

Bogdanova, N.N., Corbin, D.R., Malvar, T.M., Perlak, F.J., Roberts, J.K., Romano, C.P., 2007. Nucleotide sequences encoding insecticidal proteins. Appl. Num.PCT/US2006/033868; WO 2007/027777/A3.

Bolognesi, R., Ramaseshadri, P., Anderson, J., Bachman, P., Clinton, W., Flannagan, R., et al., 2012. Characterizing the mechanism of action of double-stranded RNA activity against western corn rootworm (Diabrotica virgifera virgifera LeConte). PLoS One. 7, e47534.

Bolter, C., Jongsma, M.A., 1997. The adaptation of insects to plant protease inhibitors. J. Insect Physiol. 43, 885–895.

Bosch, D., Schipper, B., van der Kleij, H., de Maagd, R.A., Stiekema, W.J., 1994. Recombinant *Bacillus thuringiensis* crystal proteins with new properties: possibilities for resistance management. Biotechnology (N Y). 12, 915–918.

Bosher, J.M., Labouesse, M., 2000. RNA interference: genetic wand and genetic watchdog. Nat. Cell Biol. 2, E31–E36.

Bowen, D., Rocheleau, T.A., Blackburn, M., Andreev, O., Golubeva, E., Bhartia, R., et al., 1998. Insecticidal toxins from the bacterium Photorhabdus luminescens. Science. 280, 2129–2132.

Bowen, D.J., Ensign, J.C., 1998. Purification and characterization of a high-molecular-weight insecticidal protein complex produced by the entomopathogenic bacterium *Photorhabdus luminescens*. Appl. Environ. Microbiol. 64, 3029–3035.

Bown, D.P., Wilkinson, H.S., Jongsma, M.A., Gatehouse, J.A., 2004. Characterisation of cysteine proteinases responsible for digestive proteolysis in guts of larval western corn rootworm (*Diabrotica virgifera*) by expression in the yeast *Pichia pastoris*. Insect Biochem. Mol. Biol. 34, 305–320.

Bravo, A., Soberón, M., 2008. How to cope with insect resistance to Bt toxins? Trends Biotechnol. 26, 573–579.

Bravo, A., Gill, S.S., Soberón, M., 2007. Mode of action of *Bacillus thuringiensis* Cry and Cyt toxins and their potential for insect control. Toxicon. 49, 423–435.

Bravo, A., Gomez, I., Porta, H., Garcia-Gomez, B.I., Rodriguez-Almazan, C., Pardo, L., et al., 2013. Evolution of Bacillus thuringiensis Cry toxins insecticidal activity. Microb. Biotechnol. 6, 17–26.

Brown, S.J., Mahaffey, J.P., Lorenzen, M.D., Denell, R.E., Mahaffey, J.W., 1999. Using RNAi to investigate orthologous homeotic gene function during development of distantly related insects. Evol. Dev. 1, 11–15.

Burand, J.P., Hunter, W.B., 2013. RNAi: future in insect management. J. Invertebr. Pathol. 112, S68–S74.

Canavoso, L.E., Jouni, Z.E., Karnas, K.J., Pennington, J.E., Wells, M.A., 2001. Fat metabolism in insects. Annu. Rev. Nutr. 21, 23–46.

Candas, M., Loseva, O., Oppert, B., Kosaraju, P., Bulla Jr., L.A., 2003. Insect resistance to Bacillus thuringiensis: alterations in the indianmeal moth larval gut proteome. Mol. Cell. Proteomics. 2, 19–28.

Cao, J., Ibrahim, H., Garcia, J.J., Mason, H., Granados, R.R., Earle, E.D., 2002. Transgenic tobacco plants carrying a baculovirus enhancin gene slow development and increase mortality of Trichoplusia ni larvae. Plant Cell Rep. 21, 244–250.

Carroll, J., Convents, D., Van Damme, J., Boets, A., Van Rie, J., Ellar, D.J., 1997. Intramolecular proteolytic cleavage of Bacillus thuringiensis Cry3A delta-endotoxin may facilitate its coleopteran toxicity. J. Invertebr. Pathol. 70, 41–49.

Chen, J., Hua, G., Jurat-Fuentes, J.L., Abdullah, M.A., Adang, M.J., 2007. Synergism of Bacillus thuringiensis toxins by a fragment of a toxin-binding cadherin. Proc. Natl. Acad. Sci. USA. 104, 13901–13906.

Crespo, A.L., Rodrigo-Simon, A., Siqueira, H.A., Pereira, E.J., Ferre, J., Siegfried, B.D., 2011. Cross-resistance and mechanism of resistance to Cry1Ab toxin from Bacillus thuringiensis in a field-derived strain of European corn borer, Ostrinia nubilalis. J. Invertebr. Pathol. 107, 185–192.

Crespo, A.L.B., Spencer, T.A., Alves, A.P., Hellmich, R.L., Blankenship, E.E., Magalhães, L.C., et al., 2009. On-plant survival and inheritance of resistance to Cry1Ab toxin from Bacillus thuringiensis in a field-derived strain of European corn borer, Ostrinia nubilalis. Pest. Manag. Sci. 65, 1071–1081.

Crickmore, N., Zeigler, D.R., Feitelson, J., Schnepf, E., Van Rie, J., Lereclus, D., et al., 1998. Revision of the nomenclature for the Bacillus thuringiensis pesticidal crystal proteins. Microbiol. Mol. Biol. Rev. 62, 807–813.

Crickmore, N., Baum, J., Bravo, A., Lereclus, D., Narva, K., Sampson, K., et al., 2013. Bacillus thuringiensis toxin nomenclature. http://www.btnomenclature.info/

De Barjac, H., Frachon, E., 1990. Classification of Bacillus thuringiensis strains. Entomophaga. 35, 233–240.

De Leo, F., Bonade-Bottino, M.A., Ceci, L.R., Gallerani, R., Jouanin, L., 1998. Opposite effects on Spodoptera littoralis larvae of high expression level of a trypsin proteinase inhibitor in transgenic plants. Plant. Physiol. 118, 997–1004.

de Maagd, R.A., Kwa, M.S., van der Klei, H., Yamamoto, T., Schipper, B., Vlak, J.M., et al., 1996. Domain III substitution in Bacillus thuringiensis delta-endotoxin CryIA(b) results in superior toxicity for Spodoptera exigua and altered membrane protein recognition. Appl. Environ. Microbiol. 62, 1537–1543.

de Maagd, R.A., Bravo, A., Crickmore, N., 2001. How Bacillus thuringiensis has evolved specific toxins to colonize the insect world. Trends Genet. 17, 193–199.

de Maagd, R.A., Bravo, A., Berry, C., Crickmore, N., Schnepf, H.E., 2003. Structure, diversity, and evolution of protein toxins from spore-forming enteromopathogenic bacteria. Annu. Rev. Genet. 37, 409–433.

Dean, M., Annilo, T., 2005. Evolution of the ATP-binding cassette (ABC) transporter superfamily in vertebrates. Annu. Rev. Genomics Hum. Genet. 6, 123–142.

Dennehy, T.J., Carroll, M., Head, G.P., Moar, W.J., Price, P.A., Akbar, W., et al., 2011. 2010-Season update on monitoring of resistance to Bt cotton in key Lepidopteran pests in the U.S.A. Proceedings of the Beltwide Cotton Conferences, pp. 1061–1062.

Derbyshire, D.J., Ellar, D.J., Li, J., 2001. Crystallization of the *Bacillus thuringiensis* toxin Cry1Ac and its complex with the receptor ligand N-acetyl-D-galactosamine. Acta. Crystallogr. D57, 1938—1944.

Derksen, A.C., Granados, R.R., 1988. Alteration of a lepidopteran peritrophic membrane by baculoviruses and enhancement of viral infectivity. Virology. 167, 242—250.

Dhurua, S., Gujar, G.T., 2011. Field-evolved resistance to Bt toxin Cry1Ac in the pink bollworm, Pectinophora gossypiella (Saunders) (Lepidoptera: Gelechiidae), from India. Pest. Manag. Sci. 67, 898—903.

Dos Santos, K.B., Neves, P., Meneguim, A.M., Dos Santos, R.B., Dos Santos, W.J., Boas, G.V., et al., 2009. Selection and characterization of the *Bacillus thuringiensis* strains toxic to *Spodoptera eridania* (Cramer), *Spodoptera cosmioides* (Walker) and *Spodoptera frugiperda* (Smith) (Lepidoptera: Noctuidae). Biol. Control. 50, 157—163.

Dow, J.A., Harvey, W.R., 1988. Role of midgut electrogenic K + pump potential difference in regulating lumen K + and pH in larval lepidoptera. J. Exp. Biol. 140, 455—463.

Dunse, K.M., Kaas, Q., Guarino, R.F., Barton, P.A., Craik, D.J., Anderson, M.A., 2010a. Molecular basis for the resistance of an insect chymotrypsin to a potato type II proteinase inhibitor. Proc. Natl. Acad. Sci. USA. 107, 15016—15021.

Dunse, K.M., Stevens, J.A., Lay, F.T., Gaspar, Y.M., Heath, R.L., Anderson, M.A., 2010b. Coexpression of potato type I and II proteinase inhibitors gives cotton plants protection against insect damage in the field. Proc. Natl. Acad. Sci. USA. 107, 15011—15015.

Dykxhoorn, D.M., Novina, C.D., Sharp, P.A., 2003. Killing the messenger: short RNAs that silence gene expression. Nat. Rev. Mol. Cell. Biol. 4, 457—467.

Elbashir, S.M., Lendeckel, W., Tuschl, T., 2001. RNA interference is mediated by 21- and 22-nucleotide RNAs. Genes. Dev. 15, 188—200.

Estada, U., Ferre, J., 1994. Binding of insecticidal crystal proteins of *Bacillus thuringiensis* to the Midgut brush border of the Cabbage Looper, *Trichoplusia ni* (Hübner) (Lepidoptera: Noctuidae), and selection for resistance to one of the crystal proteins. Appl. Environ. Microbiol. 60, 3840—3846.

Fernandez, L.E., Aimanova, K.G., Gill, S.S., Bravo, A., Soberon, M., 2006. A GPI-anchored alkaline phosphatase is a functional midgut receptor of Cry11Aa toxin in *Aedes aegypti* larvae. Biochem. J. 394, 77—84.

Ferré, J., Van Rie, J., 2002. Biochemistry and genetics of insect resistance to *Bacillus thuringiensis*. Annu. Rev. Entomol. 47, 501—533.

ffrench-Constant, R., Waterfield, N., 2006. An ABC guide to the bacterial toxin complexes. Adv. Appl. Microbiol. 58, 169—183.

ffrench-Constant, R., Waterfield, N., Daborn, P., Joyce, S., Bennett, H., Au, C., et al., 2003. *Photorhabdus*: towards a functional genomic analysis of a symbiont and pathogen. FEMS Microbiol. Rev. 26, 433—456.

ffrench-Constant, R.H., Dowling, A., Waterfield, N.R., 2007. Insecticidal toxins from *Photorhabdus* bacteria and their potential use in agriculture. Toxicon. 49, 436—451.

Flannagan, R.D., Yu, C.-G., Mathis, J.P., Meyer, T.E., Shi, X., Siqueira, H.A.A., et al., 2005. Identification, cloning and expression of a Cry1Ab cadherin receptor from European corn borer, *Ostrinia nubilalis* (Hübner) (Lepidoptera: Crambidae). Insect Biochem. Mol. Biol. 35, 33—40.

Forcada, C., Alcácer, E., Garcerá, M.D., Martínez, R., 1996. Differences in the midgut proteolytic activity of two *Heliothis virescens* strains, one susceptible and one resistant to Bacillus thuringiensis toxins. Arch. Insect Biochem. Physiol. 31, 257—272.

Forst, S., Dowds, B., Boemare, N., Stackebrandt, E., 1997. *Xenorhabdus* and *Photorhabdus* spp.: bugs that kill bugs. Annu. Rev. Microbiol. 51, 47—72.

Frizzi, A., Huang, S., 2010. Tapping RNA silencing pathways for plant biotechnology. Plant Biotechnol. J. 8, 655—677.

Gahan, L.J., Gould, F., Heckel, D.G., 2001. Identification of a gene associated with Bt resistance in *Heliothis virescens*. Science. 293, 857—860.

Gahan, L.J., Pauchet, Y., Vogel, H., Heckel, D.G., 2010. An ABC transporter mutation is correlated with insect resistance to *Bacillus thuringiensis* Cry1Ac toxin. PLoS Genet. 6, e1001248.

Galitsky, N., Cody, V., Wojtczak, A., Ghosh, D., Luft, J.R., Pangborn, W., et al., 2001. Structure of the insecticidal bacterial δ-endotoxin Cry3Bb1 of *Bacillus thuringiensis*. Acta Crystallogr. D57, 1101—1109.

Garczynski, S.F., Crim, J.W., Adang, M.J., 1991. Identification of putative insect brush border membrane-binding molecules specific to *Bacillus thuringiensis* δ-endotoxin by protein blot analysis. Appl. Environ. Microbiol. 57, 2816—2820.

Gassmann, A.J., Onstad, D.W., Pittendrigh, B.R., 2009. Evolutionary analysis of herbivorous insects in natural and agricultural environments. Pest. Manag. Sci. 65, 1174—1181.

Gassmann, A.J., Petzold-Maxwell, J.L., Keweshan, R.S., Dunbar, M.W., 2011. Field-evolved resistance to Bt maize by western corn rootworm. PLoS One. 6, e22629.

Gill, M., Ellar, D., 2002. Transgenic *Drosophila* reveals a functional in vivo receptor for the *Bacillus thuringiensis* toxin Cry1Ac1. Insect Mol. Biol. 11, 619—625.

Gill, S.S., Cowles, E.A., Pietrantonio, P.V., 1992. The mode of action of *Bacillus thuringiensis* endotoxins. Annu. Rev. Entomol. 37, 615—636.

Giordana, B., Sacchi, V.F., Hanozet, G.M., 1982. Intestinal amino acid adsorption in lepidoperan larvae. Biochim. Biophys. Acta. 692, 81—88.

Glare, T.R., O'Callaghan, M., 2000. *Bacillus thuringiensis*: Biology, Ecology and Safety. Wiley, West Sussex, U.K.

Gomez, I., Sanchez, J., Miranda, R., Bravo, A., Soberon, M., 2002. Cadherin-like receptor binding facilitates proteolytic cleavage of helix alpha-1 in domain I and oligomer pre-pore formation of *Bacillus thuringiensis* Cry1Ab toxin. FEBS Lett. 513, 242—246.

González-Cabrera, J., Escriche, B., Tabashnik, B.E., Ferré, J., 2003. Binding of *Bacillus thuringiensis* toxins in resistant and susceptible strains of pink bollworm (*Pectinophora gossypiella*). Insect Biochem. Mol. Biol. 33, 929—935.

Granados, R.R., Fu, Y., Corsaro, B., Hughes, P.R., 2001. Enhancement of *Bacillus thuringiensis* toxicity to lepidopterous species with the enhancin from *Trichoplusia ni* granulovirus. Biol. Control. 20, 153—159.

Griffitts, J.S., Aroian, R.V., 2005. Many roads to resistance: how invertebrates adapt to Bt toxins. Bioessays. 27, 614—624.

Grochulski, P., Masson, L., Borisova, S., Pusztai-Carey, M., Schwartz, J.-L., Brousseau, R., et al., 1995. *Bacillus thuringiensis* Cry1A(a) insecticidal toxin: crystal structure and channel formation. J. Mol. Biol. 254, 447—464.

Gunning, R.V., Dang, H.T., Kemp, F.C., Nicholson, I.C., Moores, G.D., 2005. New resistance mechanism in *Helicoverpa armigera* threatens transgenic crops expressing *Bacillus thuringiensis* Cry1Ac toxin. Appl. Environ. Microbiol. 71, 2558—2563.

Guo, S., Ye, S., Liu, Y., Wei, L., Xue, J., Wu, H., et al., 2009. Crystal structure of *Bacillus thuringiensis* Cry8Ea1: an insecticidal toxin toxic to underground pests, the larvae of *Holotrichia parallela*. J. Struct. Biol. 168, 259—266.

Habib, H., Fazili, K.M., 2007. Plant protease inhibitors: a defense strategy in plants. Biotech. Mol. Biol. Rev. 2, 68—85.

Han, S., Craig, J.A., Putnam, C.D., Carozzi, N.B., Tainer, J.A., 1999. Evolution and mechanism from structures of an ADP-ribosylating toxin and NAD complex. Nat. Struct. Biol. 6, 932—936.

Hannon, G.J., 2002. RNA interference. Nature. 418, 244—251.

Hannon, G.J., 2003. RNA interference (RNAi) and MicroRNAs. Nat. Encycl. Hum. Genome.100—108.

Heath, R.L., Barton, P.A., Simpson, R.J., Reid, G.E., Lim, G., Anderson, M.A., 1995. Characterization of the protease processing sites in a multidomain proteinase inhibitor precursor from *Nicotiana alata*. Eur. J. Biochem. 230, 250–257.

Heimpel, A.M., Angus, T.A., 1959. The site of action of crystalliferous bacteria in Lepidopteran larvae. J. Insect Path. 1, 152–170.

Hernandez-Martinez, P., Navarro-Cerrillo, G., Caccia, S., de Maagd, R.A., Moar, W.J., Ferre, J., et al., 2010. Constitutive activation of the midgut response to *Bacillus thuringiensis* in Bt-resistant *Spodoptera exigua*. PLoS One. 5.

Herrero, S., Gechev, T., Bakker, P.L., Moar, W.J., de Maagd, R.A., 2005. *Bacillus thuringiensis* Cry1Ca-resistant *Spodoptera exigua* lacks expression of one of four Aminopeptidase N genes. BMC Genomics. 6, 96–106.

Hilder, V.A., Gatehouse, A.M.R., Sheerman, S.E., Barker, R.F., Boulter, D., 1987. A novel mechanism of insect resistance engineered into tobacco. Nature. 330, 160–163.

Höfte, H., Whiteley, H.R., 1989. Insecticidal crystal proteins of *Bacillus thuringiensis*. Microbiol. Rev. 53, 242–255.

Huang, F., Andow, D.A., Buschman, L.L., 2011. Success of the high-dose/refuge resistance management strategy after 15 years of Bt crop use in North America. Entomol. Exp. Appl. 140, 1–16.

Hulpiau, P., van Roy, F., 2009. Molecular evolution of the cadherin superfamily. Int. J. Biochem. Cell Biol. 41, 349–369.

Huvenne, H., Smagghe, G., 2010. Mechanisms of dsRNA uptake in insects and potential of RNAi for pest control: a review. J. Insect Physiol. 56, 227–235.

ILSI, 2011. Problem Formulation for the Environmental Risk Assessment of RNAi Plants, Washington, DC, June 1–3, 2011. Center for Environmental Risk Assessment, International Life Sciences Institute Research Foundation, Washington, DC, pp. 1–48.

Ibrahim, M.A., Griko, N., Junker, M., Bulla, L.A., 2010. Bacillus thuringiensis: a genomics and proteomics perspective. Bioeng. Bugs. 1, 31–50.

James, C., 2009. Global status of commercialized biotech/GM crops: 2009. ISAAA Brief No. 41. ISAAA: Ithaca, NY.

Janzen, D.H., Juster, H.B., Liener, I.E., 1976. Insecticidal action of the phytohemagglutinin in black beans on a bruchid beetle. Science. 192, 795–796.

Jurat-Fuentes, J.L., Gahan, L.J., Gould, F.L., Heckel, D.G., Adang, M.J., 2004. The HevCaLP protein mediates binding specificity of the Cry1A class of *Bacillus thuringiensis* toxins in *Heliothis virescens*. Biochemistry. 43, 14299–14305.

Jurat-Fuentes, J.L., Karumbaiah, L., Jakka, S.R., Ning, C., Liu, C., Wu, K., et al., 2011. Reduced levels of membrane-bound alkaline phosphatase are common to lepidopteran strains resistant to Cry toxins from Bacillus thuringiensis. PLoS One. 6, e17606.

Keeton, T.P., Bulla Jr., L.A., 1997. Ligand specificity and affinity of BT-R1, the *Bacillus thuringiensis* Cry1A toxin receptor from *Manduca sexta*, expressed in mammalian and insect cell cultures. Appl. Environ. Microbiol. 63, 3419–3425.

Keller, M., Sneh, B., Strizhov, N., Prudovsky, E., Regev, A., Koncz, C., et al., 1996. Digestion of δ-endotoxin by gut proteases may explain reduced sensitivity of advance Instar larvae of *Spodoptera littoralis* to CryIC. Insect Biochem. Mol. Biol. 26, 365–373.

Kennerdell, J.R., Carthew, R.W., 1998. Use of dsRNA-mediated genetic interference to demonstrate that frizzled and frizzled 2 act in the wingless pathway. Cell. 95, 1017–1026.

Khajuria, C., Buschman, L.L., Chen, M.S., Siegfried, B.D., Zhu, K.Y., 2011. Identification of a novel aminopeptidase P-like gene (OnAPP) possibly involved in Bt toxicity and resistance in a major corn pest (*Ostrinia nubilalis*). PLoS One. 6, e23983.

Knowles, B.H., 1994. Mechanism of action of *Bacillus thuringiensis* insecticidal δ-endotoxins. Adv. Insect Physiol. 24, 275–308.

Kumar, P., Pandit, S.S., Baldwin, I.T., 2012. Tobacco rattle virus vector: a rapid and transient means of silencing manduca sexta genes by plant mediated RNA interference. PLoS One. 7, e31347.

Kurtz, R.W., 2010. A review of Vip3A mode of action and effects on Bt Cry protein-resistant colonies of Lepidopteran larvae. Southwest Entomol. 35, 391–394.

Labbe, R., Caveney, S., Donly, C., 2011. Genetic analysis of the xenobiotic resistance-associated ABC gene subfamilies of the Lepidoptera. Insect Mol. Biol. 20, 243–256.

Lang, A.E., Schmidt, G., Schlosser, A., Hey, T.D., Larrinua, I.M., Sheets, J.J., et al., 2010. Photorhabdus luminescens toxins ADP-ribosylate actin and RhoA to force actin clustering. Science. 327, 1139–1142.

Lawrence, P.K., Koundal, K.R., 2002. Plant protease inhibitors in control of phytophagous insects. Elect. J. Biotech. 5, 93–109.

Lecadet, M.M., Frachon, E., Dumanoir, V.C., Ripouteau, H., Hamon, S., Laurent, P., et al., 1999. Updating the H-antigen classification of Bacillus thuringiensis. J. Appl. Microbiol. 86, 660–672.

Lee, M.K., Rajamohan, F., Gould, F., Dean, D.H., 1995. Resistance to *Bacillus thuringiensis* CryIA δ-endotoxins in a laboratory-selected *Heliothis virescens* strain is related to receptor alteration. Appl. Environ. Microbiol. 61, 3836–3842.

Lee, M.K., Walters, F.S., Hart, H., Palekar, N., Chen, J.-S., 2003. The mode of action of the *Bacillus thuringiensis* vegetative insecticidal protein Vip3A differs from that of Cry1Ab δ-endotoxin. Appl. Environ. Microbiol. 69, 4648–4657.

Lee, M.K., Miles, P., Chen, J.S., 2006. Brush border membrane binding properties of *Bacillus thuringiensis* Vip3A toxin to *Heliothis virescens* and *Helicoverpa zea* midguts. Biochem. Biophys. Res. Commun. 339, 1043–1047.

Lee, S.C., Stoilova-McPhie, S., Baxter, L., Fulop, V., Henderson, J., Rodger, A., et al., 2007. Structural characterisation of the insecticidal toxin XptA1, reveals a 1.15 MDa tetramer with a cage-like structure. J. Mol. Biol. 366, 1558–1568.

Lelandais-Briere, C., Sorin, C., Declerck, M., Benslimane, A., Crespi, M., Hartmann, C., 2010. Small RNA diversity in plants and its impact in development. Curr. Genomics. 11, 14–23.

Li, H., Oppert, B., Higgins, R.A., Huang, F., Zhu, K.Y., Buschman, L.L., 2004. Comparative analysis of proteinase activities of *Bacillus thuringiensis*-resistant an d-susceptible *Ostrinia nubilalis* (Lepidoptera: Crambidae). Insect Bioch. Mol. Biol. 34, 753–762.

Li, H., Oppert, B., Higgins, R.A., Huang, F., Buschman, L.L., Zhu, K.Y., 2005. Susceptibility of dipel-resistant and -susceptible *Ostrinia nubilalis* (Lepidoptera: Crambidae) to individual *Bacillus thuringiensis* protoxins. J. Econ. Entomol. 98, 1333–1340.

Li, J., Carroll, J., Ellar, D.J., 1991. Crystal structure of insecticidal δ-endotoxin from *Bacillus thuringiensis* at 2.5 Å resolution. Nature. 353, 815–821.

Li, J., Derbyshire, D.J., Promdonkoy, B., Ellar, D.J., 2001. Structural implications for the transformation of the *Bacillus thuringiensis* δ-endotoxins from water-soluble to membrane-inserted forms. Bioch. Soc. Trans. 29, 571–577.

Liang, Y., Patel, S.S., Dean, D.H., 1995. Irreversible binding kinetics of Bacillus thuringiensis CryIA delta-endotoxins to gypsy moth brush border membrane vesicles is directly correlated to toxicity. J. Biol. Chem. 270, 24719–24724.

Liu, D., Burton, S., Glancy, T., Li, Z.S., Hampton, R., Meade, T., et al., 2003. Insect resistance conferred by 283-kDa *Photorhabdus luminescens* protein TcdA in *Arabidopsis thaliana*. Nat. Biotechnol. 21, 1222–1228.

Liu, X.S., Dean, D.H., 2006. Redesigning *Bacillus thuringiensis* Cry1Aa toxin into a mosquito toxin. Protein Eng. Des. Sel. 19, 107–111.

Loseva, O., Ibrahim, M., Candas, M., Koller, C.N., Bauer, L.S., Bulla Jr., L.A., 2002. Changes in protease activity and Cry3Aa toxin binding in the Colorado potato beetle:

implications for insect resistance to *Bacillus thuringiensis* toxins. Insect Biochem. Mol. Biol. 32, 567–577.

Ma, G., Roberts, H., Sarjan, M., Featherstone, N., Lahnstein, J., Akhurst, R., et al., 2005. Is the mature endotoxin Cry1Ac from *Bacillus thuringiensis* inactivated by a coagulation reaction in the gut lumen of resistant *Helicoverpa armigera* larvae?. Insect Biochem. Mol. Biol. 35, 729–739.

Macedo, M.R.L., Freire, M.G.M., 2011. Insect digestive enzymes as targets for pest control. Invert. Surviv. J. 8, 190–198.

Mao, Y.B., Cai, W.J., Wang, J.W., Hong, G.J., Tao, X.Y., Wang, L.J., et al., 2007. Silencing a cotton bollworm P450 monooxygenase gene by plant-mediated RNAi impairs larval tolerance of gossypol. Nat. Biotechnol. 25, 1307–1313.

Martins, E.S., Monnerat, R.G., Queiroz, P.R., Dumas, V.F., Braz, S.V., de Souza Aguiar, R.W., et al., 2010. Midgut GPI-anchored proteins with alkaline phosphatase activity from the cotton boll weevil (*Anthonomus grandis*) are putative receptors for the Cry1B protein of *Bacillus thuringiensis*. Insect Biochem. Mol. Biol. 40, 138–145.

Masson, L., Schwab, G., Mazza, A., Brousseau, R., Potvin, L., Schwartz, J.L., 2004. A novel *Bacillus thuringiensis* (PS149B1) containing a Cry34Ab1/Cry35Ab1 binary toxin specific for the western corn rootworm *Diabrotica virgifera virgifera* LeConte forms ion channels in lipid membranes. Biochemistry. 43, 12349–12357.

Matten, S.R., Head, G.P., Quemada, H.D., 2008. How governmental regulation can help or hinder the integration of Bt crops within IPM programs. In: Romeis, J., Shelton, A.M., Kennedy, G.G. (Eds.), Integration of Insect-Resistant Genetically Modified Crops within IPM Programs, vol. 5. Springer, New York, pp. 27–39.

McNall, R.J., Adang, M.J., 2003. Identification of novel *Bacillus thuringiensis* Cry1Ac binding proteins in *Manduca sexta* midgut through proteomic analysis. Insect Biochem. Mol. Biol. 33, 999–1010.

Milner, R.J., 1994. History of Bacillus thuringiensis. Agric. Ecosyt. Environ. 49, 9–13.

Moar, W.J., Pusztai-Carey, M., Van Faassen, H., Bosch, D., Frutos, R., Rang, C., et al., 1995. Development of *Bacillus thuringiensis* Cry1C Resistance by *Spodoptera exigua* (Hubner) (Lepidoptera: Noctuidae). Appl. Environ. Microbiol. 61, 2086–2092.

Moffett, D., Cummings, S., 1994. Transepithelial potential and alkalization in an in situ preparation of tobacco hornworm (*Manduca Sexta*) midgut. J. Exp. Biol. 194, 341–345.

Mohan, S., Ma, P.W., Williams, W.P., Luthe, D.S., 2008. A naturally occurring plant cysteine protease possesses remarkable toxicity against insect pests and synergizes *Bacillus thuringiensis* toxin. PLoS One. 3, e1786.

Morin, S., Biggs, R.W., Sisterson, M.S., Shriver, L., Ellers-Kirk, C., Higginson, D., et al., 2003. Three cadherin alleles associated with resistance to *Bacillus thuringiensis* in pink bollworm. Proc. Natl. Acad. Sci. USA. 100, 5004–5009.

Morse, R.J., Yamamoto, T., Stroud, R.M., 2001. Structure of Cry2Aa suggests an unexpected receptor binding epitope. Structure. 9, 409–417.

Mosolov, V.V., Valueva, T.A., 2008. Proteinase inhibitors in plant biotechnology: a review. Appl. Biochem. Microbiol. 44, 261–269.

Murdock, L.L., Brookhart, G., Dunn, P.E., Foard, D.E., Kelley, S., Kitch, L., et al., 1987. Cysteine digestive proteinases in Coleoptera. Comp. Biochem. Physiol. 87B.

Naimov, S., Weemen-Hendriks, M., Dukiandjiev, S., de Maagd, R.A., 2001. *Bacillus thuringiensis* delta-endotoxin Cry1 hybrid proteins with increased activity against the Colorado potato beetle. Appl. Environ. Microbiol. 67, 5328–5330.

Ning, C., Wu, K., Liu, C., Gao, Y., Jurat-Fuentes, J.L., Gao, X., 2010. Characterization of a Cry1Ac toxin-binding alkaline phosphatase in the midgut from *Helicoverpa armigera* (Hubner) larvae. J. Insect Physiol. 56, 666–672.

Nykanen, A., Haley, B., Zamore, P.D., 2001. ATP requirements and small interfering RNA structure in the RNA interference pathway. Cell. 107, 309–321.

Oppert, B., Kramer, K.J., Beeman, R.W., Johnson, D., McGaughey, W.H., 1997. Proteinase-mediated insect resistance to *Bacillus thuringiensis* toxins. J. Biol. Chem. 272, 23473–23476.

Pacheco, S., Gomez, I., Arenas, I., Saab-Rincon, G., Rodriguez-Almazan, C., Gill, S.S., et al., 2009. Domain II loop 3 of *Bacillus thuringiensis* Cry1Ab toxin is involved in a "ping pong" binding mechanism with *Manduca sexta* aminopeptidase-N and cadherin receptors. J. Biol. Chem. 284, 32750–32757.

Pandian, G.N., Ishikawa, T., Togashi, M., Shitomi, Y., Haginoya, K., Yamamoto, S., et al., 2008. *Bombyx mori* midgut membrane protein P252, which binds to *Bacillus thuringiensis* Cry1A, is a chlorophyllide-binding protein, and the resulting complex has antimicrobial activity. Appl. Environ. Microbiol. 74, 1324–1331.

Pardo-Lopez, L., Gomez, I., Munoz-Garay, C., Jimenez-Juarez, N., Soberon, M., Bravo, A., 2006. Structural and functional analysis of the pre-pore and membrane-inserted pore of Cry1Ab toxin. J. Invertebr. Pathol. 92, 172–177.

Park, Y., Abdullah, M.A., Taylor, M.D., Rahman, K., Adang, M.J., 2009. Enhancement of *Bacillus thuringiensis* Cry3Aa and Cry3Bb toxicities to coleopteran larvae by a toxin-binding fragment of an insect cadherin. Appl. Environ. Microbiol. 75, 3086–3092.

Pechan, T., Ye, L., Chang, Y.-M., Mitra, A., Lin, L., Davis, F.M., et al., 2000. A unique 33 kD cysteine proteinase accumulates in response to larval feeding in corn (*Zea mays* L) genotypes resistant to fall armyworm and other Lepidoptera. Plant Cell. 12, 1031–1040.

Pereira, E.J.G., Storer, N.P., Siegfried, B.D., 2008. Inheritance of Cry1F resistance in laboratory-selected European corn borer and its survival on transgenic corn expressing the Cry1F toxin. Bull. Entomol. Res. 98, 621–629.

Pigott, C.R., Ellar, D.J., 2007. Role of receptors in *Bacillus thuringiensis* crystal toxin activity. Microbiol. Mol. Biol. Rev. 71, 255–281.

Pigott, C.R., King, M.S., Ellar, D.J., 2008. Investigating the properties of *Bacillus thuringiensis* Cry proteins with novel loop replacements created using combinatorial molecular biology. Appl. Environ. Microbiol. 74, 3497–3511.

Pouvreau, L., Gruppen, H., Piersma, S.R., van den Broek, L.A., van Koningsveld, G.A., Voragen, A.G., 2001. Relative abundance and inhibitory distribution of protease inhibitors in potato juice from cv. Elkana. J. Agric. Food Chem. 49, 2864–2874.

Price, D.R., Gatehouse, J.A., 2008. RNAi-mediated crop protection against insects. Trends Biotechnol. 26, 393–400.

Rahbe, Y., Deraison, C., Bonade-Bottino, M., Girard, C., Nardon, C., Jouanin, L., 2003. Effects of the cysteine protease inhibitor oryzacystatin on different expression (OC-I) on different aphids and reduced performance of *Myzus persicae* on OC-I expressing transgenic oilseed rape. Plant Sci. 164, 441–450.

Rahman, K., Abdullah, M.A., Ambati, S., Taylor, M.D., Adang, M.J., 2012. Differential protection of Cry1Fa toxin against *Spodoptera frugiperda* larval gut proteases by cadherin orthologs correlates with increased synergism. Appl. Environ. Microbiol. 78, 354–362.

Rahman, M.M., Roberts, H.L., Sarjan, M., Asgari, S., Schmidt, O., 2004. Induction and transmission of Bacillus thuringiensis tolerance in the flour moth Ephestia kuehniella. Proc. Natl. Acad. Sci. USA. 101, 2696–2699.

Rajagopal, R., Sivakumar, S., Agrawal, N., Malhotra, P., Bhatnagar, R.K., 2002. Silencing of midgut aminopeptidase N of *Spodoptera litura* by double-stranded RNA establishes its role as *Bacillus thuringiensis* toxin receptor. J. Biol. Chem. 277, 46849–46851.

Rajagopal, R., Arora, N., Sivakumar, S., Rao, N.G., Nimbalkar, S.A., Bhatnagar, R.K., 2009. Resistance of *Helicoverpa armigera* to Cry1Ac toxin from *Bacillus thuringiensis* is due to improper processing of the protoxin. Biochem. J. 419, 309—316.

Rodriguez-Almazan, C., Reyes, E.Z., Zuniga-Navarrete, F., Munoz-Garay, C., Gomez, I., Evans, A.M., et al., 2012. Cadherin binding is not a limiting step for *Bacillus thuringiensis* subsp. israelensis Cry4Ba toxicity to *Aedes aegypti* larvae. Biochem. J. 443, 711—717.

Ryan, C.A., 1990. Protease inhibitors in plants: genes for improving defenses against insects and pathogens. Ann. Rev. Phytopathol. 28, 425—449.

Sadeghi, A., Smagghe, G., Jurado-Jacome, E., Peumans, W.J., Van Damme, E.J.M., 2009. Laboratory study of the effects of leek lectin (APA) in transgenic tobacco plants on the development of cotton leafworm *Spodoptera littoralis* (Lepidoptera: Noctuidae). Eur. J. Entomol. 106, 21—28.

Sane, V.A., Nath, P., Aminuddin, Sane, P.V., 1997. Development of insect-resistant transgenic plants using plant genes: expression of cowpea trypsin inhibitor in transgenic tobacco plants. Curr. Sci. 72, 741—747.

Saraswathy, N., Kumar, P.A., 2004. Protein enginesering of δ-endotoxins of *Bacillus thuringiensis*. Electron. J. Biotechnol. 7, 180—190.

Sayyed, A.H., Gatsi, R., Ibiza-Palacios, S., Escriche, B., Wright, D.J., Crickmore, N., 2005. Common, but complex, mode of resistance of *Plutella xylostella* to *Bacillus thuringiensis* toxins Cry1Ab and Cry1Ac. Appl. Environ. Microbiol. 71, 6863—6869.

Schnepf, E., Crickmore, N., Van Rie, J., Lereclus, D., Baum, J., Feitelson, J., et al., 1998. *Bacillus thuringiensis* and its pesticidal crystal proteins. Microbiol. Mol. Biol. Rev. 62, 775—806.

Setamou, M., Bernal, J.S., Legaspi, J.C., Mirkov, T.E., Legaspi Jr., B.C., 2002. Evaluation of lectin-expressing transgenic sugarcane against stalkborers (Lepidoptera: Pyralidae): effects on life history parameters. J. Econ. Entomol. 95, 469—477.

Sharma, H.C., Sharma, K.K., Seetharama, N., Ortiz, R., 2000. Prospects for transgenic resistance to insects. Electron. J. Biotechnol. 3, 173—179.

Sharon, N., Lis, H., 2004. History of lectins: from hemagglutinins to biological recognition molecules. Glycobiology. 14, 53R—62R.

Sheets, J.J., Hey, T.D., Fencil, K.J., Burton, S.L., Ni, W., Lang, A.E., et al., 2011. Insecticidal toxin complex proteins from *Xenorhabdus nematophilus*: structure and pore formation. J. Biol. Chem. 286, 22742—22749.

Soberon, M., Pardo-Lopez, L., Lopez, I., Gomez, I., Tabashnik, B.E., Bravo, A., 2007. Engineering modified Bt toxins to counter insect resistance. Science. 318, 1640—1642.

Soberon, M., Gill, S.S., Bravo, A., 2009. Signaling versus punching hole: how do *Bacillus thuringiensis* toxins kill insect midgut cells? Cell. Mol. Life Sci. 66, 1337—1349.

Soberon, M., Pardo, L., Munoz-Garay, C., Sanchez, J., Gomez, I., Porta, H., et al., 2010. Pore formation by cry toxins. Adv. Exp. Med. Biol. 677, 127—142.

Srinivasan, A., Giri, A.P., Gupta, V.S., 2006. Structural and functional diversities in lepidopteran serine proteases. Cell. Mol. Biol. Lett. 11, 132—154.

Storer, N.P., Babcock, J.M., Schlenz, M., Meade, T., Thompson, G.D., Bing, J.W., et al., 2010. Discovery and characterization of field resistance to Bt maize: *Spodoptera frugiperda* (Lepidoptera: Noctuidae) in Puerto Rico. J. Econ. Entomol. 103, 1031—1038.

Tabashnik, B.E., Carrière, Y., 2010. Field-evolved resistance to Bt cotton: bollworm in the U.S. and pink bollworm in India. Southwest Entomol. 35, 417—424.

Tabashnik, B.E., Finson, N., Groeters, F.R., Moar, W.J., Johnson, M.W., Luo, K., et al., 1994. Reversal of resistance to *Bacillus thuringiensis* in *Plutella xylostella*. Proc. Natl. Acad. Sci. USA. 91, 4120—4124.

Tabashnik, B.E., Liu, Y.-B., Finson, N., Masson, L., Heckel, D.G., 1997a. One gene in diamondback moth confers resistance to four *Bacillus thuringiensis* toxins. PNAS. 94, 1640−1644.

Tabashnik, B.E., Liu, Y.B., Malvar, T., Heckel, D.G., Masson, L., Ballester, V., et al., 1997b. Global variation in the genetic and biochemical basis of diamondback moth resistance to *Bacillus thuringiensis*. Proc. Natl. Acad. Sci. USA. 94, 12780−12785.

Tabashnik, B.E., Liu, Y.B., Malvar, T., Heckel, D.G., Masson, L., Ferre, J., 1998. Insect resistance to *Bacillus thuringiensis*: uniform or diverse?. Phil. Trans. Royal Soc. London-Series B: Biol. Sci. 353, 1751−1756.

Tabashnik, B.E., Johnson, K.W., Engleman, J.T., Baum, J.A., 2000. Cross-resistance to Bacillus thuringiensis toxin Cry1Ja in a strain of diamondback moth adapted to artificial diet. J. Invertebr. Pathol. 76, 81−83.

Tabashnik, B.E., Huang, F., Ghimire, M.N., Leonard, B.R., Siegfried, B.D., Rangasamy, M., et al., 2011. Efficacy of genetically modified Bt toxins against insects with different genetic mechanisms of resistance. Nat. Biotechnol. 29, 1128−1131.

Tailor, R., Tippett, J., Gibb, G., Pells, S., Pike, D., Jordan, L., et al., 1992. Identification and characterization of a novel *Bacillus thuringiensis* delta-endotoxin entomocidal to coleopteran and lepidopteran larvae. Mol. Microbiol. 6, 1211−1217.

Terra, W.R., Ferreira, C., 1994. Insect digestive enzymes: properties, compartmentalization and function. Comp. Biochem. Physiol. 109B, 1−62.

Terra, W.R., Ferreira, C., 2012. Biochemistry and molecular biology of digestion. In: Gilbert, L.I. (Ed.), Insect Biochemistry and Molecular Biology. Academic Press, London, pp. 365−418.

Tetreau, G., Bayyareddy, K., Jones, C.M., Stalinski, R., Riaz, M.A., Paris, M., et al., 2012. Larval midgut modifications associated with Bti resistance in the yellow fever mosquito using proteomic and transcriptomic approaches. BMC Genomics. 13, 248.

Tiewsiri, K., Wang, P., 2011. Differential alteration of two aminopeptidases N associated with resistance to Bacillus thuringiensis toxin Cry1Ac in cabbage looper. Proc. Natl. Acad. Sci. USA. 108, 14037−14042.

Upadhyay, S.K., Singh, P.K., 2011. Role of alkaline phosphatase in insecticidal action of Cry1Ac against *Helicoverpa armigera* larvae. Biotechnol. Lett. 33, 2027−2036.

Vachon, V., Laprade, R., Schwartz, J.L., 2012. Current models of the mode of action of *Bacillus thuringiensis* insecticidal crystal proteins: a critical review. J. Invertebr. Pathol.

Vadlamudi, R.K., Weber, E., Ji, I., Ji, T.H., Bulla Jr., L.A., 1995. Cloning and expression of a receptor for an insecticidal toxin of *Bacillus thuringiensis*. J. Biol. Chem. 270, 5490−5494.

Valaitis, A.P., 2008. *Bacillus thuringiensis* pore-forming toxins trigger massive shedding of GPI-anchored aminopeptidase N from gypsy moth midgut epithelial cells. Insect Biochem. Mol. Biol. 38, 611−618.

Valaitis, A.P., Mazza, A., Brousseau, R., Masson, L., 1997. Interaction analyses of *Bacillus thuringiensis* Cry1A toxins with two aminopeptidases from gypsy moth midgut brush border membranes. Insect Biochem. Mol. Biol. 27, 529−539.

Valaitis, A.P., Jenkins, J.L., Lee, M.K., Dean, D.H., Garner, K.J., 2001. Isolation and partial characterization of gypsy moth BTR-270, an anionic brush border membrane glycoconjugate that binds *Bacillus thuringiensis* Cry1A toxins with high affinity. Arch. Insect Biochem. Physiol. 46, 186−200.

Van Damme, E.J.M., Peumans, W.J., Barre, A., Rouge, P., 1998. Plant lectins: a composite of several distinct families of structurally and evolutionary related proteins with diverse biological roles. Crit. Rev. Plant Sci. 17, 575−692.

van Frankenhuyzen, K., 2009. Insecticidal activity of *Bacillus thuringiensis* crystal proteins. J. Invertebr. Pathol. 101, 1−16.

Van Rensburg, J.B.J., 2007. First reports of field resistance by the stem borer, *Busseola fusca* (Fuller) to Bt-transgenic maize. S. Afr. J. Plant Soil. 24, 147−151.

Vandenborre, G., Smagghe, G., Van Damme, E.J., 2011. Plant lectins as defense proteins against phytophagous insects. Phytochemistry. 72, 1538−1550.

Vaughn, T., Cavato, T., Brar, G., Coombe, T., DeGooyer, T., Ford, S., et al., 2005. A method of controlling corn rootworm feeding using a *Bacillus thuringiensis* protein expressed in transgenic maize. Crop. Sci. 45, 931−938.

Vazquez, F., Legrand, S., Windels, D., 2010. The biosynthetic pathways and biological scopes of plant small RNAs. Trends Plant. Sci. 15, 337−345.

Walters, F.S., Stacy, C.M., Lee, M.K., Palekar, N., Chen, J.S., 2008. An engineered chymotrypsin/cathepsin G site in domain I renders *Bacillus thuringiensis* Cry3A active against Western corn rootworm larvae. Appl. Environ. Microbiol. 74, 367−374.

Walters, F.S., deFontes, C.M., Hart, H., Warren, G.W., Chen, J.S., 2010. Lepidopteran-active variable-region sequence imparts coleopteran activity in eCry3.1Ab, an engineered *Bacillus thuringiensis* hybrid insecticidal protein. Appl. Environ. Microbiol. 76, 3082−3088.

Wang, P., Zhao, J.Z., Rodrigo-Simon, A., Kain, W., Janmaat, A.F., Shelton, A.M., et al., 2007. Mechanism of resistance to *Bacillus thuringiensis* toxin Cry1Ac in a greenhouse population of the cabbage looper, *Trichoplusia ni*. Appl. Environ. Microbiol. 73, 1199−1207.

Wang, Z., Zhang, K., Sun, X., Tang, K., Zhang, J., 2005. Enhancement of resistance to aphids by introducing the snowdrop lectin gene gna into maize plants. J. Biosci. 30, 627−638.

Waterfield, N., Dowling, A., Sharma, S., Daborn, P.J., Potter, U., ffrench-Constant, R. H., 2001. Oral toxicity of *Photorhabdus luminescens* W14 toxin complexes in *Escherichia coli*. Appl. Environ. Microbiol. 67, 5017−5024.

Waterfield, N., Hares, M., Yang, G., Dowling, A., ffrench-Constant, R., 2005. Potentiation and cellular phenotypes of the insecticidal Toxin complexes of *Photorhabdus* bacteria. Cell. Microbiol. 7, 373−382.

Whangbo, J.S., Hunter, C.P., 2008. Environmental RNA interference. Trends Genet. 24, 297−305.

Whyard, S., Singh, A.D., Wong, S., 2009. Ingested double-stranded RNAs can act as species-specific insecticides. Insect Biochem. Mol. Biol. 39, 824−832.

Wolfersberger, M., Luethy, P., Maurer, A., Parenti, P., Sacchi, F.V., Giordana, B., et al., 1987. Preparation and partial characterization of amino acid transporting brush border membrane vesicles from the larval midgut of the cabbage butterfly (*Pieris Brassicae*). Comp. Bioch. Physiol. 86A, 301−308.

Wu, X., Huang, F., Rogers Leonard, B., Ottea, J., 2009. Inheritance of resistance to *Bacillus thuringiensis* Cry1Ab protein in the sugarcane borer (Lepidoptera: Crambidae). J. Invertebr. Pathol. 102, 44−49.

Xu, X., Yu, L., Wu, Y., 2005. Disruption of a cadherin gene associated with resistance to Cry1Ac δ-endotoxin of *Bacillus thuringiensis* in *Helicoverpa armigera*. Appl. Environ. Microbiol. 71, 948−954.

Yang, Y., Zhu, Y.C., Ottea, J., Husseneder, C., Leonard, B.R., Abel, C., et al., 2010. Molecular characterization and RNA interference of three midgut aminopeptidase N isozymes from *Bacillus thuringiensis*-susceptible and -resistant strains of sugarcane borer, *Diatraea saccharalis*. Insect Biochem. Mol. Biol. 40, 592−603.

Yarasi, B., Sadumpati, V., Immanni, C.P., Vudem, D.R., Khareedu, V.R., 2008. Transgenic rice expressing *Allium sativum* leaf agglutinin (ASAL) exhibits high-level resistance against major sap-sucking pests. BMC Plant Biol. 8, 102.

Yu, C.G., Mullins, M.A., Warren, G.W., Koziel, M.G., Estruch, J.J., 1997. The *Bacillus thuringiensis* vegetative insecticidal protein Vip3A lyses midgut epithelium cells of susceptible insects. Appl. Environ. Microbiol. 63, 532−536.

Zamore, P.D., 2002. Ancient pathways programmed by small RNAs. Science. 296, 1265−1269.

Zeiske, W., Meyer, H., Wieczorek, H., 2002. Insect midgut K(+) secretion: concerted run-down of apical/basolateral transporters with extra-/intracellular acidity. J. Exp. Biol. 205, 463−474.

Zhang, S., Cheng, H., Gao, Y., Wang, G., Liang, G., Wu, K., 2009. Mutation of an aminopeptidase N gene is associated with *Helicoverpa armigera* resistance to *Bacillus thuringiensis* Cry1Ac toxin. Insect Biochem. Mol. Biol. 39, 421−429.

Zhang, X., Candas, M., Griko, N.B., Rose-Young, L., Bulla Jr., L.A., 2005. Cytotoxicity of Bacillus thuringiensis Cry1Ab toxin depends on specific binding of the toxin to the cadherin receptor BT-R1 expressed in insect cells. Cell Death Differ. 12, 1407−1416.

Zhang, X., Candas, M., Griko, N.B., Taussig, R., Bulla Jr., L.A., 2006. A mechanism of cell death involving an adenylyl cyclase/PKA signaling pathway is induced by the Cry1Ab toxin of *Bacillus thuringiensis*. Proc. Natl. Acad. Sci. USA. 103, 9897−9902.

Zhang, Y., Ma, F., Wang, Y., Yang, B., Chen, S., 2008. Expression of v-cath gene from HearNPV in tobacco confers an antifeedant effect against *Helicoverpa armigera*. J. Biotechnol. 138, 52−55.

Zhu, J.Q., Liu, S., Ma, Y., Zhang, J.Q., Qi, H.S., Wei, Z.J., et al., 2012. Improvement of pest resistance in transgenic tobacco plants expressing dsRNA of an insect-associated gene EcR. PLoS One. 7, e38572.

CHAPTER 5

Concepts and Complexities of Population Genetics

David W. Onstad[1] and Aaron J. Gassmann[2]
[1]DuPont Pioneer, Wilmington, DE
[2]Department of Entomology, Iowa State University, Ames, IA

Chapter Outline

Insect resistance management is, at its foundation, an application of population genetics and ecology. Population-genetic factors of dominance, fitness costs, and genetic drift interact with ecological factors, such as dispersal, mating behavior, and spatial heterogeneity, to determine the rate at which a population will evolve resistance to a management tactic. An understanding of population genetics is needed both to explain past cases of resistance and to predict the evolution of insect resistance in the future. While modeling the basic process of resistance evolution is straightforward, complexities of the natural world may lead to inaccurate predictions from overly simplistic

Insect Resistance Management
DOI: http://dx.doi.org/10.1016/B978-0-12-396955-2.00005-9

models. Effects of genetic drift, environmentally induced changes in the genetic dominance of resistance, and ecologically mediated fitness costs are examples of how biological complexity in the real world can affect resistance evolution. In this chapter, we discuss simple yet important concepts and introduce the reader to diverse research that complicates heuristic traditions and correct misconceptions.

WITHOUT NATURAL SELECTION

Unless otherwise stated, we have made the following assumption in our overview of resistance evolution. Organisms are diploid, sexually reproducing arthropods with discrete generations. Population size is very large, and consequently random genetic drift does not affect resistance-allele frequency, and genetic mutation is absent, except to initially intro-duce resistance alleles into a population. Resistance is a monogenic trait, which by definition, resides at a single locus with one allele for resistance and a second allele that confers susceptibility.

The Hardy—Weinberg Equilibrium describes the frequencies (propor-tions) of genotypes when mating is random. For p the proportion of allele s, and q the proportion of allele r, $p + q = 1$. The proportion of each genotype can be calculated as $(p + q)^2$. Thus, the frequency of ss in the population is p^2, the frequency of rr is q^2, and the frequency of rs is $2pq$. The frequencies of p and q remain constant in a population in the absence of (1) natural selec-tion, (2) genetic drift, (3) mutation, and (4) migration. If any of these factors is present, the frequency of both the alleles and the genotypes can be altered.

These relationships under Hardy—Weinberg conditions also demon-strate that the heterozygote, rs, will be the genotype that carries the most r alleles in the population as long as $q < 0.5$. (See Figure 11.4 in Chapter 11.) Homozygous resistant individuals, rr, will be rare compared to heterozygotes at low values of the resistance-allele frequency.

EVOLUTION DUE TO NATURAL SELECTION

Natural selection causes evolution when there is differential survival among genotypes. Without the presence of genetic variation (e.g., multiple alleles at a single locus), selection cannot occur. As stated in Fisher's Fundamental Theorem, assuming a constant force of natural selection, the rate of evolution is equal to the genetic variation (Frank and Slatkin, 1992). Essentially, selection creates differential fitness among genotypes and changes the average fitness of the population, by increasing the frequency of those genotypes with the highest fitness. Fitness determines, and is by definition, the number of offspring an individual produces. The number of alleles that an individual contributes to the next generation is a measure of fitness and increases as fitness becomes greater. Thus, fitness depends on survival and reproduction, and all of the life-history characteristics that in turn affect survival and reproduction.

Given a single locus with two alleles, r for resistance and s for susceptibility, the generational change in allele frequency is

$$q(t + 1) = [q(t) \times p(t) \times \mathrm{Wrs}(t) + q^2(t) \times \mathrm{Wrr}(t)]/\overline{W}(t) \qquad (5.1)$$

where Wii is the fitness of genotype ii, and t is the index for generation. The first term on the right-hand side is derived from $2pq\mathrm{Wrs}/2$ and represents the r alleles provided by the heterozygotes. The second term on the right-hand side represents the r alleles contributed by the homozygous resistant individuals. The average fitness of the entire population, $\overline{W}$, is the weighted average based on Hardy−Weinberg proportions for each genotype

$$\overline{W} = p^2\mathrm{Wss} + 2pq\,\mathrm{Wrs} + q^2\mathrm{Wrr} \qquad (5.2)$$

$\overline{W}$ also determines the weighted sum of all allele frequencies. Note that, because of selection, p, q, and $\overline{W}$ vary over generations t. Equations 5.1 and 5.2 and any underlying models are calculated iteratively to evaluate the dynamics of allele frequencies changing over a given number of generations until a constant q is found.

Table 5.1 provides an overview of the dynamics simulated with Equation 5.1. Typically, fitness values are standardized to the highest value of genotypic survival. Any values for genotypic survival may be used, provided all three genotypes are measured on the same scale (e.g., per generation, per year, proportions surviving).

When the homozygous resistant individuals are most fit and the fitness of the heterozygotes is intermediate, the resistance-allele frequency, q, eventually reaches 1.0. The opposite occurs when homozygous resistant individuals are the least fit of the three genotypes. Figure 5.1 shows how the initial frequency of the resistance allele affects the number of generations until

Table 5.1 The Evolutionary Outcomes Calculated by Equation 5.1 for the Four Basic Relationships for Genotypic Fitness Values in a Diploid Species

Relationship	Evolutionary Outcomes
$W_{ss} < W_{rs} < W_{rr}$	q fixes at 1.00
$W_{ss} > W_{rs} > W_{rr}$	q fixes at 0.00
$W_{ss} < W_{rs} > W_{rr}$	reaches an equilibrium q_e between 0.00 and 1.00
	$\qquad W_{ss} = W_{rr} : q_e = 0.5.$
	$\qquad W_{ss} < W_{rr} : 0.5 < q_e < 1.0$
	$\qquad W_{ss} > W_{rr} : 0.0 < q_e < 0.5.$
$W_{ss} > W_{rs} < W_{rr}$	a threshold exists, q_T, such that when:
	$\qquad q_0 < q_T : q$ fixes at 0.
	$\qquad q_0 > q_T : q$ fixes at 1.
	$\qquad q_0 = q_T : q$ remains at equilibrium.

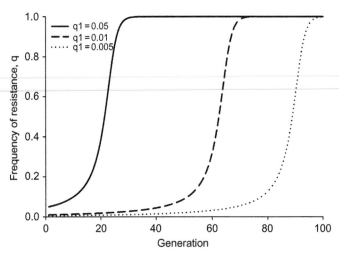

Figure 5.1 The change in resistance-allele frequency over time for a population under selection beginning with three different initial allele frequencies. Genotypic fitness is held constant at $W_{ss} = 0.5$, $W_{rs} = 0.51$, and $W_{rr} = 1.0$ with resistance expression close to recessive.

resistance evolves in the population when $W_{ss} < W_{rs} < W_{rr}$. The initial allele frequency has an inverse relationship to the number of generations required to pass through the lag phase. Regardless of the initial resistance allele-frequency (q_0), q eventually approaches 1.

Both the relative fitness of the heterozygote and the overall difference in relative fitness between homozygous resistant and homozygous susceptible genotypes determine the rate of resistance evolution. Relative fitness values are often reported in the range from 0.00 to 1.00. The range is a convenience that defines the genotype with the maximum fitness as the standard (i.e., 1.00) and the fitness of the other genotypes as proportions relative to that standard. Figure 5.2 shows how changes in relative fitness of heterozygotes influence the evolution of resistance. The curve on the far right represents Wrs = Wss, while the curve on the far left represents the scenario with Wrs = Wrr. Minor differences in fitness of the heterozygote can dramatically influence how quickly resistance evolves. Figure 5.3 presents the change in resistance-allele frequency over time for four scenarios with additive expression of resistance and Wrr = 1.0. As the relative difference between Wss and Wrr decreases (going from left to right on the figure), the rate of the evolution of resistance is slower.

The outcomes are more complicated when the fitness of the heterozygote is either the maximum or the minimum of the three genotypes (Table 5.1). Evolutionary theory indicates that an allele frequency changes to maximize fitness of the population (Speiss, 1977). A stable equilibrium of $q_e = 0.5$ is established when the heterozygote has the highest fitness (overdominance): Wss < Wrs > Wrr, and Wrr = Wss. When Wrr > Wss

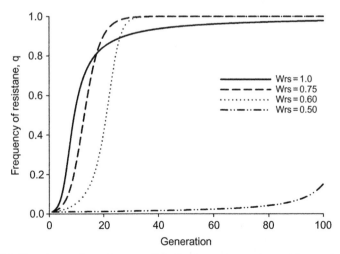

Figure 5.2 The change in resistance-allele frequency over time with constant initial q = 0.01, Wss = 0.5 and Wrr = 1.0, and incremental increases in fitness of the heterozygote, Wrs.

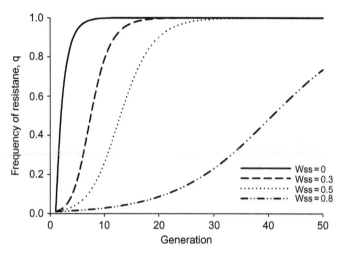

Figure 5.3 The change in resistance-allele frequency over time with constant initial q = 0.01, Wrr = 1.0, additive expression of resistance Wrs = (Wss + Wrr)/2, and incremental increases in fitness for the susceptible homozygote, Wss.

the stable equilibrium shifts such that $0.5 < q_e < 1.0$. When Wrr < Wss, the stable frequency will be $0.0 < q_e < 0.5$. In general, the equilibrium allele frequency according to Speiss (1977) is

$$q_e = (Wss - Wrs)/[(Wss - Wrs) + (Wrr - Wrs)] \qquad (5.3)$$

Equation 5.3 can also be used to find the threshold, q_T, which is an unstable equilibrium point (Table 5.1) that arises when the heterozygote has the minimum fitness (underdominance). When Wrr = Wss, $q_T = 0.5$ is the threshold allele frequency at which a resistance allele will be lost from the population or will increase and eventually become fixed. When Wrr > Wss, the threshold will be $0.0 < q_T < 0.5$; and when Wrr < Wss the threshold will be $0.5 < q_T < 1.0$. When the initial allele frequency, q_0, is below the threshold, the resistance allele is lost from the population, and when q_0 is above the threshold, the population will become fixed for resistance (i.e., all individuals in the population will eventually become homozygous resistant).

Although it is mathematically possible to have an equilibrium allele frequency, it is difficult to maintain the allele frequency at equilibrium in the real world. Fluctuations in allele frequency caused by migration and genetic drift will move the frequency from the stable point, thus causing resistance either to be lost or fixed within the population.

Of course, none of these conclusions about thresholds and equilibria tell us how fast evolution occurs. To actually explore the yearly or generational changes in resistance-allele frequencies, we must calculate the entire population genetics model.

NATURAL SELECTION IN PATCHY LANDSCAPES

A. Genotypic Fitness Constant over Time

Major Implicit Assumptions

1. Random mating within entire population from all patches.
2. Uniform distribution of offspring in landscape.
3. Survival and reproduction do not change over generations.
4. No density-dependent survival, reproduction, or behavior.

Much of population genetics and insect resistance management (IRM) emphasizes the evolution of a population inhabiting two or more patches that produce differential fitness among the genotypes. An obvious example is a set of treated fields (patch 1) and untreated refuges for susceptible pests (patch 0). Equation 5.1 can be used if the entire population mates at random and oviposits uniformly across the landscape. We assume that genotypic fitness is approximated by the survival of each genotype in each patch, $S0$ and $S1$, multiplied by fecundity in that patch, $F0$ and $F1$, and weighted by the proportional area of the patch, $P0$ and $P1$. Thus, for genotype ii in the treated patch, $W1ii = S1ii \times F1ii \times P1$. For a landscape where fecundity is equal in both patches, genotypic fitness in the whole population across the entire landscape is calculated as: $Wii = S1ii \times P1 + S0ii \times P0$.

Figure 5.4 demonstrates how evolution occurs over time in a landscape with two patches. Reproductive capacity is equal in both patches and therefore is effectively 1.0. It is assumed that fitness costs are absent; therefore, $S0ii = 1$ for all genotypes in the refuge. $S1rr = 1$, $S1rs = 0.5$, and $S1ss = 0.1$ in the treated patch. As the proportion of treated fields increases, the overall fitness of the susceptible genotype, Wss, decreases and the difference in fitness between genotypes becomes greater, resulting in a shift of the curves to the left in the figure.

B. Variable Fitness over Time and Space

Major Implicit Assumptions

1. Random mating within entire population from all patches
2. Uniform distribution of offspring in landscape

Note that Equation 5.1 contains no terms for abundance of the arthropod species being modeled. However, fitness is likely to change

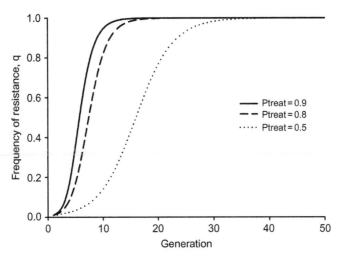

Figure 5.4 The change in resistance-allele frequency over time with constant initial allele frequency and constant genotypic fitness within treated patch and refuge patch in the landscape. Evolution is changed by altering the proportion of treated patch in the landscape.

over time and space as arthropod density changes. Given that density-dependent processes such as survival, reproduction, and dispersal are common phenomena in arthropod populations (Chapter 12), we should expect fitness to vary over time and space. Stakeholders and economists want to predict population density to better predict consequences of insect resistance management for pest damage and economic losses (Mitchell and Onstad, Chapter 2).

As described in Chapters 10 and 12, there are many ways to combine models of gene frequency with models of pest density. Modelers must make decisions about what aspects of population dynamics to include in the study of population genetics. The most important decisions are concerned with the interactions of phenotype density and behavior, in particular dispersal. Some of the first models to consider how pest density and dispersal between patches affects resistance evolution were constructed by Comins (1977), Roughgarden (1979), Taylor and Georghiou (1979), Tabashnik and Croft (1982), and Alstad and Andow (1995).

Finally, note that mean population fitness is the most appropriate measure of susceptibility in population models. Resistance-allele frequency can be adequate for models studying a single toxin and a single gene, but with more than one gene, allele frequency is a less satisfactory metric. Of course, pest density and economic measures will also be valuable for any model of integrated pest management and IRM.

GENE FLOW AND POPULATION STRUCTURE

Gene flow is the movement of alleles among populations, and it occurs over a landscape and among populations that are separated spatially or temporally. Whether space is simple or complex in a model, we need to carefully define gene flow and population structure and deal with both concepts simultaneously. The genetic structure of the population in a landscape, and in turn the dynamics of resistance evolution, is determined by natural selection and gene flow, both of which are dependent on the spatial structure of the landscape (Gassmann et al., 2009c).

Background on the population-genetics issues related to gene flow can be obtained from several publications (Hedrick, 2006; Mallet, 2001). Mallet (2001) clarifies a number of issues that often confuse nonexperts, emphasizing the actual movement of genes and genotypes in his analysis. Felsenstein (1976) reviews research concerning models of island populations and dispersal. This traditional work in population genetics still has relevance to current problems regarding patches of transgenic insecticidal crops and refuges of conventional crops. In particular, Felsenstein (1976) summarizes the relatively old studies by stating that the alleles found in island populations will depend on the amount of dispersal among the islands, with threshold levels of dispersal possibly determining the final outcome.

Kawecki (2000) created a simple model of population genetics and source–sink population dynamics in a landscape with a refuge (source) and a treated field (sink). The single locus has a mutant, nonrecessive allele that confers resistance to the insecticide while producing a fitness cost in the refuge. Kawecki (2000) concluded that both ecological and genetic aspects of dispersal and the amount of fitness conferred by one or more genes must be evaluated to determine the effects of dispersal in these landscapes. Other authors have investigated more complex systems involving gene flow in heterogeneous landscapes for insect pests under selection from insecticides (Caprio and Tabashnik, 1992; Caprio, 2001; Ives and Andow, 2002; Sisterson et al., 2005). Most authors generally conclude that an intermediate rate of movement before mating between refuge and area with insecticide delays evolution of resistance longer than low and high rates.

The number of independent origins of insecticide-resistance alleles is often debated. To deal with this issue, we must understand both arthropod movement and genetics. ffrench-Constant *et al.* (1996) offered evidence for the multiple independent origins of resistance in two *Drosophila* species, two beetle species, and the *Bemisia tabaci* complex. The repeated replacement of the same amino acid in the resistance-to-dieldrin (*Rdl*) gene, conferring resistance to cyclodiene insecticides, is a model system that can be used to study the origins of resistance alleles. They used this technique, plus an examination of flanking sequence data, to produce evidence for multiple origins of the same amino acid replacement. These results should be compared to studies of amplified esterases and insensitive acetylcholinesterase in *Culex* mosquitoes (Chapter 6). ffrench-Constant *et al.* (1996) also discussed the importance of life history in determining the likely origin and spread of different resistance alleles.

Care must be taken when using the term *gene flow* or claiming some consequence of gene flow for IRM. Gene flow depends on dispersal, and to understand the interaction of dispersal and resistance evolution, one must consider several factors. First, does dispersal occur before or after mating, or both? Second, do males and females have different dispersal rates and behaviors? Third, is gene flow unidirectional or multidirectional? If gene flow occurs in multiple directions, such as to and from a particular crop or refuge, are the rates equal or unequal? Is gene flow expressed as a proportion of the population or as the number of individuals, genotypes, or alleles that move? Fourth, is dispersal constant over time and space or does it interact with other processes such as insect density or environmental conditions? Thus, when someone claims to know how gene flow affects evolution of resistance, be prepared to ask a series of questions.

MATING

The mixing of a population for the purposes of mating is a critical process in the evolution of resistance. For example, delays in resistance to transgenic insecticidal crops predicted under the high-dose and refuge strategy are contingent on mating between susceptible insects from refuges and resistant individuals from a field planted with transgenic insecticidal crop. Mating involves a complex set of behaviors. For the purposes of IRM, details of these behaviors may need to be measured empirically and included in simulation models. Trimble *et al.* (2004) studied the effects of sublethal residues of azinphosmethyl on pheromone production, calling, female attractiveness and the ability of males to locate sources of pheromone. They compared the performances of susceptible and resistant *Choristoneura rosaceana* and found effects that depended on the phenotype and treatment. Some studies have shown a fitness cost in mating rate and male competition due to insecticide resistance (Doherty and Hales, 2002; Berticat *et al.*, 2002a; Groeters *et al.*, 1993; Boivin *et al.*, 2001; Higginson *et al.*, 2005; Anilkumar *et al.*, 2008; Crespo *et al.*, 2010; Gulzar *et al.*, 2012). In other cases, the resistant individuals have an advantage. Arnaud and Haubruge (2002) evaluated malathion-susceptible and malathion-resistant male *Tribolium* beetles in mating competition for susceptible females, and in most cases, found that resistant males had a greater reproductive success rate than susceptible males. Rolff and Kraaijeveld (2003) showed that parasitoid-resistant *Drosophila melanogaster* males achieved a higher mating success than susceptible males. Genetics and natural selection also may be important when females mate with more than one male (Haubruge *et al.*, 1997; Alyokhin and Ferro, 1999a,b; Baker *et al.*, 2005, Higginson *et al.*, 2005). Zhao *et al.* (2009) observed resistant females producing pheromone later and less frequently than susceptible females.

Traditional, simple models assume that panmixia, or random mating, occurs in populations. However, when density-dependent effects and heterogeneous landscapes are investigated, variable dispersal (and adult emergence) can lead to nonrandom mating. Spencer *et al.* (2013) demonstrate, with both empirical and computational results, the importance of isolated female adults and limited male mating capability for evolution of resistance in heterogeneous landscapes. In their system in which the maize landscape is divided into large blocks of refuge and insecticidal maize, female *D. virgifera virgifera* mostly mate before dispersing and males disperse but have a limited number of opportunities to mate.

In other models of patchy landscapes, mating can be modeled as random within a patch but nonrandom for the entire population. For example, Caprio and Hoy (1995) and Caprio (2001) evaluate random and nonrandom mating under various dispersal scenarios. They concluded that prediction about IRM must consider dispersal for mating and mating biases (assortative mating) along with dispersal that distributes offspring.

Guse *et al.* (2002) simulated mating that is influenced by irrigation of the cropland and by dispersal of males between habitats and found significant differences in the rate of resistance evolution.

RANDOM GENETIC DRIFT AND DEMOGRAPHIC ALLEE EFFECTS

Evolution (i.e., genetic change over time) differs between small and large populations in two important ways: the effects of genetic drift and the presence of Allee effects. Both of these factors can cause models based on large populations to make inaccurate predictions for small populations. Genetic drift is defined as change in gene frequency over time due to random or stochastic events. Under genetic drift, changes in allele frequency are decoupled from the effect of the allele on individual fitness. In the case of resistance evolution for small populations, genetic drift can cause rare alleles either to be lost or to become fixed solely by chance. Most IRM models do not include random drift because they are deterministic (i.e., they lack random processes) and assume that the population densities will always be very large. However, densities of arthropods can be driven to low levels in pest management (Carrière *et al.*, 2003; Hutchison *et al.*, 2010). Stochastic models, which allow chance to influence the changes in variables, are able to simulate random genetic drift (Caprio and Tabashnik, 1992; Caprio, 1994; Caprio and Hoy, 1994; Storer *et al.*, 2003; Sisterson *et al.*, 2004, 2005; Gassmann *et al.*, 2008, 2009a, 2012a).

The Allee effect is defined as a decrease in fitness for individuals in small populations. The Allee effect has been recognized in ecology for many decades (Kanarek and Webb, 2010). Even though the Allee effect can be modeled deterministically, a stochastic model also would be reasonable because some conditions and events producing the effect are stochastic, such as the inability to find a mate at low density.

The number of homozygous resistant (RR) genotypes is expected to be very rare and at extremely low densities when selection of the population begins. (See Figure 11.4 in Chapter 11.) For example, in fields of transgenic insecticidal crops, arthropod densities are expected to be very low, and most surviving genotypes are expected to be RR. At initial allele frequencies of 0.00001 to 0.001, there will be one RR individual for every million to 10 billion insects in a population. In the case of *Diabrotica virgifera*

virgifera, densities range from 1 million to 100 million per hectare in corn-fields depending on the life stage (Onstad *et al.*, 2006). Thus, for this insect, there can be fewer than 100 RR individuals per hectare at the start of selection. In this scenario, random drift and the Allee effect may influence the evolution of resistance depending on how populations mix and how long the densities remain low.

GENETIC ARCHITECTURE AND EVOLUTION

Although models of resistance typically assume monogenic resis-tance, studies from both natural and agricultural systems indicate that oli-gogenic and polygenic resistance should also be considered, and that in some cases, resistance traits are likely to involve more than one gene (Gassmann *et al.*, 2009c; Pedra *et al.*, 2004; Daborn *et al.*, 2002). For example, microevolution due to natural selection in wild species can hap-pen quickly over several generations (Hendry and Kinnison, 1999). Endler (1986) found that physiological adaptation in animals and plants involved quantitative traits in 15 species and discontinuous traits (possibly monogenic or oligogenic) in five species. For an additional seven species, the response to the same selective pressure resulted in both quantitative and discontinuous traits (Endler, 1986).

The genetic architecture, the relationship between genotypes and phe-notypes (Hansen, 2006), is often expressed in terms of monogenic versus polygenic traits (Orr and Coyne, 1992; ffrench-Constant *et al.*, 2004; Ferre *et al.*, 2008; Roush and McKenzie, 1987). By polygenic, we usually mean that the resistance is due to the combination of effects conferred by a set of genes each providing a minor contribution. Roush and McKenzie (1987) and Roush and Daly (1990) argue that field resistance to pesticides by arthropods only evolves when one or two major genes are the basis for resistance. McKenzie and Batterham (1998) used muta-genesis to demonstrate that major genes are the primary cause of resis-tance in a variety of cases. However, this simple dichotomy does not adequately represent the dynamics and complexity of the various genetic architectures expected to be encountered in studies of arthropod evolu-tion (Hansen, 2006).

A study of pyrethroid resistance in *Musca domestica* in the United States demonstrates some of this complexity (Rinkevich *et al.*, 2006). *M. domestica* can become resistant to permethrin by two major mechanisms: cytochrome P450 mediated detoxification (overtranscription/translation of *CYP6D1*) and target site insensitivity (*knockdown resistance; kdr*). Rinkevich *et al.* (2006) sampled flies from four eastern states and observed that (1) resistance-allele frequencies varied across the sites; (2) one *kdr* allele conferring a minor increase in survival could be found at frequencies as high or higher than the *kdr* allele that conferred greater survival to permethrin; and (3) an oligogenic combination of *kdr* and overexpression of *CYP6D1* is necessary for survival of insecticide treatments in the field. Rinkevich *et al.* (2006) concluded that fly populations and their genomes are adapted to local environmental conditions and that survival due to exposure to permethrin is not the only factor determining the frequency of resistance alleles.

Ferré *et al.* (2008) reviewed the literature on genetics of pest adaptation to crystal (Cry) toxins derived from *Bacillus thuringiensis*, including resistance to insecticidal crops. Ferré *et al.* (2008) reported that in 8 out of 11 cases resistance was monogenic, but in the remaining three cases resistance was polygenic. While the level of resistance was similar between cases of monogenic and polygenic resistance, more episodes of selection occurred for cases involving polygenic resistance compared with monogenic resistance. This finding is consistent with the idea that gradual selection operating in the range of existing genetic variation will lead to polygenic resistance. Some cases in which transgenic insecticidal crops do not provide high-dose control of pests include Cry1Ac cotton and maize for control of *Helicoverpa zea* and insecticidal maize for control of *Diabrotica spp.* (Tabashnik *et al.*, 2008; Siegfried *et al.*, 2005). These pests may be considered the most likely to evolve polygenic resistance to transgenic insecticidal crops.

Insecticide resistance can arise through a modification of the insecticide's target site (Pittendrigh *et al.*, Chapter 3), which may further favor the evolution of monogenic resistance (ffrench-Constant *et al.*, 2004). If, however, resistance requires greater biological complexity, then oligogenic resistance should be expected (Crow, 1954, ffrench-Constant *et al.*, 2004). One example of this is behavioral resistance to crop rotation in *Diabrotica virgifera virgifera* (Spencer *et al.*, Chapter 7). In this form of resistance, adult *D. v. virgifera* have evolved a behavior of laying eggs outside of cornfields in areas that are often rotated to corn the following season. Genetic analysis of rotation-resistant populations revealed that resistance appears to be linked with changes at three loci (Knolhoff *et al.*, 2009).

The case of the *Mayetiola destructor* and its resistance to wheat varieties with antibiosis has been the subject of several genetic studies. According to Wu *et al.* (2008), wheat varieties appear to defend against larvae through upregulation inhibitor-like genes (genes believed to interfere with digestive enzymes of the *M. destructor*) and genes involved in cell wall growth (leading to compartmentalization of larvae). Resistant *M. destructor* do not elicit these responses from wheat varieties to which they have evolved resistance, and resistance traits in *M. destructor* appear to be monogenic (El Bouhssini *et al.*, 2001; Liu *et al.*, 2007).

Adaptation to selection has been described in terms of a constant level of selection acting uniformly on a population (ffrench-Constant *et al.*, 2004), with a fixed level of selection acting either within standing genetic variation of the pest population or outside the range of existing variation. However, this simplicity is not likely to account for all possible scenarios of adaptation (Hansen, 2006). Selection pressure may be variable over space and time (Hoy *et al.*, 1998), and therefore, the concept of a fixed level should be relaxed, if not abandoned. The changes in selection over time may be random or a pattern of decreasing or increasing levels may occur.

Labbe *et al.* (2007) describe how ace-1, a gene involved in resistance to organophosphates in *Culex pipiens*, evolved during 40 years of an insecticide control program. During the early stages of resistance evolution, a major resistance allele with fitness costs spread through the population. Then a duplication combining a susceptible and a resistance ace-1 allele began to spread without replacing the original resistance allele, as it is sublethal when homozygous. More recently, a second duplication (also sublethal when homozygous) spread through the population because heterozygotes for the two duplications incur no fitness costs. Double over-dominance favoring the heterozygotes now maintains the four alleles across treated and untreated areas. Labbe *et al.* (2007) concluded that ace-1 evolution did not continue because of the steady accumulation of beneficial mutations with small effects. Instead, the observed evolution of resistance was an erratic combination of mutation, positive selection, and the rearrangement of existing genetic variation leading to a complex genetic architecture (Labbe *et al.*, 2007).

The dynamic nature of selection may be due to dynamics of management, just as over longer periods of time, coevolution of interacting species, such as plants and herbivorous insects, may display a dynamic process. As outlined by Thompson (1994), the coevolutionary dynamic of interacting species may differ from one population to the next, with selection favoring some traits in one environment and different traits in another environment. Overlaid on these contrasting selective pressures is gene flow among populations. Such evolutionary complexity, which lies outside the textbook description of a single selective force acting on a single population, is also often operating in agricultural systems. In these systems, different populations of a pest are confronted with differing management practices, but these populations are connected by some level of gene flow. Furthermore, standing genetic variation is also a dynamic condition over time due to mutation, pleiotropic effects of resistance genes, and general environmental dynamics.

Our desire for simplicity causes scientists to arbitrarily choose a gene pool at a given time as representing the standing level of genetic variation. The choice of a reference time is arbitrary, as is the practical need to disregard variation due to alleles at frequencies that are either below a typical mutation rate or too low to sample in empirical research. For example, if we assume that genetic variation in resistance to a toxin is normally distributed in a population, then from basic statistical theory we know that 3×10^{-5} of this variation is not included in studies that truncate a population at four standard deviations above the mean. Is this small proportion of the gene pool part of the standing genetic variation or not? Only when we truncate the distribution beyond five standard deviations above the mean do we include all mutations and variation occurring at frequencies higher than a typical neutral substitution rate (4×10^{-9} according to ffrench-Constant et al. (2004)). If the purpose of evolutionary biology concerning resistance is the study of rare genes and events, then it is unwise to form a habit of simplifying the genetic variation in a population.

Karasov et al. (2010) challenged the dogma concerning mutation rate and effective population size in studies of natural selection in arthropod populations. It is generally assumed that adaptation by arthropods is limited by the population-level mutation rate, which is proportional to the product of effective population size multiplied by the per-site mutation rate given the standing neutral variation. Karasov et al. (2010) observed that multiple simple and complex resistance alleles (at insecticide resistance locus Ace) evolved quickly and repeatedly within individual populations of D. melanogaster. They concluded that, contrary to dogma, short-term, effective population sizes existing during adaptation due to selection by insecticides (10–1000 generations) are essentially the same as observed population sizes (circa 10^8). Karasov et al. (2010) also concluded that the population mutation rate can be close to 1, which is much greater than the typical value of 0.01 based on traditional assumptions. They postulated that the per-site mutation rate will not limit the rate of evolution of resistance unless the arthropod population size is smaller than the reciprocal of that mutation rate. In other words, in extremely large insect populations, resistance alleles will exist at initial frequencies that permit evolution to occur within tens to hundreds of generations.

When selective pressure and genetic variation are heterogeneous and dynamic, we should expect contributions of single and multiple genes for resistance with a range of minor to major effects. Gassmann *et al.* (2009c) conclude that if model-based predictions must be made before the genetic architecture is known, the assumption of monogenic resistance will be the most conservative and cautious one to make. As scientists and regulators confront more frequent cases of oligogenic or polygenic resistance, it is likely that individual-based models will need to be created. Instead of using sparse matrices to represent the population of insects with many rare geno-types, individual-based models will replace the modeling of millions of dimensions with the modeling of millions of individuals in a landscape.

SELECTION INTENSITY AND GENETICS

In an attempt to determine the genetic basis of field-evolved resis-tance by arthropods to pesticides, Groeters and Tabashnik (2000) related published data on pesticide selection intensity to an analysis of a stochastic model. They concluded that measurements of selection intensities for nine species varied widely and that field and laboratory selection intensities are generally similar. In their stochastic model, they evaluated six unlinked loci (genes). Because the initial allele frequencies were inversely proportional to their effects on resistance, major genes had the lowest initial frequencies. Results indicated that resistance alleles with major effects dominated responses to selection no matter what the selection intensity. Resistance evolved faster in models with major genes than in models without them. The most intense selection tended to prohibit minor genes from contrib-uting. Groeters and Tabashnik (2000) concluded that knowing the inten-sity of selection is crucial for IRM predictions, but knowledge of the number of loci and their relative contributions to resistance is not. Thus, models simulating a few major genes (one or two loci) should be satisfac-tory for modeling and predicting the consequences of IRM decisions.

Tabashnik (1990) studied the influence of gene amplification at a single locus on the evolution of insecticide resistance. Gene amplification increases the number of copies of a gene per haploid genome above normal levels. When a resistance mechanism depends on the amount of biochemical products of gene expression, gene amplification will lead to greater tolerance to a toxin. Tabashnik (1990) compared simulations of a conventional two-allele model to simulations of three- and four-allele models in which additional alleles are derived from existing alleles at a rate greater than assumed mutation rates. Each subsequent allele produces more mRNA and protein to increase tolerance. Results were similar for the models when insecticide concentration was low or moderate. In contrast, when 10% of the population was not exposed in a refuge, high insecticide concentrations slowed resistance evolution in the two-allele model, but caused rapid evolution of resistance in the three- and four-allele models even at very low initial allele frequencies. Tabashnik (1990) concluded that IRM strategies based on use of a high dose of toxin are not likely to succeed if gene amplification or other mechanisms generate alleles that confer high levels of resistance.

Quantitative-genetics models of selection based on a constant heritability parameter cannot predict evolution over many generations. Heritability is the correlation between response and selection differential between two generations (Maynard Smith, 1989). It is an empirical parameter that can be calculated only for a particular set of population, generations, and environment (Maynard Smith, 1989). Heritability changes as allele frequency changes. Heritability is expected to change quickly with changes in allele frequencies when only a few genes are involved (Falconer and Macay, 1996). Therefore, heritability cannot remain constant and account for the nonlinear aspects of evolutionary dynamics such as the logarithmic or exponential increases in allele frequency often seen in resistance evolution (Maynard Smith, 1989).

DOMINANCE

Genetic dominance describes the degree to which the phenotype for a heterozygous genotype resembles the phenotype of a homozygous genotype with which it shares an allele. For a single gene, one allele is dominant over a second allele when the expression of the first allele determines the response of the heterozygote to its environment. For example, when susceptibility to a toxin is dominant, Sr, the phenotype is vulnerable to the toxin, but when the susceptibility is recessive, sR, the phenotype is resistant. Additive expression of resistance causes a genotype to produce a phenotype that is intermediate to the phenotype of either of its homozygous parents. Typically, nonrecessive resistance occurs through a gain of function; the organism can now do something it could not before. For example, detoxification enzymes are produced at a higher level, thereby allowing more of the toxin to be detoxified. One allele causing a gain in function may, therefore, provide resistance. In contrast, recessive resistance is often associated with a loss of function. For example, a change in a target site means that the toxin can no longer bind to the given receptor. However, in the heterozygous state, the interactions between the wild-type protein, still produced by the single S allele, and the toxin are enough to cause mortality. For cases with intermediate heterozygote fitness, the simplest way to model relative fitness for heterozygotes is with a function similar to

$$\text{Wrs} = (1 - h) \times (1 - \text{Mss}) + h \times (1 - \text{Mrr}) \tag{5.4}$$

where $0 \leq h \leq 1$ is the dominance level for resistance to the pest management tactic (e.g., crop rotation or application of an insecticide) and $0 \leq M \leq 1$ is the genotypic mortality. In this function, $h = 0$ represents recessive resistance, $h = 1$ represents dominant resistance, and $h = 0.5$ implies additive expression. A selection coefficient typically equals the mortality due to the pest control treatment. A simplistic conceptual model of gene expression can lead to problems, however, because the level of dominance may be affected by environment.

Although dominance is often heuristically described as a constant genetic property, this is not true for real situations (Bourguet *et al.*, 2000). The dominance of resistance depends on the environment experienced by the arthropods, including the dose of toxin (Roush and Daly, 1990). For example, several bioassays indicate that the dominance of resistance to toxins decreases as toxin concentration increases (Tabashnik *et al.*, 2004). Figure 5.5 shows how the toxin dose can alter the survival of heterozygotes (the phenotypic response), and therefore, the functional dominance of resistance to the toxin. Because toxin concentration in pesticide residues and even in transgenic insecticidal crops can vary over time, dominance and selection on the targeted pest are dynamic conditions. Onstad

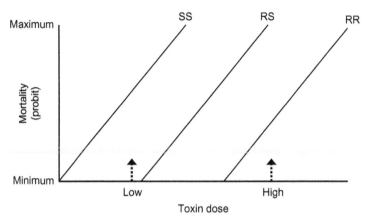

Figure 5.5 Effect of toxin concentration encountered by pest on mortality of three genotypes. Most of susceptibles (SS) and none of the heterozygotes are killed by a low dose. Resistance is identified as dominant. At a high dose, all SS and RS plus some RR individuals are killed; resistance is considered recessive.

and Gould (1998) postulated that crop senescence and reallocation of plant nitrogen from toxin to grain could cause the toxin titer to decline during much of the growing season, as has been found for transgenic insecticidal cotton (Olsen et al., 2005). Additionally, some resistant strains are capable of surviving on transgenic insecticidal corn during the reproductive state but cannot survive on younger vegetative corn (Crespo et al., 2008). Furthermore, since each life stage will respond differently to its environment and toxin, then logically dominance of resistance can also vary with the life stage of the pest that is evaluated (Bouvier et al., 2002). An example of this can be seen in the decreased susceptibility of larval insects to toxins at late versus early instars (Binning et al., 2010). For all of these reasons, claims based on a single level of dominance in a complex IRM scenario should be met with skepticism.

When selection pressure and the evolution of resistance begin, resistance alleles are rare, and the alleles occur mostly in heterozygotes. Homozygous resistant individuals are extremely rare. Therefore, the selection of heterozygotes, which is dependent on dominance, determines the early progress of resistance evolution. As such, the genetic dominance of resistance may play a central role in determining how quickly a population will evolve resistance. If transgenic insecticidal crops produce a sufficient concentration of toxin (i.e., 25 times the concentration required to kill a susceptible pest or sufficient toxin to kill 99.99% of all susceptible pests), then the crop is defined as high dose and resistance is expected to be a functionally recessive trait (Environmental Protection Agency, 1998; Tabashnik *et al.*, 2004). As the dominance of a resistance trait increases, populations are expected to evolve resistance more quickly (Tabashnik *et al.*, 2008). Indeed, several cases where pests have evolved resistance in the field to an insecticidal crop are cases where the crop was not high dose and resistance was expected to be nonrecessive (Tabashnik *et al.*, 2008; Storer *et al.*, 2010; Gassmann *et al.*, 2011, 2012b; Gassmann, 2012).

GENE INTERACTIONS

Arthropod gene interactions can be important in IRM. Epistasis is the positive (synergistic) or negative (antagonistic) interaction of genes. Gould (1986a,b) explored the antagonistic effect of resistance−gene interactions on insect fitness. He used data on *Mayetiola destructor* epistasis to calibrate an IRM model (Chapter 9). Epistasis can be measured in laboratory studies on susceptible and resistant colonies, but without these data, it is generally impossible to determine whether or not epistasis will arise among genes.

Genetic linkage occurs when two genes are on the same chromosome. When recombination is incomplete, genes may be in nonrandom associations (Groeters and Tabashnik, 2000). This is called linkage disequilibrium or gametic phase disequilibrium. Gould (1986b) explored the effects of linkage on *Mayetiola destructor* IRM. Assumptions about epistasis and linkage may be important when modeling the evolution of multiple genes in arthropod populations.

FITNESS COSTS

A fitness cost arises when individuals harboring resistance alleles have lower fitness, in the absence of the selective agent, a homozygous susceptible genotype (Gassmann *et al.*, 2009b). Fitness costs may be recessive, affecting only homozygous resistant individuals, or nonrecessive (e.g., additive or dominant) affecting both homozygous resistant and heterozygous genotypes. Fitness costs can affect a range of life-history characteristics such as fecundity, survival, and male mating success. Additionally, costs may display phenotypic plasticity, with costs arising in some environments but not others. For example, entomopathogenic nematodes can cause greater mortality of resistant than susceptible *Pectinophora gossypiella* (Gassmann *et al.*, 2006, 2008). Thus, the fitness cost of resistance to insecticidal cotton for this pest becomes greater when insect pathogenic nematodes are present in the environment. In addition to varying with ecological conditions, fitness costs may also be affected by the level of resistance, with higher levels of resistance imposing greater fitness costs (Gassmann *et al.*, 2009b).

In order for costs to affect the rate of resistance evolution, some portion of the population must be unexposed to toxin. In the case of transgenic insecticidal crops, this can arise when crops are grown with a refuge. Refuges can delay resistance because the mating of homozygous susceptible individuals from refuges with resistant individuals from insecticidal crops leads to the production of heterozygous progeny, which are often less fit on an insecticidal crop. However, the accumulation of resistance alleles in the refuge population will cause this dynamic to break down, with a rapid increase of resistance alleles occurring. Refuges are most effective when coupled with high-dose events (i.e., insecticidal crops that only allow for survival of homozygous resistant individuals). In those cases even small fitness costs can act to delay resistance. Fitness costs also can greatly prolong the efficacy of insecticidal crops if a pyramid of events is used. Gould *et al.* (2006) found that even small fitness costs coupled with a pyramided, two-toxin crop greatly delayed resistance. Significant fitness costs have been observed in arthropods evolving resistance to *Cry* toxins derived from *Bacillus thuringiensis* (Bird and Akhurst, 2005; Carrière *et al.*, 2005; Higginson *et al.*, 2005; Janmaat and Myers, 2006) and other toxins (Bourguet *et al.*, 2004). Fitness costs may be influenced by environmental stress and host-plant quality (Gassmann *et al.*, 2009b).

When resistance first evolves in a population, the majority of resistance alleles will reside in heterozygous individuals. As such, nonrecessive fitness costs, those affecting heterozygotes, will be most effective at delaying resistance. No-recessive fitness costs of resistance to *Cry* toxins have been documented, but these costs often display a high degree of ecological contingency, arising in some environments but not in others (Gassmann *et al.*, 2009b).

Fitness costs may affect symbionts or be affected by symbionts in the pest's body. Berticat *et al.* (2002b) demonstrated that *Wolbachia* density is altered by the presence of insecticide-resistant genes in the mosquito, *Culex pipiens. Wolbachia* are responsible for various alterations in host reproduction. Mosquito strains with genes conferring resistance were more infected by *Wolbachia* than by a susceptible strain. Berticat *et al.* (2002b) showed that this interaction also operates in natural populations and suggested that mosquitoes may control *Wolbachia* density less efficiently when they carry an insecticide-resistant gene and suffer a fitness cost.

Small population sizes may affect delays in resistance caused by fitness costs under the high-dose/refuge strategy for insecticidal crops. In simulation modeling, Gassmann *et al.* (2012a) found that increasing the magnitude of population bottlenecks could diminish and even eliminate delays in resistance caused by fitness costs. Because population bottlenecks create a small population size for one or more generations, resistance-allele frequency can increase or decrease from genetic drift. If alleles reach a sufficiently high level, the effects of fitness costs in delaying resistance are found to be minimal or absent.

HAPLO-DIPLOIDY

The population-genetics models described above all represent diploid species of arthropods. However, arthropod species are often haplo-diploid, which means that the males have one set of functioning chromosomes (haploid) and females have two sets. In these species, males either develop from unfertilized eggs or experience genome loss once an egg is fertilized either through inactivation or elimination of chromosomes during the early development of males. Chapter 9 describes *Mayetiola destructor*, which is haplo-diploid because of paternal genome loss. Carrière (2003) relates the evolution of haplo-diploid species to type of resistance mechanism. Pesticide resistance can result from a gain or loss of function. For example, detoxification of pesticides by enzymes is based on a gain of function when greater enzyme production reduces toxin concentration. Other examples of gain of functions involve reduced pesticide penetration or enhanced sequestration or excretion. An example of a loss of function is reduced sensitivity of target sites. According to Carrière (2003), when resistance involves a loss of function, R males and RR females should be equally tolerant of the pesticide, but with a gain of function RR females should have greater tolerance than R males. He postulated that, in most species, haploid males should be less tolerant to pesticides than diploid females. Carrière (2003) found support for his hypothesis by reviewing cases of sex-linked resistance in *Musca domestica*, *Ceratitis capitata*, *D. melanogaster*, and *Haematobia irritans*. In these diploid species, an allele associated with the female chromosome is not present or expressed on related male chromosomes.

In an associated empirical analysis, Carrière (2003) tested his hypothesis that tolerance to pesticides is lower in males than in females in haplo-diploid systems, by comparing the relative tolerance of males and females between haplo-diploid and diploid arthropods. He reviewed 16 reports pertaining to 10 haplo-diploid species involving 56 cases of pesticide tolerance observed in both males and females. He also obtained 85 cases of tolerance in both sexes from 33 reports on diploid species. Carrière (2003) found that the ratio of male to female tolerance is much smaller in haplo-diploid than in diploid arthropods (Figure 5.6), indicating that resistance alleles are not strongly upregulated (with gain of function) in haploid males. He then assessed whether factors other than ploidy affect male tolerance and discovered that males were generally less tolerant than females in both haplo-diploid and diploid arthropods. Carrière (2003) concluded that sexual size dimorphism and sex-dependent selection may account for the lower tolerance in males than in females. Therefore, the lower tolerance of males, particularly in haplo-diploid species, must be considered when developing model predictions and IRM strategies (Caprio and Hoy, 1995; Crowder et al., 2006).

By contrast, if resistance is conferred by target site insensitivity, then haploid males with a resistance allele will experience survival similar to that of a homozygous resistant female. This phenomenon has been found for sweet potato whitefly *Bemisia tabaci* with resistance to the insecticide pyriproxyfen (Crowder et al., 2008). Even at low-allele frequencies, haploid males will express the resistant phenotype, while most females, which will be heterozygous, may be phenotypically susceptible to insecticide. Under this scenario, resistance is expected to develop more quickly for haplo-diploid species than for diploid insects (Crowder and Carrière, 2009).

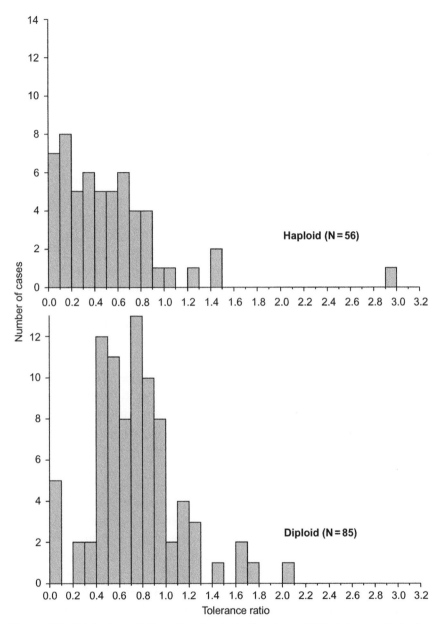

Figure 5.6 Distribution of the ratio of male to female pesticide tolerance in haplo-diploid and diploid arthropods. *From Carrière, Y., 2003,. with permission from Entomological Society of America.*

RESISTANCE EVOLUTION AND PEST GENERATION TIME

Two empirical studies have discovered an apparent influence of generation time on the rate of evolution of arthropod resistance. Both Georghiou (1980) and Tabashnik and Croft (1985) found that the shorter the generation time (greater number of arthropod generations per year), the faster the evolution of resistance. Georghiou (1980) analyzed data for 7 species of soil-dwelling crop pests selected by cyclodiene insecticides, while Tabashnik and Croft (1985) evaluated data for 24 arthropod species selected by an organophosphate pesticide in apple orchards. In a third investigation, Rosenheim and Tabashnik (1991) also identified a similar relationship in their empirical analysis of 56 pests of apple and pear.

However, in their overall analysis of 682 arthropod species in North America, Rosenheim and Tabashnik (1991) found little influence of generation time on the evolution of resistance. The highest rates of evolution were observed for arthropods with 3.5–10.5 generations per year. Thus, pests with a medium number of generations per year, including the 56 pome-fruit species (all with less than 13 generations per year), may be more likely to evolve resistance to a pesticide more quickly than those with shorter or longer generation times. Nevertheless, Rosenheim and Tabashnik (1991) provide several reasons why generation time cannot be directly influencing the rate of resistance evolution. First, secondary pests not targeted by pesticide applications will have varying numbers of applications applied during each of their generations. Thus, secondary pests may incur zero to many pesticide applications in a given generation. Second, although generation time significantly influences the intrinsic rate of increase of a population, the realized rate of population growth for many pests will vary over time and space to obscure any effects of generation time. Third, not all pest generations are the same from a management perspective. For example, crops may be vulnerable to damage and only need protection during a short period coinciding with one or two pest generations. Generations of the pest occurring at other periods may be insignificant and not receive any pesticide applications. Fourth, pests that have long generation times and also damage a crop or livestock for much of the time during each generation will likely be treated several times per generation with a pesticide; the opposite scenario can occur with pests passing through several generations without incurring even one treatment. Therefore, Rosenheim and Tabashnik (1991) concluded that no direct relationship should be expected between pest generation time and the rate of resistance evolution.

In a modeling study, Rosenheim and Tabashnik (1990) also drew the same conclusion, but noted that complex interactions between generation time and genetic, demographic, and management factors could result in some significant influence of generation time on the predicted number of years for resistance to evolve. To date, cases of pest resistance to transgenic insecticidal crops have arisen for both univoltine pests (Gassmann et al., 2011, 2012b; Gassmann, 2012) and multivoltine pests (van Rensburg, 2007; Storer et al., 2010; Dhurua and Gujar, 2011).

TEMPORAL AND SPATIAL SCALES IN HYPOTHESES

A number of ecologists have expressed concern about the lack of temporal and spatial scales in ecological hypotheses (Levandowsky and White, 1977; Allen and Starr, 1982; O'Neill et al., 1986; Peters, 1988; Roughgarden et al., 1989; Weins, 1989). Vagueness and the lack of operational definitions are indications of immature theories (Loehle, 1987; Murray, 2001; Krebs, 2006) and prevent us from evaluating predictions about evolution over time and space. Often claims are made and conclusions drawn about the conditions that promote or inhibit the evolution of arthropod resistance without reference to temporal and spatial scales. Without such scales, we do not know whether a given concept pertains to a square meter and a day or to a million square kilometers and a year. If hypotheses in population genetics are to be tested and implemented, precise concepts are needed that include general temporal and spatial scales for which the concepts are valid. Onstad (1992) evaluated this problem in epidemiology and proposed criteria for identifying appropriate scales and definitions of important terms.

Criteria for identifying temporal and spatial scales should be based on consistency of observation and ecological validity. Scales must account for differences in behavior and longevity among phenotypes. Units must be effective for both discrete and overlapping generations. Model computation or analysis may require small units of time and space to ensure proper calculation of functions and stable results, but these computational units are not the conceptual units of interest here.

Temporal and spatial units must correspond for logical reasons. The minimum time unit for analysis of evolution is clearly the generation time for the targeted arthropod. Onstad (1992) discusses the various ways to measure generation time for arthropods with discrete or overlapping generations. The minimum spatial unit should be the two- or three-dimensional space that the targeted arthropod traverses on average during a generation. The proper ecological, spatial scale depends on the behavior of a species, and therefore, must be large enough to encompass all normal movement by the average or median individuals in the phenotypes (Weins, 1976; Addicott et al., 1987; Caprio, 2001).

The maximum spatial and temporal scales should also be declared in any hypothesis. This issue is similar to the choice of time horizon, which is important for management of resistance (Mitchell and Onstad, Chapter 2). Ecological and genetic conditions underlying the hypothesis may not be valid after a certain number of arthropod generations or when a very large number of minimum spatial units are considered together. Obviously the maximum spatial scale should not exceed the existing area inhabited by a species, and even areas that are inhabited that have never and will never be treated outside a core area may not be properly included in hypotheses.

Scientists and IRM practitioners should use these minimum and maximum scales to test or implement the hypotheses. Samples should not be taken at units smaller than the minima, nor should they be taken in times or areas beyond the boundaries defined by the maxima. For example, should resistance-allele frequency be measured as the mean for a minimum spatial unit, for an area logically selected as the maximum, or for a region encompassing a species' entire geographic range? The last-named effort would likely indicate that resistance is on average very low, while at a smaller unit it may detect an increase in resistance.

CONCLUSIONS

Natural selection of arthropod populations occurs in heterogeneous landscapes where management tactics are applied in a variable manner over time and space. The intensity of selection (e.g., number of insecticide applications) is often the most important factor determining how quickly resistance evolves. Additionally, the genetic dominance of resistance, dispersal of genotypes within a landscape, and fitness costs can greatly influence the rate of resistance evolution. The greater the dominance of resistance, the faster resistance evolution occurs. On the other hand, the greater the fitness costs, the slower the evolution of resistance. In general, resistance evolution is delayed by dispersal of unselected individuals into habitats that harbor resistant individuals and are under selection.

These conditions and processes highlight the complexity of real systems in which we attempt to manage resistance. Epistasis and genetic linkage likely will have unpredictable effects on IRM, making generalizations difficult. Random genetic drift may either slow or accelerate the evolution of resistance during its early phases. Because gene flow, dispersal, and mating behaviors can all affect the rate of resistance evolution, these factors must be carefully evaluated before making predictions regarding a particular IRM strategy. The following chapters in this book describe how scientists and IRM practitioners have interpreted and conceptualized population genetics within the context of each system's complexity.

REFERENCES

Addicott, J.F., Aho, J.M., Antolin, M.F., Padilla, D.K., Richardson, J.S., Soluk, D.A., 1987. Ecological neighborhoods: scaling environmental patterns. Oikos. 49, 340–346.

Allen, T.F.H., Starr, T.B., 1982. Hierarchy. University of Chicago Press, Chicago.

Alstad, D.N., Andow, D.A., 1995. Managing the evolution of insect resistance to transgenic plants. Science. 268, 1894–1896.

Alyokhin, A.V., Ferro, D.N., 1999a. Relative fitness of Colorado potato beetle (Coleoptera: Chrysomelidae) resistant and susceptible to the *Bacillus thuringiensis* Cry3A toxin. J. Econ. Entomol. 92, 510–515.

Alyokhin, A.V., Ferro, D.N., 1999b. Electrophoretic confirmation of sperm mixing in mated Colorado potato beetles (Coleoptera: Chrysomelidae). Ann. Entomol. Soc. Am. 92, 230–235.

Anilkumar, K.J., Pusztai-Carey, M., Moar, W.J., 2008. Fitness costs associated with Cry1Ac-resistant *Helicoverpa zea* (Lepidoptera: Noctuidae): a factor countering selection for resistance to Bt cotton? J. Econ. Entomol. 101, 1421–1431.

Arnaud, L., Haubruge, E., 2002. Insecticide resistance enhances male reproductive success in a beetle. Evolution. 56, 2435–2444.

Baker, M.B., Alyokhin, A., Dastur, -S.R., Porter, -A.H., Ferro, -D.N., 2005. Sperm precedence in overwintered Colorado potato beetle (Coleoptera: Chrysomelidae) and its implications for insecticide resistance management. Ann. Entomol. Soc. Am. 98, 989–995.

Berticat, -C., Boquien, -G., Raymond, -M., Chevillon, -C., 2002a. Insecticide resistance genes induce a mating competition cost in *Culex pipiens* mosquitoes. Genet. Res. 79, 41–47.

Berticat, C., Rousset, F., Raymond, M., Berthomieu, A., Weill, M., 2002b. High Wolbachia density in insecticide-resistant mosquitoes. Proc. R. Soc. London B. 269, 1413–1416.

Binning, R.R., Lefko, S.A., Millsap, A.Y., Thompson, S.D., Nowatzki, T.M., 2010. Estimating western corn rootworm (Coleoptera: Chrysomelidae) larval susceptibility to event DAS-59122-7 maize. J. Appl. Entomol. 134, 551–561.

Bird, L.J., Akhurst, R.J., 2005. The fitness of Cry1A-resistant and -susceptible Helicoverpa armigera (Lepidoptera: Noctuidae) on transgenic cotton with reduced levels of Cry1Ac. J. Econ. Entomol. 59, 1166–1168.

Boivin, T., d'Hieres, C.C., Bouvier, J.C., Beslay, D., Sauphanor, B., 2001. Pleiotropy of insecticide resistance in the codling moth. Cydia Pomonella. Entomol. Exp. Appl. 99, 381–386.

Bourguet, D., Genissel, A., Raymond, M., 2000. Insecticide resistance and dominance levels. J. Econ. Entomol. 93, 1588–1595.

Bourguet, D., Guillemaud, T., Chevillon, C., Raymond., M., 2004. Fitness costs of insecticide resistance in natural breeding sites of the mosquito Culex pipiens. Evolution. 58, 128–135.

Bouvier, J.-C., Boivin, T., Beslay, D., Sauphanor, B., 2002. Age-dependent response to insecticides and enzymatic variation in susceptible and resistant codling moth larvae. Arch. Insect Biochem. Physiol. 51, 55–66.

Caprio, M.A., 1994. Bacillus thuringiensis gene deployment and resistance management in single- and multi-tactic environments. Biocontrol Sci. Technol. 4, 487–497.

Caprio, M.A., 2001. Source-sink dynamics between transgenic and non-transgenic habitats and their role in the evolution of resistance. J. Econ. Entomol. 94, 698–705.

Caprio, M.A., Hoy, M.A., 1994. Metapopulation dynamics affect resistance development in the predatory mite, Metaseiulus occidentalis (Acari: Phytoseiidae). J. Econ. Entomol. 87, 525–534.

Caprio, M.A., Hoy, M.A., 1995. Premating isolation in a simulation model generates frequency-dependent selection and alters establishment rates of resistant natural enemies. J. Econ. Entomol. 88, 205–212.

Caprio, M.A., Tabashnik, B.E., 1992. Gene flow accelerates local adaptation among finite populations: simulating the evolution of insecticide resistance. J. Econ. Entomol. 85, 611–620.

Carrière, Y., 2003. Haplodiploidy, sex, and the evolution of pesticide resistance. J. Econ. Entomol. 96, 1626–1640.

Carrière, Y., Ellers-Kirk, C., Sisterson, M.S., Antilla, L., Whitlow, M., Dennehy, T.J., et al., 2003. Long-term regional suppression of pink bollworm by Bacillus thuringiensis cotton. Proc. Natl. Acad. Sci. USA. 100, 1519–1523.

Carrière, Y., Ellers-Kirk, C., Biggs, R., Degain, B., Holley, D., Yafuso, C., et al., 2005. Effects of cotton cultivar on fitness costs associated with resistance of pink bollworm (Lepidoptera: Gelechiidae) to Bt cotton. J. Econ. Entomol. 98, 947–954.

Comins, H.N., 1977. The development of insecticide resistance in the presence of migration. J. Theor. Biol. 64, 177–197.

Crespo, A.L.B., Spencer, T.A., Alves, A.A., Hellmich, R.L., Blankenship, E.E., Magalhães, L.C., et al., 2008. On-plant survival and inheritance of resistance to Cry1Ab toxin from Bacillus thuringiensis in a field derived strain of European corn borer, Ostrinia nubilalis. Pest. Manag. Sci. 65, 1071–1081.

Crespo, A.L.B., Spencer, T.A., Tan, S.Y., Siegfried, B.D., 2010. Fitness costs of cry1Ab resistance in a field-derived strain of Ostrinia nubilalis (Lepidoptera: Crambidae). J. Econ. Entomol. 103, 1386–1393.

Crow, J.F., 1954. Analysis of a DDT-resistant strain of Drosophila. J. Econ. Entomol. 47, 393–398.

Crowder, D.W., Carrière, Y., 2009. Comparing the refuge strategy for managing the evolution of insect resistance under different reproductive strategies. J. Theor. Biol. 261, 423–430.

Crowder, D.W., Carrière, Y., Tabashnik, B.E., Ellsworth, P.C., Dennehy, T.J., 2006. Modeling evolution of resistance to pyriproxyfen by sweetpotato whitefly (Homoptera:Aleyrodidae). J. Econ. Entomol. 99, 1396–1406.

Crowder, D.W., Ellers-Kirk, C., Yafuso, C.M., Dennehy, T.J., Degain, B.A., Harpold, V.S., et al., 2008. Inheritance of resistance to pyriproxyfen in Bemisia tabaci (Hemiptera: Aleyrodidae) males and females (B biotype). J. Econ. Entomol. 101, 927–932.

Daborn, P.J., Yen, J.L., Bogwitz, M., LeGoff, G., Feil, E., Jeffers, S., et al., 2002. A single allele of a P450 gene is associated with insecticide resistance in Drosophila. Science. 297, 2353–2356.

Dhurua, S., Gujar, G.T., 2011. Field-evolved resistance to Bt toxin Cry1Ac in the pink bollworm, Pectinophora gossypiella (Saunders) (Lepidoptera: Gelechiidae) from India. Pest. Manag. Sci. 67, 898–903.

Doherty, H.M., Hales, D.F., 2002. Mating success and mating behaviour of the aphid, Myzus persicae (Hemiptera: Aphididae). Eur. J. Entomol. 99, 23–27.

El Bouhssini, M., Hatchett, J.H., Cox, T.S., Wilde, G.E., 2001. Genotypic interaction between resistance genes in wheat and virulence genes in the Hessian fly Mayetiola destructor (Diptera: Cecidomyiidae). Bull. Entomol. Res. 91, 327–331.

Endler, J.A., 1986. Natural selection in the wild, Monographs in Population Biology, vol. 21. Princeton University Press, Princeton, NJ.

Environmental Protection Agency, 1998. Final Report of the FIFRA Scientific Advisory Panel Subpanel on Bacillus thuringiensis (Bt) Plant-Pesticides and Resistance Management. Accessed at <http://www.epa.gov/scipoly/sap/meetings/1998/february/finalfeb.pdf>.

Falconer, D.S., Macay, T.E., 1996. Introduction to Quantitative Genetics, fourth ed. Pearson, Harlow.

Felsenstein, J., 1976. The theoretical population genetics of variable selection and migration. Annu. Rev. Genet. 10, 253–280.

Ferré, J., van Rie, J., MacIntosh, S.C., 2008. Insecticidal genetically modified crops and insect resistance management (IRM). In: Romeis, J., Shelton, A.M., Kennedy, G.G. (Eds.), Integration of Insect-Resistant Genetically Modified Crops within IPM Programs. Springer, New York.

ffrench-Constant, R.H., Anthony, N.M., Andreev, D., Aronstein, K., 1996. Single versus multiple origins of insecticide resistance: inferences from the cyclodiene resistance gene Rdl. In: Brown, T.M. (Ed.), Molecular Genetics and Evolution of Pesticide Resistance. American Chemical Society, Washington, DC, pp. 106–116, 265 pp.

ffrench-Constant, R.H., Daborn, P.J., Goff, G.L., 2004. The genetics and genomics of insecticide resistance. Trends Genet. 20, 163–170.

Frank, S.A., Slatkin, M., 1992. Fisher's fundamental theorem of natural selection. Trends Ecol. Evol. 7, 92–95.

Gassmann, A.J., 2012. Field-evolved resistance to Bt maize by western corn rootworm: predictions from the laboratory and effects in the field. J. Invertebr. Pathol. 110, 287–293.

Gassmann, A.J., Stock, S.P., Carrière, Y., Tabashnik, B.E., 2006. Effect of entomopathogenic nematodes on the fitness cost of resistance to Bt toxin Cry1Ac in pink bollworm (Lepidoptera: Gelechiidae). J. Econ. Entomol. 99, 920–926.

Gassmann, A.J., Stock, S.P., Sisterson, M.S., Carrière, Y., Tabashnik, B.E., 2008. Synergism between entomopathogenic nematodes and Bacillus thuringiensis crops: integrating biological control and resistance management. J. Appl. Ecol. 45, 957–966.

Gassmann, A.J., Fabrick, J.A., Sisterson, M.S., Hannon, E.R., Stock, S.P., Carrière, Y., et al., 2009a. Effects of pink bollworm resistance to *Bacillus thuringiensis* on phenoloxidase activity and susceptibility to entomopathogenic nematodes. J. Econ. Entomol. 102, 1224–1232.

Gassmann, A.J., Carrière, Y., Tabashnik, B.E., 2009b. Fitness costs of insect resistance to *Bacillus thuringiensis*. Annu. Rev. Entomol. 54, 147–163.

Gassmann, A.J., Onstad, D.W., Pittendrigh, B.R., 2009c. Evolutionary analysis of herbivorous insects in natural and agricultural environments. Pest. Manag. Sci. 65, 1174–1181.

Gassmann, A.J., Petzold-Maxwell, J.L., Keweshan, R.S., Dunbar, M.W., 2011. Field-evolved resistance to Bt maize by western corn rootworm. PLoS One. 6 (7), e22629. Available from: http://dx.doi.org/10.1371/journal.pone.0022629.

Gassmann, A.J., Hannon, E.R., Sisterson, M.S., Stock, S.P., Carrière, Y., Tabashnik, B.E., 2012a. Effects of entomopathogenic nematodes on the evolution of pink bollworm resistance to Bt toxin Cry1Ac. J. Econ. Entomol. 105, 994–1005.

Gassmann, A.J., Petzold-Maxwell, J.L., Keweshan, R.S., Dunbar, M.W., 2012b. Western corn rootworm and Bt maize: challenges of pest resistance in the field. GM Crops Food. 3, 235–244.

Georghiou, G.P., 1980. Insecticide resistance and prospects for its management. Residue Rev. 76, 131–145.

Gould, F., 1986a. Simulation models for predicting durability of insect-resistant germ plasm: a deterministic diploid, two-locus model. Environ. Entomol. 15, 1–10.

Gould, F., 1986b. Simulation models for predicting durability of insect-resistant germ plasm: hessian fly (Diptera: Cecidomyiidae)-resistant winter wheat. Environ. Entomol. 15, 11–23.

Gould, F., Cohen, M.B., Bentur, J.S., Kennedy, G.G., Van Duyn, J., 2006. Impact of small fitness costs on pest adaptation to crop varieties with multiple toxins: a heuristic model. J. Econ. Entomol. 99, 2091–2099.

Groeters, F.R., Tabashnik, B.E., 2000. Roles of selection intensity, major genes, and minor genes in the evolution of insecticide resistance. J. Econ. Entomol. 93, 1580–1587.

Groeters, F.R., Tabashnik, B.E., Finson, N., Johnson, M.W., 1993. Resistance to *Bacillus thuringiensis* affects mating success of the diamondback moth (Lepidoptera: Plutellidae). J. Econ. Entomol. 86, 1035–1039.

Gulzar, A., Pickett, B., Sayyed, A.H., Wright, D.J., 2012. Effect of temperature on the fitness of a Vip3a resistant population of *Heliothis virescens* (Lepidoptera: Noctuidae). J. Econ. Entomol. 105, 964–970.

Guse, C.A., Onstad, D.W., Buschman, L.L., Porter, P., Higgins, R.A., Sloderbeck, P.E., et al., 2002. Modeling the development of resistance by stalk-boring Lepidoptera (Crambidae) in areas with irrigated, transgenic corn. Environ. Entomol. 31, 676–685.

Hansen, T.F., 2006. The evolution of genetic architecture. Annu. Rev. Ecol. Evol. Syst. 37, 123–157.

Haubruge, E., Arnaud, L., Mignon, J., 1997. The impact of sperm precedence in malathion resistance transmission in populations of the red flour beetle *Tribolium castaneum* (Herbst) (Coleoptera: Tenebrionidae). J. Stored Prod. Res. 33, 143–146.

Hedrick, P.W., 2006. Genetic polymorphism in heterogeneous environments: the age of genomics. Annu. Rev. Ecol. Evol. Syst. 37, 67–93.

Hendry, A.P., Kinnison, M.T., 1999. Perspective: the pace of modern life: measuring rates of contemporary microevolution. Evolution. 53, 1637–1653.

Higginson, D.M., Morin, S., Nyboer, M., Biggs, R., Tabashnik, B.E., Carrière., Y., 2005. Evolutionary trade-offs of insect resistance to Bt crops: fitness costs affecting paternity. Evolution. 59, 915–920.

Hoy, C.W., Head, G., Hall, F.R., 1998. Spatial heterogeneity and insect adaptation to toxins. Annu. Rev. Entomol. 43, 571—594.

Hutchison, W., Burkness, E., Mitchell, P., Moon, R., Leslie, T., Fleischer, S., et al., 2010. Areawide suppression of European corn borer with Bt maize reaps savings to non-Bt maize growers. Science. 330, 222—225.

Ives, A.R., Andow, D.A., 2002. Evolution of resistance to *Bt* crops: directional selection in structured environments. Ecol. Lett. 5, 792—801.

Janmaat, A.F., Myers, J.H., 2006. The influences of host plant and genetic resistance to *Bacillus thuringiensis* on trade-offs between offspring number and growth rate in cabbage loopers, *Trichoplusia ni*. Ecol. Entomol. 31, 172—178.

Kanarek, A.R., Webb, C.T., 2010. Allee effects, adaptive evolution, and invasion success. Evol. Appl. 3, 122—135.

Karasov, T., Messer, P.W., Petrov, D.A., 2010. Evidence that adaptation in *Drosophila* is not limited by mutation at single sites. PLoS Genet. 6, e1000924.

Kawecki, T.J., 2000. Adaptation to marginal habitats: contrasting influence of the dispersal rate on the fate of alleles with small and large effects. Proc. R. Soc. Lond. B. 267, 1315—1320.

Knolhoff, L.M., Walden, K.K.O., Ratcliffe, S.T., Onstad, D.W., Robertson., H.M., 2009. Microarray analysis yields candidate markers for rotation resistance in the western corn rootworm beetle, *Diabrotica virgifera virgifera*. Evol. Appl. 3, 17—27.

Krebs, C.J., 2006. Ecology after 100 years: progress and pseudo-progress. N.Z.J. Ecol. 30, 3—11.

Labbe, P., Berticat, C., Berthomieu, A., Unal, S., Bernard, C., Weill, M., et al., 2007. Forty years of erratic insecticide resistance evolution in the mosquito *Culex pipiens*. PLoS Genet. 3, 2190—2199.

Levandowsky, M., White, B.S., 1977. Randomness, time scales, and the evolution of biological communities. Evol. Biol. 10, 69—161.

Liu, X., Bai, J., Huang, L., Zhu, L., Liu, X., Weng, N., et al., 2007. Gene expression of different wheat genotypes during attack by virulent and avirulent Hessian fly (*Mayetiola destructor*) larvae. J. Chem. Ecol. 33, 2171—2194.

Loehle, C., 1987. Hypothesis testing in ecology: psychological aspects and the importance of theory maturation. Q. Rev. Biol. 62, 397—409.

Mallet, J., 2001. Gene flow. In: Woiwood, I.P., Reynolds, D.R., Thomas, C.D. (Eds.), Insect Movement: Mechanisms and Consequences. CAB International, New York (Chapter 16).

Maynard Smith, J., 1989. Evolutionary Genetics. Oxford University Press, Oxford.

McKenzie, J.A., Batterham, P., 1998. Predicting insecticide resistance: mutagenesis, selection and response. Phil. Trans. R. Soc. Lond. B. 353, 1729—1734.

Murray, B.G., 2001. Are ecological and evolutionary theories scientific? Biol. Rev. 76, 255—289.

Olsen, K.M., Daly, J.C., Holt, H.E., Finnegan, E.J., 2005. Season-long variation in expression of Cry1Ac gene and efficacy of *Bacillus thuringiensis* toxin in transgenic cotton against *Helicoverpa armigera* (Lepidoptera: Noctuidae). J. Econ. Entomol. 98, 1007—1017.

O'Neill, R.V., DeAngelis, D.L., Waide, J.B., Allen, T.F.H., 1986. A Hierarchical Concept of Ecosystems. Princeton University Press, Princeton, NJ.

Onstad, D.W., 1992. Temporal and spatial scales in epidemiological concepts. J. Theor. Biol. 158, 495—515.

Onstad, D.W., Gould, F., 1998. Do dynamics of crop maturation and herbivorous insect life cycle influence the risk of adaptation to toxins in transgenic host plants? Environ. Entomol. 27, 517—522.

Onstad, D.W., Hibbard, B.E., Clark, T.L., Crowder, D.W., Carter., K.G., 2006. Analysis of density-dependent survival of *Diabrotica* (Coleoptera: Chrysomelidae) in cornfields. Environ. Entomol. 35, 1272–1278.

Orr, H.A., Coyne, J.A., 1992. The genetics of adaptation: a reassessment. Am. Nat. 140, 725–742.

Pedra, J.H.F., McIntyre, L.M., Scharf, M.E., Pittendrigh, B.R., 2004. Genome-wide transcription profile of field- and laboratory-selected DDT-resistant *Drosophila*. Proc. Nat. Acad. Sci. USA. 101, 7034–7039.

Peters, R.H., 1988. Some general problems for ecology illustrated by food web theory. Ecology. 69, 1673–1676.

Rinkevich, F.D., Zhang, L., Hamm, R.L., Brady, S.G., Lazzaro, B.P., Scott, J.G., 2006. Frequencies of the pyrethroid resistance alleles of Vssc1 and CYP6D1 in house flies from the eastern United States. Insect Mol. Biol. 15, 157–167.

Rolff, J., Kraaijeveld, A.R., 2003. Selection for parasitoid resistance alters mating success in *Drosophila*. Proc. Roy. Soc. London. Series B. 270, S154–S155.

Rosenheim, J.A., Tabashnik, B.E., 1990. Evolution of pesticide resistance: interactions between generation time and genetic, ecological, and operational factors. J. Econ. Entomol. 83, 1184–1193.

Rosenheim, J.A., Tabashnik, B.E., 1991. Influence of generation time on the rate of response to selection. Am. Nat. 137, 527–541.

Roughgarden, J., 1979. Theory of Population Genetics and Evolutionary Biology: An Introduction. Macmillan Publishing Co., London.

Roughgarden, J., May, R.M., Levin, S.A., 1989. Perspectives in Ecological Theory. Princeton University Press, Princeton, NJ.

Roush, R.T., Daly, J.C., 1990. The role of population genetics in resistance research and management. In: Roush, R.T., Tabashnik, B.E. (Eds.), Pesticide Resistance in Arthropods. Chapman and Hall, New York (Chapter 5).

Roush, R.T., McKenzie, J.A., 1987. Ecological genetics of insecticide and acaricide resistance. Annu. Rev. Entomol. 32, 361–380.

Siegfried, B.D., Vaughn, T.T., Spencer, T., 2005. Baseline susceptibility of western corn rootworm (Coleoptera: Chrysomelidae) to Cry3Bb1 *Bacillus thuringiensis* toxin. J. Econ. Entomol. 98, 1320–1324.

Sisterson, M.S., Antilla, L., Carrière, Y., Ellers-Kirk, C., Tabashnik, B.E., 2004. Effects of insect population size on evolution of resistance to transgenic crops. J. Econ. Entomol. 97, 1413–1424.

Sisterson, M.S., Carrière, Y., Dennehy, T.J., Tabashnik, B.E., 2005. Evolution of resistance to transgenic crops: interactions between insect movement and field distribution. J. Econ. Entomol. 98, 1751–1762.

Spencer, J., Onstad, D., Krupke, C., Hughson, S., Pan, Z., Stanley, B., et al., 2013. Isolated females and limited males: evolution of insect resistance in structured landscapes. Entomol. Exp. Appl. 146, 38–49.

Spiess, E.B., 1977. Genes in Populations. Wiley and Sons, New York.

Storer, N.P., Peck, S.L., Gould, F., Van-Duyn, J.W., Kennedy, G.G., 2003. Spatial processes in the evolution of resistance in *Helicoverpa zea* (Lepidoptera: Noctuidae) to Bt transgenic corn and cotton in a mixed agroecosystem: a biology-rich stochastic simulation model. J. Econ. Entomol. 96, 156–172.

Storer, N.P., Babcock, J.M., Schlenz, M., Meade, T., Thompson, G.D., Bing, J.W., et al., 2010. Discovery and characterization of field resistance to Bt maize: *spodoptera frugiperda* (Lepidoptera: Noctuidae) in Puerto Rico. J. Econ. Entomol. 103, 1031–1038.

Tabashnik, B.E., 1990. Implications of gene amplification for evolution and management of insecticide resistance. J. Econ. Entomol. 83, 1170–1176.

Tabashnik, B.E., Croft, B.A., 1982. Managing pesticide resistance in crop-arthropod complexes: interactions between biological and operational factors. Environ. Entomol. 11, 1137—1144.

Tabashnik, B.E., Croft, B.A., 1985. Evolution of pesticide resistance in apple pests and their natural enemies. Entomophaga. 30, 37—49.

Tabashnik, B.E., Gould, F., Carrière, Y., 2004. Delaying evolution of insect resistance to transgenic crops by decreasing dominance and heritability. J. Evol. Biol. 17, 904—912.

Tabashnik, B.E., Gassmann, A.J., Crowder, D.W., Carrière, Y., 2008. Insect resistance to Bt crops: evidence versus theory. Nat. Biotechnol. 26, 199—202.

Taylor, C.E., Georghiou, G.P., 1979. Suppression of insecticide resistance by alteration of gene dominance and migration. J. Econ. Entomol. 72, 105—109.

Thompson, J.N., 1994. The Coevolutionary Process. The University of Chicago Press, Chicago.

Trimble, R.M., El-Sayed, A.M., Pree, D.J., 2004. Impact of sub-lethal residues of azinphos-methyl on the pheromone-communication systems of insecticide-susceptible and insecticide-resistant obliquebanded leafrollers Choristoneura rosaceana (Lepidoptera: Tortricidae). Pest. Manag. Sci. 60, 660—668.

van Rensburg, J.B.J., 2007. First report of field resistance by stem borer, Busseola fusca (Fuller) to Bt-transgenic maize. S. Afr. J. Plant Soil. 24, 147—151.

Weins, J.A., 1976. Population responses to patchy environments. Ann. Rev. Ecol. Syst. 7, 81—120.

Weins, J.A., 1989. Spatial scaling in ecology. Funct. Ecol. 3, 385—397.

Wu, J., Liu, X., Zhang, S., Zhu, Y.-C., Whitworth, R.J., Chen, M.-S., 2008. Differential responses of wheat inhibitor-like genes to Hessian fly, Mayetiola destructor, attacks during compatible and incompatible interactions. J. Chem. Ecol. 34, 1005—1012.

Zhao, X.C., Wu, K.M., Liang, G.M., Guo, Y.Y., 2009. Modified female calling behaviour in Cry1Ac-resistant Helicoverpa armigera (Lepidoptera: Noctuidae). Pest. Manag. Sci. 65, 353—357.

CHAPTER 6

Resistance by Ectoparasites

Lisa M. Knolhoff[1] and David W. Onstad[2]

[1]Genective, AgReliant Genetics, Champaign, IL
[2]DuPont Pioneer, Wilmington, DE

Chapter Outline

Control of ectoparasites is important because of the inherent value of the animal host. The evolution of resistance to insecticides and acaricides in these pests is a major concern, but economic or ethical limitations leave few insect resistance management (IRM) options available. This chapter highlights the similarities between the resistance problems for the variety of pests infesting humans, livestock, pets, and domesticated bees. There has been much work on resistance mechanisms, subsequent monitoring, and even predictions of resistance, but less has been done to actively and effectively manage resistance. The economics or value of the host precludes certain steps from being taken. One cannot ethically put a value on human life, which severely limits control and IRM options for both mosquitoes and lice. For livestock and apiary pests, the case studies demonstrate that improved IRM will depend on implementing more sophisticated integrated pest management (IPM).

Insect Resistance Management
DOI: http://dx.doi.org/10.1016/B978-0-12-396955-2.00006-0

DEFINITIONS

A few descriptions of commonly used insecticides and resistance mechanisms are provided here because certain terms are used throughout the chapter. The purpose here is not to address resistance per se, but rather to inform the reader of mode of action and resistance to insecticides as they would relate to resistance management. Table 6.1 is not meant to be a comprehensive explanation of mode of action; subtle differences may occur which are not noted here. Likewise, the resistance mechanisms listed may not be the only means of adapting to a chemical control. A more complete summary of chemical modes of action can be found in Ware and Whitacre (2004), which is available online at http://ipmworld.umn.edu/chapters/ware.htm. A valuable resource of information about mode of action and how it relates to resistance management can be found at www.irac-online.org. The Insecticide Resistance Action Committee (IRAC) is an international organization dedicated to implementing appropriate resistance management strategies in agriculture and public health. Pittendrigh *et al.* (Chapter 3) present a modification of the IRAC classification system for the mode of action of insecticides.

Pyrethroids (and previously, the organochlorine dichloro-diphenyl-trichloroethane (DDT)) are commonly used for control of medical or veterinary pests because of their relatively low mammalian toxicity. They are synaptic nerve poisons; they cause sodium ion leakage from voltage-gated channels. Resistance to these compounds can occur through target site mutations in sodium channels or through metabolic detoxification. Pyrethroid insecticides are known for their rapid insecticidal effects; the major allele for target site resistance is called *kdr* for "knockdown resistance." If resistance is characteristically similar to *kdr* mutants, but the allele has not yet been identified in a particular insect, it may be referred to as "*kdr*-like." Finally, behavioral resistance may also occur because these compounds have an irritant effect.

The other major classes of insecticides have different modes of action. Organophosphate and carbamate insecticides inhibit acetylcholinesterase (AChE), leading to a buildup of acetylcholine in synapses. Resistance to these compounds usually occurs through metabolic detoxification or insensitive AChE. Cyclodiene organochlorines inhibit gamma-aminobutyric acid (GABA) receptors in neurons and therefore prevent

Table 6.1 Some Commonly Used Classes of Pesticides for Control of Ectoparasites

Class	Example	Mode of Action	Common Resistance Mechanisms (if applicable)
Pyrethroids	Permethrin, fluvalinate	Na^+ leakage in neurons	Target site, metabolic detoxification, or behavior modification
Organochlorines-1	DDT	Na^+ leakage in neurons	Target site, metabolic detoxification, or behavior modification
Organochlorines-2 (cyclodienes)	Dieldrin	Blocks GABA-gated chloride channels	Target site
Organophosphates	Malathion, coumaphos	AChE inhibition	Metabolic detoxification, target site
Carbamates	Carbaryl	AChE inhibition	Metabolic detoxification, target site
Formamidines	Amitraz	Binds to octopamine receptors	
Spinosyns	Spinosad	Nicotinic ACh receptor agonist*	
Neonicotinoids	Imidacloprid	Nicotinic ACh receptor agonist*	
Phenylpyrazoles	Fipronil	Blocks GABA-gated chloride channels	
Avermectin	Ivermectin	Activates chloride channels	

AChE: acetylcholinesterase; GABA: gamma-aminobutyric acid.
*Although the spinosyns and neonicotinoids have the same mode of action, the binding site of each to the ACh receptor is different.

chloride ion uptake. Target site resistance is most common; one example is the *Rdl* allele conferring resistance to dieldrin.

MOSQUITOES

Mosquitoes (Diptera: Culicidae) are unrivaled in their vector capability and are responsible for the transmission of a number of diseases of medical importance. Vector control is the main component of disease control programs because of its relative ease and lowered cost with respect to pathogen control. The fact that mosquitoes have different habitats at different life stages allows for more control options. Only adult females feed on blood; larvae are aquatic and nonparasitic. Preference for human hosts, anthrophily, or other animals, zoophily, also plays a role in vector control, disease transmission, and IRM strategies.

The short life cycle and high reproductive potential of mosquitoes predispose populations to evolving resistance. Resistance to at least one class of traditional synthetic insecticides is common in the major mosquito genera *Aedes*, *Anopheles*, and *Culex* that vector human disease (Hemingway and Ranson, 2000) and can occur via target site changes or metabolic mechanisms (Hemingway *et al.*, 2004). For example, resistance to multiple classes of insecticides—organochlorines, organophosphates, and pyrethroids—is found in mosquitoes in China (Cui *et al.*, 2006). Use of DDT and especially pyrethroids for control of *An. gambiae s.s.* vectors of malaria has led to widespread resistance in Africa (WHO, 2005a). Resistance to organophosphates in *Culex pipiens* is found in many parts of the world (Labbé *et al.*, 2005).

The study of resistance to organophosphates (OPs) in *C. pipiens* mosquitoes has allowed; certain aspects of selection and migration to be examined. In southern France, there is a cline of frequencies of resistance alleles for organophosphate insecticides, suggesting that mutations arose once and spread by migration of the insect (Chevillon *et al.*, 1999; Lenormand and Raymond, 2000). Duplications at the *ace-1* (acetylcholinesterase) locus conferring resistance to OPs contribute to a resistance haplotype, where subfunctionalization of paralogs could allow for compensation of costs associated with decreased acetylcholinesterase activity (Labbé *et al.*, 2007, 2009).

Multiple resistance mutations can even occur within one locus (Labbé *et al.*, 2005). The first resistance allele to appear at the *Ester* (esterase) locus in French populations of *C. pipiens* seems to have largely been replaced by a second resistance allele conferring a lower fitness cost (Labbé *et al.*, 2009). Certain alleles are favored in specific local environments, depending on both selection pressure with insecticides and natural climate/landscape conditions (Labbé *et al.*, 2005).

Insecticide resistance in mosquitoes is of special concern in areas of the world where malaria is endemic. Malaria (*Plasmodium* spp.) is a serious problem in many developing countries and is vectored by *Anopheles* mosquitoes, the most important of which is *An. gambiae* (Hemingway and Ranson, 2000). The widespread use of DDT to control malaria-vectoring populations in Africa and southern Asia has selected for resistance and caused the World Health Organization(WHO) to shift its goal from malaria eradication to malaria control (Hemingway and Ranson, 2000). WHO advocates the use of insecticide-treated bednets (ITNs) as the main intervention against malaria (WHO, 2005c, 2007a), the overall mortality and prevalence of which has decreased largely due to efforts to fully distribute nets (WHO, 2010; Lim *et al.*, 2011). Use of ITNs, though problems in implementation exist (Ordinioha, 2012), is likely to be a better IRM strategy in that control is used only where it is most needed and reduces the need for indiscriminate sprays of insecticide.

Malaria control focuses on female *Anopheles* mosquitoes because they, unlike males, require a blood meal. Most females coming in contact with an ITN searching for a host are mated (Curtis *et al.*, 1993). Host-seeking females tend to first land on the top part of the net because of the concentration of heat and carbon dioxide there and then search downwards on the net to find a human host (Guillet *et al.*, 2001).

Insecticide-Treated Bednets

ITNs covering sleeping people function like baited traps, in that host-seeking mosquitoes are either killed or repelled by the insecticide (WHO, 2006a). Irritant insecticides in ITNs are most effective in preventing bites because the mosquito will be repelled before encountering and biting a person. Some bites may occur, however, either through entry in a hole in a torn net or if the person is sleeping against the net and a mosquito bites through it. Besides ITNs, outdoor spraying and indoor residual spraying are commonly used for mosquito control. We will focus on the use of ITNs as an IRM strategy but will address these other methods later in this section.

Retreatment of ITNs is essential for their long-term efficacy, but because of difficulties in implementation, long-lasting insecticidal nets (LLINs) have been developed that offer an improvement in both structural and insecticidal durability (WHO, 2007b). LLINs are distinguished from conventionally treated ITNs in that they are factory-produced nets made of synthetic fibers impregnated with insecticide. The World Health Organization advocates the full coverage of LLINs for those populations at risk of malaria because of both increased effectiveness and economic benefit (WHO, 2007b; Yukich et al., 2008). For simplicity, and because LLINs and conventionally treated ITNs both fall under the inclusive term *ITN* (WHO, 2007a,b), we will use the ITN terminology to refer to both kinds of bednets.

Pyrethroids are the only class of insecticides currently approved for use with ITNs (WHO, 2007a). They are known for both excitorepellency, an increased tendency for mosquitoes to take off and fly, and rapid knockdown effect causing mortality of mosquitoes (Pates and Curtis, 2005). Mosquitoes have been known to evolve behavioral resistance such as exophily (Pates and Curtis, 2005), a behavior that causes mosquitoes to avoid internal walls treated with insecticide. There is no evidence yet, however, of behavioral resistance by the avoidance of ITNs (Pates and Curtis, 2005).

If using a single insecticide on an ITN, Curtis et al. (1998) noted that, because of the excitorepellency of pyrethroids, a low-dose strategy is better for IRM. With low doses, resistant heterozygotes leave ITNs before being killed by insecticide, reducing the selection pressure. With higher concentrations of insecticides, the heterozygotes incur higher mortality. In the initial stages of evolution of resistance, the frequency and fitness of heterozygotes are most critical.

Curtis (1985) modeled possible IRM strategies for mosquito control and found that the use of mixtures of insecticides was best at delaying resistance when alleles were at least partly recessive. Mixtures refer to the simultaneous use of two insecticides; it is thought that if resistance alleles are rare, then resistance to two insecticides should be especially rare. Furthermore, mixtures on an ITN decrease selection pressure on whole populations because it is assumed that only mated females are exposed to the toxins in the ITN when searching for a host (Curtis et al., 1993).

It has been hypothesized that pyrethroid resistance in anopheline mosquitoes has most likely not been selected by ITNs (Curtis et al., 1998;

Takken, 2002); it is generally not seen in areas without extensive pyrethroid use in agricultural areas (Vulule et al., 1996). However, Norris and Norris (2011a) conducted a study to test the resistance status of *Culex quinquefasciatus* after three years of ITN use in Zambia. They measured mortality and sublethal effects in response to ITNs. To quantify resistance, they compared resistance alleles between current samples and historical samples collected before ITN deployment; they also compared enzyme activity between current field samples and lab colony. This population was somewhat resistant to all insecticides tested and most of all to DDT. They did not detect the *ace-1R* allele, but they did see an increase over time in *kdr* allele frequency that coincided with introduction of ITNs. Frequency of *kdr*, however, did not perfectly correlate with pyrethroid resistance. Glutathione S-transferase (GST) enzyme activity was higher in field-collected samples compared to a laboratory colony, so resistance may occur both via target site and metabolic mechanisms. The authors did a similar survey of resistance with *Anopheles arabiensis* in Zambia after ITN introduction, and though they did not detect the *kdr* allele in these populations, they found low levels of resistance to DDT and deltamethrin (Norris and Norris, 2011b).

Pyrethroid-treated ITNs seem to remain effective where mosquitoes are resistant to pyrethroids via *kdr*, presumably because mosquitoes are not repelled before receiving a lethal dose (Chandre et al., 2000). Even in areas of pyrethroid resistance, ITNs still provide some protection (N'Guessan et al., 2007; Irish et al., 2008). The efficacy of ITNs can be improved when used in combination with other control methods. Djènontin et al. (2009, 2010) evaluated the efficacy of LLINs in combination with bendiocarb used in either indoor residual sprays or plastic sheeting in Burkina Faso in an area with resistance to pyrethroids and lower levels of resistance to carbamates in *Anopheles gambiae*. Mortality ranged from about 40 to 67% for all treatments, but LLINs and the plastic sheeting had the best residual activity. The highest level of exophily was found for the LLIN-only treatment, suggesting that pyrethroid-resistant *An. gambiae* are still deterred by the deltamethrin in the net. The most important advantage with LLINs is that any treatment that included them had significantly fewer bloodmeals. Djènontin et al. (2010) also found a high frequency of *kdr*, but it was not related to observed mortality in any treatment with LLINs. The frequency of *ace-1R* was lower, but it was significantly associated with survival in the treatments with the carbamate alone, either as an indoor spray or in the plastic sheeting (Djènontin *et al.*, 2010).

Another proposed IRM strategy was to create ITNs that function as a mixture of insecticides in that a mosquito comes in contact with two insecticides during a single attempt to feed. Guillet *et al.* (2001) tested one approach involving the use of ITNs that have been treated on the top half with a nonirritant insecticide and on the bottom half with a pyrethroid (Figure 6.1). This takes advantage of mosquito host-searching behavior; they tend to land on the top of the net and then travel down, making the ITN effectively a mixture, rather than a mosaic, of insecticide treatment (Guillet *et al.*, 2001). As a follow-up to this idea, Oxborough *et al.* (2008) tested ITNs that were treated with lambda-cyhalothrin on only one-half of the net. Their data suggested that, indeed, female mosquitoes contact multiple areas of the net, but they also noted potential difficulties in preventing cross-contamination of insecticides on different parts of the net while in storage (Oxborough *et al.*, 2008). Guillet *et al.* (2001) found the best control when ITNs are treated with a carbamate on the top half and a pyrethroid on the bottom half. Placing the nonpyrethroid insecticide on the top of the net, farther from human contact, is

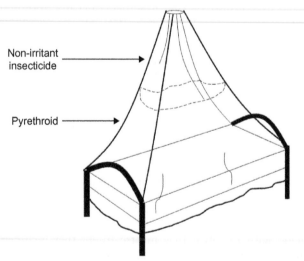

Figure 6.1 Idealized diagram of an ITN functioning as an insecticidal mixture (concept from Guillet *et al.*, 2001). Host-seeking mosquitoes land on the top part of the net and receive a dose of a non-irritant insecticide (e.g., carbamate). As the mosquitoes travel down the net to locate the host, they receive a dose of pyrethroid insecticide.

advantageous because other classes of insecticides tend to have higher mammalian toxicity.

Sustainability of ITNs (and other control methods) can be enhanced through the use of alternate hosts of mosquitoes, such as cattle, which may act both as a refuge for IRM and as a dead-end host of malaria. Kawaguchi et al. (2003) suggested that evolution of resistance can be delayed in zoophilic mosquitoes with the incorporation of cattle close to, but not within, human dwellings. Their mathematical model indicated that evolution of resistance is delayed with an increase in the number of cattle and with insecticide sprays used only in areas occupied by humans.

Indoor Residual Spraying

There has been an effort to obtain experimental data on resistance management using indoor residual spray (IRS) regimes. A field trial was set up in southern Mexico to test the effectiveness of certain IRM strategies on DDT resistance in *An. albimanus* mosquitoes (Hemingway et al., 1997). Indoor spraying regimes tested included (1) exclusive use of a pyrethroid or DDT, (2) annual rotation of three classes of insecticides (organophosphate, pyrethroid, and carbamate), and (3) spatial mosaic (within a village) of an organophosphate and pyrethroid. Metabolic detoxification by glutathione S-transferase (GST) was previously shown to be the major resistance mechanism to DDT in the mosquitoes in this area (Hemingway et al., 1997; Penilla et al., 1998). Levels of GST were measured after three years of treatment, and there was a mean decrease relative to that found in susceptible strains in every treatment except the exclusive use of DDT (Penilla et al., 2006). However, there was not a clear correlation between DDT resistance and mean GST levels in their study. This group also measured cytochrome P450 activity, and though there was a high level of variance, they found a decrease in P450 activity over the three-year time period for all IRS treatments (Penilla et al., 2007). Although P450 activity decreased over this time, there was still less susceptibility to the pyrethroid treatment, so resistance is likely to be caused be another mechanism (Penilla et al., 2007).

In addition, this group conducted a survey of villagers to determine the perceived effects of the spray programs (Rodríguez et al., 2006). They found that most people found lower numbers of mosquito bites under all treatment regimes, but that few actually associated this with a reduction in malaria incidence. Interestingly, most of the people associating reduced incidence of malaria were in the rotation regime, which had the additional perceived benefit of reducing cockroaches in the home.

This field study highlights the need for increased education about mosquito and malaria control, as well as the dynamics of resistance under certain IRS regimes. This is important because in 2006 WHO endorsed the use of IRS programs in combination with ITNs to combat malaria (WHO, 2006b). Of particular interest is their support for the use of DDT "where indicated," meaning where *Anopheles* vectors have not already evolved resistance. Previously, WHO supported the reduced reliance on DDT, and this insecticide was only allowed under special circumstances in public health. WHO recognizes the potential of the evolution of resistance in vector mosquitoes but notes in its position statement that, because DDT is not used in agricultural settings, the chance of resistance evolving in the public health sector is diminished (WHO, 2006b). Resistance to DDT and other major classes of insecticides, however, was higher in agricultural areas of several African countries surveyed (Ranson *et al.*, 2009). Ranson *et al.* (2009) also found considerable variability in resistance over short geographical distances, highlighting the importance of monitoring networks and data sharing.

Due to real concerns about DDT resistance or the potential thereof, alternative IRS methods are being tested that include different paints and binders, and their effects on residual activity (Sibanda *et al.*, 2011). A microcapsule formulation containing a mixture of two organophosphatess and an insect growth regulator (IGR) retained high efficacy on pyrethroid-resistant mosquitoes (*Anopheles gambiae* and *Culex quinquefasciatus*) after one year (Mosqueira *et al.*, 2010a,b).

There has been much effort in monitoring resistance, especially in Africa; IRS with DDT may only be effective in areas where malaria transmission is unstable and the *Anopheles* vector is susceptible, particularly in the highlands and fringes (Figure 6.2; WHO, 2005b). In light of the danger malaria poses to public health, resistance may be an acceptable risk to WHO. In 2003, WHO guidelines seemed to indicate that public health IRM simply involves monitoring for resistance, so that another product can subsequently be used when it is detected (WHO, 2003).

A unique approach to insecticidal control of mosquitoes is advocated by Koella *et al.* (2009). They suggest that delaying insecticide mortality in adult mosquitoes until the malaria parasite has matured in the female's body would reduce transmission of the disease but cause significantly less selection for resistance in the mosquito population. However, they

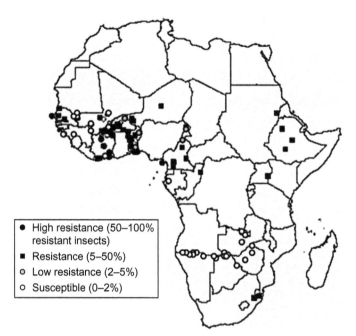

Figure 6.2 Distribution of DDT resistance in *An. gambiae s.s.* and *s.l.*, and *A. arabiensis* in Africa. *(modified from WHO, 2005b).*

limited their analysis to cases in which resistance is completely dominant in its expression in heterozygotes. Their concepts, which were derived from relatively simple models, need to be evaluated using more complex models of mosquito population dynamics, genetics, and behavior. Koella *et al.* (2009) also suggest that IRM should take advantage of fitness costs observed in some resistant mosquito populations. Experiments on DDT- and a pyrethroid-resistant colony of *An. gambiae* showed some fitness cost to resistance in that infection with a microsporidian decreased longevity (Koella *et al.*, 2012). Koella *et al.* (2012) modeled potential costs of resistance, and their results indicate that a high proportion of mosquitoes must be exposed to the infectious agent for resistance evolution to slow down. Jones *et al.* (2012) show that older mosquitoes of pyrethroid-resistant *An. gambiae*, that is, 17–19 days old as adults, are more susceptible to deltamethrin (and bendiocarb) than young mosquitoes. If that has already been the case, then how big is the concern about resistant mosquitoes? Resistance may not be too concerning, for example, if it confers decreased vector capability (Rivero *et al.*, 2010).

◢ BED BUGS

Until recently, most people in industrialized nations had little or no exposure to the common bed bug (*Cimex lectularius*, Hemiptera: Cimicidae) or its relative, the tropical bed bug (*Cimex hemipterus*). Bed bugs were a common problem until the advent of widespread use of DDT in the 1940s and 1950s in those countries (Potter, 2011). Resistance problems with DDT emerged, but when other residual organochlorines became available for use, bed bugs were all but eradicated (Potter, 2011), essentially leading to a void in published research on bed bugs since Usinger's 1966 monograph. Much new research on the basic biology of bed bugs has been spurred by its renewed pest status (Reinhardt and Siva-Jothy, 2007). Much of the recent resurgence can be attributed to increased global travel, but certain aspects of the pest biology and human behavior make infestations difficult to control. Eradication requires more than a single visit by a pest-management professional; the lack of ovicidal effect of pyrethroids, as well as alarm pheromones influencing dispersal of bed bugs, are complicating factors (Davies *et al.*, 2012). Besides the fact that bed bugs can survive a long time without feeding because of a low metabolic rate and resistance to starvation (Harlan, 2006; Polanco *et al.*, 2011b), factors associated with some modern housing also play a role. For example, increased humidity favoring bed bug growth often results from lack of air circulation (Harlan, 2006), and increased clutter and furnishings likely give bed bugs more refuge area (Potter, 2011). Social stigma associated with bed bug infestations can compel people to self-treat living areas and therefore contribute to resurgence and resistance (Moore and Miller, 2009).

Like many other case studies in this chapter, pyrethroids are often used to control bed bugs, but resistance did not take long to evolve. Yoon *et al.* (2008) and Adelman *et al.* (2011) compared two different susceptible and resistant strains that can be used to evaluate potential origins and mechanisms. Adelman *et al.* (2011) compared gene expression between a susceptible reference strain (Harlan) to a local population that exhibited over 5000-fold resistance to deltamethrin via both *kdr*-type and metabolic resistance mechanisms. A transcriptomic approach was taken as a first step in investigating metabolic resistance mechanisms, and Adelman *et al.* (2011) found increased expression of cytochrome P450s and carboxylesterases.

This resistant strain also had one of the two target site mutations identified by Yoon *et al.* (2008) in a resistant strain from New York.

Alternative IPM strategies are necessary not only because of general difficulties in extermination, but also because potential costs of resistance can be compensated. Pyrethroid-resistant bed bugs are less tolerant of starvation than susceptible strains (Polanco *et al.*, 2011a), but have a higher intrinsic rate of increase than susceptible strains due to decreased developmental time and higher survivorship (although fecundity was not compared) (Polanco *et al.*, 2011b). Renewed interest in the basic biology of bed bugs, such as responses to environmental stress (Benoit, 2011), the role of symbionts in digestion (Sakamoto and Rasgon, 2006), or dispersal ecology (Reinhardt and Siva-Jothy, 2007), will help inform new IPM strategies.

HUMAN HEAD LICE

Another pest with a history of resistance problems is the human head louse, *Pediculus humanus capitis* (Phthiraptera: Pediculidae). These insects feed on human blood and spend their entire life cycle on the scalp. They are most often considered a pest of young children because of the close proximity of students to each other in elementary school classrooms. *Pediculus. h. capitis* are most often spread by physical contact because they cannot survive for very long off of a host, but there is a great need to further investigate the basic biology of lice to understand resistance and their spread to other hosts (Burgess, 2004). Tebruegge *et al.* (2011) note a decreased efficacy in lice control products since the mid-1990s, before which it was greater than 80%. Ovediagnosis of *P. h. capitis* and subsequent unnecessary treatment is a probable factor leading to resistance; Pollack *et al.* (2000) found that most cases of infestations are misdiagnosed. Only 59% of submissions to their laboratory contained lice specimens, and of these, slightly over half contained live lice or viable eggs. They concluded that subjects without active infestations were actually more likely to receive treatments than those with active infestations.

The most common treatments available without a prescription in the United States and UK, for example, are either pyrethrins (sometimes with a synergist) or permethrin (CDC, 2010; Tebruegge *et al.*, 2011) because these compounds have lower mammalian toxicity (Mumcuoglu, 1996; Yoon *et al.*, 2004). *P. h. capitis* has therefore evolved resistance to this class of insecticides in many parts of the world because of its almost exclusive use for control and also cross-resistance to DDT, which was previously used to control *P. h. capitis* (Mumcuoglu, 1996). *Kdr*-like sodium channel mutations are likely to be the causal mechanism for resistance (Lee *et al.*, 2000), the consistency of which (Durand *et al.*, 2011) makes it possible to develop increasingly sophisticated monitoring methods (Lee *et al.*, 2010).

Resistance to other insecticides besides the most commonly used pyrethroids can still be present in some populations. Insecticides available only by a prescription from physicians in the United States include malathion, a benzyl alcohol, and lindane, which is only recommended in extreme circumstances (CDC, 2010). Carbaryl is available from a doctor in the UK, but also because of mammalian toxicity, this treatment is discouraged and should be reserved for extreme cases (Tebruegge *et al.*, 2011). Decreased efficacy of malathion has been reported (e.g., Chosidow *et al.*, 2010), and resistant populations have been found in the United States by Yoon *et al.* (2004) and in the United Kingdom by Downs *et al.* (1999). Resistance to lindane and carbaryl can be difficult to confirm (e.g., Meinking *et al.*, 2002; Yoon *et al.*, 2004), or found at lower levels (Downs *et al.*, 2002), respectively, but as there is a trend away from use of these insecticides because of negative effects on the host, it is perhaps not of great concern.

Ivermectin, was approved in 2012 for the public by the U.S. Food and Drug Administration as a treatment for control of head lice. Ivermectin does not have ovicidal effects, but is still efficacious in that nymphs treated as eggs (nits) died within 48 hours of hatching (Strycharz *et al.*, 2011). Interestingly, blood-feeding was significantly reduced in the nymphs that hatched after treatment (Strycharz *et al.*, 2011). Ameen *et al.* (2010) found 100% mortality after two treatments, but in light of the results of Strycharz *et al.* (2011), the second treatment may not have been necessary. Proper treatment should be exercised, however, because sublethal doses increase transcription of detoxification genes involved in tolerance to insecticides (Yoon *et al.*, 2011).

The main options for IRM in *P. h. capitis* could be the application of mixtures of insecticides or the use of one insecticide after another has lost its effectiveness. However, as Curtis *et al.* (1993) noted, the latter is not a strategy per se, in that this would be the normal procedure in the absence of any real plan for IRM. Mixtures of treatments are possible to implement and may cause significant mortality, but their effectiveness for IRM has yet to be studied. Many studies examine cross-resistance to pyrethroids because of their common use and control failure. Aliphatic alcohols, for example, may cause significant control of head lice, and there is no cross-resistance from permethrin-resistant lice (Mougabure Cueto *et al.*, 2002). Yoon *et al.* (2003) found that an insecticidal lotion without its active ingredient, mala-thion, still caused some mortality; this would make an effective mixture easier to create. A high-dose strategy is not likely to be effective IRM, especially when some level of resistance is already present in the population; mortality was not affected by very high doses of two pyrethroids, and it was only moderately increased with very high doses of malathion (Downs *et al.*, 2002).

An integrated approach that includes alternative or supplementary con-trol methods can provide an acceptable solution. Botanical oils and insecti-cides have received much attention because of their general safety compared to synthetic insecticides, but their efficacy is often either questionable (Tebruegge *et al.*, 2011) or nonexistent (Canyon and Speare, 2007). Cultural controls of lice include combing, though its effect needs to be better vali-dated (Tebruegge *et al.*, 2011), and the use of heat. The LouseBuster is a device with which the operator applies heated air in a controlled fashion to heat-kill and desiccate lice (Bush *et al.*, 2011; Goates *et al.*, 2006). Ease of use is another advantage; novice operators after a short training session caused the same lice mortality as experienced operators (Bush *et al.*, 2011).

Resistance management and control of *P. h. capitis* require much edu-cation and communication between parents, teachers, and health workers. Schools that are proactive in educating parents about head lice and regular head combing may lead to fewer incidents of head lice in children (Downs *et al.*, 2002). Another study indicated that keeping parents informed about head lice did not reduce the incidence, but did decrease the level of infestation in numbers of lice (Downs *et al.*, 2000). Downs *et al.* (1999) suggested that insecticidal treatments for lice be available only by prescription to slow the development of resistance. However, antibac-terial medications are also only available by prescription, and this has not completely prevented the evolution of resistance to antibiotics.

FLEAS OF CATS AND DOGS

The most important ectoparasite of cats and dogs is the flea *Ctenocephalides felis* (Siphonaptera: Pulicidae). Besides causing discomfort to pets (and owners), fleas may cause allergic dermatitis in highly sensitive animals, and they may transmit pathogens or endoparasites (Eisen and Gage, 2012). Adult fleas infest pets; females lay eggs on pet fur. The eggs soon drop from the fur, normally where the pet rests. Larvae are nonparasitic and feed on adult feces, which consists of partially digested blood.

Control of *C. felis* has relied on chemical application of both the pet and places where it rests (Rust and Dryden, 1997). Reports of resistance in *C. felis* have led to the use of newer treatments such as insect growth regulators, IGRs, and botanical compounds (Rust and Dryden, 1997). Other compounds such as fipronil, a phenylpyrazole, and imidacloprid, a neonicotinoid, are now commonly used as well (Rust and Dryden, 1997).

Resistance to the major classes of insecticides has been documented, but tests of resistance have used variable methods of detection (Bossard *et al.*, 1998). Because resistance patterns are generally unknown, the pet owner's treatment decision is usually based on price and simplicity (Rust and Dryden, 1997). Although resistance is commonly cited as the reason for control failure, this is not always certain. Some reports of resistance may actually be due to incomplete treatment (Rust and Dryden, 1997) or the inherent variability of susceptibility in flea populations or variability in the tests themselves (Bossard *et al.*, 1998). For example, Bossard *et al.* (2002) found high levels of variability in susceptibility within some of the strains they tested.

A simpler test for resistance will clarify these issues; Rust *et al.* (2002) recently developed a larval assay. Because the assay begins with flea eggs, many environmental effects (Bossard *et al.*, 1998) are minimized, and there is also no need to rear fleas until they reach adulthood. The authors determined a threshold of 3 ppm of imidacloprid at which to administer further tests to diagnose resistance; subsequent tests validated this concentration (Rust *et al.*, 2005).

Genetic and molecular studies on resistance in *C. felis* have recently been conducted. A polymerase chain reaction (PCR) assay for both *kdr* mutants (Bass *et al.*, 2004a) and *Rdl* mutants (Bass *et al.*, 2004b; Daborn *et al.*, 2004) has been developed. Both of these alleles were common in the laboratory populations tested from the United Kingdom and the United States (Bass *et al.*, 2004a,b). Most of their samples were from fleas that had been in laboratory colonies for a few years without any selection pressure. Though named for "resistance to dieldrin," an organochlorine, the *Rdl* allele also seems to confer cross-resistance to fipronil, a phenylpyrazole (Bass *et al.*, 2004b; Daborn *et al.*, 2004). When using these assays to test field-collected strains from the United States and Europe, both alleles were common, but frequencies were highly variable among the populations (Bass et al., 2004a, b). Ideally, one wants to predict molecular mechanisms of resistance before they occur. Bass et al. (2006) found protein subunits of the target site of imidacloprid, a neonicotinoid, that are likely to be involved in resistance if it evolves.

A history of resistance requires an integrated approach for control. Carlotti and Jacobs (2000) outlined the major flea control methods. To prevent resistance, they suggested using chemicals with different modes of action and integrating cultural controls such as animal grooming and vacuuming of carpets (IPM). Simply leaving an untreated refuge is generally not acceptable for medical/veterinary pests; there is a low threshold, if any, of tolerance for fleas. Temporal refuges may be implemented by only treating acute infestations, but this is also not acceptable to either the pet or its owner (Bossard et al., 1998).

MITES ON BEES

The mite *Varroa destructor* (Acari: Varroidae) is one of the most serious pests of honey bees (*Apis mellifera*) worldwide. Mites feed on the hemolymph of bees, preferentially of the drone brood, leaving them severely deformed. They may also transmit viral pathogens (Sammataro et al., 2000). If left unchecked, *V. destructor* infestations can destroy a honey bee colony in a few years.

The life history of *V. destructor* probably makes resistance easier to evolve in this species. A female mite enters a brood cell right before it is capped for the bee to pupate. She soon starts laying eggs, the first of which is a male, followed by female eggs. Nymphs feed and develop on the pupal bee, and when mature, the single male mates with all of the sisters in the cell. Haplodiploidy and sibling mating in *V. destructor* greatly increase the chance of fixation of new mutations (Cornuet et al., 2006).

Typical control of *V. destructor* has involved the use of fluvalinate, a pyrethroid—treated strips placed in the hive during times of no honey production. Intensive use of these strips has selected for resistance in some parts of Europe (Troullier, 1998), the United States (Elzen et al., 1998; Macedo et al., 2002), Israel (Mozes-Koch et al., 2000), and Mexico

(Rodríguez-Dehaibes *et al.*, 2005). The spread of pyrethroid resistance in Europe roughly follows that of the initial spread of the mite according to bee movement, suggesting that resistance evolved once and spread thereafter (Martin, 2004). Coumaphos, an organophosphate insecticide, was soon introduced for emergency use after control problems with fluvalinate, but resistance to coumaphos has already been detected in Florida (Elzen and Westervelt, 2002), Argentina (Maggi *et al.*, 2009), and northern Italy (Spreafico *et al.*, 2001).

Multiple resistance is also not uncommon (Sammataro *et al.*, 2000). Resistance to both pyrethroids and amitraz, an amidine previously used for mite control, has been reported in the United States in Minnesota (Elzen *et al.*, 2000) and in Mexico (Rodríguez-Dehaibes *et al.*, 2005, 2011). Fluvalinate and coumaphos are currently the most commonly used miticides in the United States, and there is a high level of resistance to both of them in populations from Texas and Florida (Kanga *et al.*, 2010).

Some resistance to fluvalinate seems to be partly due to detoxification via monooxygenases (Hillesheim *et al.*, 1996; Mozes-Koch *et al.*, 2000), but Johnson *et al.* (2010) found that inhibitors of monooxygenases, as well as other major classes of detoxification enzymes, had very little effect on toxicity to tolerant mites. Sodium channel target site (*kdr*-like) resistance is also possible (Wang *et al.*, 2002, 2003), which could be used to construct a PCR-based assay to monitor for resistance. On the other hand, there may be less demand for monitoring procedures. Because of widespread resistance problems and contamination by fluvalinate, the recent trend has been away from this control method and toward increased use of formic acid or mixtures of essential oils containing the monoterpenoid thymol (R. Johnson, personal communication).

Most suggestions for IRM involve either an acaricide rotation or an additional IPM method. Milani (1999) suggested the rotation of chemical and nonchemical means of control in different seasons, combined with breeding for resistant bees (Martin *et al.*, 1997), to delay resistance. In Minnesota, mites that were resistant to fluvalinate were also resistant to amitraz but not to coumaphos (Elzen *et al.*, 2000). This group suggested a rotation involving the use of coumaphos for two years followed by one year of fluvalinate (Elzen *et al.*, 2001) for IRM. However, Johnson *et al.* (2009) found synergism between these two miticides on honey bees. A combination of single toxins that are sublethal by themselves could increase honey bee mortality in a rotation scheme if residual effects from the previous treatment are present.

The relatively close phylogenetic relationship between mites and their honey bee hosts makes control and IRM difficult, obliging the use of multiple IPM practices. Breeding programs to produce commercially viable lines of honey bees that are resistant to varroa mites (Buechler *et al.*, 2010; Rinderer *et al.*, 2010) are expected to reduce the use of miticides and concomitant selection pressure. Other IPM-like methods for managing resistance incorporate trapping or heat-killing *V. destructor* mites. Webster *et al.* (2000) designed a trap in the floor of the hive preventing live fallen mites from returning to the comb. Even though their *V. destructor* colony was resistant to fluvalinate, they found larger amounts of mite-fall during fluvalinate treatment than without. They hypothesized that live mites falling in their traps had experienced sublethal doses of insecticide, and that a possible IRM strategy is to use either chemical means plus their floor trap or to use less effective essential oils in addition to the floor trap. One other method to delay resistance includes possible use of heat to kill *V. destructor*. Mites prefer the male drone brood, but drones contribute little to the bee colony. Therefore drone brood can be sacrificed for *V. destructor* control with little adverse effect (Calderone, 2005). Removing some of the drone-brood comb is done in some small apiaries in Europe, but it is labor intensive (Huang, 2001). Huang (2001) extended this idea to develop special frames for the bees to make drone-brood combs. These frames have internal heating elements that can be activated to kill mites and associated drone brood.

TICKS OF CATTLE

The tick *Rhipacephalus* (*Boophilus*) *microplus* Canestrini (Acari: Ixodidae) is a pest of cattle in Latin America and in Australia; three *Boophilus* species, including *B. microplus*, are pests in sub-Saharan Africa (Estrada-Peña *et al.*, 2006). *R. microplus* has been eradicated from the United States since the 1940s, and great care is taken to prevent its reintroduction through cattle crossing the border from Mexico. There is concern for the welfare and comfort of cattle with respect to ticks, but vector capability is also of great importance, as they may transmit protozoan or bacterial diseases. De Castro (1997) estimated an annual cost of between US$13.9 and 18.7 billion for damage and control of ticks and their diseases on cattle throughout the world.

Ticks may be categorized by the number of hosts they require to complete their life cycle. Cattle ticks are single-host ticks, meaning that they complete parasitic stages on one host. These ticks are free-living only from the time the engorged female drops from the host to lay eggs until the newly hatched larvae locate another host. There are likely to be a mixture of developmental stages on the host at any one time.

Most cattle producers rely solely on acaricides for tick control, which may be done in the form of dips, sprays, or pour-on formulas (George, 2000). Amitraz (an amidine), coumaphos (an organophosphate), and cypermethrin (a pyrethroid) are the main acaricides used (Jonsson et al., 2001; Rodríguez-Vivas et al., 2006). Many tick control schemes involve regular prophylactic dipping, which has led to some cases of resistance (Sangster, 2001). Also, not surprisingly, an increase in the number of acaricide treatments leads to higher incidence of resistance (Jonsson et al., 2000; Rodríguez-Vivas et al., 2006). Concern over resistance and animal health has led to research on botanical compounds, repellents, and vaccines for cattle (Kiss et al., 2012; Guerrero et al., 2012; De La Fuente and Kocan, 2006).

Because of intensive use of acaricides, resistance has evolved in many parts of the world. Some areas in Kenya have experienced resistance to amitraz (Kamidi and Kamidi, 2005). In Mexico, resistance is commonly seen with more than one class of acaricide (Foil et al., 2004), and multiclass resistance is widespread in Brazil (Graf et al., 2004). In Australia, Sutherst et al. (1979) estimated that widespread resistance to an acaricide occurs in 4 to 7 years after a new product is introduced. Aguirre et al. 1986 detected resistance to coumaphos in 7 years, while Foil et al. (2004) gave an average of 12 years for evolution of resistance to amitraz.

Despite some problems with resistance, amitraz is still the preferred control of ticks on dairy cattle in Australia because of its efficacy compared to cost and reduced toxicity on mammals and the environment (Jonsson and Hope, 2007). In a review of amitraz resistance, these authors suggested that movement of cattle due to sales is responsible for the spread of amitraz resistance in Australia, rather than multiple independent origins. As would be expected, Rodriguez-Vivas et al. (2007) reported that an increase in amitraz application frequency was most significantly associated with amitraz resistance in southeastern Mexico.

Rotation strategies are often studied not as an a priori IRM strategy, but rather as a way of restoring the efficacy of one or multiple controls where ticks have already evolved some level of resistance. Jonsson et al.

(2010) conducted a multiyear study testing the rotation of amitraz with spinosad. They established populations initiated with equal numbers of ticks from a susceptible strain and an amitraz- and pyrethroid-resistant strain. Treatments consisted of acaricides applied when thresholds where reached: spinosad, amitraz, or a spinosad-amitraz rotation every two months, which was meant to approximate one tick generation in that region of Australia. While spinosad was found to be generally less efficacious than amitraz, rotation of these two acaricides almost restored full susceptibility to amitraz, which is also a less expensive treatment.

Kamidi and Kamidi (2005) tested the rapid rotation of amitraz and chlorfenvinphos (an organophosphate), with each being used once a week, as an IRM strategy for ticks showing resistance to each of these acaricides at different times. However, they measured incidence of tick-borne diseases, rather than any data on the ticks themselves. Disease incidence drastically decreased after deployment of the rapid rotation of acaricides, and the authors claimed that this strategy was still effective after 2.5 years.

Mixtures are often used as an IRM strategy because of relative ease. In Australia organophosphates are used to synergize the toxicity of pyrethoids used in tick control, which probably not only delayed resistance, but also provided a more economical use of pyrethroids (George *et al.*, 2004). George *et al.* (2004) pointed out that a possible added benefit to the use of these insecticide mixtures is control of horn fly (*Haematobia irritans*). However, widespread resistance to multiple classes of insecticide, such as in Brazil, would most likely render mixture strategies ineffective (Graf *et al.*, 2004). Cross-resistance studies are also necessary, while keeping in mind the potential for different resistance mechanisms or patterns in different regions of the world (Graf *et al.*, 2004, Li *et al.*, 2004, 2007ab; Miller *et al.*, 2007; Rodriguez-Vivas *et al.*, 2007).

Simply increasing the dose is a short-term strategy that could be recommended under special conditions such as the tick quarantine zone to prevent reintroduction into the United States from cattle being traded from Mexico. Davey *et al.* (2004) found that three intervals of coumaphos, an organophosphate treatment, did not completely kill all organophosphate-resistant ticks, but it did lower the reproductive potential sufficiently that they felt resurgence was not likely to occur. A subsequent study by this group revealed that a repeated treatment of a high dose of organophosphate insecticide effectively controlled resistant ticks (Miller *et al.*, 2005). For ticks from Mexico that were already resistant to amitraz, however, the level necessary to attain 99% control is probably cost-prohibitive and potentially toxic to cattle (Davey et al., 2008).

George et al. (2004) noted that even though integrated tick control measures have been suggested (Sutherst et al., 1979; Norton et al., 1983), little has been done to help producers implement them. They advocate means of educating both producers and regulatory groups of the benefits and problems of tick control and resistance management. However, evidence is severely lacking as to which plan would be best for resistance management (Willadsen, 2006).

Chemical companies involved in tick control have an interest in IRM. Graf et al. (2004) suggested that resistance has possibly arisen due to rotation of trade names, rather than rotation of chemical classes. They highlighted the need for resistance management and claim that chemical controls should not be viewed as a renewable resource. High investment risks are present for chemical companies (Witty, 1999), especially because the market for insecticidal products for livestock pests tends to be shrinking in comparison to that market for companion animal pests (Graf et al., 2004). Regulatory procedures are one factor that may be preventing effective control products from entering the market. Exact thresholds vary among nations, but many require an average mortality in the target pest of 95–98% for introduction of a control product (Graf et al., 2004). Graf et al. (2004) applauded changes to regulatory guidelines in Australia that allowed lower efficacy products on the market that "aid in the control of" target pests. Repellents (e.g., Khallaayoune et al., 2009) or plant extracts (e.g., Srivastava et al., 2008) may find greatest utility as components of an IPM program or components of treatment mixtures. Pérez De Leon et al. (2010) advocate more research into slower acting compounds such as growth regulators, to be used in combination with other acaricides.

Because of a history of resistance problems in Australia (George et al., 2004), an effort has been made to take an integrated approach in tick control, including methods of dipping cattle in acaricides, pasture spelling/rotation, tick vaccines, and raising tick-resistant cattle (Angus, 1996). Pasture spelling is the practice of removing cattle from pastureland for a period of time to prevent ticks from locating a host until most ticks die. Graf et al. (2004) suggested a combination of pasture rotation with acaricidal treatment to delay evolution of resistance. Sutherst et al. (1979) found that the combination of the use of tick-resistant breeds of cattle and pasture spelling offered the best sustainable control of B. microplus in a mathematical model. Concerns about organic beef and improved productivity have renewed interest in breeding for tick-resistant cattle (Frisch *et al.*, 2000). Genomic technologies are being used to improve the search for and use of molecular markers in breeding programs (Porto Neto *et al.*, 2011).

BLOW FLY IN SHEEP

Cutaneous myiasis in sheep is caused by the Australian sheep blow fly, *Lucilia cuprina* (Diptera: Calliphoridae). *L. cuprina* is not a constant menace, but "flywaves" may be anticipated according to weather predictions and monitoring of populations (Levot, 1995). Females oviposit in wet fleece, namely, around the backside of the animal, and often when a bacterial infection (called fleece rot) is already present. Rainy weather predisposes the sheep to *L. cuprina* infestation and associated fleece rot.

Larvae feed on the flesh of sheep and can cause painful wounds. They drop off the host to pupate in the soil. *L. cuprina* infestations can be prevented through chemical or cultural means. Cultural controls include mulesing and tail docking. Mulesing is the process of cutting off large patches of skin on the backside of the animal to remove the skin folds there that are conducive to *L. cuprina* infestations. Research is ongoing to develop either nonsurgical or sheep breeding alternatives to this controversial practice (James, 2005). Dipping or spraying sheep with insecticide is the most common control; insect growth regulators (IGRs) are used most often because of control failures from other insecticides (Levot and Sales, 2004).

Lucilia cuprina has a history of evolving resistance to the major classes of insecticides used against it. Organochlorines were used from 1948 to 1958 when resistance problems surfaced (Levot, 1995), and in 1962 they were withdrawn from the market (Hughes and McKenzie, 1987). Resistance to dieldrin, an organochlorine, took only three years to evolve, probably because of previous use of lindane, another organochlorine, for sheep lice (*Bovicola ovis*) control (Hughes and McKenzie, 1987). Organochlorines were soon replaced by organophosphates, which also later failed to control *L. cuprina*. Lack of other control options in the late 1960s and 1970s drove organophosphate resistance to fixation (Hughes and McKenzie, 1987; Levot, 1995). It took five years for the frequency of resistance to diazinon, an organophosphate, to reach 95% in field populations from the time that it was first detected (Levot, 1995). Despite widespread resistance, organophosphates are still used for *L. cuprina* control because they still provide quick protection for the short term (Levot, 1995). Cyromazine is an IGR that is effective for longer-term control (about 14 weeks), but it is slow acting and does not provide much protection for active infestations (Levot, 1995). Resistance has not yet evolved in the field since cyromazine was introduced to producers in 1979 (Levot, 1995). Resistance has been recorded, however, to diflubenzuron (another IGR) in some parts of Australia (Levot and Sales, 2002, 2008).

Because of this insect's relative ease in rearing and because resistance mechanisms are known, certain IRM assumptions have been examined. Strains resistant to dieldrin, diazinon (an organophosphate), or malathion (another organophosphate), respectively, created in the laboratory by artificial mutagenesis had identical mutations in resistance alleles as those found in field populations (McKenzie and Batterham, 1998). Artificial mutagenesis was also used to study and predict possible resistance to cyromazine and four loci were found to confer resistance, all at low resistance ratios (Yen *et al.*, 1996). A resistance ratio is a comparison of the insecticide concentration required to cause mortality of resistant to susceptible insects. Only one of these alleles was likely to make a viable homozygote, so based on these results they predicted a low chance of evolution of resistance to cyromazine (Yen *et al.*, 1996). Levot and Sales (2004) also found low resistance ratios to cyromazine in a laboratory-selected population, so they too concluded that the chance of evolution of resistance is low. However, this laboratory-selected resistance may not effectively mimic that found in the field. Levot and Sales's (2004) study included other IGRs as well, and they found cross-resistance to dicyclanil in diflubenzuron-selected laboratory populations, but this cross-resistance was not found in resistant populations collected from the field. This surely complicates any extensions from laboratory work to field application. Roush and McKenzie (1987) outlined some of the problems with comparisons between laboratory-selected and field-selected resistant strains of insects, namely, that polygenic resistance tends to be selected in the laboratory, and monogenic resistance tends to be selected in the field.

Establishment of resistance alleles depends both on initial frequency and concentration of selective agent, that is, selection pressure; see Scott *et al.* (2000). If selection occurs at a range outside the normal phenotypic variation of susceptible insects, then monogenic resistance is expected to occur (McKenzie, 2000). The combination of knowing the resistance gene and applying the chemical control above the LC_{100} of susceptibles allows prediction of and hopefully prevention of resistance. Because initial development of resistance is dependent on the fitness of heterozygotes, McKenzie and Batterham (1998) advocate the use of square-wave decay curves to determine a concentration lethal to resistant heterozygotes. This is analogous to a high-dose strategy for IRM.

Other IRM strategies include the use of a refuge or the use of two insecticides in a mixture or rotation. In their model of dieldrin resistance, Goss and McKenzie (1996) found that even a small increase in the amount of insects in a refuge has a large impact on the time for resistance to evolve. The feasibility of a refuge strategy depends on the size of the population and if there is some level of tolerance to fly infestations. The genetic bases of resistance to two different IGRs, diflubenzuron and cyromazine, seem to be independent (Batterham *et al.*, 2006), so this has potential for use in a mixture or rotation IRM strategy.

HORN FLY ON CATTLE

The horn fly, *Haematobia irritans* (Diptera: Muscidae), is mainly a pest of cattle, although it may infest other types of livestock as well. Adults look similar to house flies (*Musca domestica*), but are about half the size and have piercing mouthparts to feed on the blood of cattle. Economic damage occurs in lost productivity of cattle due to blood loss and stress on the animals. Females oviposit in fresh manure; larvae may remain there to pupate or may first migrate to the soil. *Haematobia irritans* damage to cattle in the United States alone is estimated to cost US\$876 million annually (Kunz *et al.*, 1991).

A common method of control of *H. irritans* is the use of ear tags that have been impregnated with pyrethroid or organophosphate insecticides (Byford and Sparks, 1987; Foil *et al.*, 2005). Resistance to pyrethroids is widespread because it can evolve in as little as two years with the use of ear tags on cattle (Quisenberry *et al.*, 1984). Barros *et al.* (2001) found that resistance had evolved to all ear tags with organophosphates they tested in less than nine years. Resistance is common because of the high reproductive potential and mobility of this insect (Byford *et al.*, 1999).

Even though resistance to pyrethroids via the *kdr* mutation might confer some fitness cost (Scott *et al.*, 1997), resistance is persistent even without selection. Use of diazinon ear tags for 18 weeks on farms with at least three consecutive years of permethrin tags decreased, but did not eliminate, pyrethroid resistance (Cilek and Knapp, 1993). Weinzierl *et al.* (1990) found no reversion of pyrethroid efficacy after two years of disuse. Likewise, Jamroz *et al.* (1998) found no decrease in *kdr* allele frequency in a wild population with no pyrethroid selection. Guglielmone *et al.* (2002) reported some decrease in frequency of resistant pyrethroid alleles (*kdr*) after cessation of selection pressure, but this decrease was not enough to restore susceptibility to cypermethrin. It is likely that a combination of resistance mechanisms to pyrethroids is present in *H. irritans* because the level of resistance to pyrethroids does not seem to be highly correlated to *kdr* frequency (Jamroz *et al.*, 1998). A combination of resistance mechanisms may contribute to the persistence of pyrethroid resistance observed in field populations.

Like the other case studies in this chapter, IRM has been difficult for this insect. Insecticide-free refuges are essentially nonexistent, due to the host-specific nature of the insect, widespread use of ear tags, and the high mobility of horn flies (Byford *et al.*, 1999). Because ear tags are easy for producers to implement, many studies have focused on examining mixtures or rotations of insecticides in ear tags for IRM. Barros *et al.* (2002) found high levels of susceptibility to the organophosphate diazinon in populations with a history of pyrethroid resistance. This is in contrast to the studies of Barros *et al.* (1999) and Guerrero *et al.* (2002), which found that resistance evolved to both insecticides in yearly rotations of pyrethroid and organophosphate ear tags. In a laboratory study, McKenzie and Byford (1993) found that mixtures and rotations of pyrethroids and organophosphates delayed and reduced the magnitude of resistance to each insecticide, but did not prevent it from evolving. In a field study by Byford *et al.* (1999), mosaic treatment of pyrethroid and organophosphate was not shown to select for resistance to either insecticide in a three-year study. A mosaic strategy is possible within a farm, but probably is not feasible across the landscape.

Because of the constant selection pressure from insecticide-impregnated ear tags, Sparks *et al.* (1985) recommended a return to control methods that deliver discrete doses of insecticide, such as sprays or dust bags, for IRM. Farmers in central Chile mostly use pour-ons with either cypermethrin or permethrin plus a synergist, and sampled horn fly populations showed high levels of resistance (Oyarzún *et al.*, 2011). One farm that showed high pyrethroid resistance had no insecticide use for the previous five years, which highlights the importance of understanding pest mobility for IRM (Oyarzún *et al.*, 2011). Because of control problems with pyrethroid pour-ons, some farmers surveyed by Oyarzún *et al.* (2011) developed their own IRM strategy of rotating these pour-ons with diazinon ear tags.

MUSCA DOMESTICA

Muscoid flies are common pests of livestock. Their similar life histories and association with production animals allow a more comprehensive study of IRM for veterinary pests. *M. domestica* flies are a worldwide nuisance pest of both humans and production animals. They lay eggs in manure or other refuse and feed in these areas as well, highlighting the importance of cultural controls such as sanitation of feedlots. Although they cannot bite, adult flies may attempt to feed on moist areas of cattle, and as such, they are sometimes implicated in disease transmission. *M. domestica* is a notorious and well-known pest with a long history of evolving resistance to chemical controls, and a large amount of research has been done on resistance mechanisms. For this reason, the approach taken in this section will be to apply what is currently known about resistance to a broader context of IRM principles.

That many ectoparasites are flies, and with relatively short generation time, allows one to use *Drosophila melanogaster* as a model for resistance studies (Perry *et al.*, 2011, Pittendrigh *et al.*, Chapter 3). However, in a similar fashion, fly species profiled in this chapter can also be used to evaluate expectations about resistance evolution in a way that can be tied to field observations. Synthetic pyrethroids are often used to control ectoparasites because of their low mammalian toxicity, but frequent application of one mode of action rapidly selects for resistance. As illustrated in many of the other case studies in this chapter, resistance to pyrethroid insecticides can occur via either target site modification or metabolic detoxification, or a combination of both mechanisms. That *kdr*-type resistance via sodium channel mutations is found in all of the insect case studies in this chapter, and sometimes remains in populations without selection, suggests that variability in this locus exists and that it has little to no fitness cost, as reflected in Rinkevich *et al.* (2012), for example, who found independent origins of *kdr* in house flies from different continents. Prediction of resistance to other classes of compounds can also be made. Resistance to organophosphates, among others as well, often occurs via a mechanism involving carboxylesterases; seven orders of insects sequenced and profiled had two mutations in common (Cui *et al.*, 2011).

M. *domestica* has become a model of the study of resistance because it has evolved resistance to every major class of insecticide used against it. Keiding (1999) provided an extensive review of resistance in *M. domestica*; highlights are provided here. Widespread resistance to DDT and other organochlorines quickly evolved in the 1940s and 1950s because of their ubiquitous use worldwide. Resistance to DDT and other organochlorines, such as lindane and dieldrin, in *M. domestica* is still present, despite the discontinued use of these products in most countries by the 1970s because of toxic effects on the environment. After the control failures of the organochlorines, organophosphates were commonly used and resistance also evolved to many of these as well. Resistance to carbamates has been documented; where it occurs, it is associated with either cross-resistance to organophosphates or intensive use of baits with methomyl, a carbamate. Resistance to pyrethroids is common in North America and Europe, again due to widespread use. It was slow to evolve in Asia because of a low initial frequency of resistance alleles, but it is now widespread.

Monitoring populations for resistance is important in determining the frequency of resistance alleles, a parameter considered important in most IRM strategies. Molecular methods to determine resistance mechanisms and frequencies of resistance alleles are being developed with ever-increasing efficiency. As a model system, resistance mechanisms and their respective alleles have been intensely studied in *M. domestica*, leading to molecular diagnostic assays. PCR-based assays have been developed for alleles conferring target site knockdown resistance (Williamson *et al.*, 1996; Huang *et al.*, 2004). These assays have the potential to be used for quicker results on initial frequencies of resistant alleles before a control method is used or one for which some level of cross-resistance is suspected. For example, though Scott and Wen (1997) found cross-resistance to fipronil in *Rdl* house flies, Gao *et al.* (2007) screened flies from three distant areas in the United States for that allele. Because they could not detect the *Rdl* allele and were unsuccessful in their efforts to select a highly resistant line, they concluded that fipronil should present a low risk for resistance (Gao *et al.*, 2007). Kozaki *et al.* (2009), in contrast, observed high frequencies of *Ace* variants, *Vssc*, and *CYP6D1* in flies collected from dairies in New York and Florida, indicating multiple resistance. These genes refer to an acetylcholinesterase, a sodium channel variant *kdr*, and a cytochrome P450 monooxygenase, respectively, and they were found to be still under selection (Kozaki *et al.*, 2009). Phylogenetic analysis of *Ace* alleles identified from 13 different strains revealed that two of the five resistance alleles had multiple origins (Kozaki *et al.*, 2009). Interestingly, Seifert and Scott (2002) found the same cytochrome P450 variant, Cyp6d1v1, in pyrethroid-resistant strains from both New York and Georgia. The same mutation found in such geographically distant locations implies either more dispersal than once thought, or evolvability at this locus, as there were six alleles found among a few populations.

Because of the prevalence of resistance to most insecticides used against *M. domestica*, researchers are now looking for methods of control that do not confer cross-resistance or perhaps have negative cross-resistance. These compounds could be used in possible mixtures or rotations with commonly used insecticides. Insect growth regulators, such as cyromazine, are used for control of *M. domestica*, but resistance has developed in some locations in the United States to this compound, probably due to low doses added to chicken feed (Tang *et al.*, 2002). Tang *et al.* (2002) found that resistance is probably due to three loci, and they recommend a high-dose sprayed on manure to combat it. A low level of resistance to cyromazine was documented on a pig farm in the United Kingdom (Bell *et al.*, 2010), so a high dose may be a short-term solution. For newer insecticidal compounds such as fipronil (a phenylpyrazole) or spinosad (a spinosyn), possible resistance interactions are investigated. Liu and Yue (2000) found that spinosad was effective against one pyrethroid-resistant strain. Scott (1998) came to the same conclusion but noted the slow action of spinosad. A pyrethroid-resistant strain was found to exhibit cross-resistance to fipronil (Wen and Scott, 1999). These results may present a warning in the use of these compounds for control of other muscoid flies. As previously noted, *L. cuprina* has not yet evolved resistance to cyromazine, and it may be unlikely (Yen *et al.*, 1996; Levot and Sales, 2004), but it is possible to inadvertently select for it (Tang *et al.*, 2002).

Widespread resistance in *M. domestica* provides lessons in IRM strategies that may be applied to other medical/veterinary pests. The tolerance threshold is quite low for ectoparasites, and so intensive use of a single compound for control quickly selects for resistance. Like most other insects in the case studies documented here, major resistance problems force people to take an IPM-like approach to combating resistance. Crespo *et al.* (2002) found that populations of *M. domestica* in poultry farms were reduced most effectively using a combination of chemical control (cyromazine), cultural control (using lime to dry the manure), and biological control (parasitic wasps).

DISCUSSION

Given that arthropods have evolved resistance to crops, crop rotation, and insecticides applied to crops, it is not surprising that they have evolved resistance to insecticides and acaricides used to manage pests on animals we consider very valuable: pets, livestock, and ourselves. Ethical considerations and social realities constrain options for control and research, especially when human lives or livelihoods are affected. Even pests that are not considered vectors can pose limitations. For example, Moore and Miller (2009) found that cooperation in control apartments declined when they learned authors were only counting bed bugs, and tenants frequently disposed of infested material. Susceptibility can be difficult to maintain when certain IRM strategies are precluded. Resistance can be directly selected from control methods of ectoparasites or indirectly selected from outside sources.

Georghiou (1990) reviewed the problem of crop-protection chemicals contributing to the evolution of resistance in vectors of animal diseases. Broad-spectrum insecticides used for pest control on crops can reduce the fitness of mosquitoes infesting the same or nearby habitats. Certain fungicides and herbicides used on crops can even synergistically promote the efficacy of insecticides targeted against mosquitoes (Georghiou, 1990). Four types of evidence support the claim of a link between agricultural pesticide use and resistance in mosquitoes: (1) mosquito resistance is often higher in agricultural than in nonagricultural areas; (2) mosquito resistance can be observed before insecticides have been targeted against the vectors of disease; (3) there often is a correlation between intensity of insecticide use on crops and the level of resistance in mosquitoes; and (4) fluctuations in mosquito resistance have been observed in synchrony with seasonal fluctuations of crop spraying. In general, these studies demonstrate that IRM must account for selection pressure from all sources in a pest's habitat to be effective. Georghiou (1990) proposed two guidelines for mosquito IRM in agricultural areas. First, agricultural and public health agencies and industries must collaborate on mosquito IRM. Second, comprehensive IPM for crops must be supported by collaborators to reduce dependency on insecticides.

The link between public health and agricultural use of insecticides helps explain the persistence of resistance to insecticides. Dusfour *et al.* (2010) found a direct correlation between resistance to malathion in populations of *An. albimanus* with proximity to agricultural areas. The same link to DDT and permethrin resistance was not observed, but the authors note that malathion is much more water-soluble as compared to the other two classes, which tend to bind to soil particles (Dusfour *et al.*, 2010). Persistence of resistance to DDT, though this compound has not been used for decades, is common in all or most of the pests described in this chapter, suggesting little to no fitness cost. A second reason is the prevalence of pyrethroid use for control, but resistance may remain in spite of the disuse of either compound. Persistence of pyrethroid resistance in the absence of selection has been shown in *H. irritans* in field conditions and in *C. felis* in the laboratory. There is generally little or no fitness cost to resistant genotypes (Roush and McKenzie, 1987); one example is pyrethroid resistance in *V. destructor* (Martin *et al.*, 2002). Another reason for persistence of resistance to DDT and pyrethroids is that multiple mechanisms of resistance exist. Resistance to pyrethroids in *H. irritans* may occur in the form of the *kdr* mutation (Guerrero *et al.*, 1997), metabolic detoxification (Sparks *et al.*, 1990), or behavioral mechanisms (Lockwood *et al.*, 1985; Byford and Sparks, 1987). Multiple mechanisms of resistance to pyrethroids (Brogdon *et al.*, 1999) and DDT (Penilla *et al.*, 2006) are also seen in *Anopheles* mosquitoes, *P. h. capitis* (Lee *et al.*, 2000), and *M. domestica* (Liu and Yue, 2000).

Persistent resistance to other insecticides is less common than to pyrethroids but nevertheless is still present. Resistance to organochlorines is widespread in mosquitoes, even though those compounds have not been used for decades (Hemingway and Ranson, 2000). Laboratory colonies of *C. felis* fleas had variable frequencies of *Rdl* alleles without selection by dieldrin or any other insecticidal compound (Bass *et al.*, 2004b). IRM strategies should account for persistent resistance to certain compounds and the possibility of cross-resistance to another compound that may be used.

Because of commonalities that exist in control and subsequent resistance evolution in ectoparasites, there are already a few candidate mechanisms to explore. Certain sodium channel mutations are frequently found in individuals or populations that are resistant to DDT and pyrethroids (Davies *et al.*, 2007). However, although there tends to be a relationship between *kdr* and DDT or pyrethroid resistance status in mosquitoes, there are concerns about assuming this status based solely on *kdr* genotype (Brooke, 2008; Donnelly *et al.*, 2009). Resistance to pyrethroids can also occur via metabolic mechanisms, which may also be responsible for cross-resistance to other classes of compounds. Metabolic resistance usually involves increased expression of at least one enzyme responsible for detoxification of xenobiotics, such as cytochrome P450 monooxygenases, GSTs, and esterases, but can also occur via target site mutations leading to a more efficient enzyme (Li *et al.*, 2007). Gene amplification (duplication) is the most common way of achieving increased enzyme production resulting in resistance, but increases in transcription can also occur via coding mutations in some region contributing to its regulation (Bass and Field, 2011). With ever increasing efficiency of sequencing and gene expression technologies, one can use these methods as an initial approach to investigate resistance mechanisms.

Monitoring and prediction of resistance is recognized as an important first step in its mitigation (Stanley, Chapter 15). Resistance mechanisms have been intensely studied in *M. domestica* and mosquitoes, most notably (Hemingway, 1998). The WHO has developed bioassays for resistance in a number of pests of public health importance. Furthermore, as the cases presented here indicate, molecular methods for diagnosing resistance have been developed (or are in development) for a variety of pests. PCR allele-specific amplification, or PASA, is a method of genotyping individuals when a single-base mutation has been identified, and it has been used for monitoring *kdr*-type and *Rdl* resistance. Newer technologies may also be utilized to take advantage of reduced inputs or higher throughput. Lee *et al.* (2010) outline a three-tier method for molecular monitoring of populations in human head lice that are resistant to permethrin via *kdr*-type resistance. When the responsible resistance mutations are known, quantitative sequencing of pools of individuals can be used to estimate allele frequency in pest populations (Kwon *et al.*, 2008; Seong *et al.*, 2010). Because this method tends to have a lower detection limit (between 2 and 8%), the authors recommend its use for routine monitoring or initial phases of resistance screening. A more sensitive method, though more expensive and labor-intensive, is to use real-time PCR to quantify copy numbers of given alleles in pools of individuals (Lee *et al.*, 2009). Finally, if more resolution is warranted, individual genotyping can be accomplished by a high-throughput SNP detection system (Kim *et al.*, 2004).

Genomic resources as a tool for either resistance monitoring or research into resistance mechanisms are greatly enhanced by coordination of scientists and stakeholders. VectorBase (www.vectorbase.org) is an online database of currently nine genomes of arthropod vectors of humans, with tools and resources for gene ontologies, population genomics, and insecticide resistance (Lawson *et al.*, 2009; Megy *et al.*, 2012). More broadly, an ambitious project to sequence 5000 insect and other arthropod genomes is becoming reality (http://www.arthropodgenomes. org/wiki/i5K), the priorities of which include pests of medical and veterinary importance.

There is much attention on predicting and monitoring resistance, but little has been done to actually manage resistance or to change practices according to predictions. This may be due to the inherent problematic nature of IRM for ectoparasites because of the value of the host. The value of the hosts of these pest arthropods—pets, livestock, and ourselves—contributes to the difficulty in researching or implementing long-term IRM solutions. The creation of untreated refuges to increase the number of susceptible individuals is common in agricultural settings. However, managed refuges are generally not acceptable for medical pests because there is usually no threshold of tolerance of these insects relating to either vector capability or comfort level. Managed refuges may not even be possible in certain arthropods spending all or most of their life cycle on or in close proximity to the host. Use of a refuge depends on whether the pest is holometabolous or hemimetabolous. A holometabolous insect is likely to be only a temporary parasite, usually (but not necessarily) during the adult stage; a refuge could consist of nonparasitic larvae. Conversely, a different control method may be used on the larval stage, essentially functioning as a rotation strategy for IRM. Hemimetabolous insects or acarines are usually either permanent parasites or spend very little time off of a host. An untreated refuge may not be possible in these situations, so other strategies such as insecticide mixtures or rotations may be useful. Finally, a pest's fidelity to a host further complicates the use of a refuge. As discussed above, some mosquitoes may feed on an alternative host, but examples such as these are probably the exception rather than the rule.

Another typical IRM strategy is the use of high doses of insecticide to ensure no heterozygotes survive treatment, but this too proves to be difficult for ectoparasites. Detriment to the host is of great concern, as exemplified by the *V. destructor/A. mellifera* association. Because both are arthropods, they are likely to be susceptible to similar compounds. Finally, a high-dose strategy may not even yield effective control. For example, a higher dose generally does not cause higher mortality in *P. h. capitis* lice (Downs *et al.*, 2002).

Coordination and education among all affected parties are crucial in mitigating resistance; one cannot do it alone. There have been calls for more communication in almost every case study here. A concern for public health gives greater impetus to control disease vectors and manage resistance. Even though there is a need for more coordination in distribution of ITNs and educational materials about malaria, it is one example where there is some level of centralization of mosquito control. WHO has a great influence on policy-makers of different nations, and they have made great strides in education about malaria, and consequently mosquito resistance. Another organization dedicated to coordination on resistance management activities is IRAC (Insecticide Resistance Action Committee). This international group consists of industry leaders and some academics, and they advise regulatory bodies on policy issues relating to sustainable agriculture and public health. Information on resistance and resistance management issues is available on their website (www.irac-online.org).

Other case studies show examples of difficulty in coordinating IRM efforts. For example, *R. (B.) microplus* can be easily spread between herds of cattle, and George *et al.* (2004) lament the disconnect between researchers and producers in integrated tick control. Some pests are so ubiquitous that a coordinated effort to manage resistance would be almost impossible. A great number of people own pets, and a great number of these pets will encounter fleas at some point in time. The regulation of flea control methods through veterinarians could slow the evolution of resistance. However, as mentioned above, resistance to antibiotics has not been prevented by exclusive prescription by physicians.

Finally, the role of behavior in resistance, and therefore resistance management, is one that should not be overlooked. The behavior of an ectoparasite could be exploited in its control and managing resistance; this is seen in the possible use of ITNs that are treated with two insecticides, one on the top and one on the bottom (Figure 6.1). These nets function as a mixture strategy for IRM because of mosquito host-seeking behavior. Conversely, a parasite may evolve resistance by behavioral means. Behavioral resistance has evolved in a number of the arthropods mentioned here, namely to pyrethroid insecticides (or DDT) because of their excitorepellent effect. A notable example of behavioral resistance is certain *Anopheles* mosquitoes that change their resting places in response to indoor residual spraying. Some species naturally rest indoors, but have adapted to indoor sprays by resting outdoors where insecticide is not sprayed, that is, exophily. Behavioral avoidance of pyrethroids is also seen in *H. irritans*, where they have been noted to rest on cattle on the areas where they are farthest from the insecticide-treated ear tag, namely, the underside and backside (Byford *et al.*, 1987). Resistance by a temporal change in behavior is another possibility. There is concern that mosquitoes could change preferred feeding times to hours of the day when people are not likely to be sleeping under an ITN (Pates and Curtis, 2005).

Cooperband and Allan (2009) provide a good discussion of behavioral terms with regard to the effect of pyrethroids on three different mosquito species. They revisit terminology from Dethier *et al.* (1960), in which behavioral terms such as *attractant*, *repellent*, and *arrestant* are defined. Dethier (1956) further categorized repellency into two categories: one involving a directed avoidance response and one that stimulates general locomotor activity. The latter definition would refer to the contact irritancy that DDT and pyrethroids effect and should be classified as locomotor stimulants, rather than by the term *excitorepellent* (Cooperband and Allan, 2009).

There are many difficulties in devising and implementing IRM strategies for ectoparasites. The valuable nature of the host, or even ethical reasons, may preclude certain types of strategies from being tested. Models greatly assist in determining the best strategy, but they clearly need to be built on reliable biological data, which may be difficult to obtain due to host specificity of the ectoparasite. Some people may view the magnitude and severity of a disease to outweigh the very real risk of evolution of resistance by the vectors of the disease. Because of these challenges, many have adopted IPM-like methods of control to manage resistance.

REFERENCES

Adelman, Z.N., Kilcullen, K.A., Koganemaru, R., Anderson, M.A.E., Anderson, T.D., Miller, D.M., 2011. Deep sequencing of pyrethroid-resistant bed bugs reveals multiple mechanisms of resistance within a single population. PLoS ONE. 6.

Aguirre, J., Sobrino, A., Santamaria, M., Aburto, A., Roman, S., Hernandez, M., et al., 1986. Resistancia de garrapatas en Mexico. In: Cavazzani, A.H., Garcia, Z. (Eds.), Seminario Internacional de Parasitologia Animal. Cuernavaca, Morelos, Mexico, pp. 282—306.

Ameen, M., Arenas, R., Villanueva-Reyes, J., Ruiz-Esmenjaud, J., Millar, D., Domínguez-Dueñas, F., et al., 2010. Oral ivermectin for treatment of pediculosis capitis. Pediatr. Infect. Dis. J. 29, 991—993.

Angus, B.M., 1996. The history of the cattle tick Boophilus microplus in Australia and achievements in its control. Int. J. Parasitol. 26, 1341—1355.

Barros, A.T.M., Alison Jr., M.W., Foil, L.D., 1999. Evaluation of a yearly insecticidal ear tag rotation for control of pyrethroid-resistant horn flies (Diptera: Muscidae). Vet. Parasitol. 82, 317—325.

Barros, A.T.M., Ottea, J., Sanson, D., Foil, L.D., 2001. Horn fly (Diptera: Muscidae) resistance to organophosphate insecticides. Vet. Parasitol. 96, 243—256.

Barros, A.T.M., Gomes, A., Ismael, A.P.K., Koller, W.W., 2002. Susceptibility to diazinon in populations of the horn fly, Haematobia irritans (Diptera: Muscidae), in central Brazil. Mem. Inst. Oswaldo Cruz. 97, 905—907.

Bass, C., Field, L.M., 2011. Gene amplification and insecticide resistance. Pest Manag. Sci. 67, 886—890.

Bass, C., Schroeder, I., Turberg, A., Field, L.M., Williamson, M.S., 2004a. Identification of mutations associated with pyrethroid resistance in the para-type sodium channel of the cat flea, Ctenocephalides felis. Ins. Biochem. Mol. Biol. 34, 1305—1313.

Bass, C., Schroeder, I., Turberg, A., Field, L.M., Williamson, M.S., 2004b. Identification of the Rdl mutation in laboratory and field strains of the cat flea, Ctenocephalides felis (Siphonaptera: Pulicidae). Pest manag. Sci. 60, 1157—1162.

Bass, C., Lansdell, S.J., Millar, N.S., Schroeder, I., Turberg, A., Field, L.M., et al., 2006. Molecular characterisation of nicotinic acetylcholine receptor subunits from the cat flea, Ctenocephalides felis (Siphonaptera: Pulicidae). Ins. Biochem. Mol. Biol. 36, 86—96.

Batterham, P., Hill-Williams, A., Levot, G., Sales, N., McKenzie, J.A., 2006. The genetic bases of high-level resistance to diflubenzuron and low-level resistance to cyromazine in a field strain of the Australian sheep blowfly, Lucilia cuprina (Wiedemann) (Diptera : Calliphoridae). Aust. J. Entomol. 45, 87—90.

Bell, H.A., Robinson, K.A., Weaver, R.J., 2010. First report of cyromazine resistance in a population of UK house fly (Musca domestica) associated with intensive livestock production. Pest Manag. Sci. 66, 693—695.

Benoit, J.B., 2011. Stress tolerance of bed bugs: a review of factors that cause trauma to Cimex lectularius and C. hemipterus. Insects. 2, 151—172.

Bossard, R.L., Hinkle, N.C., Rust, M.K., 1998. Review of insecticide resistance in cat fleas (Siphonaptera: Pulicidae). J. Med. Entomol. 35, 415—422.

Bossard, R.L., Bossard, R.L., Broce, A.B., 2002. Insecticide susceptibilities of cat fleas (Siphonaptera: Pulicidae) from several regions of the United States. J. Med. Entomol. 39, 742—746.

Brogdon, W.G., McAllister, J.A., Corwin, A.M., Cordon-Rosales, C., 1999. Independent selection of multiple mechanisms for pyrethroid resistance in Guatemalan Anopheles albimanus (Diptera: Culicidae). J. Econ. Entomol. 92, 298—302.

Brooke, B.D., 2008. kdr: can a single mutation produce an entire insecticide resistance phenotype? Trans. R. Soc. Trop. Med. Hyg. 102, 524—525.

Buechler, R., Berg, S., Le Conte, Y., 2010. Breeding for resistance to Varroa destructor in Europe. Apidologie. 41, 393—408.

Burgess, I.F., 2004. Human lice and their control. Annu. Rev. Entomol. 49, 457—481.

Bush, S.E., Rock, A.N., Jones, S.L., Malenke, J.R., Clayton, D.H., 2011. Efficacy of LouseBuster, a new medical device for treating head lice (Anoplura: Pediculidae). J. Med. Entomol. 48, 67−72.

Byford, R.L., Sparks, T.C., 1987. Chemical approaches to the management of resistant horn fly, Haematobia irritans (L.) populations. In: Ford, M.G., Holloman, D.W., Khambay, B.P.S., Sawicki, R.M. (Eds.), Combating Resistance to Xenobiotics: Biological and Chemical Approaches. Ellis Horwood, London, pp. 178−191.

Byford, R.L., Lockwood, J.A., Smith, S.M., Franke, D.E., 1987. Redistribution of behaviorally resistant horn flies (Diptera: Muscidae) on cattle treated with pyrethroid-impregnated ear tags. Environ. Entomol. 16, 467−470.

Byford, R.L., Craig, M.E., Derouen, S.M., Kimball, M.D., Morrison, D.G., Wyatt, W.E., et al., 1999. Influence of permethrin, diazinon and ivermectin treatments on insecticide resistance in the horn fly (Diptera: Muscidae). Int. J. Parasitol. 29, 125−135.

Calderone, N.W., 2005. Evaluation of drone brood removal for management of Varroa destructor (Acari: Varroidae) in colonies of Apis mellifera (Hymenoptera: Apidae) in the Northeastern United States. J. Econ. Entomol. 98, 645−650.

Canyon, D.V., Speare, R., 2007. A comparison of botanical and synthetic substances commonly used to prevent head lice (Pediculus humanus var. capitis) infestation. Int. J. Dermatol. 46, 422−426.

Carlotti, D.N., Jacobs, D.E., 2000. Therapy, control, and prevention of flea allergy dermatitis in dogs and cats. Vet. Dermatol. 11, 83−98.

Centers for Disease Control and Prevention (CDC, United States), 2010. <http://www.cdc.gov/parasites/lice/head/treatment.html>. Page (last updated November 2, 2010).

Chandre, F., Darriet, F., Duchon, S., Finot, L., Manguin, S., Carnevale, P., et al., 2000. Modifications of pyrethroid effects associated with kdr mutation in Anopheles gambiae. Med. Vet. Entomol. 14, 81−88.

Chevillon, C., Raymond, M., Guillemaud, T., Lenormand, T., Pasteur, N., 1999. Population genetics of insecticide resistance in the mosquito Culex pipiens. Biol. J. Linn. Soc. 68, 147−157.

Chosidow, O., Giraudeau, B., Cottrell, J., Izri, A., Hofman, R., Mann, S.G., et al., 2010. Oral ivermectin versus malathion lotion for difficult-to-treat head lice. N. Engl. J. Med. 362, 896−905.

Cilek, J.E., Knapp, F.W., 1993. Enhanced diazinon susceptibility in pyrethroid-resistant horn flies (Diptera: Muscidae): potential for insecticide resistance management. J. Econ. Entomol. 86, 1303−1307.

Cooperband, M.F., Allan, S.A., 2009. Effects of different pyrethroids on landing behavior of female Aedes aegypti, Anopheles quadrimaculatus, and Culex quinquefasciatus mosquitoes (Diptera: Culicidae). J. Med. Entomol. 46, 292−306.

Cornuet, J.M., Beaumont, M.A., Estoup, A., Solignac, M., 2006. Inference on microsatellite mutation processes in the invasive mite, Varroa destructor, using reversible jump Markov chain Monte Carlo. Theor. Pop. Biol. 69, 129−144.

Crespo, D.C., Lecuona, R.E., Hogsette, J.A., 2002. Strategies for controlling house fly populations resistant to cyromazine. Neotrop. Entomol. 31, 141−147.

Cui, F., Raymond, M., Qiao, C.-L., 2006. Insecticide resistance in vector mosquitoes in China. Pest Manag. Sci. 62, 1013−1022.

Cui, F., Lin, Z., Wang, H., Liu, S., Chang, H., Reeck, G., et al., 2011. Two single mutations commonly cause qualitative change of nonspecific carboxylesterases in insects. Insect Biochem. Mol. Biol. 41, 1−8.

Curtis, C.F., 1985. Theoretical models of the use of insecticide mixtures for the management of resistance. Bull. Entomol. Res. 75, 259−265.

Curtis, C.F., Hill, N., Kasim, S.H., 1993. Are there effective resistance management strategies for vectors of human disease? Biol. J. Linn. Soc. 48, 3−18.

Curtis, C.F., Miller, J.E., Hodjati, M.H., Kolaczinski, J.H., Kasumba, I., 1998. Can anything be done to maintain the effectiveness of pyrethroid-impregnated bednets against
   malaria vectors? Phil. Trans. R. Soc. Lond. B. 353, 1769−1775.
Daborn, P., McCart, C., Woods, D., ffrench-Constant, R.H., 2004. Detection of insecticide resistance-associated mutations in cat flea Rdl by TaqMan-allele specific amplification. Pestic. Biochem. Physiol. 79, 25−30.
Davey, R.B., George, J.E., Miller, R.J., 2004. Control of an organophosphate-resistant
   strain of Boophilus microplus (Acari: Ixodidae) infested on cattle after a series of dips in
   coumaphos applied at different treatment intervals. J. Med. Entomol. 41, 524−528.
Davey, R.B., Miller, R.J., George, J.E., 2008. Efficacy of amitraz applied as a dip against
   an amitraz-resistant strain of Rhipicephalus (Boophilus) microplus (Acari: Ixodidae)
   infested on cattle. Vet. Parasitol. 152, 127−135.
Davies, T.G.E., Field, L.M., Usherwood, P.N.R., Williamson, M.S., 2007. A comparative
   study of voltage-gated sodium channels in the Insecta: implications for pyrethroid resistance in Anopheline and other Neopteran species. Insect Mol. Biol. 16, 361−375.
Davies, T.G.E., Field, L.M., Williamson, M.S., 2012. The re-emergence of the bed bug
   as a nuisance pest: implications of resistance to the pyrethroid insecticides. Med. Vet.
   Entomol. 26, 241−254.
De Castro, J.J., 1997. Sustainable tick and tickborne disease control in livestock improve-
   ment in developing countries. Vet. Parasitol. 71, 77−97.
De La Fuente, J., Kocan, K.M., 2006. Strategies for development of vaccines for control
   of ixodid tick species. Parasite Immunol. 28, 275−283.
Dethier, V.G., 1956. Repellents. Annu. Rev. Entomol. 1, 181−202.
Dethier, V.G., Barton Browne, L., Smith, C.N., 1960. The designation of chemicals in
   terms of the responses they elicit from insects. J. Econ. Entomol. 53, 134−136.
Djènonton, A., Chabi, J., Baldet, T., Irish, S., Pennetier, C., Hougard, J.-M., et al., 2009.
   Managing insecticide resistance in malaria vectors by combining carbamate-treated
   plastic wall sheeting, and pyrethroid-treated bed nets. Malar. J. 8, 233.
Djènonton, A., Chandre, F., Dabiré, K.R., Chabi, J., N'Guessan, R., Baldet, T., et al.,
   2010. Indoor use of plastic sheeting impregnated with carbamate combined with
   long-lasting insecticidal mosquito nets for the control of pyrethroid-resistant malaria
   vectors. Am. J. Trop. Med. Hyg. 83, 266−270.
Donnelly, M.J., Corbel, V., Weetman, D., Wilding, C.S., Williamson, M.S., Black IV, W.
   C., 2009. Does kdr genotype predict insecticide-resistance phenotype in mosquitoes?.
   Trends Parasitol. 25, 213−219.
Downs, A.M.R., Stafford, K.A., Coles, G.C., 1999. Head lice: prevalence in schoolchil-
   dren and insecticide resistance. Parasitol. Today. 15, 1−4.
Downs, A.M.R., Stafford, K.A., Coles, G.C., 2000. Factors that may be influencing the
   prevalence of head lice among British school children. Paediatr. Dermatol. 17,
   72−74.
Downs, A.M.R., Stafford, K.A., Hunt, L.P., Ravenscroft, J.C., Coles, G.C., 2002.
   Widespread insecticide resistance in head lice to the over-the-counter pediculosides
   in England, and the emergence of carbaryl resistance. Brit. J. Dermat. 146, 88−93.
Durand, R., Bouvresse, S., Andriantsoanirina, V., Berdjane, Z., Chowsidow, O., Izri, A.,
   2011. High frequency of mutations associated with head lice pyrethroid resistance in
   schoolchildren from Bobigny, France. J. Med. Entomol. 48, 73−75.
Dusfour, I., Achee, N.L., Briceno, I., King, R., Grieco, J.P., 2010. Comparative data on
   the insecticide resistance of Anopheles albimanus in relation to agricultural practices in
   northern Belize, CA. J. Pestic. Sci. 83, 41−46.
Eisen, R.J., Gage, K.L., 2012. Transmission of flea-borne zoonotic agents. Annu. Rev.
   Entomol. 57, 61−82.

Elzen, P.J., Westervelt, D., 2002. Detection of coumaphos resistance in *Varroa destructor* in Florida. Am. Bee J. 142, 291–292.

Elzen, P.J., Eischen, F.A., Baxter, J.B., Pettis, J., Elzen, G.W., Wilson, W.T., 1998. Fluvalinate resistance in *Varroa jacobsoni* from several geographic locations. Am. Bee J. 138, 674–676.

Elzen, P.J., Baxter, J.R., Spivak, M., Wilson, W.T., 2000. Control of *Varroa jacobsoni* Oud resistant to fluvalinate and amitraz using coumaphos. Apidologie. 31, 437–441.

Elzen, P.J., Baxter, J.R., Westervelt, D., Causey, D., Randall, C., Cutts, L., et al., 2001. Acaricide rotation plan for control of varroa. Am. Bee J. 141, 412.

Estrada-Peña, A., Bouattour, A., Camicas, J.-L., Guglielmone, A., Horak, I., Jongejan, F., et al., 2006. The known distribution and ecological preferences of the tick subgenus *Boophilus* (Acari: Ixodidae) in Africa and Latin America. Exp. Appl. Acarol. 38, 219–235.

Foil, L.D., Coleman, P., Eisler, M., Fragoso-Sanchez, H., Garcia-Vazquez, Z., Guerrero, F.D., et al., 2004. Factors that influence the prevalence of acaricide resistance and tick-borne diseases. Vet. Parasitol. 125, 163–181.

Foil, L.D., Guerrero, F., Alison, M.W., Kimball, M.D., 2005. Association of the *kdr* and *superkdr* sodium channel mutations with resistance to pyrethroids in Louisiana populations of the horn fly, *Haematobia irritans irritans* (L.). Vet. Parasitol. 129, 149–158.

Frisch, J.E., O'Neill, C.J., Kelly, M.J., 2000. Using genetics to control cattle parasites — the Rockhampton experience. Int. J. Parasitol. 30, 253–264.

Gao, J.-R., Kozaki, T., Leichter, C.A., Rinkevich, F.D., Shono, T., Scott, J.G., 2007. The A302S mutation in *Rdl* that confers resistance to cyclodienes and limited cross-resistance to fipronil is undetectable in field populations of house flies from the USA. Pestic. Biochem. Physiol. 88, 66–70.

George, J.E., 2000. Present and future technologies for tick control. Ann. N. Y. Acad. Sci. 916, 583–588.

George, J.E., Pound, J.M., Davey, R.B., 2004. Chemical control of ticks on cattle and the resistance of these parasites to acaricides. Parasitol. 129, S353–S366.

Georghiou, G.P., 1990. The effect of agro-chemicals on vector populations. In: Roush, R.T., Tabashnik, B.E. (Eds.), Pesticide Resistance in Arthropods. Chapman and Hall, New York, pp. 183–202. , Chapter 7.

Goates, B.M., Atkin, J.S., Wilding, K.G., Birch, K.G., Cottam, M.R., Bush, S.E., et al., 2006. An effective nonchemical treatment for head lice: a lot of hot air. Pediatrics. 118, 1962–1970.

Goss, P.J.E., McKenzie, J.A., 1996. Selection, refugia, and migration: simulation of evolution of dieldrin resistance in *Lucilia cuprina* (Diptera: Calliphoridae). J. Econ. Entomol. 89, 288–301.

Graf, J.-F., Gogolewski, R., Leach-Bing, N., Sabatini, G.A., Molento, M.B., Bordin, E. L., et al., 2004. Tick control: an industry point of view. Parasitol. 129, S427–S442.

Guerrero, F.D., Alison Jr., M.W., Kammlah, D.M., Foil, L.D., 2002. Use of the polymerase chain reaction to investigate the dynamics of pyrethroid resistance in Haematobia irritans irritans (Diptera: Muscidae). J. Med. Entomol. 39, 747–754.

Guerrero, F.D., Miller, R.J., Pérez de León, A.A., 2012. Cattle tick vaccines: Many candidate antigens, but will a commercially viable product emerge? Int. J. Parasitol. 42, 421–427.

Guerrero, R.D., Jamroz, R.C., Kammlah, D., Kunz, S.E., 1997. Toxicological and molecular characterization of pyrethroid-resistant horn flies, *Haematobia irritans*: identification of *kdr* and *super-kdr* point mutations. Ins. Biochem. Mol. Biol. 27, 745–755.

Guglielmone, A.A., Castelli, M.E., Volpogni, M.M., Anziani, O.S., Mangold, A.J., 2002. Dynamics of cypermethrin resistance in the field in the horn fly, *Haematobia irritans*. Med. Vet. Entomol. 16, 310–315.

Guillet, P.N., Guessan, R., Darriet, F., Traore-Lamizana, M., Chandre, F., Carnevale, P., 2001. Combined pyrethroid and carbamate "two-in-one" treated mosquito nets: field efficacy against pyrethroid-resistant *Anopheles gambiae* and *Culex quiquefasciatus*. Med. Vet. Entomol. 15, 105−112.

Harlan, H.J., 2006. Bed bugs 101: the basics of *Cimex lectularius*. Am. Entomol. 52, 99−101.

Hemingway J., 1998. Techniques to detect insecticide resistance mechanisms (a field and laboratory manual). Document WHO/CDS/CPC/MAL/98.6. World Health Organization, Geneva <http://www.who.int/whopes/resistance/en/>.

Hemingway, J., Ranson, H., 2000. Insecticide resistance in insect vectors of human disease. Annu. Rev. Entomol. 45, 371−391.

Hemingway, J., Penilla, R.P., Rodríguez, A.D., James, B.M., Edge, W., Rogers, H., et al., 1997. Resistance management strategies in malaria vector mosquito control. A large-scale field trial in Southern Mexico. Pestic. Sci. 51, 375−382.

Hemingway, J., Hawkes, N.J., McCarroll, L., Ranson, H., 2004. The molecular basis of insecticide resistance in mosquitoes. Insect Biochem. Mol. Biol. 34, 653−665.

Hillesheim, E., Ritter, W., Bassand, D., 1996. First data on resistance mechanisms of *Varroa jacobsoni* (Oud.) against tau fluvalinate. Exp. Appl. Acarol. 20, 283−296.

Huang, J., Kristensen, M., Qiao, C.L., Jespersen, J.B., 2004. Frequency of *kdr* gene in house fly field populations: correlation pyrethroid resistance and *kdr* frequency. J. Econ. Entomol. 97, 1036−1041.

Huang, Z., 2001. Mite zapper-a new and effective method for varroa mite control. Am. Bee J. 141, 730−732.

Hughes, P.B., McKenzie, J.A., 1987. Insecticide resistance in the Australian sheep blowfly, *Lucilia cuprina*: speculation, science, and strategies. In: Ford, M.G., Holloman, D.W., Khambay, B.P.S., Sawicki, R.M. (Eds.), Combating Resistance to Xenobiotics: Biological and Chemical Approaches. Ellis Horwood, London, pp. 162−177.

Irish, S.R., N'Guessan, R., Boko, P.M., Metonnou, C., Odjo, A., Akogbeto, M., et al., 2008. Loss of protection with insecticide-treated nets against pyrethroid-resistant *Culex quinquefasciatus* mosquitoes once nets become holed: an experimental hut study. Parasit Vectors. 1, 17.

James, P.J., 2005. Genetic alternatives to mulesing and tail docking in sheep: a review. Aust. J. Exp. Agric. 46, 1−18.

Jamroz, R.C., Guerrero, F.D., Kammlah, D.M., Kunz, S.E., 1998. Role of the *kdr* and *super-kdr* sodium channel mutations in pyrethroid resistance: correlation of allelic frequency to resistance level in wild and laboratory populations of horn flies (*Haematobia irritans*). Ins. Biochem. Mol. Biol. 28, 1031−1037.

Johnson, R.M., Pollock, H.S., Berenbaum, M.R., 2009. Synergistic interactions between in-hive miticides in *Apis mellifera*. J. Econ. Entomol. 102, 474−479.

Johnson, R.M., Huang, Z.Y., Berenbaum, M.R., 2010. Role of detoxification in *Varroa destructor* (Acari: Parasitidae). tolerance of the miticide tau-fluvalinate. Int. J. Acarol. 36, 1−6.

Jones, C.M., Sanou, A., Guelbeogo, W.M., Sagnon, N., Johnson, P.C.D., Ranson, H., 2012. Aging partially restores the efficacy of malaria vector control in insecticide-resistant populations of *Anopheles gambiae s.l.* from Burkina Faso. Malar. J. 11, 24.

Jonsson, N.N., Hope, M., 2007. Progress in the epidemiology and diagnosis of amitraz resistance in the cattle tick *Boophilus microplus*. Vet. Parasitol. 146, 193−198.

Jonsson, N.N., Mayer, D.G., Green, P.E., 2000. Possible risk factors on Queensland dairy farms for acaricide resistance in cattle tick (*Boophilus microplus*). Vet. Parasitol. 88, 79−92.

Jonsson, N.N., Davis, R., De Witt, M., 2001. An estimate of the economic effects of cattle tick (*Boophilus microplus*) infestation on Queensland dairy farms. Aust. Vet. J. 79, 826−831.

Jonsson, N.N., Miller, R.J., Kemp, D.H., Knowles, A., Ardila, A.E., Verrall, R.G., et al., 2010. Rotation of treatments between spinosad and amitraz for the control of *Rhipicephalus (Boophilus) microplus* populations with amitraz resistance. Vet. Parasitol. 169, 157−164.

Kamidi, R.E., Kamidi, M.K., 2005. Effects of a novel pesticide resistance management strategy on tick control in a smallholding exotic-breed dairy herd in Kenya. Trop. Anim. Health Prod. 37, 469−478.

Kanga, L.H.B., Adamczyk, J., Marshall, K., Cox, R., 2010. Monitoring for resistance to organophosphorus and pyrethroid insecticides in *Varroa* mite populations. J. Econ. Entomol. 103, 1797−1802.

Kawaguchi, I., Sasaki, A., Mogi, M., 2003. Combining zooprophylaxis and insecticide spraying: a malaria-control strategy limiting the development of insecticide resistance in vector mosquitoes. Proc. R. Soc. Lond. B. 271, 301−309.

Keiding, J., 1999. Review of the global status and recent development of insecticide resistance in field populations of the housefly, *Musca domestica* (Diptera: Muscidae). Bull. Entomol. Res. 89 (Suppl. 1), 67.

Khallaayoune, K., Biron, J.M., Chaoui, A., Duvallet, G., 2009. Efficacy of 1% geraniol (Fulltec®) as a tick repellent. Parasite. 16, 223−226.

Kim, H.J., Symington, S.B., Lee, S.H., Clark, J.M., 2004. Serial invasive signal amplification reaction for genotyping permethrin-resistant (*kdr*-like) human head lice, *Pediculus capitis*. Pestic. Biochem. Physiol. 80, 173−182.

Kiss, T., Cadar, D., Spînu, M., 2012. Tick prevention at a crossroad: new and renewed solutions. Vet. Parasitol. 187, 357−366.

Koella, J.C., Lynch, P.A., Thomas, M.B., Read, A.F., 2009. Towards evolution-proof malaria control with insecticides. Evol. Appl. 2, 469−480.

Koella, J.C., Saddler, A., Karacs, T.P.S., 2012. Blocking the evolution of insecticide-resistant malaria vectors with a microsporidian. Evol. Appl. 5, 283−292.

Kozaki, T., Brady, S.G., Scott, J.G., 2009. Frequencies and evolution of organophosphate insensitive acetylcholinesterase alleles in laboratory and field populations of the house fly, *Musca domestica* L. Pestic. Biochem. Physiol. 95, 6−11.

Kunz, S.E., Murrell, K.D., Lamber, G., James, L.F., Terrill, C.E., 1991. Estimated losses of livestock to pests.. In: Pimental, D. (Ed.), Handbook of Pest Management in Agriculture, vol. 1. CRC Press, Boca Raton, FL, pp. 69−98.

Kwon, D.H., Yoon, K.S., Strycharz, J.P., Clark, J.M., Lee, S.H., 2008. Determination of permethrin resistance allele frequency of human head louse populations by quantitative sequencing. J. Med. Entomol. 45, 912−920.

Labbé, P., Lenormand, T., Raymond, M., 2005. On the worldwide selection of an insecticide resistance gene: a role for local selection. J. Evol. Biol. 18, 1471−1484.

Labbé, P., Berthomieu, A., Berticat, C., Alout, H., Raymond, M., Lenormand, T., et al., 2007. Independent duplications of the acetylcholinesterase gene conferring insecticide resistance in the mosquito *Culex pipiens*. Mol. Biol. Evol. 24, 1056−1067.

Labbé, P., Sidos, N., Raymond, M., Lenormand, T., 2009. Resistance gene replacement in the mosquito *Culex pipiens*: fitness estimation from long-term cline series. Genetics. 182, 303−312.

Lawson, D., Arensburger, P., Atkinson, P., Besansky, N.J., Bruggner, R.V., Butler, R., et al., 2009. VectorBase: a data resource for invertebrate vector genomics. Nucleic Acids Res. 37, D583−D587.

Lee, S.H., Yoon, K.S., Williamson, M.S., Goodson, S.J., Takano-Lee, M., Edman, J.D., et al., 2000. Molecular analysis of *kdr*-like resistance in permethrin-resistant strains of head lice, *Pediculus capitis*. Pestic. Biochem. Physiol. 66, 130−143.

Lee, S.H., Clark, J.M., Yoon, K.S., Kwon, D.H., Hodgen, H.E., Seong, K.M., 2009. Resistance Management of the Human Head Louse Using Molecular Tools, ACS Book Series.

Lee, S.H., Clark, J.M., Ahn, Y.J., Lee, W.-J., Yoon, K.S., Kwon, D.H., et al., 2010. Molecular mechanisms and monitoring of permethrin resistance in human head lice. Pestic. Biochem. Physiol. 97, 109−114.

Lenormand, T., Raymond, M., 2000. Analysis of clines with variable selection and variable migration. Am. Nat. 155, 70−82.

Levot, G., Sales, N., 2004. Insect growth regulator cross-resistance studies in field- and laboratory-selected strains of the Australian sheep blowfly, Lucilia cuprina (Wiedemann) (Diptera: Calliphoridae). Aust. J. Entomol. 43, 374−377.

Levot, G., Sales, N., 2008. In vitro effectiveness of ivermectin, and spinosad flystrike treatments against larvae of the Australian sheep blowfly Lucilia cuprina (Wiedemann) (Diptera: Calliphoridae). Aust. J. Entomol. 47, 365−369.

Levot, G.W., 1995. Resistance and the control of sheep ectoparasites. Int. J. Parasitol. 25, 1355−1362.

Levot, G.W., Sales, N., 2002. New high level resistance to diflubenzuron detected in the Australian sheep blowfly, Lucilia cuprina (Wiedemann) (Diptera: Calliphoridae). Gen. Appl. Entomol. 31, 43−45.

Li, A.Y., Davey, R.B., Miller, R.J., George, J.E., 2004. Detection, and characterization of amitraz resistance in the Southern cattle tick, Boophilus microplus (Acari: Ixodidae). J. Med. Entomol. 41, 193−200.

Li, A.Y., Chen, A.C., Miller, R.J., Davey, R.B., George, J.E., 2007a. Acaricide resistance and synergism between permethrin, and amitraz against susceptible and resistant strains of Boophilus microplus (Acari: Ixodidae). Pest Manag. Sci. 63, 882−889.

Li, A.Y., Guerrero, F.D., Pruett, J.H., 2007b. Involvement of esterases in diazinon resistance, and biphasic effects of piperonyl butoxide on diazinon toxicity to Haematobia irritans irritans (Diptera: Muscidae). Pestic. Biochem. Physiol. 87, 147−155.

Li, X., Schuler, M.A., Berenbaum, M.R., 2007. Molecular mechanisms of metabolic resistance to synthetic and natural xenobiotics. Annu. Rev. Entomol. 52, 231−253.

Lim, S.S., Fullman, N., Stokes, A., Ravishankar, N., Masiye, F., Murray, C.J.L., et al., 2011. Net benefits: A multicountry analysis of observational data examining associations between insecticide-treated mosquito nets and health outcomes. PLoS Med.8.

Liu, N., Yue, X., 2000. Insecticide resistance and cross-resistance in the house fly (Diptera: Muscidae). J. Econ. Entomol. 93, 1269−1275.

Lockwood, J.A., Byford, R.L., Story, R.N., Sparks, T.C., Quisenberry, S.S., 1985. Behavioral resistance to pyrethroids in the horn fly (Diptera: Muscidae). Environ. Entomol. 14, 873−890.

Macedo, P.A., Ellis, M.D., Siegfried, B.D., 2002. Detection and quantification of fluvalinate resistance in varroa mites in Nebraska. Am. Bee J. 142, 17−25.

Maggi, M.D., Ruffinengo, S.R., Damiani, N., Sardella, N.H., Eguaras, M.J., 2009. First detection of Varroa destructor resistance to coumaphos in Argentina. Exp. Appl. Acarol. 47, 317−320.

Martin, S., Holland, K., Murray, M., 1997. Non-reproduction in the honeybee mite Varroa jacobsoni. Exp. Appl. Acarol. 21, 539−549.

Martin, S.J., 2004. Acaricide (pyrethroid) resistance in Varroa destructor. Bee World. 85, 67−69.

Martin, S.J., Elzen, P.J., Rubink, W.R., 2002. Effect of acaricide resistance on reproductive ability of the honey bee mite Varroa destructor. Exp. Appl. Acarol. 27, 195−207.

McKenzie, C.L., Byford, R.L., 1993. Continuous, alternating, and mixed insecticides affect development of resistance in the horn fly (Diptera: Muscidae). J. Econ. Entomol. 86, 1040−1048.

McKenzie, J.A., 2000. The character or the variation: the genetic analysis of the insecticide-resistance phenotype. Bull. Entomol. Res. 90, 3−7.

McKenzie, J.A., Batterham, P., 1998. Predicting insecticide resistance: mutagenesis, selection and response. Phil. Trans. R. Soc. Lond. B. 353, 1729—1734.

Megy, K., Emrich, S.J., Lawson, D., Campbell, D., Dialynas, E., Hughes, D.S.T., et al., 2012. VectorBase: improvements to a bioinformatics resource for invertebrate vector genomics. Nucleic Acids Res. 40, D729—D734.

Meinking, T.L., Serrano, L., Hard, B., Entzel, P., Lemard, G., Rivera, E., et al., 2002. Comparative in vitro pediculicidal efficacy of treatments in a resistant head lice population in the United States. Arch. Dermat. 138, 220—224.

Milani, N., 1999. The resistance of Varroa jacobsoni Oud. to acaricides. Apidologie. 30, 229—234.

Miller, R.J., Davey, R.B., George, J.E., 2005. First report of organophosphate-resistant Boophilus microplus (Acari: Ixodidae) within the United States.. J. Med. Entomol. 42, 912—917.

Miller, R.J., Esparza Rentaria, J.A., Quiroz Martinex, H., George, J.E., 2007. Characterization of permethrin-resistant Boophilus microplus (Acari: Ixodidae) collected from the state of Coahuila, Mexico. J. Med. Entomol. 44, 895—897.

Moore, D.J., Miller, D.M., 2009. Field evaluations of insecticide treatment regimens for control of the common bed bug, Cimex lectularius (L.). Pest Manag. Sci. 65, 332—338.

Mosqueira, B., Duchon, S., Chandre, F., Hougard, J.-M., Carnevale, P., Mas-Coma, S., 2010a. Efficacy of an insecticide paint against insecticide susceptible and resistant mosquitoes - Part 1: laboratory evaluation. Malar. J. 9.

Mosqueira, B., Chabi, J., Chandre, F., Akogbéto, M., Hougard, J.-M., Carnevale, P., et al., 2010b. Efficacy of an insecticide paint against malaria vectors and nuisance in West Africa - Part 2: field evaluation. Malar. J. 9.

Mougabure Cueto, G., Gonzalez Audino, P., Vassena, C.V., Picollo, M.I., Zerba, E.N., 2002. Toxic effect of aliphatic alcohols against susceptible and permethrin-resistant Pediculus humanus capitis (Anoplura: Pediculidae). J. Med. Entomol. 39, 457—460.

Mozes-Koch, R., Slabezki, Y., Efrat, H., Kalevi, H., Kamer, Y., Yakobson, B.A., et al., 2000. First detection in Israel of fluvalinate resistance in the varroa mite using bioassay and biochemical methods. Exp. App. Acarol. 24, 35—43.

Mumcuoglu, K.Y., 1996. Control of human lice (Anoplura: Pediculidae) infestations: past and present. Am. Entomol. 42, 175—178.

N'Guessan, R., Corbel, V., Akogbéto, M., Rowland, M., 2007. Reduced efficacy of insecticide-treated nets, and indoor residual spraying for malaria control in pyrethroid resistance area, Benin. Emerging Infect. Dis. 13, 199—206.

Norris, L.C., Norris, D.E., 2011a. Insecticide resistance in Culex quinquefasciatus mosquitoes after the introduction of insecticide-treated bed nets in Macha, Zambia. J. Vector Ecol. 36, 411—420.

Norris, L.C., Norris, D.E., 2011b. Efficacy of long-lasting insecticidal nets in use in Macha, Zambia, against the local Anopheles arabiensis population. Malar. J. 10, 254.

Norton, G.A., Sutherst, R.W., Maywald, G.F., 1983. A framework for integrating control methods against the cattle tick, Boophilus microplus in Australia. J. Appl. Biol. 20, 489—505.

Ordinioha, B., 2012. The use and misuse of mass distributed free insecticide-treated bed nets in a semi-urban community in Rivers State, Nigeria. Ann. Afr. Med. 11, 163—168.

Oxborough, R.M., Mosha, F.W., Matowo, J., Mndeme, R., Feston, E., Hemingway, J., et al., 2008. Mosquitoes and bednets: testing the spatial positioning of insecticide on nets and the rationale behind combination insecticide treatments. Ann. Trop. Med. Parasit. 102, 717—727.

Oyarzún, M.P., Li, A.Y., Figueroa, C.C., 2011. High levels of insecticide resistance in introduced horn fly (Diptera: Muscidae) populations and implications for management. J. Econ. Entomol. 104, 258–265.

Pates, H., Curtis, C., 2005. Mosquito behavior and vector control. Annu. Rev. Entomol. 50, 53–70.

Penilla, R.P., Rodríguez, A.D., Hemingway, J., Torres, J.L., Arredondo-Jiménez, J.I., Rodríguez, M.H., 1998. Resistance management strategies in malaria vector mosquito control. Baseline data for a large-scale field trial against Anopheles albimanus in Mexico. Med. Vet. Entomol. 12, 217–233.

Penilla, R.P., Rodríguez, A.D., Hemingway, J., Torres, J.L., Solis, R., Rodríguez, M.H., 2006. Changes in glutathione S-transferase activity in DDT resistant natural Mexican populations of Anopheles albimanus under different insecticide resistance management strategies. Pestic. Biochem. Physiol. 86, 63–71.

Penilla, R.P., Rodriguez, A.D., Hemingway, J., Trejo, A., Lopez, A.D., Rodriguez, M.H., 2007. Cytochrome P-450-based resistance mechanism and pyrethroid resistance in the field Anopheles albimanus resistance management trial. Pestic. Biochem. Physiol. 89, 111–117.

Pérez De Leon, A.A., Strickman, D.A., Knowles, D.P., Fish, D., Thacker, E., De La Fuente, J., et al., 2010. One Health approach to identify research needs in bovine and human babesioses: workshop report. Parasit Vectors. 3, 36.

Perry, T., Batterham, P., Daborn, P.J., 2011. The biology of insecticidal activity and resistance. Insect Biochem. Mol. Biol. 41, 411–422.

Polanco, A.M., Brewster, C.C., Miller, D.M., 2011a. Population growth potential of the bed bug, Cimex lectularius L.: a life table analysis. Insects. 2, 173–185.

Polanco, A.M., Miller, D.M., Brewster, C.C., 2011b. Survivorship during starvation for Cimex lectularius L. Insects. 2, 232–242.

Pollack, R.J., Kiszewski, A.E., Spielman, A., 2000. Overdiagnosis and consequent mismanagement of head louse infestations in North America. Pediatr. Infect. Dis. 19, 689–693.

Porto Neto, L.R., Jonsson, N.N., D'Occhio, M.J., Barendse, W., 2011. Molecular genetic approaches for identifying the basis of variation in resistance to tick infestation in cattle. Vet. Parasitol. 180, 165–172.

Potter, M.F., 2011. The history of bed bug management-with lessons from the past. Am. Entomol. 57, 14–25.

Quisenberry, S.S., Lockwood, J.A., Byford, R.L., Wilson, H.K., Sparks, T.C., 1984. Pyrethroid resistance in the horn fly Haematobia irritans (L.) (Diptera: Muscidae). J. Econ. Entomol. 77, 1095–1098.

Ranson, H., Abdallah, H., Badolo, A., Guelbeogo, W.M., Kerah-Hinzoumbé, C., Yangalbé-Kalnoné, E., et al., 2009. Insecticide resistance in Anopheles gambiae: data from the first year of a multi-country study highlight the extent of the problem. Malar. J. 8, 299.

Reinhardt, K., Siva-Jothy, M.T., 2007. Biology of the bed bugs (Cimicidae). Annu. Rev. Entomol. 52, 351–374.

Rinderer, T.E., Harris, J.W., Hunt, G.J., de Guzman, L.I., 2010. Breeding for resistance to Varroa destructor in North America. Apidologie. 41, 409–424.

Rinkevich, F.D., Hedtke, S.M., Leichter, C.A., Harris, S.A., Su, C., Brady, S.G., et al., 2012. Multiple origins of kdr-type resistance in the house fly, Musca domestica. PLoS ONE. 7, e52761.

Rivero, A., Vezilier, J., Weill, M., Read, A.F., Gandon, S., 2010. Insecticide control of vector-borne diseases: when is insecticide resistance a problem? PLoS Pathog. 6, 8.

Rodríguez, A.D., Penilla, R.P., Rodríguez, M.H., Hemingway, J., Trejo, A., Hernández-Avila, J.E., 2006. Acceptability and perceived side effects of insecticide indoor residual spraying under different resistance management strategies. Salud Publica Mex. 48, 317–324.

Rodríguez-Dehaibes, S.R., Otero-Colina, G., Pardo-Sedas, V., Villanueva-Jimenez, J.A., 2005. Resistance to amitraz and flumethrin in *Varroa destructor* populations from Veracruz, Mexico. J. Apic. Res. 44, 124–125.

Rodriguez-Dehaibes, S.R., Otero-Colina, G., Villanueva-Jimenez, J.A., Corcuera, P., 2011. Susceptibility of *Varroa destructor* (Gamasida: Varroidae) to four pesticides used in three Mexican apicultural regions under two different management systems. Int. J. Acarol. 37, 441–447.

Rodríguez-Vivas, R.I., Alonso-Díaz, M.A., Rodríguez-Arevalo, F., Fragoso-Sanchez, H., Santamaria, V.M., Rosario-Cruz, R., 2006. Prevalence and potential risk factors for organophosphate and pyrethroid resistance in *Boophilus microplus* ticks on cattle ranches from the State of Yucatan, Mexico. Vet. Parasitol. 136, 335–342.

Rodriguez-Vivas, R.I., Rivas, A.L., Chowell, G., Fragoso, S.H., Rosario, C.R., Garcia, Z., et al., 2007. Spatial distribution of acaricide profiles (*Boophilus microplus* strains susceptible or resistant to acaricides) in southeastern Mexico. Vet. Parasitol. 146, 158–169.

Roush, R.T., McKenzie, J.A., 1987. Ecological genetics of insecticide and acaricide resistance. Annu. Rev. Entomol. 32, 361–380.

Rust, M.K., Dryden, M.W., 1997. The biology, ecology, and management of the cat flea. Annu. Rev. Entomol. 42, 451–473.

Rust, M.K., Waggoner, M., Hinkle, N.C., Mencke, N., Hansen, O., Vaughn, M., et al., 2002. Development of a larval bioassay for susceptibility of cat fleas (Siphonaptera: Pulicidae) to imidacloprid. J. Med. Entomol. 39, 671–674.

Rust, M.K., Denholm, I., Dryden, M.W., Payne, P., Blagburn, B.L., Jacobs, D.E., et al., 2005. Determining a diagnostic dose for imidacloprid susceptibility testing of field-collected isolates of cat fleas (Siphonaptera: Pulicidae). J. Med. Entomol. 42, 631–636.

Sakamoto, J.M., Rasgon, J.L., 2006. Endosymbiotic bacteria of bed bugs: evolution, ecology and genetics. Am. Entomol. 52, 119–122.

Sammataro, D., Gerson, U., Needham, G., 2000. Parasitic mites of honey bees: life history, implications, and impact. Annu. Rev. Entomol. 45, 519–548.

Sangster, N.C., 2001. Managing parasiticide resistance. Vet. Parasitol. 98, 89–109.

Scott, J.A., Plapp Jr., F.W., Bay, D.E., 1997. Pyrethroid resistance associated with decreased biotic fitness in horn flies (Diptera: Muscidae). Southwest. Entomol. 22, 405–410.

Scott, J.G., 1998. Toxicity of spinosad to susceptible and resistant strains of house flies, *Musca domestica*. Pestic. Sci. 54, 131–133.

Scott, J.G., Wen, Z., 1997. Toxicity of fipronil to susceptible and resistant strains of German cockroaches (Dictyoptera: Blattellidae) and house flies (Diptera: Muscidae). J. Econ. Entomol. 90, 1152–1156.

Scott, M., Diwell, K., McKenzie, J.A., 2000. Dieldrin resistance in *Lucilia cuprina* (the Australian sheep blowfly): chance, selection and response. Hered. 84, 599–604.

Seifert, J., Scott, J.G., 2002. The CYP6D1v1 allele is associated with pyrethroid resistance in the house fly, *Musca domestica*. Pestic. Biochem. Physiol. 72, 40–44.

Seong, K.M., Lee, D.Y., Yoon, K.S., Kwon, D.H., Kim, H.C., Klein, T.A., et al., 2010. Establishment of quantitative sequencing, and filter contact vial bioassay for monitoring pyrethroid resistance in the common bed bug, *Cimex lectularius*. J. Med. Entomol. 47, 592–599.

Sibanda, M.M., Focke, W.W., Labuschagne, F.J.W.J., Moyo, L., Nhlapo, N.S., Maity, A., et al., 2011. Degradation of insecticides used for indoor spraying in malaria control and possible solutions. Malar. J. 10, 307.

Sparks, T.C., Quisenberry, S.S., Lockwood, J.A., Byford, R.L., Roush, R.T., 1985. Insecticide resistance in the horn fly, *Haematobia irritans*. J. Agric. Entomol. 2, 217–233.

Sparks, T.C., Byford, R.L., Craig, M.E., Crosby, B.L., McKenzie, C., 1990. Permethrin metabolism in pyrethroid-resistant adults of the horn fly (Diptera: Muscidae). J. Econ. Entomol. 83, 662–665.

Spreafico, M., Eordegh, F.R., Bernardinelli, I., Colombo, M., 2001. First detection of strains of Varroa destructor resistant to coumaphos. Results of laboratory tests and field trials. Apidologie. 32, 49–55.

Srivastava, R., Ghosh, S., Mandal, D.B., Azhahianambi, P., Singhal, P.S., Pandey, N.N., et al., 2008. Efficacy of Azadirachta indica extracts against Boophilus microplus. Parasitol. Res. 104, 149–153.

Strycharz, J.P., Berge, N.M., Alves, A.-M., Clark, J.M., 2011. Ivermectin acts as a posteclosion nymphicide by reducing blood feeding of human head lice (Anoplura: Pediculidae) that hatched from treated eggs. J. Med. Entomol. 48, 1174–1182.

Sutherst, R.W., Norton, G.A., Barlow, N.D., Conway, G.R., Birley, M., Comins, H.N., 1979. An analysis of management strategies for cattle tick (Boophilus microplus) control in Australia. J. Appl. Ecol. 16, 359–382.

Takken, W., 2002. Do insecticide-treated bednets have an effect on malaria vectors?. Trop. Med. Int. Hlth. 7, 1022–1030.

Tang, J.D., Caprio, M.A., Sheppard, D.C., Gaydon, D.M., 2002. Genetics and fitness costs of cyromazine resistance in the house fly (Diptera: Muscidae). J. Econ. Entomol. 95, 1251–1260.

Tebruegge, M., Pantazidou, A., Curtis, N., 2011. What's bugging you? An update on the treatment of head lice infestation. Arch. Dis. Child.: Educ. Prac. Ed. 96, 2–8.

Troullier, J., 1998. Monitoring Varroa jacobsoni resistance to pyrethroids in western Europe. Apidologie. 29, 537–546.

Usinger, R.L., 1966. Monograph of Cimicidae (Hemiptera-Heteroptera). Entomological Society of America, College Park, MD, The Thomas Say Foundation, Vol. VII.

Vulule, J.M., Beach, R.F., Atieli, F.K., Mount, D.L., Roberts, J.M., Mwangi, R.W., 1996. Long-term use of permethrin-impregnated nets does not increase Anopheles gambiae permethrin tolerance. Med. Vet. Entomol. 10, 71–79.

Wang, R., Liu, Z., Dong, K., Elzen, P.J., Pettis, J., Huang, Z.Y., 2002. Association of novel mutations in a sodium channel gene with fluvalinate resistance in the mite, Varroa destructor. J. Apic. Res. 41, 17–25.

Wang, R., Huang, Z.Y., Dong, K., 2003. Molecular characterization of an arachnid sodium channel gene from the varroa mite (Varroa destructor). Ins. Biochem. Mol. Biol. 33, 733–739.

Ware, G.W., Whitacre, D.M., 2004. An introduction to insecticides, The Pesticide Book. sixth ed. MeisterPro Information Resources, Willoughby, Ohio.

Webster, T.C., Thacker, E.M., Vorisek, F.E., 2000. Live Varroa jacobsoni (Mesostigmata: Varroidae) fallen from honey bee (Hymenoptera: Apidae) colonies. J. Econ. Entomol. 93, 1596–1601.

Weinzierl, R.A., Schmidt, C.D., Faulkner, D.B., Cmarik, G.F., Zinn, G.D., 1990. Chronology of permethrin resistance in a southern Illinois population of the horn fly (Diptera: Muscidae) during and after selection by pyrethroid use. J. Econ. Entomol. 83, 690–697.

Wen, Z., Scott, J.G., 1999. Genetic and biochemical mechanisms limiting fipronil toxicity in the LPR strain of house fly, Musca domestica. Pestic. Sci. 55, 988–992.

Willadsen, P., 2006. Tick control: thoughts on a research agenda. Vet. Parasitol. 138, 161–168.

Williamson, M.S., Martinez-Torres, D., Hick, C.A., Devonshire, A.L., 1996. Identification of mutations in the housefly para-type sodium channel gene associated with knockdown resistance (kdr) to pyrethroid insecticides. Mol. Gen. Genet. 252, 51–60.

Witty, M.J., 1999. Current strategies in the search for novel antiparasitic agents. Int. J. Parasitol. 29, 95–103.

World Health Organization, 2003. Draft: guidelines on the management of public health pesticides: report of the WHO Interregional Consultation, Chiang Mai, Thailand 25—28 February 2003. World Health Organization. Department of Communicable Disease Prevention, Control and Eradication. <http://whqlibdoc.who.int/hq/2003/WHO_CDS_WHOPES_2003.7.pdf>.

World Health Organization, 2005a. The work of the African network on vector resistance to insecticides 2000—2004. <http://www.afro.who.int/des/phe/publications/>.

World Health Organization, 2005b. Atlas of insecticide resistance in malaria vectors of the WHO African region <http://www.afro.who.int/des/phe/publications/>.

World Health Organization, Working Group for Scaling-up Insecticide-treated Netting, Roll Back Malaria partnership, 2005c. A strategic framework for coordinated national action: scaling up insecticide-treated netting programmes in Africa. <http://www.who.int/malaria/publications/atoz/winitn_strategicframework/en/>.

World Health Organization, 2006a. Malaria vector control and personal protection: report of a WHO study group. WHO Technical Report Series 936. Geneva, World Health Organization. <http://whqlibdoc.who.int/trs/WHO_TRS_936_eng.pdf>.

World Health Organization, 2006b. Indoor residual spraying: use of indoor residual spraying for scaling up global malaria control and elimination: WHO position statement. World Health Organization, Global Malaria Programme. <http://whqlibdoc.who.int/hq/2006/WHO_HTM_MAL_2006.1112_eng.pdf>.

World Health Organization, 2007a. Insecticide-treated mosquito nets: a position statement. <http://www.who.int/malaria/publications/atoz/itnspospaperfinal/en/index.html>.

World Health Organization, 2007b. Long-lasting insecticidal nets for malaria prevention. Trial edition: A manual for malaria programme managers. <http://www.who.int/malaria/publications/atoz/insecticidal_nets_malaria/en/index.html>.

World Health Organization, 2010. World Malaria Report 2010. <http://www.who.int/malaria/world_malaria_report_2010/en/index.html>.

Yen, J.L., Batterham, P., Gelder, B., McKenzie, J.A., 1996. Predicting resistance and managing susceptibility to cyromazine in the Australian sheep blowfly Lucilia cuprina. Aust. J. Exp. Agric. 36, 413—420.

Yoon, K.S., Gao, J.-R., Lee, S.H., Clark, J.M., Brown, L., Taplin, D., 2003. Permethrin-resistant human head lice, Pediculus capitis, and their treatment. Arch. Dermatol. 139, 994—1000.

Yoon, K.S., Gao, J.-R., Lee, S.H., Coles, G.C., Meinking, T.L., Taplin, D., et al., 2004. Resistance and cross-resistance to insecticides in human head lice from Florida and California. Pest. Biochem. Physiol. 80, 192—201.

Yoon, K.S., Kwon, D.H., Strycharz, J.P., Hollingsworth, C.S., Lee, S.H., Clark, J.M., 2008. Biochemical and molecular analysis of deltamethrin resistance in the common bed bug (Hemiptera: Cimicidae). J. Med. Entomol. 45, 1092—1101.

Yoon, K.S., Strycharz, J.P., Baek, J.H., Sun, W., Kim, J.H., Kang, J.S., et al., 2011. Brief exposures of human body lice to sublethal amounts of ivermectin over-transcribes detoxification genes involved in tolerance. Insect Mol. Biol. 20, 687—699.

Yukich, J.O., Lengeler, C., Tediosi, F., Brown, N., Mulligan, J.A., Chavasse, D., et al., 2008. Costs and consequences of large-scale vector control for malaria. Malar. J. 7, 258.

CHAPTER 7

Insect Resistance to Crop Rotation

Joseph L. Spencer[1], Sarah A. Hughson[2] and Eli Levine[3]

[1]Illinois Natural History Survey, Prairie Research Institute, University of Illinois, Champaign, IL
[2]Department of Entomology, University of Illinois Urbana, IL
[3]Illinois Natural History Survey, Prairie Research Institute, University of Illinois, Champaign, IL

Chapter Outline

BACKGROUND

Crop rotation is the agricultural practice of sequentially growing a series of plant species on the same land (Yates, 1954). Rotational cycles may be as short as two years, like the corn–soybean rotation that dominates the United States (U.S.) Corn Belt, or as long as 10–12 years for

Insect Resistance Management
DOI: http://dx.doi.org/10.1016/B978-0-12-396955-2.00007-2
233

the grass-legume pasture and cash-grain rotation that is found in Argentina (Bullock, 1992). The origins of crop rotation as an agricultural practice lie in antiquity; MacRae and Mehuys (1985) report that it was in use more than 3000 years ago in Han Dynasty China. Historically, crop rotation was a remedy for poor soil productivity and involved a series of crops that incorporated legumes with cereal crops (Karlen *et al.*, 1994). Although the mechanism responsible for the agronomic benefit of rotation was not understood, the benefits of incorporating a leguminous crop in a cropping sequence were known during Roman times and reported by a number of contemporary historians (Karlen *et al.*, 1994). The modern concept of crop rotation goes back to the four-crop Norfolk rotation (turnip, barley, clover, and wheat) that was popular in eighteenth-century England. In the United States, Thomas Jefferson and George Washington experimented with a variety of rotational schemes on their farms. Rotation, along with the artificial addition of lime and soil minerals, became nearly universal in England by the middle of the nineteenth century; immigrant farmers brought these practices to the United States where many different rotational systems were developed and used extensively during the nineteenth century. The discovery that a mutualism between leguminous plants (e.g., alfalfa, soybean, and clover) and a common soil bacterium *Rhizobium* fixed atmospheric nitrogen (converting unusable N_2 gas into biologically useful NH_3) that was then available to successive crops provided an explanation for some of the benefits of crop rotation. Corn grown in a two-year rotation with soybean yields 5 to 20% more than corn grown in a continuous cultivation (Heichel, 1987; Power, 1987; Pierce and Rice, 1988; Bullock, 1992). In addition to the fertility benefits of increased available soil nitrogen and reduced need for commercial nitrogen fertilizers, soil organic matter can be increased and soil structure may be improved when crop rotation is practiced (Bullock, 1992).

At current prices, producing corn in an annual rotation with soybeans or in a corn–corn–soybean rotation has the greatest profit potential for Illinois and Indiana growers (Dobbins *et al.*, 2011; Schnitkey, 2012). However, there are circumstances associated with continuous corn cultivation for which the costs of inputs may be more than offset by local demand, market forces, or premiums, making continuous corn an attractive option. Where the added costs for continuous production of one crop cannot be justified, crop rotation offers a simple, economical, and environmentally benign alternative to reliance on chemical inputs.

Specialist Pests are Vulnerable to Crop Rotation

The intentional use of crop rotation as a tool for managing pests is a relatively recent development. Along with other cultural controls such as alteration of planting date to disrupt pest—host synchrony or field sanitation to remove overwintering sites, crop rotation for pest management requires an understanding of pest ecology. Not all pest organisms can be managed with crop rotation; vulnerable pests will (1) have a life stage(s) that is relatively immobile, (2) have a narrow feeding relationship requiring consumption of host-plant tissue for development, and (3) not survive for long periods in the absence of its host plant. Soil- and root-dwelling nematodes, soilborne pathogens, and certain weed species are examples of noninsect pests that may be controlled with rotation (Bullock, 1992).

One of the best examples of crop rotation used for insect pest management involves diabroticite corn-rootworm beetles (Coleoptera: Chrysomelidae). In the U.S. Corn Belt, 75.2% of surveyed farmers used crop rotation for corn-rootworm management (Wilson *et al.*, 2005). Nationwide, 67% of cornfields were in a corn—soybean rotation (USDA ERS, 2003), and 75% of corn was grown in some type of rotation (USDA ERS, 2006). Since the first reports of corn rootworm injury to corn, entomologists have recommended crop rotation to control corn rootworms. Stephen A. Forbes (1883) suggested that because the northern corn rootworm (NCR), *Diabrotica barberi* [then known as *Diabrotica longicornis*], depended on the availability of corn roots as food for the relatively immobile larvae, rotating production of corn with a plant that NCR larvae could not eat would destroy rootworm populations by depriving NCR larvae of corn roots. The strong host fidelity that was the basis for Forbes's NCR management recommendation was also later applied to management of the related western corn rootworm (WCR), *Diabrotica virgifera virgifera*. Although WCR did not become a pest until the twentieth century, today it surpasses the NCR in its overall importance as a corn pest. Together the NCR and WCR are arguably the most economically significant pests of corn. The 1992 discovery of WCR in Europe and their subsequent spread have only enhanced the status of corn rootworms as important pests. It has been estimated that annual U.S. losses and management expenditures related to corn rootworm amount to US$1.17 billion (Sappington *et al.*, 2006). The projected cost of managing economic infestations of WCR across the current European range has been estimated at €472 million (Wesseler and Fall, 2010). The Commonwealth Agricultural Bureaux International (CABI) recently designated the WCR as the world's most expensive pest to control (Cock, 2011). Because crop rotation was so effective in reducing their populations, there are now populations of NCR and WCR that are resistant to rotation. The history of corn-rootworm management reminds us that any management practice that affects pest survival or reproduction can select for resistance, even a seemingly "unbeatable" cultural control.

CORN PRODUCTION, CORN ROOTWORM, AND INSECTICIDES

In the rich, fertile soils of the U.S. Corn Belt, rotations were not commonly accepted during the early twentieth century, even though higher yields were demonstrated for rotated corn (also referred to as first-year corn). Superior yields of corn in rotation could not compensate for the poor economic return from the other rotated crops; thus, many farmers continued to grow continuous corn at the expense of depleted soil fertility. The needs of draft animals in the period before mechanization of agriculture also mandated that significant acreage be devoted to continuous corn production for on-farm use as animal feed.

An agricultural transformation began after the U.S. Civil War and bloomed in the early twentieth century thanks to development of hybrid corn and improved farming techniques that dramatically altered the landscape of crop production (Allen and Rajotte, 1990). The availability of hybrid corn, fertilizers, pesticides, irrigation, and mechanization simplified farming operations, reduced costs, and enabled corn to be produced over a vast area with increased yield (Bullock, 1992). Expansion of corn cultivation to the West and onto the prairie during and after the U.S. Civil War brought the NCR and WCR (a once sparsely distributed leaf beetle) into contact with an abundant host plant. Adoption of cultivated corn as a host plant by corn rootworms as well as human reliance on insecticides to control corn rootworms were responsible for their rise as the most serious insect threat to U.S. corn; *Diabrotica* are truly "human-made" pests (Metcalf, 1986).

History of *Diabrotica*

Both the NCR and WCR are species indigenous to North America (Chiang, 1973). The NCR was described by Thomas Say as *Galleruca longicornis* from specimens collected in 1820 from Colorado (Say, 1824). Branson and Krysan (1981) argue that the NCR invaded areas of the United States before the introduction of corn (circa A.D. 700; Galinat, 1965) and only later switched to corn after extensive corn cultivation reached the central plains following the Civil War. The NCR was first reported as a pest of corn in July 1880 in Stark County, Illinois, by Cyrus Thomas in his Fifth Annual Report of the State Entomologist (Thomas, 1881). Even in this early correspondence, crop rotation was suggested as "the most feasible means of dealing with the insect." In his sixth report, Thomas (1882) noted that C. V. Riley, State Entomologist of Missouri, had already reported the same rootworm species feeding in corn roots from Missouri in 1878 as part of Riley's March 1879 report to the Commissioner of Agriculture. In his First Annual Report of the State Entomologist covering 1882, Forbes (1883) presents a detailed account of suspected NCR injury predating the Missouri records, along with a thorough treatment of NCR biology and suggestions for controlling NCR, which prominently features crop rotation.

The WCR was first collected in 1867 and described by LeConte (1868), who collected two specimens from wild gourd (probably *Cucurbita foetidissima*) near Fort Wallace, Kansas. Gillette (1912) was first to report WCR as a pest of cultivated corn (sweet corn) in 1909 and 1910 near Fort Collins and Loveland, Colorado, respectively. Smith (1966) suggested that the native hosts of WCR populations in the Colorado–New Mexico–Arizona region of the United States were likely *Tripsacum* (any of approximately 15 species of perennial grass closely related to corn). However, Krysan *et al.* (1977) and Branson and Krysan (1981) suggested that WCR likely evolved in the same region as corn, and, as a corn specialist, it followed corn as it was moved into the southwestern United States after A.D. 700.

Evidence of an even earlier host association is evident in diabroticite chemical ecology. The corn rootworm's compulsion to feed on species of Cucurbitaceae (squashes) containing cucurbitacins B and E (Metcalf, 1979) is offered as evidence of its coevolution with cucurbitaceous host plants prior to a host shift onto graminaceous hosts (Metcalf, 1983; Tallamy *et al.*, 2005).

The North American range of the WCR has expanded dramatically since the 1950s (Gray *et al.*, 2009). Currently, WCR can be found from Colorado and Wyoming east to the Atlantic coast, south into northern portions of Alabama, Georgia, and the Texas Panhandle, and north to near the Canadian border in the West and through the southern portion of the Canadian province of Ontario. The distribution of NCR overlaps with that of the WCR; in northern reaches, the NCR tends to be dominant in interactions; NCRs are less common in the East (Chiang, 1973).

The WCR range was very dramatically expanded when it was inadvertently introduced into central Europe from North America in the late 1980s. The initial discovery in Serbia near the Belgrade airport in 1992, has been followed by additional introductions from North America (Miller et al., 2005) and rapid spread within Europe (via stratified dispersal (Ciosi et al., 2011)) to more than 15 countries as of fall 2011 (Gray et al., 2009; Wesseler and Fall, 2010; Edwards and Kiss, 2011). Analyses of European and North American WCR failed to identify a definitive North American source location for European WCR (Miller et al., 2005). European WCR sampled from two of five introductions bore other signatures of a North American field population; they were resistant to the cyclodiene insecticide, aldrin, but susceptible to the organophosphate, methyl parathion (Ciosi et al., 2009).

Corn-Rootworm Biology

Corn-rootworm biology and ecology have been extensively reviewed (Chiang, 1973; Krysan, 1986; Levine and Oloumi-Sadeghi, 1991; Meinke et al., 2009; Spencer et al., 2009; Gray et al., 2009). The history of corn-rootworm management and adaptation in Illinois is reviewed in Spencer (2010). Univoltine adults of NCR and WCR are present in cornfields beginning in late June or early July through frost where they feed on corn foliage, pollen, silks, and developing kernels, and in the case of NCR, the pollen of other plants (Spencer et al., 2009). Numerous feeding adults (usually more than five WCR per plant) may interfere with corn pollination due to silk clipping (Levine and Oloumi-Sadeghi, 1991). Oviposition has traditionally taken place almost exclusively in cornfields from late July through mid-September (Shaw et al., 1978; Levine and Oloumi-Sadeghi, 1991). WCR females will accept almost any moist particulate substrate for oviposition, a circumstance that facilitates egg laying in a variety of crop habitats (Kirk et al., 1968, Gustin, 1979; Kirk, 1979). Drought cracks, open earthworm burrows, and soil cracks around corn roots provide access to suitable damp locations (Kirk, 1979, 1981a,b).

WCRs are capable of depositing more than 1000 eggs, though 440 is a more realistic figure (reviewed in Spencer et al., 2009), whereas NCRs may deposit 300 eggs in their lifetime. Diapausing eggs overwinter in the soil and begin to hatch once soils warm in the spring; after emergence, neonate larvae locate corn roots and begin to feed (Levine and Oloumi-Sadeghi, 1991). Larvae of both species can survive only on the roots of corn and a limited number of grassy weeds (Branson and Ortman, 1967, 1970; Oyediran et al., 2004a,b; Ellsbury et al., 2005), including the second-generation biomass crop, *Miscanthus x giganteus*, which is a suitable oviposition site and larval host for WCR (Spencer and Raghu, 2009; Gloyna et al., 2011).

In corn, larval feeding disrupts root system function, which can reduce grain yield (Levine and Oloumi-Sadeghi, 1991). The feeding injury may facilitate infection by root and stalk rot fungi, resulting in further damage. Extensive root injury makes plants more susceptible to lodging (i.e., the plants fall over), resulting in yield losses from the difficulty of harvesting fallen plants. After completion of three larval instars, WCR pupate in soil near the corn roots and emerge as adults 5 to 10 days later (Fisher, 1986). WCR is a protandrous species; adult males emerge *circa* 5 days before females (Branson, 1987); adult males require about 5 to 7 days of development to reach sexual maturity (Guss, 1976).

Adult females are sexually mature upon emergence; they quickly release a pheromone and broadcast their readiness to accept a mate (Hammack, 1995). Mate-seeking males rapidly locate and attempt to mate with calling females on nearby plants (Hill, 1975; Lew and Ball, 1979). Mating lasts 3–4 hours, during which the male transfers a large spermatophore to the female. Most females mate only once, though subsequent matings do occur (Branson et al., 1977). Larger females have a greater likelihood of mating than small females (Kang and Krupke, 2009a). Male WCRs may mate multiple times (Hill, 1975); however, Kang and Krupke (2009b) report that the average male may mate only two to three times. The limited mating capacity may be related to declining mating ability with age (Kang and Krupke, 2009b) or the high cost of the male spermatophore, which is protein rich (Murphy and Krupke, 2011) and has a mass equal to about 9% of total male mass (Quiring and Timmins, 1990). Spermatophore nutrients can be detected within developing eggs, suggesting the male's contribution has fecundity-enhancing qualities (Branson et al., 1977; Sherwood and Levine, 1993; Spencer et al., 2009; Murphy and Krupke, 2011).

Insecticides and Corn-Rootworm Control

Following World War II, a new era in pest control began, as synthetic organic insecticides became widely available (Aspelin, 2003). Soil-applied cyclodiene insecticides soon became important management tools for protecting corn roots from corn-rootworm larval feeding injury (Hill et al., 1948). The first WCR control failures were reported in 1959; by 1961, central Nebraska WCRs were 100-fold more resistant to cyclodiene insecticides than susceptible populations elsewhere in the state (Ball and Weekman, 1962; Metcalf, 1983). The rise of insecticide-resistant corn rootworms coincided with expansion in the WCR and NCR geographic ranges, though to a much lower extent among NCRs (Metcalf, 1983; Gray et al., 2009).

Prior to this period, during the 1920s–1940s, the distribution of WCR had only slowly expanded eastward across the western corn-growing region [reviewed by Metcalf, 1983; Gray et al., 2009; Meinke et al., 2009]. An apparent acceleration in the eastward spread of WCR coincided with the rise of widespread cyclodiene resistance in the early 1960s (Metcalf, 1982, 1983). The cyclodiene-resistant WCR populations that had overspread the Corn Belt by 1979 (retaining 1000–2500 fold levels of cyclodiene resistance at the expanding front) reached corn-producing areas of the eastern states by the mid-1980s (Tallamy et al., 2005). Metcalf (1983) hypothesized that the rapid range expansion was related to the increased fitness among cyclodiene-resistant beetles; however, stratified dispersal, where discontinuous satellite populations disperse and grow far ahead of the expanding front leading to rapid colonization is a more likely explanation (Gray et al., 2009; Meinke et al., 2009). The WCR's current, rapid invasion of Central and Southeastern (CSE) Europe is also consistent with stratified dispersal (Ciosi et al., 2011).

After the failure of the cyclodiene insecticides, organophosphate and carbamate insecticides were introduced for rootworm control. Methyl parathion (an organophosphate) experienced failures by the early 1990s. Meinke et al. (1998) documented organophosphate and carbamate resistance in Nebraska where insecticides in these classes were in heavy use.

Systemic neonicotinoid seed treatments have been applied to seed corn since 1999 (Wilson et al., 2005). These products are intended primarily for management of secondary insect pests on seedling maize, though at high rates they may provide some root protection under low rootworm larval pressure. Unlike Bt-transgenic crops (see below), there are no refuge requirements for the ubiquitously applied neonicotinoids (they are present on nearly every corn and soybean seed planted in the Corn Belt); resistance to these products may be inevitable (Gray, 2011a).

RESISTANCE TO CROP ROTATION
NCR and Prolonged Diapause

Although crop rotation usually prevented NCR problems, instances of rootworm injury to corn grown in rotation with other crops have been reported (Levine and Oloumi-Sadeghi, 1991). In the early 1930s, Bigger (1932) noted that growing corn after oats or sweet clover failed to control this pest. Lilly (1956) reported severe NCR injury in cornfields that were planted to oats or soybean the previous year but observed that two years without corn always provided control. Branson and Krysan (1981) and Hill and Mayo (1980) suggested that NCR infestations in rotated corn could be explained by oviposition in fields planted in crops other than corn the preceding season, but other studies failed to support this hypothesis. Chiang (1965a) sampled NCR eggs in fields other than corn where adults were feeding and concluded that few eggs were laid in those sites. In Illinois, Shaw *et al.* (1978) found that oviposition in soybean fields and injury to corn the following year were negligible where soybean fields were free of volunteer corn; root injury did not reach economic levels even when corn was planted after weedy soybeans. In South Dakota, Gustin (1984) found that NCR adults laid the majority of their eggs in corn plots, regardless of the maturation stage of corn, rather than in small grain stubble. Boetel *et al.* (1992) studied the oviposition habits of NCR in South Dakota, concluding that while adults were often found feeding on weeds and noncorn crops, they returned to corn to lay a majority of their eggs.

An alternate explanation for root injury to rotated corn was that NCR eggs may undergo prolonged diapause; that is, eggs pass through two or more winters before hatching rather than the typical one-year pattern (Levine and Oloumi-Sadeghi, 1991) (Figure 7.1). Diapause is a state of arrested development occurring in one stage of the life cycle that allows an insect to survive seasonally recurring periods of adverse conditions. Prolonged diapause (also referred to as extended diapause) is common among insects, and it may spread emergence over as many as 12 or

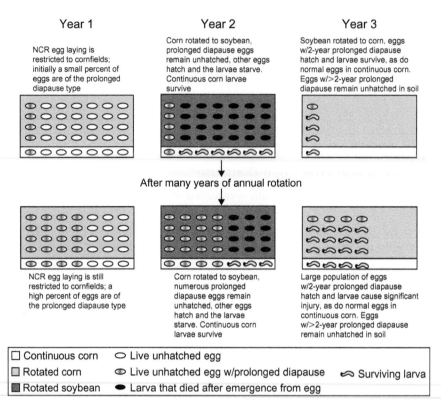

Figure 7.1 NCR prolonged diapause. Selection scenario and fate of northern corn rootworm (NCR) eggs (with and without prolonged diapause) deposited in continuous and rotated cornfields. Only eggs deposited in Year 1 are followed. Lower section depicts outcome after years of crop rotation-imposed selection for prolonged diapause.

more years for certain species (Tauber *et al.*, 1986; Danks, 1992). Usually, only a small percentage of individuals remain in diapause while the major portion of the population becomes active. Chiang (1965b) was first to document the fact that NCR eggs could remain in diapause for more than one winter, but concluded that the percentage of eggs with this trait (0.3%) was too small to be of economic consequence. Krysan *et al.* (1984, 1986), however, reported that about 40% of the NCR eggs from a South Dakota population collected in 1981 underwent prolonged diapause; such a large percentage of prolonged diapause could potentially cause significant injury to corn in an annual rotation. Indeed, reports of corn rootworm injury to corn planted after soybeans became more prevalent in the early 1980s, particularly in the northwestern region of the Corn Belt (Iowa, Minnesota,

Nebraska, and South Dakota). Prolonged diapause has been confirmed for NCR eggs from South Dakota (Krysan *et al.*, 1984, 1986), Minnesota (Krysan *et al.*, 1986), Illinois (Levine *et al.*, 1992b), Michigan (Landis *et al.*, 1992), and North Dakota (Levine and Weiss, unpublished data). A 1988 survey of randomly sampled northwestern Iowa cornfields suggested that the prolonged diapause trait was generally distributed throughout the NCR population infesting rotated corn (Tollefson, 1988).

Levine *et al.* (1992b) showed that diapause in NCR eggs was quite variable in length, ranging from one to four years in Illinois and South Dakota populations. Levine *et al.* (1992b) collected NCR eggs from Champaign, Illinois, females, among the 777 eggs that survived to hatch during the four-year study period, 50.6%, 41.2%, 8.0%, and 0.3% hatched after one, two, three, and four simulated winters, respectively (Levine *et al.*, 1992b). Similarly, of 311 eggs that hatched from a Madison, South Dakota, population that overwintered 20 cm deep in the field, 48.9%, 20.6%, 20.9%, and 9.6% hatched after one, two, three, and four winters, respectively (Levine *et al.*, 1992b). The population differences may reflect adaptation to field cropping patterns at the adult collection sites. Collections from Illinois were from an area that was annually rotated between corn and soybean. The South Dakota farm used many different rotations during the 25 years prior to the collections. These data support the hypothesis that NCRs are adapting to site-specific cropping practices. The less varied rotational patterns used by Illinois farmers may explain the large percentage of egg hatch that occurred following two winters; the more varied cropping patterns in South Dakota may explain the nearly equal hatch of eggs after two or three winters (Levine *et al.*, 1992b). An analysis of eggs from females collected around Illinois revealed a highly significant correlation between the percentage of NCRs with prolonged diapause in a given county and the percentage of rotated corn in that county (Levine *et al.*, 1992b). This finding is supported by reports of a greater incidence of prolonged diapause in areas of South Dakota and Minnesota where corn is rotated annually than in areas where corn is not rotated (Krysan *et al.*, 1986; Ostlie, 1987). Consistent patterns in the percentage of eggs showing prolonged diapause among individual female clutches from single locations suggests that there is a genetic component to prolonged diapause (Levine, unpublished data). Annual rotation of corn with another crop provides intense selection pressure for NCR eggs to remain in diapause for two years. Larvae that emerge from eggs that remain in diapause for two winters before hatching have a greater chance for survival and are more likely to pass these traits to their progeny under such a cropping pattern (Levine et al., 1992b). Early reports by Bigger (1932) and Lilly (1956) of NCR injury to rotated corn are now best explained by our knowledge of prolonged diapause as a type of resistance to crop rotation.

The NCR can be considered to have two phenotypes: univoltine (individuals with the typical one-winter egg diapause) and semivoltine (individuals with a prolonged egg diapause). Both phenotypes live in a heterogeneous host-plant environment. In continuous cornfields, semi-voltine beetles are at a distinct reproductive disadvantage compared with univoltine beetles because their reproductive rate is essentially half that of univoltine beetles. Conversely, in fields where corn is regularly rotated, univoltine beetles are at a disadvantage. Because both types of corn are often planted in the same area, beetles are subject to disruptive selection (Krafsur, 1995). Krysan (1993) suggested that where planting practices in individual fields tend to remain the same for several years, it is possible for NCR populations to adapt to the cropping practices of *individual* growers. Local adaptation, however, could be counteracted by random intermating (Krafsur, 1995).

WCR and Behavioral Resistance to Crop Rotation

WCR first entered northwest Illinois in 1964 (Petty, 1965). Management of WCR (and NCR) in continuous corn traditionally involved a planting-time application of soil insecticide to protect corn roots from lar-val feeding or application of foliar insecticides to protect silks from adult feeding (Levine and Oloumi-Sadeghi, 1991). Excellent management of WCR could also be achieved with crop rotation. Following WCR entry into Illinois, the use of soil-applied insecticide in cornfields increased; through the late 1960s and 1970s, more than 60% and 50%, respectively, of all cornfields were treated for corn rootworms (Pike and Gray, 1992). Thereafter, insecticide use steadily declined into the 1990s as education succeeded in reducing insecticide use on rotated corn. In 1990, 80% of continuous corn and only about 13% of rotated corn in Illinois were trea-ted with insecticide (Pike and Gray, 1992). The 1960s–1990s shift from widespread prophylactic insecticide application to reliance on cultural control was a victory for integrated pest management. Ironically, insecti-cide use at this level was probably still excessive. Steffey et al. (1992) sur-veyed root injury in first-year corn around Illinois from 1986 to 1989, and found that only 1.7% of rotated cornfields experienced injury that exceeded theoretical economic injury levels. Steffey *et al.* (1992) concluded that Illinois producers rarely needed to apply soil insecticides to prevent rootworm injury in corn rotated with soybean. At the time, most of the economic injury reported by Steffey *et al.* (1992) was attributed to NCR prolonged diapause. Areas with high adoption of crop rotation (northeast, central, and east-central Illinois) were most at risk.

In 1987, the first evidence of WCR behavioral adaptation to crop rotation was observed in isolated seed-corn fields in Ford County, Illinois (Levine and Oloumi-Sadeghi, 1996); the roots of corn plants in rotated cornfields suffered serious larval feeding injury. More than 95% of the adults collected in the fields were WCR, suggesting the problem was not related to NCR prolonged diapause. Follow-up studies in 1988 ruled out corn-rootworm oviposition around volunteer corn or grassy weeds; the fields were free of volunteer corn and grassy weeds during the previous years. Because these were seed cornfields, it was suspected that pyrethroid insecticides, used to control *Helicoverpa zea*, had repelled adult WCR beetles into nearby soybean fields where they laid eggs (Levine and Oloumi-Sadeghi, 1996). Recovery of WCR eggs from soil samples in soybean fields and subsequent adult emergence from rotated corn supported this hypothesis.

The repellency hypothesis was abandoned beginning in 1993 when increasing reports of serious WCR larval injury to Illinois and Indiana first-year corn included many commercial cornfields that were far from pyrethroid-treated fields. Growers who had successfully controlled WCR with only crop rotation suffered serious crop losses (Levine *et al.*, 2002). Studies ruled out a WCR prolonged diapause; none of the eggs examined from "problem areas" expressed the prolonged diapause trait (Levine and Oloumi-Sadeghi, 1996), which is rare ($<$ 0.21%) in WCR (Levine *et al.*, 1992a).

The possibility that WCR females deposited large numbers of eggs outside of cornfields was counter to their well-understood biology and almost inconceivable. However, by 1995 what was once an isolated curiosity only a few years earlier had spread to nine east-central Illinois counties and 15 nearby Indiana counties where growers suffered devastating losses due to severe WCR larval injury in rotated corn (Levine *et al.*, 2002). The unlikely conclusion became inescapable: A behavioral change had occurred, and WCR beetles in east-central Illinois and northwestern Indiana were leaving cornfields to lay their eggs in neighboring soybean fields and other crops (e.g., alfalfa, wheat, and oats) as well as in corn.

Enthusiastic adoption of crop rotation over a broad area [e.g., 95—98% of corn in east-central Illinois was rotated, usually with soybean (Onstad *et al.*, 1999, 2003a)], combined with the great efficacy of the technique, created a strong selection that favored an existing, but presumably uncommon, WCR phenotype with reduced ovipositional (egg-laying) fidelity to cornfields (Onstad *et al.*, 2001) (Figure 7.2). Records of root injury for first-year corn from the late 1970s to late 1980s may be

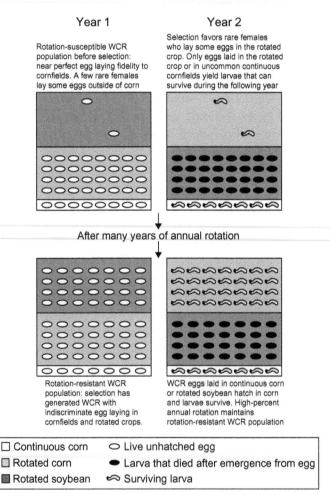

Figure 7.2 WCR rotation resistance. Selection scenario and fate of western corn rootworm (WCR) eggs deposited in cornfields (rotated and continuous) and rotated crops (i.e., soybean). Only eggs deposited in Year 1 are followed. Lower section depicts outcome after years of crop rotation-imposed selection for rotation resistance.

evidence of early expression of rotation-resistant behavior (Shaw *et al.*, 1978; Steffey *et al.*, 1992). Unexpected injury in rotated corn was frequently attributed to the presence of volunteer corn or grassy weeds, which can make rotated soybean fields attractive sites for rootworm feeding and egg laying (Shaw *et al.*, 1978). So strong was the belief that WCR populations could not overcome crop rotation that as late as 1993 it was suggested that "it is highly unlikely that the WCR could become adapted to crop rotation by oviposition in the alternate crop" (Krysan, 1993). An appreciation for the intensity of selection imposed by widespread crop rotation on WCR was lacking.

In a landscape dominated by an annual corn—soybean crop rotation, females with reduced ovipositional fidelity to cornfields enjoy a reproductive advantage over females with perfect fidelity to cornfields. This is because soybean fields are nearly always rotated to corn in the following year but only a small percentage of cornfields are planted with corn for two consecutive years. Females that lay at least some eggs in soybean (or another crop in annual rotation with corn) have a much greater chance that their eggs will hatch in a cornfield the following year than females who laid eggs in a cornfield. Over time, the reproductive advantage accruing to the once-uncommon females that laid some of their eggs outside of cornfields made them more common in rotated cornfields. It is hypothesized that this selection generated a WCR population in which a high proportion of females have some propensity to exit cornfields and oviposit in soybean fields and other locations in addition to cornfields. Modeling by Onstad *et al.* (2001) suggests that a high level of annual crop rotation in the landscape ($\geq 80\%$) is necessary for evolution of behavioral resistance; below 80% rotation, the contribution of rotation-susceptible alleles from WCR produced in continuous corn prevents rapid evolution of rotation resistance. The agricultural landscape of east-central Illinois presented conditions that strongly favored evolution of rotation resistance as envisaged by Onstad *et al.* (2001).

Today, rotation-resistant WCRs are found across large portions of Illinois, Indiana, and smaller portions of Michigan, Ohio, Wisconsin (Spencer *et al.*, 2005), eastern Iowa (Prasifka *et al.*, 2006; Dunbar, 2012b), and the Canadian province of Ontario (Meloche and Hermans, 2004). The rate of gradual west- and northward expansion of rotation resistance initially progressed faster than predicted by Onstad *et al.* (1999); however, Dunbar (2012b) found that rotation resistance was rare in eastern Iowa and absent from central Iowa. Greater landscape diversity in the eastern

Corn Belt may have slowed the eastward expansion (Onstad et al., 2003a). Curzi et al. (2012) document high landscape diversity in specific counties hosting WCR populations that are located to the west and south of where rotation resistance is known.

The Biology of Rotation-Resistant WCR

Frequent, season-long adult WCR movement outside of cornfields and presence in rotated crops are characteristic of rotation-resistant WCR behavior (O'Neal et al., 1999; Isard et al., 2000, 2004; Levine et al., 2002; Pierce, 2003; Rondon and Gray, 2003; Spencer et al., 2005; Pierce and Gray, 2006). Movement of female adults plays a key role in the phenomenon of WCR rotation resistance and in WCR biology generally. In the area threatened by WCR rotation resistance, WCR eggs may be recovered as commonly in the soil of soybean fields as in the soil of cornfields (Levine et al., 2002; Pierce, 2003; Rondon and Gray, 2004; Pierce and Gray, 2006). Measurement of adult WCR abundance in crops rotated with corn is the basis for assessing the risk of economic WCR larval injury to rotated corn (O'Neal et al., 2001).

The pre-mating and mating behavior of rotation-resistant WCR males and females appears to be similar to that historically reported for rotation-susceptible populations (Ball, 1957; Hill, 1975; Guss, 1976; Bartelt and Chiang, 1977; Branson et al., 1977; Lew and Ball, 1979, 1980), though the populations have not been directly compared. Likewise, the post-mating behavior of all WCR females likely involves a period of feeding on corn silks and pollen for as long as a week before many fly out of their natal field and disperse or migrate an unknown distance to another cornfield. Spencer et al. (2005) collected flying females as they ascended from an Illinois rotated cornfield; 84% of the females contained a spermatophore—an indication that they were young and recently mated. Coats et al. (1986) measured an impressive capacity for long- and short-distance flight among young WCR females; however, long-distance sustained migratory flights were not observed after females were nine days old (Coats et al., 1986).

During the post-mating, pre-ovipositional dispersal period, favorable atmospheric conditions (e.g., instability due to heating of air near the ground and the passage of summertime convective storms) and light winds promote flight and favor ascent from cornfields (Witkowski et al., 1975; VanWoerkom et al., 1983). A diel periodicity in flight tendency is reflected in peaks of flight during early-late morning and early evening; Isard et al. (1999, 2000) review many factors that influence flight. During the passage of summertime storms, some WCR adults may be drawn into storms and carried tens of miles before being washed out of the storm in rain (Grant and Seevers, 1989, 1990). Startling evidence of storm transport has been documented in the upper Midwest, where astonishing numbers of WCR beetles occasionally wash up along the waterline of Lake Michigan after the passage of summertime storms (Grant and Seevers, 1989). After short- or long-distance post-mating dispersal from their natal field, most WCRs will locate a new cornfield, where they continue to feed and provision a first clutch of about 100 eggs. It is after this post-mating dispersal that rotation-resistant individuals commence frequent interfield movement between corn and soybean fields and that the behavioral differences between rotation-resistant and susceptible populations become more evident.

Earlier work on WCR movement (Hill and Mayo, 1980; Lance et al., 1989; Naranjo, 1991) emphasized a reluctance among WCRs to leave cornfields; when interfield movement occurred, it was nonrandom and oriented toward flowering corn (Naranjo, 1994). Frequent interfield movement leading to high WCR abundance in the rotated crop is a hallmark of rotation-resistant WCR (Levine et al., 2002; Spencer et al., 2005). The strong periodicity of interfield movement by rotation-resistant WCRs (Isard et al., 2000) is influenced by predictable changes in local atmospheric conditions (Isard et al., 2004). Where rotation-resistant populations are present, large numbers of WCR adults become noticeable in soybean and other nonhost fields adjacent to cornfields within about one week after adult females are first noted in cornfields; they remain abundant in rotated crops throughout the growing season (O'Neal et al., 1999; Isard et al., 2000; Levine et al., 2002; Rondon and Gray, 2003; Pierce and Gray, 2006). Where WCRs are still susceptible to crop rotation, few or only modest numbers of WCR adults are detected in rotated crops, even when there is a high abundance in adjacent corn (Figure 7.3) (Levine et al., 2002; Spencer et al., 2005; Pierce and Gray, 2006).

Movement and the Mechanism of Behavioral Resistance to Crop Rotation

Most (about 60%) of the rotation-resistant WCRs in soybean fields or collected while moving between cornfields and other rotated crops are female (Levine et al., 2002; O'Neal et al., 1999; Rondon and Gray, 2003).

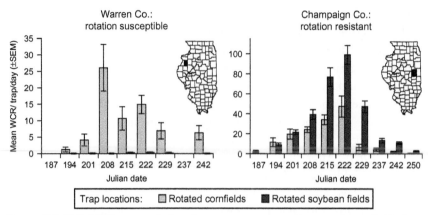

Figure 7.3 2000 WCR seasonal abundance patterns in soybean and cornfields from eastern (Monmouth, IL in Warren Co.; rotation-resistant population) and western (Urbana, IL in Champaign Co.; rotation-susceptible population) Illinois. Each bar represents a mean daily capture rate (± SEM) for cucurbitacin + insecticide-baited vial traps positioned at ear height in corn ($n = 5$) or at the top of the plant canopy in soybean ($n = 5$) at each location.

This proportion is reminiscent of the proportion of females that was previously typical of first-year corn prior to rotation resistance (Godfrey and Turpin, 1983). After rotation-resistant WCRs leave a cornfield and arrive in a soybean field, many feed on soybean tissue [despite a lack of nutritive value to WCR (Mabry and Spencer, 2003)]. During the growing season, about 55% of rotation-resistant WCR females in soybean contain identifiable soybean tissue in their gut contents (Levine *et al.*, 2002; Gray *et al.*, 2009; Seiter *et al.*, 2010). By late July when rotation–resistant WCR abundance in Illinois soybean fields can exceed 200 beetles per 100 sweeps, 86% of females have soybean tissue in their gut contents (Spencer *et al.*, 2005). Similar proportions of soybean feeding can be found among rotation-susceptible WCRs in soybean fields, an indication that soybean herbivory is not unique to rotation-resistant population. A 10- to 30-fold greater abundance of WCRs in soybean is diagnostic of areas where rotation-resistant WCR areas present versus areas where they are absent. Although readily eaten, soybean tissue does not support WCR egg development (Mabry and Spencer, 2003). In the laboratory, few field-collected WCRs that eat only soybean tissue live more than one week. However, WCR adults that eat a 50:50 mixed diet that alternates daily between corn and soybean tissue, survive and reproduce as well as WCR fed on a continual diet of corn plant tissues (Mabry *et al.*, 2004). When the interval between corn and soybean diet alternation was lengthened to four days, Dunbar and Gassmann (2012a) found that females on alternating diets had lower fecundity than those on a continual corn tissue diet. Frequent inter-field movement between corn and soybean fields may facilitate beneficial diet mixing in the field.

Gravid females or those capable of maturing some eggs account for only 20% of the females that fly into soybean fields from cornfields (Mabry and Spencer, 2003). The season-long presence of many females without mature eggs in soybean fields suggests that egg laying is not the sole reason females enter soybean fields. Clearly, the 80% of females that enter soybean fields without sufficient reserves to mature eggs must return to a cornfield to feed before they can lay eggs (Mabry et al., 2004). Laboratory assays indicate that soybean herbivory significantly increases WCR activity and the likelihood of egg laying (Mabry, 2002; Mabry et al., 2004; Knolhoff et al., 2010a). Exposure to soybean roots was found to increase oviposition by rotation-resistant populations, but not by rotation-susceptible populations (Knolhoff et al., 2010a). Using a behavioral assay, Knolhoff et al. (2006) found that female WCRs from rotation-resistant populations were faster to escape an arena than rotation-susceptible populations; greater general activity levels may explain the abundance of WCR moving between corn and soybean. A predisposition toward greater activity combined with behavioral effects of soybean herbivory may pro-vide the proximate mechanism behind the movement of rotation-resistant WCR from soybean fields back into adjacent cornfields (Mabry et al., 2004; Spencer et al., 2005). Subsequent microarray analyses of WCR gene expression identified some candidate genes (chiefly antimicrobial genes) that were differentially expressed in rotation-resistant and—susceptible females (Knolhoff et al., 2010b), though no rotation resistance "gene" was found.

Comparative analyses of rotation-susceptible and rotation-resistant WCR from multiple locations have documented differences in behavior, physiology, protease activity, and gene expression in the guts of rotation-resistant WCR adults (Curzi et al., 2012). Three- to fourfold increases in Cathepsin L activity along with higher baseline activity among rotation resistant versus susceptible beetles suggest that they are adapted to tolerate soybean defenses longer than beetles from susceptible populations. These changes are components of the mechanism that enables individuals from rotation-resistant populations to enter soybean fields, circumvent soybean defenses, and, thus, crop rotation.

Although abundant rotation-resistant WCRs readily visit and deposit eggs in corn, soybean, and other rotated crops (Rondon and Gray, 2003, 2004; Schroeder *et al.*, 2005), the number of eggs laid in corn may be greater (Rondon and Gray, 2004; Schroeder *et al.*, 2005) or fewer than the number laid in soybean (Pierce and Gray, 2006) or other rotated crops. Pierce and Gray (2006) measured season-long patterns of WCR adult abundance, oviposition, and crop phenology in corn and soybean fields near the east-central Illinois epicenter of rotation resistance. They found that differences in relative corn phenology can lead to significantly greater numbers of WCR eggs laid in late-planted corn versus early-planted corn in east-central Illinois (Pierce and Gray, 2006). They also found that WCRs were present in soybean fields before corn was mature and that egg laying outside of corn occurred throughout the season. O'Neal *et al.* (2002, 2004) hypothesized that early corn planting might play a critical role in the phenomenon of rotation resistance and perhaps even be the mechanism behind the phenomenon if WCR left maturing corn to oviposit in still-green soybean fields.

Pierce and Gray (2006) tested whether phenology differences between corn and soybean alone could lead to WCR egg laying in soybean under field conditions in an area *without* rotation resistance. Pierce and Gray (2006) planted corn on dates that were one month apart to create extreme differences in corn and soybean phenology in Champaign County, Illinois, and in an area 220 km northwest of Champaign that was free from rotation-resistant WCR (Warren County, Illinois). The number of eggs laid by WCR in Champaign County soybean plots was uniformly high, and twice as many eggs were laid in the late corn (planted May 15) than in early corn (planted April 15). In Warren County, no WCR eggs were recovered from soybean field soil, and there were no differences in the number of eggs recovered from cornfields planted on different dates. While differences in corn phenology can influence egg laying by rotation-resistant WCR in corn, a wide phenology difference was not sufficient to stimulate egg laying by WCR in soybean fields from an area where rotation-resistant WCR were not known to be present.

Interfield WCR movement rates between corn and soybean fields, or other rotated crops like wheat, suggest a possible explanation for crop-to-crop variability in WCR egg laying. Spencer *et al.* (2003–2005 field data) found that interfield movement from corn into soybean was slower than movement from corn into wheat. WCR abundance (measured with canopy level sweep net and above canopy aerial net samples and on

Pherocon® AM sticky traps) was significantly greater in soybean than in wheat or wheat double-cropped with soybean. Corn in rotation with wheat escaped yield-reducing root injury from WCR larvae, while corn after soybean or wheat double-cropped with soybean suffered injury likely to reduce yields. Using the same crop treatments, Schroeder et al. (2005) also report the least root injury in corn after wheat; however, none of the injury in rotated corn had the potential to reduce yields. The generally hotter and drier conditions in wheat and wheat stubble (compared to soybean canopy) may promote more rapid movement [about 6.8–7.2 m/day from corn into wheat versus 4.8–5.1 m/day into soybean (Spencer unpublished)]. Consequently, fewer insects accumulate there, and gravid females have less opportunity to lay eggs in wheat fields. Average adult WCR intrafield movement rates (within cornfields) range from 4.9–17 m/day (Spencer et al., 2003, 2009) to 11.9 m/day between corn and rotated crops (Spencer unpublished field data). Impressive movement is possible; in a European mark–release–recapture study conducted in a grassy steppe in Hungary, Toepfer et al. (2006) reported 2.8% of WCR adults moved 300 m from a release site to a cornfield.

MANAGING ROTATION-RESISTANT CORN ROOTWORMS

Although crop rotation is no longer an effective pest-management tool against some WCR and NCR populations, developing rootworm larvae still cannot survive on soybean roots. When eggs of either species are deposited in cornfields that are rotated to soybean, the larvae starve and die soon after emergence. Crop rotation remains the primary recommended management option where rotation-resistant WCR and NCR populations are absent. Where rotation resistance is present, application of a planting time soil insecticide and selection of a transgenic insecticidal corn hybrid are options for producers of first-year corn.

Monitoring Rotation-Resistant WCR

The O'Neal et al. (2001) economic threshold for adult WCR abundance in rotated soybean fields provided growers with a tool to assess risk and to guide decision making about use of soil insecticides or planting of transgenic insecticidal corn to protect rotated cornfields. The O'Neal et al.

(2001) protocol calls for deployment of 12 widely spaced Pherocon® AM yellow sticky traps (unbaited) (Great Lakes IPM, Vestaburg, Michigan) in a soybean field for four weeks from the last week of July through the third week of August. Traps are changed weekly to count the trapped WCR and determine the average number of WCR captured per trap per day. The cost of monitoring is about $71 per field (traps cost approximately $1.50 each). Seasonal averages of ≥ 5 WCR/trap/day indicate that economic injury [i.e., mean root injury scores of ≥ 0.25 on the 0−3 Node Injury Scale (Oleson et al., 2005) or mean root ratings of ≥ 2.5 on the 1−6 scale (Hills and Peters, 1971)] to first-year corn planted in that field is likely during the following year.

Although the threat of economic injury due to WCR larval feeding is present each year in both continuous and first-year cornfields in many areas with rotation-resistant WCR populations, surveys show that economic injury does not occur in all fields (Schroeder and Ratcliffe, 2003, 2004; Steffey et al., 2003, 2004). Outside of the known rotation resistance problem area and locally within the affected region there are areas where WCR abundance and egg laying in soybean is too low to cause economic injury to rotated corn (Gray and Steffey, 2004). However, few growers do any WCR monitoring in soybean to assess their risk. Gray and Steffey (2004) indicate that the sticky trap method was primarily used by growers from areas where rotation-resistant WCR populations were newly arrived. They note that in regions of Illinois where rotation-resistant WCRs are well established, "use of Pherocon® AM traps ... is noticeably lacking" (Gray and Steffey, 2004). Given that the cost of traps and labor for Pherocon® AM monitoring in one field is about $71, it is difficult to understand why growers would reflexively apply insecticide at a cost of $47−$70/A (Dobbins et al., 2011; Schnitkey, 2012), when they could monitor and make an informed decision for what it costs to treat just one acre with insecticide.

Cullen et al. (2008) evaluated farmers' perspectives on scouting for rotation-resistant WCR. When asked to indicate all factors that might prevent adoption of scouting, growers most commonly chose the following (the percentage choosing is indicated): (1) too time consuming—50.9%, (2) too labor intensive−28.4%, and (3) the extra financial expense—23.3%. As discouraging as the disinterest in monitoring was among the majority, it was encouraging that 24.1% of respondents indicated that they would want to monitor WCR if present in their county. Along similar lines, Gray (2011a) documents a troubling disregard for

basic IPM principles among some growers. In a set of surveys adminis-
tered during late fall meetings, Gray (2011a) found that 73 to 79% of
growers had planted *Bt* hybrids knowing that corn-rootworm and corn-
borer levels were anticipated to be low. Gray's (2011a) data show that
growers are willing to neglect ecological principles fundamental to IPM
and resort to insurance-based approaches to pest management.

Root injury surveys (Schroeder and Ratcliffe, 2003, 2004) offer strong
evidence in support of monitoring WCR rather than simply assuming
the worst (or the best). Schroeder and Ratcliffe (2003) surveyed cornfields
in 36 Illinois counties, including 5 (Bureau, Lee, Marshall, Ogle, and
Stark) where a rotation-resistance problem was widely suspected. Average
root injury (on the Iowa State 1−6 scale) was ≥ 3.0 (i.e., above the economic injury level) in just 10 of 50 (20%) randomly sampled fields (5 roots
were dug from each of 10 fields/county; ratings ≥ 3.0 indicate economic
injury is possible). In the following year, Schroeder and Ratcliffe (2004)
sampled in 24 counties outside the area of established rotation resistance.
They reported that 8.4% of randomly collected first-year corn roots from
these counties rated ≥ 3.0 (some of the injury may also be due to NCR
feeding which cannot be distinguished from that of WCR). The
Schroeder and Ratcliffe (2003, 2004) surveys illustrate just how risky
assumptions can be. Based on the county-level "assumptions" about the
status of the rotation-resistant WCR threat, up to 80% of the 2003 fields
would have been treated unnecessarily and 8.4% of the 2004 roots from
the "unthreatened" counties would have suffered economic injury
because high WCR populations were not detected in time to take action.
Monitoring WCR abundance in soybean fields requires a modest financial and time commitment; however, when it saves a grower from making
a management error, the savings can far exceed the perfield cost of
monitoring.

Monitoring WCR in Europe

WCR rotation resistance is a concern to European growers because crop
rotation is a primary management technique for WCR in Europe (along
with soil insecticides, seed treatments, and foliar sprays; practical biological control is not yet commercially available) (Meissle *et al.*, 2011).
Introduction of rotation-resistant WCR would increase the economic
impact of WCR in Europe (Ciosi *et al.*, 2008). Extensive monitoring of
WCR population abundance and distribution across Europe (Keszthelyi,

2005; Kiss et al., 2005a; Igrc Barcic et al., 2007; Sivcev et al., 2009; Szalai et al., 2011) has helped document the spread and may also detect the presence of rotation-resistant behavior should it be introduced or evolve.

While larval injury to rotated corn occurs in Europe (Kiss et al., 2005b; Igrc Barcic et al., 2007), factors such as previous close proximity to adjacent WCR-infested cornfields and generally smaller field sizes (which will tend to exaggerate the seeming significance of edge effects because of high edge-to-area ratios) may account for limited injury in rotated cornfields.

At current WCR infestation levels, which are generally low outside of the central and south- eastern European epicenter of WCR activity, abundance that exceeded economic thresholds was found only after three years of continuous corn (Sivcev et al., 2009).

At present there is little or no WCR pressure in rotated corn (Sivcev et al., 2009), and there is no indication that European populations are adapted to any particular crop rotation (Kiss et al., 2005a,b). Maintaining diverse crop rotations is important to avoid selecting a European strain of rotation-resistant WCR (Meissle et al., 2011). However, WCR introductions into Europe from North America are continuing (Ciosi et al., 2011), and there is an increasing representation of the total genetic variation from North American populations present in European WCRs. Because of this factor, it should be expected that variability associated with various types of North American WCR resistance will eventually be detected in Europe (Ciosi et al., 2011).

Insecticides

Where monitoring indicates the potential for economic injury due to WCR or NCR rotation resistance exists, growers should consider management options. Protection of corn roots from larval injury was once typically accomplished with soil insecticides applied as granules or liquids in the furrow or in a narrow (about 18 cm wide) band over the surface at planting. In Europe (where Bt-corn hybrids are unavailable to nearly all growers), insecticide use and crop rotation remain primary methods of managing WCR injury (Meissle et al., 2011).

Rootworm emergence from soil-insecticide-treated corn may actually be higher than that from untreated corn. The insecticide only penetrates a short distance from the application point; however, in this zone it protects the developing root allowing it to grow to a large size. Eventually, the root growth extends outside of the protected area and can be exploited by rootworm larvae. Because a larger root is eventually produced, more larvae may develop on a root that was treated with soil insecticide (Gray *et al.*, 1992). The juxtaposition of insecticide-treated and untreated areas around every plant may explain why resistance to modern soil insecticides used against rootworms has not occurred—rootworms emerging from around a single plant will include adults that developed with and without exposure to the soil insecticide. In effect, each plant includes a built-in refuge for susceptible insects (Gray *et al.*, 1992).

Since 1999, corn hybrids have been available with a seed-applied neonicotinoid insecticide known as a seed treatment. Seed treatments applied to *Bt*-corn hybrids controlling rootworms are marketed as providing protection against secondary insect pests that were once controlled by broad-spectrum insecticides, but are not affected by the rootworm-specific toxins in transgenic insecticidal hybrids. Neonicotinoid seed treatments offer some advantages over insecticide application (e.g., reduced human exposure and toxicity versus traditional soil insecticides, targeted application reduces active ingredient per ha, water-soluble treatments are easily absorbed by the plants, and no special equipment is needed to apply them). At low to moderate rootworm pressure, seed treatments can effectively contribute to corn pest management; their efficacy is questionable under heavy WCR pest pressure (Gray *et al.*, 2006).

Foliar sprays are also used for corn rootworm management. These sprays are applied to prevent silk-clipping. Plants may be too tall to allow application from in-field sprayers when pollination is occurring and foliar sprays are applied by applicators using airplanes. Owing to rapid silk growth and abundant pollen, high rootworm densities are necessary to justify foliar sprays. Adult WCR are the targets for WCR suppression programs in portions of Nebraska. Use of scouting with aerial application of methyl parathion targets egg-laying females to reduce larval injury to continuous corn during the following year. In these areas, there is a problem with resistance to methyl parathion (organophosphate) and carbaryl (carbamate) insecticides (Meinke *et al.*, 1998).

Overall insecticide use is higher in Corn Belt states that face more consistent threats from rotation-resistant WCR. In Illinois and Indiana, 28% and 14% of planted cornfield acres received some insecticide treatment, respectively, in 2010. Meanwhile in Iowa, where the rotation-resistant WCR threat is limited, only 8% of planted cornfield acres received some insecticide treatment (USDA NASS, 2010). Dunbar and

Gassmann (2012b) suggested that rotation-resistant WCRs were rare in northeast Iowa.

Owing to the periodicity of WCR movement in and out of soybean fields, specifically targeting WCR for adult control is impractical. However, when soybean fields are treated for the soybean aphid, *Aphis glycines*, WCR in those fields will be inadvertently exposed to foliar insecticides and/or soybean aphid-resistant soybean cultivars expressing *Rag* (Dunbar and Gassmann, 2012a). While the presence of *rag1* and *rag1/rag3* will not likely impose selection for or against rotation-resistant WCR, "nontarget" exposure to insecticide in soybean fields may actually select against rotation resistance (Dunbar and Gassmann, 2012a).

Transgenic Insecticidal Corn

In 2003, rootworm-protected transgenic insecticidal corn hybrids expressing a *Bacillus thuringiensis* (*Bt*) toxin (*Bt* corn) became available to U.S. corn growers. The toxin expressed in *Bt* corn root tissue kills neonate rootworm larvae as they attempt to colonize the roots (Vaughn *et al.*, 2005; Storer *et al.*, 2006). *Bt*-corn hybrids are an effective alternative to soil insecticides and stand-alone seed treatments at a comparable price, and when used as part of an IPM program, they offer a variety of benefits compared to soil insecticides (Rice, 2004; Vaughn *et al.*, 2005). Unlike insecticide treatment, there is little risk of grower exposure to insecticide during handling and application. Also, specialized insecticide boxes and spray equipment are not needed to "apply" the protection. The high species–specificity of the *Bt* toxin also means that most nontarget species, including beneficial insects, are unharmed in the process of protecting corn from rootworms. However, like any management tool that kills pests, its use carries a risk for resistance development. In an IPM framework, planting *Bt* corn for rootworm management should be justified by monitoring that indicates there is a risk of economic injury.

The use of refuges is a key element of the insect resistance management (IRM) plans required for transgenic corn in the United States and Canada. The requirement that a portion of each cornfield be set aside as a refuge (EPA Office of Pesticide Programs, 2003) allows a significant WCR population to develop without exposure to the specific *Bt* protein present in the transgenic area of a field. The amount of refuge that is necessary depends on many factors, including the toxicity of the *Bt* trait, the number of insecticidal traits expressed, and the biology of the target pest.

Acceptable minimum refuge sizes are hybrid specific and range from 5% to 20% of the area of *Bt* corn. For some dual mode of action hybrids, the appropriate percentages of refuge and *Bt* seed are available to growers pre-mixed in one bag as a seed blend. Deploying refuge in a seed blend ensures refuge compliance and greatly simplifies grower planting operations. Confusion surrounding refuge deployment options for various hybrids may play a role in the trend toward declining refuge compliance (Jaffe, 2009). Use of non-*Bt* refuges as part of an IRM plan is grounded on the assumption that mate-seeking refuge males will thoroughly disperse from refuges to find and mate with potentially resistant females emerging from transgenic corn. If assumptions about the movement and mating of males and females in the refuge and transgenic portions of a field are wrong, the likelihood of rapid resistance development may be greater than assumed.

Previous anticipatory modeling of corn rootworm resistance to *Bt*-corn hybrids was difficult because the allele frequencies for *Bt* resistance in wild populations were unknown. Baseline *Bt*-toxin susceptibility studies indicated that significant variation existed among wild WCR populations (Siegfried *et al.*, 2005). Studying *Bt*-protein resistance in laboratory-selected colonies may improve resistance monitoring strategies and early detection (Siegfried *et al.*, 2005). Laboratory selection experiments with Cry3Bb1 toxin showed that resistance could be rapidly selected (Meihls *et al.*, 2008; Meihls, 2010; Oswald *et al.*, 2011). The circumstances surrounding the 2009 appearance of field-evolved resistance to Cry3Bb1 toxin and its documentation by Gassmann *et al.* (2011) were consistent with the laboratory selection studies (Gassmann, 2012; Gassmann *et al.*, 2012). Field resistance is perhaps not surprising given mounting evidence of declining grower compliance with refuge requirements, disregard for ecological principles, and grower reluctance to fully commit to IPM for rootworm management in *Bt* corn (Cullen *et al.*, 2008; Jaffe, 2009; Gray, 2011a).

The importance of preserving susceptibility to transgenic crops may be of special concern in regions where other types of resistance are already present. For example, when WCR larvae consumed diet treated with the same *Bt* toxin expressed in a *Bt*-corn hybrid, neonate larvae of the rotation-resistant WCR had the lowest susceptibility to growth inhibition of any tested population and one of the highest rates of survival (Siegfried *et al.*, 2005). If traits associated with rotation resistance confer some advantage to WCR encountering *Bt* protein, it may be particularly important that *Bt*-corn hybrids be used judiciously in areas inhabited by rotation-resistant populations. Special attention may be needed for areas of Iowa and Illinois where resistance to both Cry3Bb1 *and* crop rotation may now occur (Gassmann *et al.*, 2011; Gray, 2011b).

Gray (2000) suggested that WCR susceptibility to transgenic insecticidal corn constitutes a natural resource and proposed that transgenic insecticidal corn be used only in fields where monitoring indicates its use is justified. However, in a modeling analysis, Crowder *et al.* (2006) concluded that economic thresholds would not be valuable when transgenic insecticidal corn is very effective and has a low price premium. Prescriptive use of *Bt* corn would help guarantee that susceptibility to a particular *Bt* toxin is not squandered.

Prediction and Rootworm IRM

Evidence supporting the claim that resistance by western corn rootworm to crop rotation is based on genetic changes was published by Onstad *et al.* (1999). They clearly showed that rotation resistance has been spreading away from its initial focus in east central Illinois in the mid-1980s. By 2001, the geographic spread of the resistant phenotype had extended into Michigan and Ohio (Onstad *et al.*, 2003a). Onstad *et al.* (1999, 2003a) created a simple model implemented with a geographic information system to describe the spread of the resistant phenotype from one county in Illinois using meteorological and behavioral information. Figure 7.4 shows one prediction versus the observations through 2001. Since then the resistance has spread into Wisconsin, Iowa, and perhaps Ontario. It is hoped that the genetic basis for the evolution of rotation resistance will be confirmed using satisfactory bioassays to distinguish wild-type and rotation-resistant phenotypes (Knolhoff *et al.*, 2006) or by genomic analysis.

Onstad *et al.* (2001, 2003b) demonstrated that a simple model of population dynamics, behavior, and genetics could account for the evolution of rotation resistance by the western corn rootworm from its invasion of eastern Illinois as a wild type around 1970 to the discovery of its damage in first-year corn in 1987 (Figure 7.5). They also showed that landscape diversity could slow the evolution of resistance to crop rotation (Figure 7.6). In this case, landscape diversity means the proportion of the vegetated area that is not planted to corn or the crop rotated with corn (e.g., soybean). This work caused Onstad *et al.* (2003a) to modify their model of geographic spread to include a variable for landscape diversity.

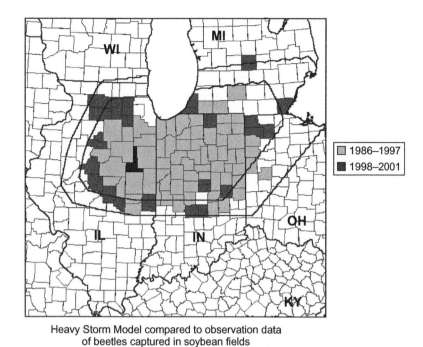

Heavy Storm Model compared to observation data
of beetles captured in soybean fields

Figure 7.4 Comparison of basic heavy storm model results to observations (20 beetles/100 sweeps or 2.0 beetles/trap/day), with the dark contours representing the 12th (1997) and 16th (2001) years of the simulations.

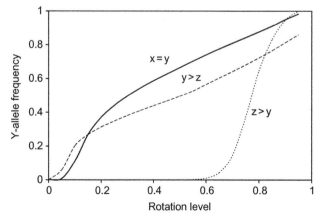

Figure 7.5 Resistance-allele (Y) frequency in Year 15 as a function of the level of crop rotation, with 5% extra vegetation, with X dominant (X > y), X and Y additive (x = y), or Y dominant (Y > x). *Onstad* et al., *2003b, reprinted by permission of Entomological Society of America.*

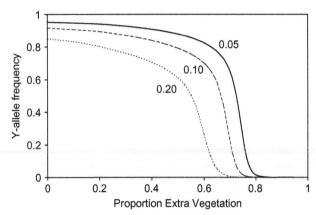

Figure 7.6 Resistance-allele frequency in year 15 produced with additive gene expression as a function of the proportion of extra vegetation when fraction of area in continuous corn is 0.05, 0.10, or 0.20. *Onstad* et al., *2003b, reprinted by permission of Entomological Society of America.*

This modification permitted the model to more accurately simulate the observed slowing of the spread of rotation-resistance in regions with greater landscape diversity (Figure 7.7).

Onstad *et al.* (2003b) expanded their simple model to explain how to manage the western corn rootworm in a landscape of corn, soybean, and winter wheat where evolution of rotation resistance may occur. They modeled six alternative IRM strategies over a 15-year time horizon (Table 7.1), as well as the typical scenario involving a two-year rotation of corn and soybean in 85% of the landscape, to investigate their effectiveness from both a biological and economic perspective.

Each of the alternative IRM strategies has different effects on western corn rootworm survival and behavior relative to the standard two-year crop rotation (2 yr). Management strategies A–C (Table 7.1) alter the proportions of the landscape in which rootworms will survive, but retain the movement parameters of the typical two-year crop rotation. The remaining three strategies (D–F, Table 7.1) alter the behavior of the western corn rootworm, thereby increasing the proportion of eggs laid in locations that will not be rotated to corn the following year. Note that strategies D–F are hypothetical, given current technology. In strategy D, the rotation-resistant beetles that are repelled from soybeans lay their eggs randomly throughout the rest of the landscape. This strategy will increase the number of adults that lay eggs in corn while decreasing the number that lay

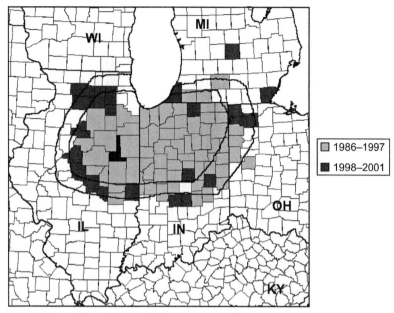

Heavy Storm Model with the maximum distance a
beetle can travel in any direction reduced by
a factor of landscape diversity (1−MEV)

Figure 7.7 Comparison of heavy storm model results to observations (20 beetles/
100 sweeps or 2.0 beetles/trap/day) with the dark contours representing the 12th
(1997) and 16th (2001) years of the simulations. Rate of spread reduced by landscape
diversity in each county. *[Modified from Onstad* et al. *(2003a)].*

Table 7.1 Six Alternative IRM Strategies Simulated by Onstad *et al.* (2003b). All Have
Two-Year Rotation of Corn (42.5%) and Soybean (42.5%) with 10% Continuous
Corn Unless Indicated Otherwise. Labels for Figures 7.8−7.9 are Given at End of Each
Description

A. Precede three-year rotation with corn (30%) by either soybean (30%) or
   other vegetation (30%), 3 yr
B. Plant transgenic insecticidal corn in rotated fields (90% of neonates die in this
   rotated corn), Trans
C. Plant more continuous corn in landscape (increase to 35% of landscape),
   MCC
D. Plant repellant soybean (repels 90% of rotation-resistant phenotypes), RSoy
E. Plant rotated corn that is more attractive (attracts 90% of all phenotypes),
   Att. Corn
F. Precede three-year rotation with corn by a crop that repels 90% of rotation-
   resistant phenotypes, 3-yr UE

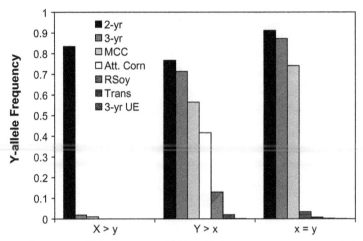

Figure 7.8 Resistance-allele frequency in Year 15 for two-year rotation and six alternative management strategies with three types of gene expression. *Onstad* et al., *2003b, reprinted by permission of Entomological Society of America.*

eggs in soybeans. Strategy E causes all beetles to lay more eggs in rotated corn than in other parts of the landscape. With this strategy the number of beetles laying eggs in both continuous corn and soybeans decreases. In strategy F, use of a three-year rotation with a less attractive crop, such as winter wheat, attempts to prevent rotation resistance by affecting both the survival of larvae and the movement rates of adults. Onstad *et al.* (2003b) investigated a winter wheat crop that repels rotation-resistant beetles, forcing them to lay eggs randomly throughout the rest of the landscape. This increases the proportion of beetles emerging in continuous corn while decreasing those emerging in rotated corn.

Each strategy (Table 7.1) was evaluated according to its effects on resistance-allele (Y) frequency, 15-year average larval densities, and the economic costs and benefits of each approach. Generally, resistance to crop rotation evolves in fewer than 15 years, and the rate of evolution increases as the level of rotated landscape (selection pressure) increases. When resistance was recessive, all six alternative strategies were effective at preventing evolution of rotation resistance. The two most successful strategies were the use of transgenic rotated corn in a two-year rotation and a three-year rotation of corn, soybean, and wheat with unattractive wheat (for oviposition) preceding corn (Figure 7.8). Economically, three alternative strategies were robust solutions to the problem, if technology

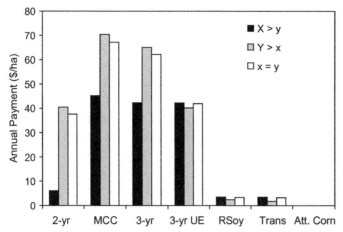

Figure 7.9 Annual payment ($/ha) that would equate the annualized net present value for each management strategy to the strategy with the greatest annualized net present value for each type of gene expression. *(Onstad* et al., *2003b, reprinted by permission of Entomological Society of America).*

fees were not too high. Repellant soybeans, attractive rotated corn, and transgenic rotated corn, all in two-year rotations, were economically valuable approaches (Figure 7.9). Onstad and Carriere (Chapter 10 in this volume) discusses the simultaneous management of resistance to crop rotation and transgenic insecticidal corn.

Even the typical two-year rotation was economical as an IRM strategy when resistance was recessive, and it took 14 years for the resistance-allele frequency to reach 50% (Figure 7.9). This result occurs because even though the frequency of the resistance allele grows rapidly, population densities do not increase for several years. As a result, yield losses on rotated corn in excess of 1% do not occur until years 14 and 15, when losses are 4.2 and 11.7%, respectively. However, since these losses occur far in the future, they are substantially discounted by the net present value criterion, implying that farmers concerned only with economic returns will prefer to do nothing about the evolution of rotation resistance. This result gives insight into the economic logic that underlies the evolution of rotation resistance among WCRs.

It is difficult, if not practically impossible, to halt the evolution of rotation resistance once the resistance-allele frequency nears 1%. Thus, the uncertainty in the timing of invasion by the rotation-resistant variant and the initial resistance gene frequency make it more difficult to choose a good strategy. Furthermore, reductions in farmer returns are not the only costs of resistance. If soil insecticide use will greatly increase as a result of rotation resistance, it may be desirable from a social perspective to subsidize farmers in the present, so that they have an incentive to change practices now and can delay the development of resistance in the future.

Onstad *et al.* (2003b) determined that under a standard set of assumptions, several alternatives, including transgenic insecticidal corn in rotation, are superior to a typical two-year rotation with regard to rootworm IPM in areas where rotation resistance is a serious problem. Unless a third crop can be found that is less attractive for rootworm egg laying and economically competitive with corn and soybeans, a three-year rotation does not appear to be a practical IPM solution in existing problem areas.

Incorporating WCR Biology into IRM

As of 2013, transgenic and other biotechnological options for corn insect pest management have been widely adopted in the United States: 76% of cornfields were planted with transgenic insecticidal corn (includes hybrids expressing lepidopteran and/or coleopteran resistance traits), and 90% of cornfields were planted with some type of genetically engineered corn (USDA-ERS, 2013). The sustainability of these solutions will depend on accurate information about pest biology. In a future in which the transgenic insecticidal crops available to producers will express a range of *Bt* toxins, be deployed in new ways (e.g., seed blends), and may include new technologies like RNA interference (Baum *et al.*, 2007; Pittendrigh *et al.*, Chapter 3 in this volume), continuing studies of rootworm ecology will be necessary to ensure that expectations for refuge function are realized in the field.

In addition to understanding rootworm movement and mating dynamics between refuge and *Bt* corn, the consequences of long-distance transport should be considered. Local and long-distance movement patterns have particular relevance for containing and managing the spread of rootworm resistance to *Bt* toxins just as they affected the spread of rotation resistance (Onstad *et al.*, 1999). Local variation in crop phenology, local and regional weather patterns, and the simultaneous presence of multiple types of resistance—all may affect rootworm dispersal and the spatial distribution of resistance genes.

FUTURE RESISTANCE

Resistance to crop rotation by the NCR and WCR was unexpected and happened in the absence of any planned strategy to avoid it. If the history of worldwide insecticide resistance over the last 60 years teaches us anything, it is that there are many possible routes to resistance. For corn growers already dealing with two modes of rotation resistance in rootworms and facing the reality of WCR resistance to *Bt* corn, preserving pest susceptibility to new technologies is of paramount importance because their set of management tools is limited. The continuing spread of rotation-resistant WCR into the western Corn Belt will place more pressure on remaining IPM methods and underscore the need to make judicious use of management tools.

Seed treatments are a popular vehicle to protect many crops, including corn and soybean. The presence of a systemic, neonicotinoid-insecticide seed treatment on corn (targeting secondary pests) and soybean (targeting *Aphis glycines*) exposes two different life stages of rotation-resistant WCR populations to neonicotinoids—once as larvae in corn and again as soybean-feeding adults. This dual exposure occurs over a vast area of the U.S. Corn Belt. Exposure of rotation-resistant WCR to insecticide applications targeting *Aphis glycines* in early-reproductive stage soybean (Myers *et al.*, 2005) adds another level of resistance risk for Corn Belt WCR.

With of the evolution of WCR resistance to one *Bt* toxin, Cry3Bb1 (Gassmann *et al.*, 2011), there should be a reevaluation of resistance management plans (Porter *et al.*, 2012). Compliance with refuge requirements must be improved to avoid placing additional selection pressure on remaining *Bt* rootworm traits. Reliance on repeated deployment of a single *Bt* trait in continuous corn is an unsustainable method of rootworm management and likely contributed to the field evolution of Cry3Bb1 resistance in WCR (Gassmann *et al.*, 2011, 2012). By using a *Bt* hybrid only when monitoring indicates that it is justified, and adhering to refuge requirements, even growers of continuous corn will preserve valuable sources of susceptible alleles (Crowder *et al.*, 2005). The use of *Bt* corn in rotated cornfields has been proposed by Crowder *et al.* (2005) as a strong strategy to manage both *Bt*- and rotation-resistant WCR populations.

Given that WCR resistance to *Cry3Bb1* toxin exists, the potential for cross-resistance with other rootworm resistance traits expressed in pyramided *Bt* hybrids is a genuine concern. However, laboratory experiments with suspected Cry3Bb1-resistant WCR by Gassmann *et al.* (2011) found no evidence of cross-resistance between two *Bt* toxins that are commonly deployed against WCR in stacked (i.e., expressing a single rootworm *Bt* toxin along with a lepidopteran *Bt* toxin) and pyramided (i.e., expressing multiple different rootworm *Bt* toxins along with multiple lepidopteran *Bt* toxins) hybrids: Cry3Bb1 and Cry34/35Ab1. On this basis, it was recommended that growers who have observed damage in *Bt* corn and suspect *Bt* resistance is present in their fields should plant *Bt* hybrids expressing multiple traits targeting rootworms (EPA Office of Pesticide Programs, 2012).

There are inescapable similarities between rotation resistance, insecticide resistance, and resistance to *Bt* corn in the role played by dependence on a single management technique. In these examples, resistance was (1) inconceivable (rotation resistance), (2) unexpected but predictable (insecticide resistance), and (3) anticipated, simulated, and regulated (resistance to *Bt* corn). Regardless of the scale of our preparations, resistance to each of these technologies occurred because we treated them as silver bullet solutions. Each example teaches the same lesson: We must be cautious about overreliance on a single solution or management method. The history of pest management and corn rootworm management in particular suggests that simple solutions are often not simple, and once solved, many problems don't stay solved forever!

REFERENCES

Allen, W.A., Rajotte, E.G., 1990. The changing role of extension entomology in the IPM era. Ann. Rev. Entomol. 35, 379–397.

Aspelin, A.L., 2003. Pesticide usage in the United States: Trends during the 20th century. NSF CIPM Tech. Bull. 105, Feb 2003.

Ball, H.J., 1957. On the biology and egg-laying habits of the western corn rootworm. J. Econ. Entomol. 50, 126–128.

Ball, H.J., Weekman, G.T., 1962. Insecticide resistance in the adult western corn rootworm in Nebraska. J. Econ. Entomol. 55, 4439–4441.

Bartelt, R.J., Chiang, H.C., 1977. Field studies involving the sex-attractant pheromones of the western and northern corn rootworm beetles. Environ. Entomol. 6, 853–861.

Baum, J.A., Bogaert, T., Clinton, W., Heck, G.R., Feldmann, P., Ilagan, O., et al., 2007. Control of coleopteran insect pests through RNA interference. Nat. Biotechnol. 25, 1322–1326.

Bigger, J.H., 1932. Short rotation fails to prevent attack of *Diabrotica longicornis* say. J. Econ. Entomol. 25, 196–199.

Boetel, M.A., Walgenbach, D.D., Hein, G.L., Fuller, B.W., Gray, M.E., 1992. Oviposition site selection of the northern corn rootworm (Coleoptera: Chrysomelidae). J. Econ. Entomol. 85, 246–249.

Branson, T.F., 1987. The contribution of prehatch and posthatch development to protandry in the chrysomelid, *Diabrotica virgifera virgifera*. Entomol. Exp. Appl. 43, 205–208.

Branson, T.F., Krysan, J.L., 1981. Feeding and oviposition behavior and life cycle strategies of *Diabrotica*: an evolutionary view with implications for pest management. Environ. Entomol. 10, 826–831.

Branson, T.F., Ortman, E.E., 1967. Host range of larvae of the western corn rootworm. J. Econ. Entomol. 60, 201–203.

Branson, T.F., Ortman, E.E., 1970. Host range of larvae of the western corn rootworm: further studies. J. Econ. Entomol. 68, 800–803.

Branson, T.F., Guss, P.L., Jackson, J.J., 1977. Mating frequency of the western corn rootworm. Ann. Entomol. Soc. Am. 70, 506–508.

Bullock, D.G., 1992. Crop rotation. Crit. Rev. Plant Sci. 11, 309–326.

Chiang, H.C., 1965a. Research on corn rootworms. Minn. Farm Home Sci. 23, 10–13.

Chiang, H.C., 1965b. Survival of northern corn rootworm eggs through one and two winters. J. Econ. Entomol. 58, 470–472.

Chiang, H.C., 1973. Bionomics of the northern and western corn rootworms. Ann. Rev. Entomol. 18, 47–72.

Ciosi, M., Miller, N.J., Kim, K.S., Giordano, R., Estoup, A., Guillemaud, T., 2008. Invasion of Europe by the western corn rootworm, *Diabrotica virgifera virgifera*: multiple transatlantic introductions with various reductions of genetic diversity. Mole. Ecol. 17, 3614–3627.

Ciosi, M., Toepfer, S., Li, H., Haye, T., Kuhlmann, U., Wang, H., et al., 2009. European populations of *Diabrotica virgifera virgifera* are resistant to aldrin, but not methylparathion. J. Appl. Entomol. 133, 307–314.

Ciosi, M., Miller, N.J., Toepfer, S., Estoup, A., Guillemaud, T., 2011. Stratified dispersal and increasing genetic variation during the invasion of Central Europe by the western corn rootworm, *Diabrotica virgifera virgifera*. Evol. Appl. 4, 54–70.

Coats, S.A., Tollefson, J.J., Mutchmor, J.A., 1986. Study of migratory flight in western corn rootworm (Coleoptera: Chrysomelidae). Environ. Entomol. 15, 620–625.

Cock, M., 2011. Plant pests: the biggest threats to food security? BBC News: Sci. Environ. 8, Nov. 2011. <http://www.bbc.co.uk/news/science-environment-15623490>.

Crowder, D.W., Onstad, D.W., Gray, M.E., Pierce, C.M.F., Hagar, A.G., Ratcliffe, S.T., et al., 2005. Analysis of the dynamics of adaptation to transgenic for and crop rotation by western corn rootworm (Coleoptera: Chrysomelidae) using a daily time-step model. J. Econ. Entomol. 98, 534–551.

Crowder, D.W., Onstad, D.W., Gray, M.E., 2006. Planting transgenic insecticidal crops based on economic thresholds: consequences for integrated pest management and insect resistance management. J. Econ. Entomol. 99, 899–907.

Cullen, E.M., Stute, J.K., Raymond, K.L., Boyd, H.H., 2008. Farmers' perspectives on IPM field scouting during a period of insect pest range expansion: a case study of variant western corn rootworm (Coleoptera: Chrysomelidae) in Wisconsin. Am. Entomol. 54 (3), 170–178.

Curzi, M., Zavala, J., Spencer, J.L., Seufferheld, M.J., 2012. Abnormally high digestive enzyme activity and gene expression explains the contemporary evolution of a *Diabrotica* biotype able to feed on soybeans. Ecol. Evol. 2, 2005–2017. Available from: http://dx.doi.org/10.1002/ece3.331.

Danks, H.V., 1992. Long life cycles in insects. Can. Entomol. 124, 167–187.

Dobbins, C.L., Miller, W.A., Nielsen, B., Vyn, T.J., Casteel, S., Johnson, B., et al., 2011. 2012 Purdue Crop Costs & Return Guide: October 2011 Estimates. Purdue Extension (ID-166-W), Purdue University, West Lafayette, IN. <http://www.agecon.purdue.edu/extension/pubs/id166_2012_AUG29_2011_final.pdf> (accessed 11.04.12.)

Dunbar, M.W., Gassmann, A.J., 2012a. Effect of soybean varieties on survival and fecundity of western corn rootworm. J. Econ. Entomol. 105, 625–631.

Dunbar, M.W., Gassmann, A.J., 2012b. Abundance and distribution of western and northern corn rootworm (*Diabrotica* spp.) and prevalence of rotation resistance in eastern Iowa. J. Econ. Entomol. [In Press].

EPA Office of Pesticide Programs, 2003. Biopesticides Registration Action Document: Event MON863 Bacillus thuringiensis Cry3Bb1 Corn. April 2003. <http://www.epa.gov/pesticides/biopesticides/ingredients/tech_docs/cry3bb1/>.

EPA Office of Pesticide Programs, 2012. Comment submitted by Patrick Porter, North Central Coordinating Committee NCCC46. March 2012. <http://www.regulations.gov/#!searchResults;rpp=10;so=DESC;sb=postedDate;po=0;s=EPA-HQ-OPP-2011-0922> (accessed 15.04.12.)

Edwards, C.R., Kiss, J., 2011. Diabrotica virgifera virgifera LeConte in Europe 2011. Online IWGO map: <http://extension.entm.purdue.edu/wcr/> (accessed 11.04.12.)

Ellsbury, M.M., Banken, K.R., Clay, S.A., Forcella, F., 2005. Interactions among western corn rootworm (Coleoptera: Chrysomelidae), yellow foxtail, and corn. Environ. Entomol. 34, 627–634.

Fisher, J.R., 1986. Development and survival of pupae of *Diabrotica virgifera virgifera* and *D. undecimpunctata howardi* (Coleoptera: Chrysomelidae) at constant temperatures and humidities. Environ. Entomol. 15, 626–630.

Forbes, S.A., 1883. The corn root-worm. (*Diabrotica longicornis*, Say) order coleoptera. Family chrysomelidae. Ill. State Entomologist Annu. Rep. 12, 10–31.

Galinat, W.C., 1965. The evolution of corn and culture in North America. Econ. Bot. 19, 350–357.

Gassmann, A.J., 2012. Field-evolved resistance to Bt maize by western corn rootworm: predictions from the laboratory and effects in the field. J. Invertebr. Pathol. 110, 287–293.

Gassmann, A.J., Petzold-Maxwell, J.L., Keweshan, R.S., Dunbar, M.W., 2011. Field-evolved resistance to Bt maize by western corn rootworm. PLoS One. 6 (7), e22629. Available from: http://dx.doi.org/10.1371/journal.pone.0022629.

Gassmann, A.J., Petzold-Maxwell, J.L., Keweshan, R.S., Dunbar, M.W., 2012. Western corn rootworm and Bt maize: challenges of pest resistance in the field. GM Crops Food. 3, 235–244.

Gillette, C.P., 1912. *Diabrotica virgifera* Lec. as a corn root-worm. J. Econ. Entomol. 5, 364–366.

Gloyna, K., Thieme, T., Zellner, M., 2011. Miscanthus, a host for larvae of a European poulation of *Diabrotica v. virgifera*. J. Appl. Entomol. 135, 780–785.

Godfrey, L.D., Turpin, F.T., 1983. Comparison of western corn rootworm (Coleoptera: Chrysomelidae) adult populations and economic thresholds in first-year and continuous cornfields. J. Econ. Entomol. 76, 1028–1032.

Grant, R.H., Seevers, K.P., 1989. Local and long-range movement of adult western corn rootworm (Coleoptera: Chrysomelidae) as evidenced by washup along southern Lake Michigan shores. Environ. Entomol. 18, 266–272.

Grant, R.H., Seevers, K.P., 1990. The vertical movement of adult western corn rootworms (*Diabrotica virgifera virgifera*) relative to the transport of momentum and heat. Agr. For. Meteorol. 49, 191–203.

Gray, M.E., 2000. Prescriptive use of transgenic hybrids for corn rootworms: an ominous cloud on the horizon? In: Proceedings of the Crop Protection Technology Conference. University of Illinois, Urbana-Champaign, pp. 97–103.

Gray, M.E., 2011a. Relevance of Traditional Integrated Pest Management (IPM) Strategies for Commercial Corn Producers in a Transgenic Agroecosystem: A Bygone Era? J. Agric. Food Chem. 59, 5852–5858. Available from: http://dx.doi.org/10.1021/jf102673s.

Gray, M.E., 2011b. Severe root damage to Bt corn observed in Northwestern Illinois. The Bulletin: pest management and crop development information for Illinois, No. 20, August 26, 2011.

Gray, M.E., Steffey, K.L., 2004. Development of economic thresholds: show us the science. The Bulletin: pest management and crop development information for Illinois, No. 1, March 18, 2004.

Gray, M.E., Felsot, A.S., Steffey, K.L., Levine, E., 1992. Planting time application of soil insecticides and western corn rootworm (Coleoptera: Chrysomelidae) emergence: implications for long-term management programs. J. Econ. Entomol. 85, 544–553.

Gray, M.E., Steffey, K.L., Estes, R. , Schroeder, J.B., Bakken, D.M., 2006. Transgenic corn rootworm hybrids, soil insecticides, and seed treatments: does anything work on the variant western corn rootworm? In: Proceedings of the Crop Protection Technology Conference. University of Illinois, Urbana-Champaign, pp. 54–61.

Gray, M.E., Sappington, T.W., Miller, N.J., Moeser, J., Bohn, M.O., 2009. Adaptation and invasiveness of western corn rootworm: intensifying research on a worsening pest. Ann. Rev. Entomol. 54, 303–321.

Guss, P.L., 1976. The sex pheromone of the western corn rootworm (*Diabrotica virgifera*). Environ. Entomol. 5, 219–223.

Gustin, R.D., 1979. Effect of two moisture and populaton levels on oviposition of the western corn rootworm. Environ. Entomol. 8, 406–407.

Gustin, R.D., 1984. Effect of crop cover on oviposition of the northern corn rootworm, *Diabrotica longicornis barberi* Smith and Lawrence. J. Kansas Entomol. Soc. 57, 515–516.

Hammack, L., 1995. Calling behavior in female western corn rootworm beetles (Coleoptera: Chrysomelidae). Ann. Entomol. Soc. Am. 88, 562–569.

Heichel, G.H., 1987. Legumes as a source of nitrogen in conservation tillage systems. In: Power J.F. (Ed.) The role of legumes in conservation tillage systems. Proceedings of the National Conservation Tillage Conference., Athens, GA. 27–29 April. Soil Conservation Society of America. Ankeny, IA, p. 29–35.

Hill, R.E., 1975. Mating, oviposition patterns, fecundity and longevity of the western corn rootworm. J. Econ. Entomol. 68, 311–315.

Hill, R.E., Mayo, Z.B., 1980. Distribution and abundance of corn rootworm species as influenced by topography and crop rotation in eastern Nebraska. Environ. Entomol. 9, 122–127.

Hill, R.E., Hixson, E., Muma, M.H., 1948. Corn rootworm control tests with benzene hexachloride, DDT, nitrogen fertilizers and crop rotations. J. Econ. Entomol. 41, 393–401.

Hills, T.M., Peters, D.C., 1971. A method of evaluating post-planting insecticide treatments for control of western corn rootworm larvae. J. Econ. Entomol. 64, 764–765.

Igrc Barcic, J., Bazok, R., Edwards, C.R., Kos, T., 2007. Western corn rootworm adult movement and possible egg laying in fields bordering maize. J. Appl. Entomol. 131, 400–405.

Isard, S.A., Nasser, M.A., Spencer, J.L., Levine, E., 1999. The influence of weather on western corn rootworm flight activity at the borders of a soybean field in east central Illinois. Aerobiologia. 15, 95–104.

Isard, S.A., Spencer, J.L., Nasser, M.A., Levine., E., 2000. Aerial movement of western corn rootworm, *Diabrotica virgifera virgifera* (Coleoptera: Chrysomelidae): Diel periodicity of flight activity in soybean fields. Environ. Entomol. 29, 226–234.

Isard, S.A., Spencer, J.L., Mabry, T.R., Levine, E., 2004. Influence of atmospheric conditions on high elevation flight of western corn rootworm (Coleoptera: Chrysomelidae). Environ. Entomol. 33, 650–656.

Jaffe, G., 2009. Complacency on the farm: significant noncompliance with EPA's refuge requirements threatens the future effectiveness of genetically engineered pest-protected corn. Center for Science in the Public Interest. Washington, D.C., pp. 19 <www.cspinet.org>.

Kang, J., Krupke, C.H., 2009a. Influence of weight of male and female western corn rootworm (Coleoptera: Chrysomelidae) on mating behaviors. Ann. Entomol. Soc. Am. 102, 326–332.

Kang, J., Krupke, C.H., 2009b. Likelihood of Multiple Mating in Diabrotica virgifera virgifera (Coleoptera: Chrysomelidae). J. Econ. Entomol. 102, 2096–2100.

Karlen, D.L., Varvel, G.E., Bullock, D.G., Cruse, R.M., 1994. Crop rotations for the 21st century. Adv. Agron. 53, 1–45.

Keszthelyi, S., 2005. Immigration of western corn rootworm (Diabrotica virgifera virgifera LeConte) adults into first year corn in Somogy county 2004. Cereal Res. Commun. 33, 747–754.

Kirk, V.M., 1979. Drought cracks as oviposition sites for western and northern corn root-worms (Diabrotica: Coleoptera). J. Kansas Entomol. Soc. 52, 769–776.

Kirk, V.M., 1981a. Earthworm burrows as oviposition sites for western and northern corn rootworms (Diabrotica: Coleoptera). J. Kansas Entomol. Soc. 54, 68–74.

Kirk, V.M., 1981b. Base of corn stalks as oviposition sites for western and northern corn rootworms (Diabrotica: Coleoptera). J. Kansas Entomol. Soc. 54, 255–262.

Kirk, V.M., Calkins, C.O., Post, F.J., 1968. Oviposition preferences of western corn root-worms for various soil surface conditions. J. Econ. Entomol. 61, 1322–1324.

Kiss, J., Edwards, C.R., Berger, H.K., Cate, P., Cean, M., Cheek, S., et al., 2005a. Monitoring of western corn rootworm (Diabrotica virgifera virgifera LeConte) in Europe 1992–2003. In: Vidal, S., Kuhlmann, U., Edwards, C.R. (Eds.), Western Corn Rootworm: Ecology and Management. CABI Publishing, Wallingford, Oxfordshire, UK, pp. 29–39.

Kiss, J., Komaromi, J., Bayar, K., Edwards, C.R., Hatala-Zseller, I., 2005b. Western corn rootworm (Diabrotica virgifera virgifera LeConte) and the crop rotation systems in Europe. In: Vidal, S., Kuhlmann, U., Edwards, C.R. (Eds.), Western Corn Rootworm: Ecology and Management. CABI Publishing, Wallingford, Oxfordshire, UK, pp. 189–220.

Knolhoff, L.M., Onstad, D.W., Spencer, J.L., Levine., E., 2006. Behavioral differences between rotation-resistant and wild-type Diabrotica virgifera virgifera (Coleoptera: Chrysomelidae). Environ. Entomol. 35, 1049–1057.

Knolhoff, L.M., Glass, J.J., Spencer, J.L., Berenbaum, M.R., 2010a. Oviposition behaviors in relation to rotation resistance in the western corn rootworm. Environ. Entomol. 39 (6), 1922–1928.

Knolhoff, L.M., Walden, K.O., Ratcliffe, S.T., Onstad, D.W., Robertson, H.M., 2010b. Microarray analysis yields candidate markers for rotation resistance in the western corn rootworm beetle, Diabrotica virgifera virgifera. Evol. Appl. 3 (1), 17–27.

Krafsur, E.S., 1995. Gene flow between univoltine and semivoltine northern corn root-worm (Coleoptera: Chrysomelidae) populations. Ann. Entomol. Soc. Am. 88, 699–704.

Krysan, J.L., 1986. Introduction: biology, distribution, and identification of pest Diabrotica. In: Krysan, J.L., Miller, T.A. (Eds.), Methods for the study of pest Diabrotica. Springer, New York, pp. 1–24.

Krysan, J.L., 1993. Adaptations of Diabrotica to habitat manipulations. In: Kim, K.C., McPheron, B.A. (Eds.), Evolution of Insect Pests. Wiley, New York, pp. 361–373.

Krysan, J.L., Branson, T.F., Castro, G.D., 1977. Diapause in Diabrotica virgifera (Coleoptera: Chrysomelidae): a comparison of eggs from temperate and subtropical climates. Entomol. Exp. Appl. 22, 81–89.

Krysan, J.L., Jackson, J.J., Lew, A.C., 1984. Field termination of egg diapause in *Diabrotica* with new evidence of extended diapause in *D. barberi* (Coleoptera: Chrysomelidae). Environ. Entomol. 13, 1237—1240.

Krysan, J.L., Foster, D.E., Branson, T.F., Ostlie, K.R., Cranshaw, W.S., 1986. Two years before the hatch: rootworms adapt to crop rotation. Bull. Entomol. Soc. Am. 32, 250—253.

Lance, D.R., Elliott, N.C., Hein, G.L., 1989. Flight activity of *Diabrotica* spp. at borders of cornfields and its relation to ovarian stage in *D. barberi*. Entomol. Experiment. Appl. 50, 61—67.

Landis, D.A., Levine, E., Haas, M.J., Meints, V., 1992. Detection of prolonged diapause of northern corn rootworm in Michigan (Coleoptera: Chrysomelidae). Great Lakes Entomol. 25, 215—222.

LeConte, J.L., 1868. New Coleoptera collected on the survey for the extension of the Union Pacific Railway, E.D. from Kansas to Fort Craig, New Mexico. Trans. Am. Entomol. Soc. 2, 49—59.

Levine, E., Oloumi-Sadeghi, H., 1991. Management of diabroticite rootworms in corn. Ann. Rev. Entomol. 36, 229—255.

Levine, E., Oloumi-Sadeghi, H., 1996. Western corn rootworm (Coleoptera: Chrysomelidae) larval injury to corn grown for seed production following soybeans grown for seed production. J. Econ. Entomol. 89, 1010—1016.

Levine, E., Oloumi-Sadeghi, H., Ellis, C.R., 1992a. Thermal requirements, hatching patterns, and prolonged diapause in western corn rootworm (Coleoptera: Chrysomelidae) eggs. J. Econ. Entomol. 85, 2425—2432.

Levine, E., Oloumi-Sadeghi, H., Fisher, J.R., 1992b. Discovery of multiyear diapause in Illinois and South Dakota northern corn rootworm (Coleoptera: Chrysomelidae) eggs and incidence of the prolonged diapause trait in Illinois. J. Econ. Entomol. 85, 262—267.

Levine, E., Spencer, J.L., Isard, S.A., Onstad, D.W., Gray, M.E., 2002. Adaptation of the western corn rootworm to crop rotation: evolution of a new strain in response to a management practice. AM Entomol. 48, 94—107.

Lew, A.C., Ball, H.J., 1979. The mating behavior of the western corn rootworm *Diabrotica virgifera virgifera* (Coleoptera: Chrysomelidae). Ann. Entomol. Soc. Am. 72, 391—393.

Lew, A.C., Ball, H.J., 1980. Effect of copulation time on spermatozoan transfer of *Diabrotica virgifera* (Coleoptera: Chrysomelidae). Ann. Entomol. Soc. Am. 73, 360—361.

Lilly, J.H., 1956. Soil insects and their control. Ann. Rev. Entomol. 1, 203—222.

Mabry, T.R., 2002. The effects of soybean herbivory on the behavior and ecology of the western corn rootworm (Diabrotica virgifera virgifera LeConte) variant. MSc thesis, University of Illinois, Urbana, IL.

Mabry, T.R., Spencer, J.L., 2003. Survival and oviposition of a western corn rootworm variant feeding on soybean. Entomol. Experiment. Appl. 109, 113—121.

Mabry, T.R., Spencer, J.L., Levine, E., Isard, S.A., 2004. Western corn rootworm (Coleoptera: Chrysomelidae) behavior is affected by alternating diets of corn and soybean. Environ. Entomol. 33, 860—871.

MacRae, R.J., Mehuys, G.R., 1985. The effect of green manuring on the physical properties of temperate-area soils. Adv. Soil Sci. 3, 71—94.

Meihls, L.N., 2010. Development and characterization of resistance to transgenic corn in western corn rootworm. Ph.D. Dissertation, University of Missouri, Columbia.

Meihls, L.N., Higdon, M.L., Siegfried, B.D., Miller, N.J., Sappington, T.W., Ellersieck, M.R., et al., 2008. Increased survival of western corn rootworm on transgenic corn within three generations of on-plant greenhouse selection. Proc. Nat. Acad. Sci. USA. 105, 19177—19182.

Meinke, L.J., Siegfried, B.D., Wright, R.J., Chandler, L.D., 1998. Adult susceptibility of Nebraska western corn rootworm (Coleoptera: Chrysomelidae) populations to selected insecticides. J. Econ. Entomol. 91, 594−600.

Meinke, L.J., Sappington, T.W., Onstad, D.W., Guillemaud, T., Miller, N.J., Komaromi, J., et al., 2009. Western corn rootworm (*Diabrotica virgifera virgifera* LeConte) population dynamics. Agricult. For. Entomol. 11, 29−46.

Meissle, M., Romeis, J., Bigler, F., 2011. Bt maize and integrated pest management—a European perspective. Pest. Manag. Sci. 67, 1049−1058.

Meloche, F., Hermans, P., 2004. Eastward expansion and discovery of the soybean biotype of western corn rootworm (*Diabrotica virgifera virgifera* LeConte) in Canada. Can. J. Plant Sci. 84, 305−309.

Metcalf, R.L., 1979. Plants, chemicals, and insects: some aspects of coevolution. Bull. Entomol Soc. Am. 25, 30−35.

Metcalf, R.L., 1982. Insecticides in pest management. In: Metcalf, R.L., Luckmann, W. H. (Eds.), Introduction to Insect Pest Management, second ed. Wiley-Interscience, New York, pp. 217−277.

Metcalf, R.L., 1983. Implications and prognosis of resistance to insecticides. In: Georghiou, G.P., Saito, T. (Eds.), Pest Resistance to Pesticides. Plenum Press, New York, pp. 703−733.

Metcalf, R.L., 1986. Foreword. In: Krysan, J.L., Miller, T.A. (Eds.), Methods for the Study of Pest Diabrotica. Springer-Verlag, New York, pp. vii−xv.

Miller, N., Estoup, A., Toepfer, S., Bourguet, D., Lapchin, L., Derridj, S., et al., 2005. Multiple transatlantic introductions of the western corn rootworm. Science. 310, 992.

Murphy, A.F., Krupke, C.H., 2011. Mating success and spermatophore composition in western corn rootworm (Coleoptera: Chrysomelidae). Environ. Entomol. 40, 1585−1594.

Myers, S.W., Hogg, D.B., Wedberg, J.L., 2005. Determining the optimal timing of foliar insecticide applications for control of soybean aphid (Hemiptera: Aphididae) on soybean. J. Econ. Entomol. 98, 2006−2012.

Naranjo, S.E., 1991. Movement of corn rootworm beetles, *Diabrotica spp.* (Coleoptera: Chrysomelidae), at cornfield boundaries in relation to sex, reproductive status, and crop phenology. Environ. Entomol. 20, 230−240.

Naranjo, S.E., 1994. Flight orientation of *Diabrotica virgifera virgifera* and *D. barberi* (Coleoptera: Chrysomelidae) at habitat interfaces. Ann. Entomol. Soc. Am. 87, 383−394.

Oleson, J.D., Park, Y.-L., Nowatzki, T.M., Tollefson, J.J., 2005. Node-injury scale to evaluate root injury by corn rootworms (Coleoptera: Chrysomelidae). J. Econ. Entomol. 98, 1−8.

O'Neal, M.E., Gray, M.E., Smyth, C.A., 1999. Population characteristics of a western corn rootworm (Coleoptera: Chrysomelidae) strain in east-central Illinois corn and soybean fields. J. Econ. Entomol. 92, 1301−1310.

O'Neal, M.E., Gray, M.E., Ratcliffe, S., Steffey., K.L., 2001. Predicting western corn rootworm (Coleoptera: Chrysomelidae) larval injury to rotated corn with Pherocon AM traps in soybeans. J. Econ. Entomol. 94, 98−105.

O'Neal, M.E., DiFonzo, C.D., Landis, D.A., 2002. Western corn rootworm (Coleoptera: Chrysomelidae) feeding on corn and soybean leaves affected by corn phenology. Environ. Entomol. 31, 285−292.

Onstad, D.W., Joselyn, M.G., Isard, S.A., Levine, E., Spencer, J.L., Bledsoe, L.W., et al., 1999. Modeling the spread of western corn rootworm (Coleoptera: Chrysomelidae) populations adapting to soybean-corn rotation. Environ. Entomol. 28 (2), 188−194.

Onstad, D.W., Spencer, J.L., Guse, C.A., Levine, E., Isard, S., 2001. Modeling evolution of behavioral resistance by an insect to crop rotation. Entomol. Experiment. Appl. 100, 195−201.

Onstad, D.W., Crowder, D.W., Isard, S.A., Levine, E., Spencer, J.L., O'Neal, M.E., et al., 2003a. Does landscape diversity slow the spread of rotation-resistant western corn rootworm (Coleoptera: Chrysomelidae)?. Environ. Entomol. 32, 992–1001.

Onstad, D.W., Crowder, D.W., Mitchell, P.D., Guse, C.A., Spencer, J.L., Levine, E., et al., 2003b. Economics versus alleles: balancing IPM and IRM for rotation-resistant western corn rootworm (Coleoptera: Chrysomelidae). J. Econ. Entomol. 96, 1872–1885.

Ostlie, K.R., 1987. Extended diapause: northern corn rootworm adapts to crop rotation. Crops Soils Mag. 39, 23–25.

Oswald, K.J., French, B.W., Nielson, C., Bagley, M., 2011. Selection for Cry3Bb1 resistance in a genetically diverse population of nondiapausing western corn rootworm (Coleoptera: Chrysomelidae). J. Econ. Entomol. 104, 1038–1044.

Oyediran, I.O., Hibbard, B.E., Clark, T.L., 2004a. Prairie grasses as alternate hosts of the western corn rootworm (Coleoptera: Chrysomelidae). Environ. Entomol. 33, 740–747.

Oyediran, I.O., Hibbard, B.E., Clark, T.L., French, B.W., 2004b. Selected grassy weeds as alternate host of northern corn rootworm (Coleoptera: Chrysomelidae). Environ. Entomol. 33, 1497–1504.

Petty, H.B., 1965. Insect situation, 1965, In Proceedings of the Seventeenth Illinois Custom Spray Operators' Training School. Extension Service, University of Illinois College of Agriculture, Urbana-Champaign, pp. 48–59.

Pierce, C.M.F., 2003. Case study of a variant of western corn rootworm, Diabrotica virgifera virgifera LeConte, in east central Illinois. Ph.D. Dissertation, University of Illinois, Champaign, IL. p. 216.

Pierce, C.M.F., Gray, M.E., 2006. Seasonal oviposition of a western corn rootworm, Diabrotica virgifera virgifera LeConte (Coleoptera: Chrysomelidae), variant in east central Illinois commercial maize and soybean fields. Environ. Entomol. 35, 676–683.

Pierce, J.J., Rice, C.W., 1988. Crop rotation and its impact on efficiency of water and nitrogen use. In: Hargrove, W.L. (Ed.), Cropping Strategies for Efficient Use of Water and Nitrogen. Amer. Soc. Agron., Madison, Wisconsin, pp. 21–42. ASA Spec. Publ. 51.

Pike, D.R., Gray, M.E., 1992. A history of pesticide use in Illinois. In Proceedings of Eighteenth Annual Illinois Crop Protection Workshop. University of Illinois, Urbana-Champaign, pp. 43–52.

Porter, P., Cullen, E., Sappington, T., Schaafsma A., Pueppke, S., Andow, D., et al., 2012. Comment submitted by Patrick Porter, North Central Coordinating Committee NCCC46. <http://www.regulations.gov/#!documentDetail;D=EPA-HQ-OPP-2011-0922-0013>.

Power, J.F., 1987. Legumes: their potential role in agricultural production. Am. J. Altern. Agr. 2 (2), 69–73.

Prasifka, P.L., Tollefson, J.J., Rice, M.E. (2006) Rotation-resistant corn rootworms in Iowa. Iowa State University Integrated Crop Manage ment Newsletter, 496, 209–211 [WWW document. URL <http://www.ipm.iastate.edu/ipm/icm//ipm/icm//ipm/icm/issue/6086> (accessed on 11.04.12.)

Quiring, D.T., Timmins, P.R., 1990. Influence of reproductive ecology on feasibility of mass trapping Diabrotica virgifera virgifera (Coleoptera: Chrysomelidae). J. Appl. Ecol. 27, 965–982.

Rice, M.E., 2004. Transgenic rootworm corn: Assessing potential agronomic, economic, and environmental benefits. Online. Plant Health Progress. Available from: http://dx.doi.org/10.1094/PHP-2004-0301-01-RV.

Rondon, S.I., Gray, M.E., 2003. Captures of western corn rootworm (Coleoptera: Chrysomelidae) adults with Pherocon AM and vial traps in four crops in east central Illinois. J. Econ. Entomol. 96, 737–747.

Rondon, S.I., Gray, M.E., 2004. Ovarian development and ovipositional preference of the western corn rootworm (Coleoptera: Chrysomelidae) in east central Illinois. J. Econ. Entomol. 97, 390–396.

Sappington, T.W., Siegfried, B.D., Guillemaud, T., 2006. Coordinated *Diabrotica* genetics research: accelerating progress on an urgent insect pest problem. Am. Entomol. 52, 90–97.

Say, T.H., 1824. Descriptions of coleopterous insects collected in the late expedition to the Rocky Mountains, performed by order of Mr. Calhoun, Secretary of War, under the command of Major Long. J. Acad. Sci. Phila. 3, 403–462.

Schnitkey, G., 2012. Crop Budgets, Illinois, 2012. Department of Agricultural and Consumer Economics, University of Illinois, Urbana-Champaign, IL. <www.farmdoc.illinois.edu/manage/2012_crop_budgets.pdf> (accessed 11.04.12.)

Schroeder, J.B., Ratcliffe, S.T., 2003. 2003 variant western corn rootworm on-farm survey. <http://www.ipm.uiuc.edu/wcrsurvey/2003.html> (accessed 08.08.12.)

Schroeder, J.B., Ratcliffe, S.T., 2004. 2004 variant western corn rootworm on-farm survey. <http://www.ipm.uiuc.edu/wcrsurvey/index.html> (accessed 08.08.12.)

Schroeder, J.B., Ratcliffe, S.T., Gray, M.E., 2005. Effect of four cropping systems on variant western corn rootworm (Coleoptera: Chrysomelidae) adult and egg densities and subsequent larval injury in rotated maize. J. Econ. Entomol. 98, 1587–1593.

Seiter, N.J., Richmond, D.S., Holland, J.D., Krupke, C.H., 2010. A novel method for estimating soybean herbivory in western corn rootworm (Coleoptera: Chrysomelidae). J. Econ. Entomol. 103 (4), 1464–1473.

Shaw, J.T., Paullus, J.H., Luckmann, W.H., 1978. Corn rootworm oviposition in soybeans. J. Econ. Entomol. 71, 189–191.

Sherwood, D.R., Levine, E., 1993. Copulation and its duration affects female weight, oviposition, hatching patterns, and ovarian development in the western corn rootworm (Coleoptera: Chrysomelidae). J. Econ. Entomol. 86, 1664–1671.

Siegfried, B.D., Vaughn, T.T., Spencer, T., 2005. Baseline susceptibility of western corn rootworm (Coleoptera: Chrysomelidae) to Cry3Bb1 *Bacillus thuringiensis* toxin. J. Econ. Entomol. 98, 1320–1324.

Sivcev, I., Stankovic, S., Kostic, M., Lakic, N., Popovic, Z., 2009. Population density of *Diabrotica virgifera virgifera* LeConte beetles in Serbia first year and continuous maize fields. J. Appl. Entomol. 133, 430–437.

Smith, R.F., 1966. Distributional patterns of selected western North American insects. The distribution of diabroticites in western North America. Bull. Entomol. Soc. Am. 12, 108–110.

Spencer, J.L., 2010. "What's past is prologue": lessons from the history of corn rootworm management in Illinois and the U.S. In: Rodríguez-del-Bosque, L.A., Morón, M.A. (Eds.), Ecología y Control de Plagas Edaficolas. Publicación especial del Instituto de Ecología, A.C. México, pp. 223–243, p. 329.

Spencer, J.L., Raghu, S., 2009. Refuge or reservoir? The potential impacts of the biofuel crop *Miscanthus* x *giganteus* on a major pest of maize. PLoS One. 4 (12), e8336. Available from: http://dx.doi.org/10.1371/journal.pone.0008336, December 16, 2009.

Spencer, J.L., Mabry, T.R., Vaughn, T., 2003. Use of transgenic plants to measure insect herbivore movement. J. Econ. Entomol. 96, 1738–1749.

Spencer, J.L., Mabry, T.R., Levine, E., Isard, S.A., 2005. Movement, dispersal, and behavior of western corn rootworm adults in rotated corn and soybean fields. In: Vidal, S., Kuhlmann, U., Edwards, C.R. (Eds.), Western Corn Rootworm: Ecology and Management. CABI Publishing, Wallingford, Oxfordshire, UK, pp. 121–144.

Spencer, J.L., Hibbard, B.E., Moeser, J., Onstad, D.W., 2009. Behaviour and ecology of the western corn rootworm (*Diabrotica virgifera virgifera* LeConte). Agr. For. Entomol. 11, 9—27.

Steffey, K.L., Gray, M.E., Kuhlman, D.E., 1992. Extent of corn rootworm (Coleoptera: Chrysomelidae) larval damage in corn after soybeans: Search for the expression of the prolonged diapause trait in Illinois. J. Econ. Entomol. 85, 268—275.

Steffey, K.L., Ratcliffe, S.T., Schroeder, J.B., 2003. Results from 2003 on-farm survey for rootworm larval damage in first-year corn. The Bulletin: pest management and crop development information for Illinois No. 23, October 3, 2003.

Steffey, K.L., Gray, M.E., Cook, K., 2004. Results of variant western corn rootworm larval-injury survey. The Bulletin: pest management and crop development information for Illinois No. 23, October 8, 2004.

Storer, N.P., Babcock, J.M., Edwards, J.M., 2006. Field measures of western corn rootworm (Coleoptera: Chrysomelidae) mortality caused by Cry34/35Ab1 proteins expressed in maize event 59122 asnd implications for trait durability. J. Econ. Entomol. 99, 1381—1387.

Szalai, M., Koszegi, J., Toepfer, S., Kiss, J., 2011. Colonisation of first-year maize fields by western corn rootworm (*Diabrotica virgifera virgifera* LeConte) from adjacent infested maize fields. Acta Phytophathol. Hun. 46, 213—223.

Tallamy, D.W., Hibbard, B.E., Clark, T.L., Gillespie, J.J., 2005. Western corn rootworm, cucurbits and cucurbitacins. In: Vidal, S., Kuhlmann, U., Edwards, C.R. (Eds.), Western Corn Rootworm: Ecology and Management. CABI Publishing, Wallingford, Oxfordshire, UK, pp. 67—93.

Tauber, M.J., Tauber, C.A., Masaki, S., 1986. Variability and genetics of seasonal adaptations. Seasonal Adaptations of Insects. Oxford University Press, New York, pp. 192—217.

Thomas, C., 1881. New corn insect—*Diabrotica longicornis* Say. Ill. State Entomologist Annu. Rep. 10, 44—46.

Thomas, C., 1882. The corn-root worm (*Diabrotica longicornis*, Say). Ill. State Entomologist Annu. Rep. 11, 65—72.

Toepfer, S., Levay, N., Kiss, J., 2006. Adult movement of newly introduced alien *Diabrotica virgifera virgifera* (Coleptera: Chrysomelidae) from non-host habitats. Bull. Entomol. Res. 96, 327—335.

Tollefson, J.J., 1988. A pest adapts to the cultural control of crop rotation. Proc. Br. Crop Prot. Conf. 3, 1029—1033.

United States Department of Agriculture Economic Research Service, 2003. ARMS briefing room. Online. <http://www.e.usda.gov/Briefing/ARMS>.

United States Department of Agriculture Economic Research Service, 2006, Soil Management and Conservation (Section 4.2) In Agricultural Resources and Environmental Indicators. Wiebe, K. Gollehon N. (Eds.). Economic Information Bulletin No. (EIB-16). <http://www.ers.usda.gov/publications/arei/eib16/> (accessed 11.04.12.)

United States Department of Agriculture National Agricultural Statistics Service, 2010. Field crop chemical insecticide treatments as percent of planted corn acres. <http://quickstats.nass.usda.gov/?sector_desc=ENVIRONMENTAL> (accessed 13.04.12.)

United States Department of Agriculture Economic Research Service, 2013. Adoption of genetically engineered crops in the U.S.: Corn. Online. <http://www.ers.usda.gov/data-products/adoption-of-genetically-engineered-crops-in-the-us.aspx#.UfwWelNK J0Q>(accessed 8.2.13).

VanWoerkom, G.J., Turpin, F.T., Barret Jr., J.R., 1983. Wind effect on western corn rootworm (Coleoptera: Chrysomelidae) flight behavior. Environ. Entomol. 12, 196—200.

Vaughn, T., Cavato, T., Brar, G., Coombe, T., DeGooyer, T., Ford, S., et al., 2005. A method of controlling corn rootworm feeding using a *Bacillus thuringiensis* protein expressed in transgenic maize. Crop. Sci. 45, 931–938.

Wesseler, J., Fall, E.H., 2010. Potential damage costs of *Diabrotica virgifera virgifera* infestation in Europe—the 'no control' scenario. J. Appl. Entomol. 134, 385–394.

Wilson, T.A., Rice, M.E., Tollefson, J.J., Pilcher, C.D., 2005. Transgenic corn for control of the European corn borer and corn rootworms: a survey of Midwestern farmers' practices and perceptions. J. Econ. Entomol. 98, 237–247.

Witkowski, J.F., Owens, J.C., Tollefson, J.J., 1975. Diel activity and vertical flight distribution of adult western corn rootworms in Iowa cornfields. J. Econ. Entomol. 68, 351–352.

Yates, F., 1954. The analysis of experiments containing different crops. Biometrics. 10, 324–346.

CHAPTER 8

Resistance to Pathogens and Parasitic Invertebrates

David W. Onstad
DuPont Pioneer, Wilmington, DE

Chapter Outline

Pathologists and parasitologists believe that every multicellular species has at least one other species parasitizing it, and current evidence suggests that half of all species are parasites (Poulin, 2011). However, estimates of the number of parasite species will likely increase as biologists explore symbiotic relationships more carefully and as molecular and genomic technologies are used to identify species with greater precision. Braxton *et al.* (2003) described 2285 pathogen species that infect 4454 arthropod species (primarily insects), creating 9400 parasitic associations between pathogen and host. Most of the described entomopathogens are protozoa, microsporidia, viruses, and fungi. Humber (2008) recently provided a synthesis of knowledge regarding the evolution of pathogenicity by entomopathogenic fungi. Stireman *et al.* (2006) reviewed parasitism by Tachinidae. Others have focused more broadly on the variety of parasitic strategies that have evolved in many parasites of animals, including parasitoids (Kuris and Lafferty, 2000; Poulin, 2011).

We should expect any kind of effective pest management to cause the evolution of insect resistance. Therefore, insect resistance management (IRM) strategies must be considered for any highly effective biological control utilizing parasites. This chapter describes the variety of cases in which insects have evolved resistance to parasites, such as pathogens and parasitoids. Note, however, that determining the genetic basis for resistance to parasitic

Insect Resistance Management
DOI: http://dx.doi.org/10.1016/B978-0-12-396955-2.00008-4

natural enemies can be complicated by the possibility of host populations carrying symbiotic microbes that protect the hosts (Oliver et al., 2005; Von Burg et al., 2008). In Chapter 12, Onstad et al. discuss how natural enemies directly or indirectly influence the selection for insect resistance to host-plant resistance or chemical control.

RESISTANCE TO PATHOGENS

Although the arthropod's resistance to infectious pathogens causing contagious diseases is not considered a serious issue and is rarely observed outside the laboratory (Shelton et al., 2007), scientists should understand the potential for resistance evolution for several reasons. First, microbial insecticides consisting of viruses, fungi, or bacteria, other than insecticidal Bacillus thuringiensis, are increasingly being studied and developed (Moscardi, 1999; Lacey and Kaya, 2000; Butt et al., 2001). Second, as more microbial insecticides are used, selection pressure may increase, resulting in a higher probability of resistance evolution. Third, management of domesticated beneficial insects (silkworm, Bombyx mori, and bees such as Apis mellifera) may require populations that are resistant to natural pathogens (Briese, 1981; Stephen and Fichter, 1990).

Briese (1981) was one of the first to summarize the state of knowledge concerning insect resistance to viruses, bacteria, fungi, microsporidia, and nematodes. Most of the cases were identified in laboratory colonies, bee hives, or silkworm populations. Several additional studies published since 1981 exemplify the ability of insects to evolve resistance to infectious pathogens (Milner, 1982; Ignoffo et al., 1985; Stephen and Fichter, 1990). For example, Briese and Mende (1983) observed a 140-fold increase in LD_{50} after serial exposure of a field-collected population of Phthorimaea operculella to granulosis virus over six generations in the laboratory.

Fuxa (2004) reviewed the cases of insect resistance to nucleopolyhedroviruses and discussed cross-resistance to several pathogens. He also stated that there are several similarities between arthropod resistance to nucleopolyhedroviruses and resistance to chemical pesticides. However, he noted that there can be important differences in mechanisms and that the potential for coevolution exists between insects and pathogens.

Bindu et al. (2012) investigated the roles of body color (melanism) and host density in the resistance by Hyblaea puera to H. puera nucleopolyhedrovirus. At low densities the larvae are lighter in color, while at high densities they are darker. Larvae reared under crowded conditions were slightly more resistant to the virus compared to larvae reared individually. Bindu *et al.* (2012) concluded that there is a link between density, melanism, and resistance to the virus.

Dubovskiy *et al.* (2013) studied adaptations and mechanisms of resistance in a melanic population of *Galleria mellonella*. After 25 generations of constant selection by the entomopathogenic fungus *Beauveria bassiana*, host larvae exhibited significantly greater resistance, which was specific to this pathogen and not to another pathogenic fungus, *Metarhizium anisopliae*. The authors concluded that, during *B. bassiana* infection, systemic immune defenses are suppressed in favor of a more limited but targeted set of enhanced defenses in the cuticle and epidermis of the integument.

Vijendravarma *et al.* (2009) reared *Drosophila melanogaster* for 61 weeks along with the microsporidian *Tubulinosema kingi*. Compared to infected controls, selected cohorts had higher early-life fecundity and increased longevity when infected, suggesting successful selection for resistance or tolerance. Fitness costs in the absence of the pathogen included lower fecundity when reared under stressful conditions. Selected cohorts were also less competitive for food than controls. Vijendravarma *et al.* (2009) also observed that resistant larvae had relatively high levels of hemocytes, which are components of the cellular immune system.

Franklin *et al.* (2012) describe the history of their work on *Malacosoma californicum pluviale* and its variable susceptibility to *M. c. pluviale nucleopolyhedrovirus* in forests. This insect is a primitively social species in which individuals in families work gregariously to construct web tents on tree branches and mark food trails with chemical cues to attract larvae of the same family. Genetic variability and polyandry in the host may be related to resistance to the virus. Franklin *et al.* (2012) also contrast their findings with those derived from research on disease resistance in eusocial insects (Baer and Schmid-Hempel, 2001; Whitehorn *et al.*, 2011; Calleri *et al.*, 2006; Tarpy, 2003; Schmid-Hempel, 1994; Rosengaus and Traniello, 2001). Resistance in social insects often involves behaviors that either prevent entrance of pathogens to colonies or remove them from surfaces of insects and nests.

It is likely that many arthropods can evolve behavioral resistance to pathogens. American foulbrood is the bacterial disease caused by *Bacillus larvae* infecting *A. mellifera*. Incidence of the disease in hives is determined by behaviors such as the speed with which diseased bee larvae are detected and removed by bees. Rothenbuhler (1964a,b) demonstrated that these behaviors are genetically controlled. DeJong (1976) observed a similar scenario with behavior by *A. mellifera* and the fungal disease caused by *Ascosphaera apis* (chalkbrood). Although no genetics were evaluated, Inglis *et al.* (1996), Villani *et al.* (2002), and Thompson and Brandenburg (2005) observed behaviors by insects that allowed individuals to reduce or prevent infection by fungi. These insect behaviors included thermoregulation to change body temperature as well as movements and tunneling behavior to avoid contact with the pathogen.

Many isolates of the bacterium, *Bacillus thuringiensis*, produce a crystalline protein that is toxic to insects that ingest it (Nelson and Alves, Chapter 4). In the midgut, the crystals are solubilized and digested into the delta-endotoxin, which perforates the midgut membrane, killing the insect. Although there are multiple means of resistance to Bt, the most common mechanism involves some modification of the binding site in the midgut (Chapter 4).

Because infection or parasitism by the bacterium is not the cause of death, insect resistance to *B. thuringiensis* is discussed in more detail by Spencer *et al.* (Chapter 7) and Onstad and Knolhoff (Chapter 9) than in this chapter. These proteins are commonly used as insecticides applied to foliage or expressed in transgenic insecticidal crops. Each crystalline protein from *B. thuringiensis* tends to have a limited range of species that it harms.

Several insect species have evolved resistance to insecticidal sprays consisting of proteins from *B. thuringiensis* (McGaughey and Whalon, 1992). Several insect species have been selected for resistance in the laboratory (Tabashnik, 1994). In addition, two lepidopteran pests of crucifers have evolved resistance to applications of *B. thuringiensis*: greenhouse populations of *Trichoplusia ni* (Janmaat and Myers, 2003) and field populations of *Plutella xylostella* (Tabashnik *et al.*, 1990). *P. xylostella* evolved resistance on farms continuously growing watercress (*Nasturtium offinicale*) under "organic" conditions after local populations received over 50 treatments of *B. thuringiensis* toxin during at least 50 pest generations in a five-year period (Tabashnik *et al.*, 1990).

Cydia pomonella Resistance to a Granulovirus

Chemical insecticides, pheromones for mating disruption, and viral insecticides are used to control *Cydia pomonella*, a major orchard pest found in most temperate regions of the world. Eberle and Jehle (2006) describe the first documented case of field resistance by an insect to a commercially applied baculovirus. The *C. pomonella granulovirus* (CpGV) is a species-specific and extremely virulent pathogen (Lacey *et al.*, 2008).

Several commercial CpGV products have been used for microbial control on 100,000 ha of apple orchards (*Malus domestica*) in Europe since the mid-1990s (Eberle and Jehle, 2006). However, by 2004, reduced suscepti- bility to CpGV was observed in several populations inhabiting organic apple orchards where control failed despite intensive CpGV application. Since then, the resistance has become common in several countries in Europe (Schmitt *et al.*, 2013, Zichova *et al.*, 2013). In a monitoring program, Schulze-Bopp and Jehle (2013) found resistance in 7 out of 10 different orchards in Germany, Austria, Switzerland, Italy, and the Netherlands.

Several studies have investigated the genetics of the resistance. Asser- Kaiser *et al.* (2007) determined that a strain of *C. pomonella* with 1000- fold resistance carries a dominant, sex-linked major resistance gene. Berling *et al.* (2013) found similar characteristics in a different resistant strain of the insect that exhibited 7000-fold more resistance than a sus- ceptible strain. However, Berling *et al.* (2013) suggest that other genes may be involved. Zichova *et al.* (2013) also found a resistant strain with a sex-linked gene expressing dominant resistance.

Resistance to CpGV seems to be stable and lacking in fitness costs. Eberle and Jehle (2006) originally postulated that this resistance is stable after selection by virus is removed from a colony. Undorf-Spahn *et al.* (2012) tested this hypothesis by observing a colony that is 1000 times less susceptible than a nonselected strain for more than 60 genera- tions in the laboratory. Without virus selection pressure, the high level of resistance remained stable for more than 30 generations and declined only by a factor of 10 after 60 generations. When cohorts were exposed to CpGV again after more than 30 generations, the resistance level increased to a factor of >1,000,000 compared to a susceptible colony. Undorf- Spahn *et al.* (2012) observed no fitness costs under laboratory conditions in terms of fecundity and fertility. Asser-Kaiser *et al.* (2011) experimen- tally determined that CpGV can invade but not replicate in the cells of resistant *C. pomonella* larvae. They also demonstrated that a modified peri- trophic membrane, a modified midgut receptor, or a change in the innate immune response could not be possible resistance mechanisms.

Gund *et al.* (2012) investigated the population-genetic structure of resistant and susceptible *C. pomonella* populations using microsatellite and mitochondrial DNA markers. Their structure analysis found neither population-level nor individual correlations associated with CpGV resis- tance. Gund *et al.* (2012) concluded that they could not determine whether there has been only one or multiple independent origins of resistance in Europe.

Resistance management for *C. pomonella* has mostly emphasized finding new isolates of CpGV. Eberle *et al.* (2008), Berling *et al.* (2009), and Zichova *et al.* (2013) demonstrated that larvae of their resistant strains of *C. pomonella* can be killed by newer isolates and formulations of CpGV. Now scientists must develop multiple types of microbial insecticides or other control tactics that can be used in combination with CpGV. Eberle *et al.* (2008) concluded, based on bioassays, that CpGV-resistant *C. pomonella* would not likely be cross-resistant to toxins from *B. thuringiensis*.

RESISTANCE TO PARASITIC INVERTEBRATES

Over the past ten years, a variety of studies have focused on the genetics and ecology of resistance by invertebrates to their multicellular (invertebrate) parasites (Carton *et al.*, 2005). Much can be learned from observations on host species that are not arthropods. This section discusses some of the important work on insect resistance to their invertebrate parasites. Although most studies have been performed under laboratory conditions, valuable conceptual lessons can still be learned from these investigations. A good conceptual overview was provided by Kraaijeveld (2004), who attempted to explain why resistance to parasitoids rarely, if ever, evolves under biological control.

Nagel *et al.* (2010) studied mite parasitism of damselflies over seven seasons in their natural environment. The mechanism of resistance is melanotic encapsulation of mite feeding tubes. Although parasite prevalence in newly emerged damselflies was greater than 77% each year, no resistance to parasitism was observed in the early years of study. Nagel *et al.* (2010) observed resistance developing during the year in which there was an unusual increase in the number of mites on newly emerged damselflies. Resistance was correlated with mite prevalence and intensity throughout the seven-year study. However, the percentage of resistant hosts only ranged from 0 to 13%. Resistance was not correlated with air temperature or with timing of damselfly emergence.

In a population of *Anopheles* mosquitoes, Woodard and Fukuda (1977) found that larvae from a strain selected in the laboratory for resistance to nematodes were much more active and defended themselves against attacking nematodes. Thus, behavioral resistance is also a possibility when insects evolve to pressure from invertebrate parasites.

Herzog *et al.* (2007) studied genetically diverse clones of *Myzus persicae* reared together either in the presence of two species of parasitoid wasps or without parasitoids. *M. persicae* was allowed to evolve for several generations in the two systems. In the absence of parasitoids, clones with high rates of population growth became more prevalent than other clones. In the other system, the parasitoid *Diaeretiella rapae* had little effect on host densities but preferred to attack a highly susceptible clone and not an entirely resistant clone. This preference changed the outcome of competition among the clones. *Aphidius colemani* significantly reduced host densities and significantly changed host clonal frequencies. The most resistant clone, which was not dominant without parasitoids, became totally dominant. No fitness costs for resistance to parasitism were found in a second study (Von Burg *et al.*, 2008).

Drosophila melanogaster Resistance to a Parasitoid

Drosophila is parasitized by several parasitoids, especially those in the genera *Asobara* and *Leptopilina* (Kraaijeveld and Godfray, 1999). In Europe, *A. tabida* attacks larvae of several species of *Drosophila*. Parasitism by *A. tabida* is lowest in northern Europe and highest in southern Europe, where it mostly attacks *D. melanogaster*. *D. melanogaster* resistance to parasitism is highest in central and southern Europe and lower elsewhere. Larvae of *D. melanogaster* defend themselves against parasitoids by encapsulating the parasitoid egg. Kraaijeveld and Godfray (1997) developed a strain of *D. melanogaster* resistant to *A. tabida*. However, this strain had several fitness costs in unparasitized larvae. Resistance by *D. melanogaster* results in a reduction in adult size and fecundity. Another fitness cost is the reduction in larval competitive ability that is important only when food is limited (Kraaijeveld *et al.*, 2002). However, selection for higher competitive ability under crowded conditions for eight generations simultaneously selected for increased resistance to *A. tabida* parasitism. These changes were associated with elevated levels of hemocytes in young larvae of *D. melanogaster* (Sanders *et al.*, 2005).

Kraaijeveld *et al.* (2009) determined that critical aspects of the evolutionary interaction between *D. melanogaster* and *A. tabida* are the population-dependent and density-dependent fitness costs for both resistance and parasitic virulence. These fitness costs maintain significant levels of genetic variation in the two species. Kraaijeveld *et al.* (2009) suggest that the genes conferring the actual immune response by *D. melanogaster* against the parasitoids are different from the genes involved in the evolution of resistance.

Green *et al.* (2000) designed an experiment to study changes in resistance in replicate populations of *D. melanogaster* and *A. tabida* reared together for 10 generations of the host (five parasitoid generations). The experiment had three treatments: no parasitoids, outbred parasitoids, or partially inbred parasitoids. Host resistance increased similarly in both treatments that included parasitoids. Green *et al.* (2000) found no evidence for unique local adaptation, as hosts from the different replicates of the treatment with partially inbred parasitoids responded in the same way as hosts in all other replicates. The authors did not observe evolution of behavioral traits in the host that could reduce attacks nor any change in parasitoid virulence.

Fellowes *et al.* (1999) showed that there is cross-resistance in *D. melanogaster* for defense against larval parasitoids. Larvae selected for resistance to either *A. tabida* or *Leptopilina boulardi* were cross-resistant to attack by parasitoids in the other genus. Fellowes *et al.* (1998) demonstrated that there is negative cross-resistance in *D. melanogaster* that are resistant to larval parasitism by *A. tabida*. Those larvae that successfully resisted parasitism by encapsulating the parasitoid's egg became more likely to be parasitized as pupae by a different species of parasitoid.

CONCLUSIONS

Clearly, arthropods can evolve resistance to their parasitic natural enemies under conditions controlled by humans. However, under more natural conditions, parasitism is not consistently high enough over both space and time to select for resistance. Parasitism levels fluctuate with host density (Onstad, 1993). Encounters between parasites and hosts are limited in heterogeneous environments where hosts can find refuges (Onstad and Carruthers, 1990). In general, pathogens and parasitoids are limited in their efficiency to attack most individuals in a host population under natural conditions (Onstad and McManus, 1996). To delay or prevent evolution of resistance to parasitic biological-control agents, entomologists and practitioners of integrated pest management (IPM) should avoid trying to make microbial insecticides and inundative releases of parasitoids 100% effective in every generation of a pest. The more effective biological control is, the less frequently it should be used. Or it must be integrated with different IPM tactics to limit the ability of rare resistant individuals to increase in numbers.

Thus, the threat of insect resistance to biological control by parasites means that chemical control, biological control, cultural control, and host-plant resistance must be integrated to preserve all tactics. As the cases of resistance to CpGV in *C. pomonella* and resistance to *B. thuringiensis* by *P. xylostella* demonstrate, even farmers using "organic" agriculture can cause pests to evolve resistance. Thus, the lessons of IPM and its philosophy of integration can be lost to farmers who believe that a natural toxin is different from a human synthesized compound with regard to IRM. Simplicity and convenience usually trump integration and complexity with negative consequences for long-term IPM.

Furthermore, all entomologists (not just those focusing on chemical control) must understand the concepts and practices of IRM. Scientists studying and promoting biological control must learn to recognize mechanisms of resistance that can be toxicological, behavioral, or even due to changes in maturation. This will be especially important as microbial control becomes more common in the future.

REFERENCES

Asser-Kaiser, S., Fritsch, E., Undorf-Spahn, K., Kienzle, J., Eberle, K.E., Gund, N.A., et al., 2007. Rapid emergence of baculovirus resistance in codling moth due to dominant, sex-linked inheritance. Science. 317, 1916–1918.

Asser-Kaiser, S., Radtke, P., El-Salamouny, S., Winstanley, D., Jehle, J.A., 2011. Baculovirus resistance in codling moth (*Cydia pomonella* L.) caused by early block of virus replication. Virology. 410, 360–367.

Baer, B., Schmid-Hempel, P., 2001. Unexpected consequences of polyandry for parasitism and fitness in the bumblebee *B. terrestris*. Evolution. 55, 1639–1643.

Berling, M., Rey, J.-B., Ondet, S.-J., Tallot, Y., Soubabere, O., Bonhomme, A., et al., 2009. Field trials of CpGV virus isolates overcoming resistance to CpGV-M. Virol. Sin. 24, 470–477.

Berling, M., Sauphanor, B., Bonhomme, A., Siegwart, M., Lopez-Ferber, M., 2013. A single sex-linked dominant gene does not fully explain the codling moth's resistance to granulovirus. Pest Manag. Sci. [In press].

Bindu, T.N., Balakrishnan, P., Sudheendrakumar, V.V., Sajeev, T.V., 2012. Density-dependent polyphenism and baculovirus resistance in teak defoliator, *Hyblaea puera* (Cramer). Ecol. Entomol. 37, 536–540.

Braxton, S.M., Onstad, D.W., Dockter, D.E., Giordano, R., Larsson, R., Humber, R.A., 2003. Description and analysis of two Internet based pathogen databases: EDWIP and VIDIL. J. Invert. Pathol. 83, 185–195.

Briese, D.T., 1981. Resistance of insect species to microbial pathogens. In: Davidson, E. W. (Ed.), Pathogenesis of Invertebrate Microbial Diseases. Allanheld, Osmun & Co. Publishers, New Jersey, p. 562, Chapter 18.

Briese, D.T., Mende, H.A., 1983. Selection for increased resistance to a granulosis virus in the potato moth, *Phthorimaea operculella* (Zeller) (Lepidoptera: Gelechiidae). Bull. Ent. Res. 73, 1–9.

Butt, T.M., Jackson, C.W., Magan, N., 2001. Fungi as biocontrol agents. Progress, Problems and Potential. CABI Publishing, New York.

Calleri II, D.V., McGrail Reid, E., Rosengaus, R.B., Vargo, E.L., Traniello, J.F.A., 2006. Inbreeding and disease resistance in a social insect: effects of heterozygosity on immunocompetence in the termite *Zootermopsis angusticollis*. Proc. R. Soc. Lond. B. 273, 2633–2640.

Carton, Y., Nappi, A.J., Poirie, M., 2005. Genetics of anti-parasite resistance in invertebrates. Dev. Comp. Immunol. 29, 9–32.

DeJong, D., 1976. Experimental enhancement of chalkbrood infections. Bee World. 57, 114–115.

Dubovskiy, I.M., Whitten, M.M.A., Yaroslavtseva, O.N., Greig, C., Kryukov, V.Y., Grizanova, E.V., et al., 2013. Can insects develop resistance to insect pathogenic Fungi? PLoS One. Available from: http://dx.doi.org/10.1371/journal.pone.0060248VL 8, AR e60248.

Eberle, K.E., Jehle, J.A., 2006. Field resistance of codling moth against *Cydia pomonella granulovirus* (CpGV) is autosomal and incompletely dominant inherited. J. Invert. Pathol. 93, 201–206.

Eberle, K.E., Asser-Kaiser, S., Sayed, S.M., Nguyen, H.T., Jehle, J.A., 2008. Overcoming the resistance of codling moth against conventional *Cydia pomonella granulovirus* (CpGV-M) by a new isolate CpGV-I12. J. Invert. Pathol. 98, 293–298.

Fellowes, M.D.E., Masnatta, P., Kraaijeveld, A.R., Godfray, H.C.J., 1998. Pupal parasitoid attack influences the relative fitness of *Drosophila* that have encapsulated larval parasitoids. Ecol. Entomol. 23, 281–284.

Fellowes, M.D.E., Kraaijeveld, A.R., Godfray, H.C.J., 1999. Cross-resistance following artificial selection for increased defense against parasitoids in Drosophila melanogaster. Evolution. 53, 966–972.

Franklin, M.T., Ritland, C.E., Myers, J.H., Cory, J.S., 2012. Multiple mating and family structure of the western tent caterpillar, *Malacosoma californicum pluviale*: impact on disease resistance. PLoS One. 7, e37472. Available from: http://dx.doi.org/10.1371/journal.pone.0037472.

Fuxa, J.R., 2004. Ecology of insect nucleopolyhedroviruses. Agric. Ecosys. & Environ. 103, 27–43.

Green, D.M., Kraaijeveld, A.R., Godfray, H.C.J., 2000. Evolutionary interactions between Drosophila melanogaster and its parasitoid Asobara tabida. Heredity. 85, 450–458.

Gund, N.A., Wagner, A., Timm, A.E., Schulze-Bopp, S., Jehle, J.A., Johannesen, J., et al., 2012. Genetic analysis of *Cydia pomonella* (Lepidoptera: Tortricidae) populations with different levels of sensitivity towards the *Cydia pomonella granulovirus* (CpGV). Genetica. 140, 235–247.

Herzog, J., Muller, C.B., Vorburger, C., 2007. Strong parasitoid-mediated selection in experimental populations of aphids. Biol. Lett. 3, 667–668.

Humber, R.A., 2008. Evolution of entomopathogenicity in fungi. J. Invert. Pathol. 98, 262–266.

Ignoffo, C.M., Huettel, M.D., McIntosh, A.H., Garcia, C., Wilkening, P., 1985. Genetics of resistance of *Heliothis subflexa* (Lepidoptera: Noctuidae) to baculovirus *Heliothis*. Ann. Entomol. Soc. Amer. 78, 468–473.

Inglis, G.D., Johnson, D.L., Goettel, M.S., 1996. Effects of temperature and thermo-regulation on mycosis by *Beauveria bassiana* in grasshoppers. Biol. control. 7, 131–139.

Janmaat, A.F., Myers, J.H., 2003. Rapid evolution and the cost of resistance to *Bacillus thuringiensis* in greenhouse populations of cabbage loopers, *Trichoplusia ni*. Proc. Roy. Soc. Lond. B. 270, 2263–2270.

Kraaijeveld, A.R., 2004. Experimental evolution in host-parasitoid interactions. In: Ehler, L.E., Sforza, R., Mateille, T. (Eds.), Genetics, Evolution and Biological Control. CAB International Publishing, Oxon, UK, Chapter 8.

Kraaijeveld, A.R., Godfray, H.C.J., 1997. Trade-off between parasitoid resistance and lar-val competitive ability in Drosophila melanogaster. Nature. 389, 278–280.

Kraaijeveld, A.R., Godfray, H.C.J., 1999. Geographic patterns in the evolution of resistance and virulence in *Drosophila* and its parasitoids. Am. Nat. 153, 61–74.

Kraaijeveld, A.R., Ferrari, J., Godfray, H.C.J., 2002. Costs of resistance in insect-parasite and insect-parasitoid interactions. Parasitology. 125, S71–S82.

Kraaijeveld, A.R., Godfray, H.C.J., Prevost, G., 2009. Evolution of host resistance and parasitoid counter-resistance. Adv. Parasitol. 70, 257–280.

Kuris, A.M., Lafferty, K.D., 2000. Parasite–host modelling meets reality: adaptive peaks and their ecological attributes. In: Poulin, R., Morand, S., Skorping, A. (Eds.), Evolutionary Biology of Host–Parasite Relationships: Theory Meets Reality. Elsevier Science, Amsterdam, pp. 9–26.

Lacey, L.A., Kaya, H.K., 2000. Field Manual of Techniques in Invertebrate Pathology: Application and Evaluation of Pathogens for Control of Insects and Other Invertebrate Pests. Kluwer Academic Publishers, Dordrecht, p. 911.

Lacey, L.A., Thomson, D., Vincent, C., Arthurs, S.P., 2008. Codling moth granulovirus: a comprehensive review. Biocontrol Sci Technol. 18, 639–663.

McGaughey, W.H., Whalon, M.E., 1992. Managing insect resistance to *Bacillus thuringien-sis* toxins. Science. 258, 1451–1455.

Milner, R.J., 1982. On the occurrence of pea aphids, *Acyrthosiphon pisum*, resistant to iso-lates of the fungal pathogen *Erynia neoaphidis*. Entomol. Exp. Appl. 32, 23–27.

Moscardi, F., 1999. Assessment of the application of baculoviruses for control of Lepidoptera. Annu. Rev. Entomol. 44, 257–289.

Nagel, L., Robb, T., Forbes, M.R., 2010. Inter-annual variation in prevalence and inten-sity of mite parasitism relates to appearance and expression of damselfly resistance. BMC Ecol. 10. 10.1186/1472-6785-10-5.

Oliver, K.M., Moran, N.A., Hunter, M.S., 2005. Variation in resistance to parasitism in aphids is due to symbionts not host genotype. Proc. Nat. Acad. Sci. USA. 102, 12795–12800.

Onstad, D.W., 1993. Thresholds and density dependence: the roles of pathogen and insect densities in disease dynamics. Biol Control. 3, 353–356.

Onstad, D.W., Carruthers., R.I., 1990. Epizootiological models of insect diseases. Annu. Rev. Entomol. 35, 399–419.

Onstad, D.W., McManus., M.L., 1996. Risks of host-range expansion by insect-parasitic biocontrol agents. Bio Sci. 46, 430–435.

Poulin, R., 2011. The Many Roads to Parasitism: A Tale of Convergence. Adv. Parasitol. 74, 1−40.

Rosengaus, R.B., Traniello, J.F.A., 2001. Disease susceptibility and the adaptive nature of colony demography in the dampwood termite *Zootermipsis angusticollis*. Behav. Ecol. Sociobiol. 50, 546−556.

Rothenbuhler, W.C., 1964a. Behaviour genetics of nest-cleaning in honeybees. I. Responses of four inbred lines to disease-killed larvae. Anim. Behav. 12, 578−583.

Rothenbuhler, W.C., 1964b. Behaviour genetics of nest-cleaning in honeybees. IV. Responses of F_1 and backcross generations to disease-killed brood. Am. Zool. 4, 111−123.

Sanders, A.E., Scarborough, C., Layen, S.J., Kraaijeveld, A.R., Godfray, H.C.J., 2005. Evolutionary change in parasitoid resistance under crowded conditions in *Drosophila melanogaster*. Evolution. 59, 1292−1299.

Schmid-Hempel, P., 1994. Infection and colony variability in social insects. Phil. Trans. R. Soc. Lond. B. 346, 313−321.

Schmitt, A., Bisutti, I.L., Ladurner, E., Benuzzi, M., Sauphanor, B., Kienzle, J., et al., 2013. The occurrence and distribution of resistance of codling moth to Cydia pomonella granulovirus in Europe. J. Appl. Entomol.

Schulze-Bopp, S., Jehle, J.A., 2013. Development of a direct test of baculovirus resistance in wild codling moth populations. J. Appl. Entomol. 137, 153−160.

Shelton, A.M., Roush, R.T., Wang, P., Zhao, J.Z., 2007. Resistance to insect pathogens and strategies to manage resistance: an update. In: Lacey, L.A., Kaya., H.K. (Eds.), Field Manual of Techniques in Invertebrate Pathology: Application and Evaluation of Pathogens for Control of Insects and Other Invertebrate Pests. Kluwer Academic Publishers, Dordrecht, pp. 793−811, p. 911.

Stephen, W.P., Fichter, B.L., 1990. Chalkbrood (*Ascosphaera aggregata*) resistance in the leaf-cutting bee (*Megachile rotundata*) 1. Challenge of selected lines. Apidologie. 21, 209−219.

Stireman, J.O., O'Hara, J.E., Wood, D.M., 2006. Tachinidae: Evolution, behavior, and ecology. Ann. Rev. Entomol. 51, 525−555.

Tabashnik, B.E., 1994. Evolution of resistance to *Bacillus thuringiensis*. Annu. Rev. Entomol. 39, 47−79.

Tabashnik, B.E., Cushing, N.L., Finson, N., Johnson, M.W., 1990. Field development of resistance to *Bacillus thuringiensis* in diamondback moth (Lepidoptera: Plutellidae). J. Econ. Entomol. 83, 1671−1676.

Tarpy, D.R., 2003. Genetic diversity within honeybee colonies prevents severe infections and promotes colony growth. Proc. R. Soc. Lond. B. 270, 99−103.

Thompson, S.R., Brandenburg, R.L., 2005. Tunneling responses of mole crickets (Orthoptera: Gryllotalpidae) to the entomopathogenic fungus, Beauveria bassiana. Environ. Entomol. 34, 140−147.

Undorf-Spahn, K., Fritsch, E., Huber, J., Kienzle, J., Zebitz, C.P.W., Jehle, J.A., 2012. High stability and no fitness costs of the resistance of codling moth to Cydia pomonella granulovirus (CpGV-M). J Invert Pathol. 111, 136−142.

Vijendravarma, R.K., Kraaijeveld, A.R., Godfray, H.C.J., 2009. Experimental evolution shows Drosophila melanogaster resistance to a microsporidian pathogen has fitness costs. Evolution. 63, 104−114.

Villani, M.G., Allee, L.L., Preston-Wilsey, L., Consolie, N., Xia, Y., Brandenburg, R.L., 2002. Use of radiography and tunnel castings for observing mole cricket (Orthoptera: Gryllotalpidae) behavior in soil. Am. Entomol. 48, 42−50.

Von Burg, S., Ferrari, J., Muller, C.B., Vorburger, C., 2008. Genetic variation and covariation of susceptibility to parasitoids in the aphid Myzus persicae: no evidence for trade-offs. Proc. Royal Soc. B: Biol Sci. 275, 1089−1094.

Whitehorn, P.R., Tinsley, M.C., Brown, M.J.F., Darvill, B., Goulson, D., 2011. Genetic diversity, parasite prevalence and immunity in wild bumblebees. Proc. R. Soc. Lond. B. 278, 1195–1202.

Woodard, D.B., Fukuda, T., 1977. Laboratory resistance of the mosquito Anopheles quadrimaculatus to the mermithid nematode Diximermis peterseni. Mosq. News. 37, 192–195.

Zichova, T., Stara, J., Kundu, J.K., Eberle, K.E., Jehle, J.A., 2013. Resistance to Cydia pomonella granulovirus follows a geographically widely distributed inheritance type within Europe. BioControl. pp. 1–10.

FURTHER READING

Pereira, R.M., 2003. Areawide suppression of fire ant populations in pastures: Project update. J. Agric. Urban Entomol. 20, 123–130.

CHAPTER 9

Arthropod Resistance to Crops

D.W. Onstad[1] and Lisa Knolhoff[2]

[1]DuPont Pioneer, Wilmington, DE
[2]DuPont Pioneer, Johnston, IA

Chapter Outline

Humans have been managing crops for thousands of years, often choosing plants that are more or less resistant to pests. And for millions of years, arthropods have been evolving mechanisms to counteract defenses in plants (Berenbaum, 2001; Gassmann *et al.*, 2009). This chapter describes attempts by agriculturalists and entomologists to manage the resistance by host plants to deal with the real or potential evolution of counteracting resistance by arthropods.

Host–plant resistance (HPR) is one of the tactical pillars of IPM (Smith, 2005). The standard definition of HPR offered by Painter (1951) is "the relative amount of heritable qualities possessed by the plant which influence the ultimate degree of damage done by the insect in the field." Panda and Khush (1995) further describe it as "any degree of host reaction less than full immunity." Host-plant resistance under genetic control can be divided into three categories: tolerance, antixenosis, and antibiosis (Panda and Khush, 1995). Plant tolerance reduces the amount of yield-

Insect Resistance Management
DOI: http://dx.doi.org/10.1016/B978-0-12-396955-2.00009-6

loss or loss of plant fitness per unit of pest injury. Antixenosis is a mechanism by which the plant deters herbivores or reduces their colonization by affecting their behavior (e.g., leaf trichomes). Antibiosis either kills the herbivore or negatively affects its development after feeding. Note that in some cases, it may be hard to differentiate between antixenosis and antibiosis if the plant exhibits both types of effects (Panda and Khush, 1995). To avoid confusion in the use of the term *resistance*, throughout this chapter we will use the abbreviation HPR for resistance by plants to insects and reserve the term *resistance* for the case of resistance by insects to crops.

Recently, Stout (2012) proposed a new categorization for types of HPR. He believes that the new description will allow more kinds of HPR to be incorporated into the categories. The new categories will also facilitate better linkages between basic and applied-science approaches to herbivore-plant interactions. Because antixenosis and antibiosis are often difficult to separate and therefore to avoid prematurely implying a mechanism, a main change is the combining of these terms into one category called "resistance." The resulting dichotomous division defines resistance as traits that reduce injury to plants and maintains the traditional definition of tolerance. In addition, Stout (2012) provided subcategories for resistance: constitutive or inducible and direct or indirect. Constitutive traits are expressed throughout a plant all the time; inducible traits are expressed in response to injury. Resistance traits can directly block or harm herbivores, or they can indirectly affect them through the plant's influences on the herbivore's natural enemies. In this proposed scheme, arthropods can evolve resistance to Stout's "resistance" HPR traits, but not to tolerance traits.

As with most approaches to integrated pest management (IPM), the emphasis in HPR work has been on incorporating a mechanism that, by itself or in conjunction with other IPM tactics, can either decrease the pest population, reduce damage to the crop, or both. Long-term durability of a cultivar is only one of many characteristics of the crop that must be considered, weighed, and balanced by plant breeders in a breeding program (Kennedy et al., 1987). Gould (1983) provides a valuable discussion of the evolutionary biology of plant–herbivore interactions. Kennedy et al. (1987) promote the use of landscape ecology and modeling of IPM and insect resistance management (IRM) during early stages of breeding programs to make breeding decisions more efficient.

From an insect's perspective, there is no difference between crops that have been bred using a traditional approach and those that have been genetically engineered. Evolution of resistance to either technology can occur. The extensive planting of transgenic insecticidal crops in agro-ecosystems does force us to deal with many issues (Onstad and Carrière, Chapter 10), but for now we will focus on the similarities of IRM for the two types of HPR crops.

In the following sections, we describe a variety of case studies concerning the evolution and management of insect resistance to HPR. Each section is labeled according to the arthropod pest that has evolved resistance to the crop. First, we discuss crops developed using traditional methods. Then we describe case studies concerning transgenic insecticidal crops. In the discussion, we address some lessons that can be learned from resistance to crops.

TRADITIONAL CROPS
Liriomyza trifolii

Hawthorne (1998) used two simple models to predict the evolution of resistance by the leafminer *Liriomyza trifolii* (Diptera: Agromyzidae) to a resistant cultivar of chrysanthemum, *Dendranthema grandiflora*. For each insect generation, each model calculates a change in insect performance based on selection intensity, S, and heritability, h^2 (Response = Sh^2). In the first model, heritability was held constant throughout the 10-generation simulation. In the second model, heritability declined due to the effects of linkage disequilibrium. For three types of leafminer populations that differed in their history of host-plant use, Hawthorne (1998) compared model predictions of larval survivorship against data from a selection experiment performed on the same leafminer populations.

In general, the predicted evolutionary trajectories of the leafminer populations on the resistant host plant were very similar to the trajectories observed in the selection experiment (Hawthorne, 1998). Two of the populations had very good match between the predicted and the observed trajectories. The third simulated population evolved too quickly to the resistant crop, due in part to an overestimate of its genetic variance (heritability). Simulations that accounted for reduced heritability, because of factors such as linkage disequilibrium, produced better predictions of all three observed patterns (Hawthorne, 1998). Hawthorne predicted that a resistant cultivar of chrysanthemum would not remain resistant to leafminers in Florida greenhouses. Hawthorne's results indicate that laboratory studies of selection intensity and genetic variance can be used to make rational IRM decisions regarding the commercialization of crops with HPR.

Tetranychus Urticae

The two-spotted spider mite, *Tetranychus urticae* (Acari: Tetranychidae), is a generalist herbivore that feeds on many crop and ornamental plants. Hot, dry weather is conducive to spider mite outbreaks. Damage can be seen as chlorosis of the leaves where the mites have been feeding. Its short life cycle and high reproductive potential predispose this mite to evolving resistance to many chemical control methods, so some growers may opt to use HPR plants. The fact that these mites are polyphagous has many implications for devising a resistance management strategy with HPR hosts.

One host of *T. urticae* is cucumber (*Cucumis sativus*). Gould (1978a) found that adaptation to HPR cultivars of cucumber expressing antibiosis could occur in as little as nine generations. Antixenosis is not a factor in HPR because resistant and susceptible cultivars were equally attractive (Gould, 1979).

Resistance to HPR cucumbers promoted resistance by *T. urticae* to other plant or insecticidal compounds. Once mites were adapted to an HPR cultivar of cucumber, they were predisposed to utilize tobacco (*Nicotiana tabacum*) and potato (*Solanum tuberosum*) as hosts (Gould, 1979), which are both taxonomically and chemically unrelated to cucumbers. In addition, Gould *et al.* (1982) found that mites adapted to HPR cucumber had significantly higher survivorship when exposed to three organophosphate insecticides. The reverse effect, however, was not observed; mites resistant to several insecticides did not have higher survivorship on resistant cucumber varieties than the susceptible mites did.

While adaptation to cucumbers seems to confer advantages with respect to other hosts or chemical controls, there may be either no effect or even a cost associated with adaptation to another host. Fry (1992) found that adaptation to tomato (*Solanum lycopersicum*) did not significantly increase or decrease ability to survive on tobacco and cucumber. Fry (1990) reported no difference in survival or fecundity on lima bean, a highly preferred host, when comparing bean- and tomato-adapted mites. However, lines of mites that were originally adapted to cucumber and tomato gradually lost the ability to utilize these hosts after acclimation to an attractive host, such as lima bean (Gould, 1979; Fry, 1990; Agrawal, 2000). Reversion of resistance could indicate a possible fitness cost associated with it. Gould (1979) found a small but significant difference in fitness on the original lima bean host after adaptation to cucumber, but Fry (1990) found no difference in survival or fecundity on lima bean after adaptation to tomato.

Antixenosis as an HPR mechanism is likely to affect the evolution of resistance. One thing to consider is whether the pest will feed upon resistant cultivars or merely be repelled; resistance is thought to evolve more slowly if the pest simply avoids the resistant cultivar over the susceptible one (Cantelo and Sanford, 1984). If a more favorable alternative host is present and the pest can access it, this should weaken selection for resistant pests (Cantelo and Sanford, 1984). HPR of tomatoes and broccoli seems to be both behavioral and toxicological, in that mites tended to disperse from these plants and had high mortality on them (Fry, 1989). HPR may take place as an antixenotic mechanism because of the morphological features of these hosts: trichomes and wax, respectively (Fry, 1988, 1989). Fry (1989) reported that it took 21 weeks for mites to diverge in survival on broccoli and only 7 weeks for divergence on tomato. Certain morphological features may have a larger effect on the evolution of resistance.

Factors affecting economic decisions by the producer cannot be ignored. Environmental effects, such as amount of water or natural enemies, must be considered in a resistance management strategy because certain regions may experience climatic conditions for which HPR expression is compromised. Gould (1978b) found that mites that were not adapted to HPR cucumbers still destroyed susceptible varieties, but they did no noticeable damage to water-stressed seedlings. In contrast, the resistant mites destroyed HPR seedlings regardless of water stress, but did have lower survivorship on stressed plants (Gould, 1978b).

In determining the adaptation to HPR in the presence of natural enemies, one must ascertain whether those natural enemies will increase or decrease the fitness differential (Gould et al., 1991). Generally, adaptation to host plant resistance occurs more slowly with a combination of low HPR and natural enemies than a high level of HPR alone (Gould et al., 1991). Therefore, an IPM-like approach to resistance management could be beneficial.

Finally, one important aspect of IRM is the initial proportion of resistance alleles in a population. Resistance alleles are usually assumed to be rare because they seem to have some cost associated with them. Gould (1978a, 1979) found that the genetic variation in survivorship on cucumber cultivars was present within a small area, meaning that it is more likely that resistant individuals will encounter each other to mate. With respect to resistance management, Gould (1978a) highlighted the need to test multiple populations of the insect target and to look at population size, mobility, and whether there is mono- or polygenic inheritance of resistance. Fry (1988) found large differences in survivorship on tomato in populations of mites. The genetic variability with respect to resistance seems to be common in many populations.

Mayetiola Destructor

The Hessian fly, *Mayetiola destructor* (Diptera: Cecidomyiidae), is an important pest of wheat, *Triticum aestivum*. While it also feeds on barley, rye, and some grasses, wheat is the host of most economic interest (Painter, 1951). Injury is caused by larval feeding, resulting in stunted plants with weakened stems. A major cultural control of *M. destructor* is the planting of winter wheat late enough in the fall after most *M. destructor* will have been killed by the frost. Besides planting after the "fly-free" date, one of the main tactics for control of *M. destructor* is the use of antibiotic wheat varieties.

Thirty-two genes conferring wheat HPR to *M. destructor* have been identified (Sardesai et al., 2005), and all except one are either dominant or partially dominant (Formusoh et al., 1996). Approximately a third of the identified HPR genes have been deployed for commercial use (Williams et al., 2003). A gene-for-gene relationship exists between wheat antibiosis and *M. destructor* resistance (Hatchett and Gallun, 1970). Resistance by the insect (usually designated as *virulence*) is essentially the ability to overcome the HPR and to alter the plant's physiology as it would a susceptible plant (Stuart et al., 2012).

HPR resistance in wheat (for 3 R genes) exists both as a constitutive defense and induced response to *M. destructor* attack (Anderson and Harris, 2006), and for the 3 HPR genes that were examined, there appears to be no fitness cost associated with them (Anderson et al., 2011). However, infestation with a resistant (virulent) *M. destructor* strain on its respective HPR cultivar made those plants more susceptible to a subsequent larval attack with a strain that is considered avirulent (Baluch et al., 2012).

The extensive use of HPR in wheat has selected for resistant biotypes in the field (Ratcliffe *et al.*, 1994). Resistance to HPR wheat varieties is controlled by single-locus recessive alleles in *M. destructor* (Hatchett and Gallun, 1970; El Bouhssini *et al.*, 2001). Biotypes of *M. destructor* are identified by their ability to survive on antibiotic varieties of wheat (Gallun, 1977). The Great Plains (GP) biotype is susceptible to all resistant wheat varieties, and the others (A through O) have resistance to one or some combination of HPR genes in wheat.

Figure 9.1 shows the distribution of biotypes in regions of the United States based on data from Ratcliffe *et al.* (1994, 2000). Insects were collected during 1989–1992 (Figure 9.1a) and 1996–1999 (Figure 9.1b). Biotype L is resistant to all deployed wheat HPR genes, except the more

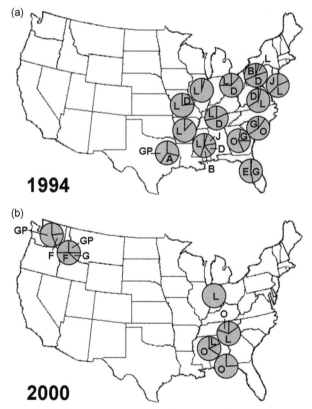

Figure 9.1 Distribution of *Mayetiola destructor* biotypes in (a) 1994 and (b) 2000. Unmarked portions of pie graphs refer to a mix of other biotypes that comprised 10% or less of the total biotype composition for a given area.

recently deployed H13 (Ratcliffe, 2000), and it is often the reference bio-type of breeding experiments for new sources of resistance in wheat. Biotype L is the predominant biotype in the eastern half of the United States above 33°N latitude, including the Midwest (Chen et al., 1990; Ratcliffe et al., 1994, 2000). Biotype O is the main biotype south of 33°N latitude in the eastern United States (Ratcliffe et al., 2000). Populations from the northwest are more variable, with frequencies of the susceptible GP biotype ranging from 25 to 75% and biotypes E, G, F, and O comprising the other sizable portions (Ratcliffe et al., 2000).

With an increasing number of HPR genes that are or will be deployed, the biotype designations can become unwieldy and labor-inten-sive, leading to classification by a "virulence test" (Chen et al., 2009, Cambron et al., 2010). Screening itself is similar; the difference is more in the classification, in that results are reported as a table of resistant propor-tions, rather than using the proportion of resistant individuals to assign a biotype. In a virulence survey of 20 populations over five states in the southeastern United States, Cambron et al. (2010) found homogeneity among M. destructor populations in their response to the 21 HPR genes in wheat. That southeastern populations of M. destructor are highly similar was reflected in a population genetics study using microsatellite markers (Morton et al., 2011).

Because there are numerous antibiosis genes for M. destructor in wheat, these genes can be deployed in a number of ways. Cox and Hatchett (1986) suggested that sequential release of HPR genes would be simple and effective, but Gould (1986b) recommended the use of wheat cultivars with pyramided HPR genes in addition to a 20% refuge of susceptible plants. Sequential release is the deployment of a single HPR gene until it loses efficacy, after which it is replaced by another single HPR gene. Pyramided HPR genes simply refer to multiple genes within a cultivar. Under optimal conditions, the pyramid plus refuge strategy was expected by Gould (1986b) to last at least 400 fly generations. This strategy is simi-lar to that recommended for transgenic insecticidal crops (see the section in this chapter, Transgenic Insecticidal Crops). While Foster et al. (1991) noted that a pyramid strategy would be quite useful, they discussed potential difficulties in executing it, particularly the time necessary to develop the pyramided cultivars. Because wheat cultivars are generally only grown for up to 10 years, the time required to develop a pyramid may preclude its use (Foster et al., 1991). The authors suggested that if a pyramid is developed, it should be incorporated into newer cultivars with higher yields. HPR gene pyramiding in wheat has not yet been accomplished (Harris *et al.*, 2003).

The utility of a refuge is dependent on the dispersal of adult females seeking oviposition sites. The short adult life span (a few days at most) of *M. destructor* makes dispersal and movement of mated females easier to quantify. Zeiss *et al.* (1993) concluded that any interactions with grasses outside of the tribe Triticeae would be negligible. Indeed, Withers *et al.* (1997) observed the rates of dispersal of mated females in different hosts, and they predicted more than a doubling of dispersal in plots with non-hosts compared to plots with hosts. However, during oviposition, female *M. destructor* do not differentiate by tillage (Del Conte *et al.*, 2005) or between resistant and susceptible cultivars (Harris *et al.*, 2001). Note that oviposition preference does not necessarily correlate with the success of offspring. For example, while triticale (Durum wheat x rye hybrid) experienced a high rate of oviposition, there was low survival of larvae (Harris *et al.*, 2001). Additionally, while females do not have a preference for plants with conspecific eggs, they do avoid plants with already infested larvae at a rate proportional to the larval density (Kanno and Harris, 2002). This is probably due to chemical cues from the stressed plants rather than the established larvae themselves (Harris *et al.*, 2006). Females also prefer to oviposit on taller rather than shorter plants (Kanno and Harris, 2002). These aspects of female *M. destructor* behavior allow predictions for a refuge strategy.

The potential of a pest to develop resistance to a control method should not be underestimated. Ratcliffe *et al.* (1994) found some level of resistance to almost all HPR genes in all of the field-collected *M. destructor* populations that they studied in the laboratory. This may mean that there is a relatively high initial frequency of resistance alleles (Gould, 1983). Another factor that contributes to rapid adaptation to antibiotic cultivars of wheat is paternal gene loss in this insect. During spermatogenesis, the paternal chromosomes are eliminated, meaning that males transmit only maternal genes (Gallun and Hatchett, 1969). This reduces the time for *M. destructor* to evolve resistance to antibiotic cultivars (Gould, 1986b). The *M. destructor* genome consists of two pairs of autosomes and two pairs of sex chromosomes (Stuart and Hatchett, 1988a,b), meaning that linkage of resistance genes may likely occur.

Finally, it is important to consider IRM for *M. destructor* to be part of regionally based IPM. Certainly, economics influences the decisions made by farmers. In an economic analysis of management of *M. destructor* in the southeastern United States, Buntin *et al.* (1992) concluded that when infestations were light, net returns were greater with the use of susceptible cultivars (with or without insecticide treatment). However, when infestations were high, it was desirable to instead plant resistant cultivars. Because it is difficult to predict *M. destructor* outbreaks, Buntin *et al.* (1992) and Buntin (1999) advocate the use of a resistant cultivar with high yield or a susceptible cultivar with preemptive insecticide treatment. In much of the United States, winter wheat planted after the "fly-free" date (determined by latitude) prevents most damage. In the far South, delayed planting is of limited use, as the winter is not sufficiently cold to kill all insects (Buntin *et al.*, 1990). Climate affects IRM in another way; certain HPR genes are less effective at high temperatures (Sosa, 1979; Ratanatham and Gallun, 1986; El Bouhssini *et al.*, 1999).

Sitodiplosis Mosellana

The wheat midge, *Sitodiplosis mosellana* (Diptera: Cecidomyiidae), is another important gall midge attacking wheat (*Triticum aestivum*) around the world. Females oviposit eggs on the wheat head directly before anthesis. After hatching, the larvae migrate to and feed on the developing seeds, causing direct economic damage. Recently, an HPR gene called *Sm*1 has been successfully incorporated into commercial cultivars of wheat in Canada. This gene controls an inducible hypersensitive reaction in the seed surface that causes antibiosis against young larvae (Smith *et al.*, 2004). Some cultivars also express oviposition deterrence against the wheat midge, but this HPR is less effective and more difficult to incorporate into wheat than *Sm*1 (Lamb *et al.*, 2002).

In a field study, Smith *et al.* (2004) evaluated the use of a seed mixture of resistant and susceptible wheat cultivars to maintain susceptibility in the *S. mosellana* population and to conserve a parasitoid that is an effective biological control agent of the *S. mosellana*. They anticipated the evolution of resistance by the midge on HPR wheat because of the high mortality caused by *Sm*1. Smith *et al.* (2004) also believe that a seed mixture creating an interspersed refuge will succeed as an IRM strategy because larvae do not move from the natal plant and adults mate before dispersal. In HPR wheat, few larvae completed development, 2% or less compared with about 80% in susceptible wheat. The densities of mature midge larvae and parasitoids were in proportion to the size of the refuge.

A 5% refuge produced about 41 mature parasitoid larvae for each mature *S. mosellana* larva from the resistant wheat. Smith *et al.* (2004) concluded that a seed mixture is a promising strategy for sustaining HPR conferred by *Sm*1 and biocontrol of the *S. mosellana*. Mixtures may be extremely useful in cases such as these, where the target pest has low dispersal or mobility.

Schizaphis Graminum

The greenbug aphid, *Schizaphis graminum* (Hemiptera: Aphididae), is a pest on wheat and sorghum (*Sorghum bicolor*), but its crop hosts also include barley, oats, and rye (Puterka and Peters, 1990). At least 70 other noncultivated grasses have also been reported as hosts for *S. graminum* (Michels, 1986). Alternate hosts such as wild grasses may sustain *S. graminum* populations when preferred hosts such as wheat and sorghum are not present (Sambaraju and Pendleton, 2005). *S. graminum* damages crop plants by feeding on the phloem, resulting in chlorosis of the leaves, and they may transmit plant viruses. Control is usually done through HPR crops, insecticidal applications, and conserving natural enemies. Resistance to both HPR varieties and organophosphate insecticides has been found.

HPR to *S. graminum* exists in wheat and sorghum as antibiosis, antixenosis, and tolerance. Most sources of plant resistance seem to exhibit some combination of HPR mechanisms. For example, the sorghum cultivars tested by Dixon *et al.* (1990) and Schuster and Starks (1973) exhibited all three. Bowling and Wilde (1996) and Teetes *et al.* (1974) found levels of antibiosis and antixenosis in several sorghum cultivars. The HPR wheat cultivars tested by Sambaraju and Pendleton (2005) exhibited tolerance and some antibiosis. Fritts *et al.* (2000) found that it may be difficult to distinguish between antibiosis and antixenosis in some HPR wheat cultivars.

There are 11 documented *S. graminum* biotypes, designated by letters from A to K, although only 8 have any relation to HPR varieties (Porter *et al.*, 1997) and 3 evolved in the field (Shufran *et al.*, 2000). The term *biotype* here will be used to designate strains of insects differing in their capability of infesting certain HPR varieties (Diehl and Bush, 1984). There is considerable variation even within aphid clones for ability to damage certain sorghum cultivars, which may explain why *S. graminum* adapts so quickly to new cultivars (Wilhoit and Mittler, 1991). Various markers have not shown any discrimination among biotype designation (Anstead *et al.*, 2002; Shufran 2011; Weng *et al.*, 2010), but these studies have shown a common pattern of three major evolutionary clades (Shufran *et al.*, 2000).

There are two schools of thought concerning the evolution of *S. graminum* biotypes. The traditional hypothesis maintains that the use of resistant plant varieties has selected for the biotypes, while the more recent hypothesis claims that biotypes are artifacts of population-level host selection on noncultivated grasses (Porter *et al.*, 1997). Kindler and Hays (1999) suggested that the development of biotype F was due to maintenance of *S. graminum* populations on certain native grasses. Anstead *et al.* (2003) reported a larger host range of noncultivated grasses with biotype I, which is resistant to a number of HPR varieties of sorghum and wheat. The deployment of HPR wheat varieties has not apparently caused the evolution of new biotypes, because HPR wheat genes were released to growers after their respective biotypes were reported (Porter *et al.*, 1997). On the other hand, Bowling *et al.* (1998) noted that as area planted with resistant sorghum cultivars increased in the southern United States, so did the proportion of insects in the resistant biotypes. Deployment of HPR cultivars may have caused an increase in resistant alleles already present in a given population of *S. graminum*.

One possible reason for the perceived disconnect between HPR varieties and resistant biotypes is that HPR is scored by plant response to *S. graminum* (Burd and Porter, 2006), rather than a difference in the insect itself. Furthermore, sorghum HPR to *S. graminum* is measured in the length of time it takes for the insect to cause economic damage to the variety (Porter *et al.*, 1997). Finally, because cultivars usually have more than one HPR mechanism, it may make determination of the origin of *S. graminum* biotypes more difficult to ascertain.

For these reasons, resistance management of *S. graminum* continues to be a challenge. Because of the time and effort required to pyramid HPR genes into a cultivar, sequential deployment has been practiced. Porter *et al.* (2000) found no benefit of pyramiding resistance genes in wheat for control of *S. graminum*, because the HPR genes tested were already susceptible to at least one biotype. Bush *et al.* (1991) found that mixtures of 3:1 resistant:susceptible wheat cultivars reduced *S. graminum* damage. This study, however, did not address the sustainability of this practice, as it lasted for only one season.

TRANSGENIC INSECTICIDAL CROPS

Transgenic insecticidal crops have been available to farmers since 1996 (Frutos *et al.*, 1999). The first generation of crops expressed genes from the soil bacterium *Bacillus thuringiensis*, which was the source of a variety of manufactured insecticides for much of the twentieth century. Nelson and Alves (Chapter 4) review plant-incorporated protectants that have been considered for use in insecticidal crops. Given its status as the most common insecticide incorporated into crops, Nelson and Alves emphasize *Bacillus thuringiensis* (Bt) in their review. Bates *et al.* (2005b) and Christou *et al.* (2006) review the history of transgenic insecticidal crops.

In all countries where they are grown, governments regulate transgenic insecticidal crops, though this is rarely the case with traditionally bred HPR crops. Thus, legal restrictions have been placed on the sale, planting, and cultivation of this modern technology. In most countries, corporations producing the transgenic insecticidal crops and selling the seeds are required to implement an IRM strategy for these cultivars. Hurley and Mitchell (Chapter 13) discuss some of the economic aspects of these regulations.

A refuge strategy is the preferred approach to managing resistance on transgenic insecticidal crops if a high dose of the toxin can be consistently expressed by the plants (Gould, 1998). The high dose plus refuge strategy is based on the work of Comins (1977), Taylor and Georghiou (1979), and Tabashnik and Croft (1982). A high dose is defined as one that kills all susceptible homozygotes and most, if not all, heterozygotes. This makes resistance functionally recessive. A refuge of conventional plants is a source of susceptible homozygotes that can mate with any surviving heterozygotes to prevent the production of homozygotes that are resistant to the toxin. The resistance allele must be initially rare to ensure the effectiveness of this approach. Liu and Tabashnik (1997) were the first to experimentally demonstrate the value of a refuge in delaying resistance to a Bt toxin not incorporated into a plant. Onstad and Carrière (Chapter 10) describe the complexities of refuge strategies in real landscapes.

Several cases of field-evolved resistance have occurred in a few crops and a few countries (Van Rensburg ,2007; Tabashnik *et al.*, 2008; Dhurua and Gujar, 2011; Gassmann *et al.*, 2011; Storer *et al.*, 2012). Field-evolved resistance to Bt toxins has caused field control failures in some but not all cases. In this section, we highlight four case studies concerning insecticidal crops that differ significantly in agricultural and regulatory history. For example, the first case involves a registered crop, Bt potato, that failed to succeed commercially because of marketing concerns by buyers and competition from new insecticides. The third case pertains to broccoli (*Brassica oleracea*) that has not yet been commercialized or registered by the U.S. Environmental Protection Agency.

Leptinotarsa Decemlineata

The Colorado potato beetle, *Leptinotarsa decemlineata* (Coleoptera: Chrysomelidae), is the most devastating defoliator of potato (*Solanum tuberosum*) around the world (Hare, 1990). Multiple generations may occur per year depending on the latitude and climate. Adults overwinter in the soil in diapause. Both larvae and adults feed on potato leaves. Even though IPM alternatives exist, most potato farmers rely on insecticides (Feldman and Stone, 1997). *L. decemlineata* has a long history of evolving resistance to insecticides (Hare, 1990).

In 1995, transgenic insecticidal potatoes created by Monsanto became the first genetically engineered HPR crop to be registered by the U.S. Environmental Protection Agency for commercial use (Feldman and Stone, 1997). The Bt cultivars were also registered for commercial use in Canada and Russia. The potato expresses a *Bt* protein toxic to some Coleoptera including *L. decemlineata*. This was the first time that an IRM plan had been submitted to the U.S. Environmental Protection Agency during the registration process for an insecticidal crop. Thus, Monsanto's IRM plan for transgenic insecticidal potato was the first to be developed prior to market introduction (Feldman and Stone, 1997).

The IRM strategy was incorporated within an overall IPM plan (Feldman and Stone, 1997; Hoy, 1999). This permitted potato farmers to reduce insecticide use and take advantage of biological control. Monsanto recognized the importance of adjusting the IRM strategy according to local conditions. The strategy included expression of a high dose in the Bt potatoes, planting of a refuge of susceptible potatoes, and monitoring *L. decemlineata* for survival and resistance. The high dose was important because the Bt potatoes would have to be competitive in the marketplace with the very effective systemic insecticide imidacloprid. Selection for resistance to the *Bt* toxin (not the transgenic insecticidal plant) had already occurred in a laboratory study (Whalon *et al.*, 1993).

The choice of refuge configuration was the most difficult problem during the development of Monsanto's IRM strategy. Monsanto realized that research was needed to investigate the complexity and uncertainty of larval and adult behavior and to learn how farmers would deal with Bt potato under realistic conditions. In addition to a refuge block near a Bt potato field, Monsanto considered using a seed mixture to randomly insert the refuge within the *Bt* potato field to ensure grower compliance with the IRM strategy.

Field observations, laboratory experiments, and modeling contributed to the decision making regarding seed mixtures. Both older larvae and adults move frequently between potato plants. Hoy and Head (1995) measured a positive genetic correlation between larval avoidance and larval tolerance to the Bt toxin that could lead to more rapid evolution of resistance. In addition, Ferro (1993) evaluated the effects of seed mixtures with a high-dose Bt potato and susceptible potato as well as a low-dose strategy that could delay maturation of *L. decemlineata*. He concluded that a seed mixture would not be effective in delaying resistance evolution. He advocated the use of a low-dose Bt potato crop that would reduce the number of generations of *L. decemlineata* per year, but warned that non-random mating among susceptible and resistant beetles could occur because of the delays in maturation. Therefore, for several reasons, the seed mixture idea was abandoned.

Monsanto's IRM plan required the use of Bt potato in an annual rotation with a non-Bt crop (temporal refuge) and a 20% spatial refuge. Potato growers could not apply a foliar *Bt* application for *L. decemlineata* control on refuge fields, but they could treat the refuge with other insecticides to prevent damage by *L. decemlineata* according to local IPM recommendations. Monsanto's transgenic insecticidal potatoes were annually planted on less than 20,000 hectares from 1995 to 2001. At its maximum, Bt potato accounted for less than 5% of total potato production in United States. The Bt cultivars never became broadly established in the IPM programs imagined by Monsanto.

Pectinophora Gossypiella

Pectinophora gossypiella is a very destructive, cosmopolitan pest and the most serious lepidopteran pest of cotton (*Gossypium hirsutum*) in the southwestern United States (Henneberry and Naranjo, 1998). Female moths lay eggs on cotton plants, and larvae bore into cotton bolls where they eat cotton seeds. At least four generations per year occur in the United States. Larvae do not move from plant to plant (Carrière et al., 2005), and they overwinter in diapause. Although P. gossypiella can infest 70 plant species in seven families, it feeds almost exclusively on cotton in Arizona (Henneberry and Naranjo, 1998).

Transgenic insecticidal cotton expressing Bt toxin was first grown on a large scale in Arizona in 1996. Since then, a large team of Arizona scientists has been studying the evolution of resistance by P. gossypiella to Bt cotton. A coordinated research and educational effort called the Arizona Bt Cotton Working Group includes the University of Arizona, the Arizona Department of Agriculture, the U.S. Department of Agriculture's Western Cotton Research Laboratory, the Arizona Cotton Research and Protection Council, and the Arizona Cotton Growers Association (Carrière et al., 2001a).

Before Bt cotton was commercialized in the United States, the U.S. Environmental Protection Agency granted a registration to Monsanto that included IRM requirements based on a refuge strategy. The percentage of cotton fields required to be refuges depends on whether refuges are planted inside or outside of the Bt cotton fields (Carrière et al., 2005). For refuges outside of Bt cotton fields, refuge size is either 5% non-Bt cotton if the refuge is not sprayed with insecticides effective against P. gossypiella or 20% non-Bt cotton if the refuge is sprayed with such insecticides. Refuges inside Bt cotton fields must be at least 5% of the field, and they can be planted as one row of conventional cotton for every six to ten rows of Bt cotton (Carrière et al., 2005). The internal refuge block and Bt cotton block can be treated with any insecticides, except Bt, as long as the refuge and Bt cotton are both treated.

The risk of resistance by P. gossypiella to Bt cotton was initially considered high in Arizona for several reasons. Lab selection with Bt toxin in an artificial diet quickly produced strains resistant to Bt cotton. From 1998 to 2003 in most of the major cotton areas of Arizona, Bt cotton was planted in about 60 to 80% of the fields (Carrière et al., 2005). Bt cotton produces a high dose of toxin that kills almost 100% of susceptible P. gossypiella larvae that ingest it. In Arizona, P. gossypiella has no other host plants and has up to five generations per year. All of these conditions favor the rapid evolution of resistance.

Subsequent observations on the P. gossypiella and Bt cotton system challenged the early predictions of fast evolution of resistance. For example, on non-Bt cotton, the Arizona team observed significant fitness costs for homozygous resistant individuals relative to homozygous susceptible ones (Carrière *et al.*, 2001b, c, 2004; Williams *et al.*, 2011). Also, survival of homozygous resistant individuals is lower on Bt cotton than on conventional cotton (Carrière and Tabashnik, 2001).

The Arizona IRM team monitored resistance to Bt cotton by collecting infested bolls from 10 to 17 cotton fields in Arizona per year (Tabashnik *et al.*, 2005b). The progeny of field-collected *P. gossypiella* from each site were tested using bioassays involving an average of 2541 larvae per year. Neonates were tested on an artificial diet containing a dose of toxin that kills all larvae but the homozygous resistant ones (Tabashnik *et al.*, 2000, 2002, 2005a). In 1999 and 2000, no larvae survived on diet treated with a diagnostic concentration of Bt toxin (Tabashnik *et al.*, 2005b). Figure 9.2 presents the results of this monitoring program and more recent monitoring programs. The highest frequency of resistance to the Bt toxin occurred in 1997 with a mean resistance-allele frequency of 0.16. Five of ten cotton fields sampled in 1997 yielded one or more resistant larvae that survived exposure to the diagnostic concentration of toxin (Tabashnik *et al.*, 2000). The resistance-allele frequency decreased in 1998 to 0.0070 and then varied over the next six years, with no net increase from 1997 to 2008 (Figure 9.2).

Tabashnik *et al.* (2006) also screened the *P. gossypiella* population for resistance to Bt cotton with a DNA-based method. The technique uses polymerase chain reaction primers that specifically amplify three mutant alleles of a cadherin gene linked with resistance to Bt cotton to detect single resistance alleles in heterozygotes. Tabashnik *et al.* (2006) found no resistance alleles in 5571 insects derived from 59 cotton fields in three states during 2001–2005.

Additional data support the conclusion that *P. gossypiella* resistance to Bt cotton in Arizona did not increase from 1997 to 2004. No control failures caused by resistance have been reported, and regional declines of *P. gossypiella* have occurred in areas of Arizona with high use of Bt cotton (Carrière *et al.*, 2003). In six major cotton-growing counties of Arizona, non-Bt cotton refuges on the average accounted for 14 to 78% of the fields planted with cotton per county from 2000 to 2003 (Figure 9.3). Furthermore, 88% of the fields, in five out of six years appeared to be in compliance with refuge requirements (Carrière *et al.* 2005). Therefore, the Arizona IRM team concluded that the high dose/refuge strategy helped to delay *P. gossypiella* resistance to Bt cotton (Tabashnik *et al.*, 2005b).

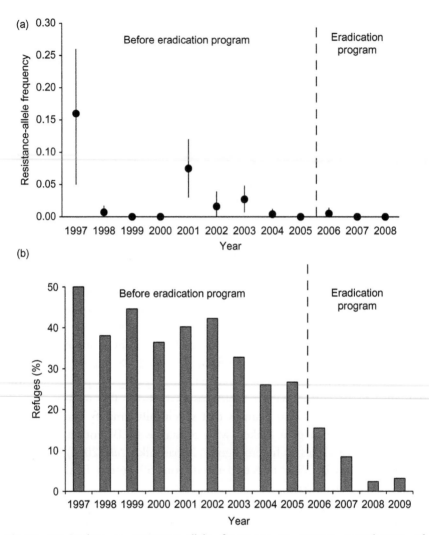

Figure 9.2 Decline in resistance-allele frequency in Arizona populations of *Pectinophora gossypiella* and (b) changes in cotton refuge levels over time in Arizona. *Reprinted from Tabashnik* et al. *(2012) by permission of Landes Bioscience.*

Unfortunately, the success with IRM in Arizona was not repeated in India and China. Resistance increased in populations of *P. gossypiella* (Dhurua and Gujar, 2011; Wan *et al.*, 2012b). The primary problem may have been the lack of refuge planting in Asia (Tabashnik *et al.*, 2012). It is also possible that Asian cotton plants expressed lower doses of insecticidal proteins than plants in Arizona (Tabashnik *et al.*, 2012).

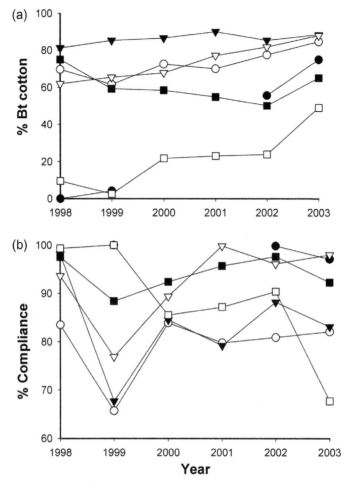

Figure 9.3 Percentage of Bt cotton grown in six counties (a) and percentage compliance (b) during 1998–2003. *From Carriere* et al., *2005 Copyright Society of Chemical Industry. Reproduced with permission.*

Because Wan *et al.* (2012a) identified regional population suppression as a valuable consequence of planting Bt cotton in 94% of cotton fields in one area of China, they advocated increased refuge plantings to delay evolution of resistance in *P. gossypiella* and prolong the economic benefits of Bt cotton.

After its introduction to Arizona, Bt cotton was quickly adopted by growers; over 93% adopted the technology by 2007 when the requirement for a refuge was eliminated (Naranjo and Ellsworth, 2010). From

2006 to 2011, refuge levels declined in Arizona during the period when an integrated-control eradication program was carried out (Figure 9.2). The strategic program included the mass release of sterile *P. gossypiella*, reduction in non-Bt cotton refuge percentage, increased percentage of pyramided Bt cotton expressing two toxins, and early-season pheromone treatments to disrupt mating in non-Bt cotton fields. No resistance to Bt cotton was detected during this period (Figure 9.2; Tabashnik *et al.*, 2010, 2012). The use of Bt cotton caused *P. gossypiella* abundance to decrease by > 99%, while eliminating insecticide sprays against this pest (Tabashnik *et al.*, 2010). In Arizona cotton fields, overall insecticide use (number of sprays per hectare) has decreased by 88% since 1995 (Naranjo and Ellsworth, 2010). Thus, IRM and IPM combined to improve management of *P. gossypiella* and other cotton pests.

Plutella xylostella

The diamondback moth, *Plutella xylostella* (Lepidoptera: Plutellidae), is a major pest of crucifers (Brassicaceae) in many parts of the world. Damage is caused by larval feeding on the leaves and around buds. The insect passes through many generations per year. *P. xylostella* has evolved resistance to numerous synthetic insecticides (Talekar and Shelton, 1993), so it has become the focus of many IRM studies. Because *P. xylostella* was the first insect to evolve resistance to Bt sprays in the field (Shelton *et al.*, 1993; Tabashnik, 1994) and because it can be studied under greenhouse conditions, a team at Cornell University used the insect as a model system to examine the assumptions underlying IRM strategies in transgenic insecticidal crops.

Before they could perform the IRM experiments, the Cornell University team had to develop both transgenic insecticidal broccoli (*Brassica oleracea*) and laboratory strains of *P. xylostella* resistant to the transgenic Bt cultivars. The team developed transgenic Bt cultivars with high expression of one or two Bt toxins (Metz *et al.*, 1995; Cao *et al.*, 1999, 2002). Resistant *P. xylostella* were collected from areas where control with Bt products was failing. Colonies derived from these insects were subsequently selected on transgenic Bt broccoli plants (Zhao *et al.*, 2002).

The resistant strains of *P. xylostella* can be used to study the dynamics of resistance to Bt transgenic crops. Resistance to Bt toxin is an autosomal recessive trait in the moth (Zhao *et al.*, 2000). Zhao *et al.* (2002) concluded that initial frequencies of resistance alleles are underestimated when using transgenic broccoli with high *Bt* expression as a screen; they recommended a diet assay that would yield fewer false negatives. Tang *et al.* (1997) found that after selection for resistance to *Bt* is stopped, there is an initial drop in resistance, followed by stabilization. Resistance quickly rose to initial levels after selection was resumed (Tang *et al.*, 1997; Tabashnik, 1994).

The Cornell University team was the first to experimentally evaluate the refuge strategy for IRM for transgenic insecticidal crops under realistic conditions. The team varied the size and arrangement of the refuge in the field and in the greenhouse. Field tests were possible because *P. xylostella* does not overwinter in New York; released resistant insects would die in the fall. Each field plot contained 300 broccoli plants, and the initial resistance-allele frequency was either 0.12 or 0.8 (Shelton *et al.*, 2000). In the greenhouse, Tang *et al.* (2001) used cloth cages containing 30 potted broccoli plants. In the greenhouse experiments, the initial resistance-allele frequencies were either 0.007−0.0125 (Tang *et al.*, 2001) or 0.10−0.34 (Zhao *et al.*, 2003, 2005).

All of the experiments demonstrated the effectiveness of separate refuges to delay the evolution of insect resistance to *Bt* broccoli (Shelton *et al.*, 2000; Tang *et al.*, 2001; Zhao *et al.*, 2003, 2005). There is no ovipositional preference between Bt and non-Bt broccoli cultivars (Tang *et al.*, 1999). *P. xylostella* larvae tend to disperse more readily from *Bt* plants than from non-*Bt* broccoli plants (Tang *et al.*, 2001), increasing the chance that larvae will receive a sublethal dose of the toxin. This means that refuge plants should be separated from (rather than randomly mixed with) *Bt* broccoli plants to prevent or delay resistance (Shelton *et al.*, 2000; Tang *et al.*, 2001).

Zhao *et al.* (2003, 2005) evaluated the deployment of refuges plus the pyramiding of Bt genes within plants (Cao *et al.*, 2002). They concluded that use of pyramids of insecticidal genes in *Bt* broccoli is the most effective strategy for delaying resistance (Zhao *et al.*, 2003), but that this benefit is lost in the presence of transgenic insecticidal crops containing only one of the genes in the pyramid (Zhao *et al.*, 2005).

Shelton (2012) describes the social, economic, and political forces that have delayed the commercialization of Bt broccoli and all other Bt vegetables other than sweet corn. It is clear that resistance management is not the only issue that must be addressed in the diverse set of countries that will likely consider using transgenic insecticidal crops to feed their citizens.

Diabrotica virgifera virgifera

This insect is a serious pest of corn (*Zea mays*) in North America and Europe (Spencer *et al.*, Chapter 7). It has a long history of evolving resistance to various IPM tactics (Onstad, Chapter 14; Gassmann *et al.*, 2011). Onstad and Meinke (2010) used a model to evaluate insecticidal corn expressing one or two toxin traits and IRM strategies deploying block refuges. In hypothetical scenarios, results demonstrated that evolution is generally delayed by pyramids compared to deployment of a single-toxin corn hybrid. However, soil insecticide use in the block refuge reduced this delay and quickened the evolution of resistance. Results were sensitive to the degree of male beetle dispersal prior to mating and to the effectiveness of both toxins in the pyramid. Resistance evolved slower as fecundity declined for survivors of insecticidal corn. Onstad and Meinke (2010) concluded that effects on fecundity must be measured to predict which resistance management plans will work well. Evolution of resistance also occurred faster if the survival rate due to exposure to the two toxins was not calculated by multiplication of two independent survival rates (one for each insect gene) but was equivalent to the minimum of the two (no redundant killing). Furthermore, when single-trait and pyramided corn hybrids were planted within adult-dispersal distance of each other, the toxin traits lost efficacy more quickly than they did in scenarios without single-trait insecticidal corn.

Pan *et al.* (2011) created a different model of *D. virgifera virgifera* to evaluate the refuge strategies for single-trait insecticidal corn. Because female adults usually mate before dispersing and because male adults are limited in the distances that they fly during their lifetimes, block refuges may not be valuable for IRM for *D. virgifera virgifera*. Model results indicated that blended refuge delays evolution of resistance longer than block refuges of equal size, if the blocks are relocated in the cornfields every year. If growers can be encouraged to plant a refuge at the same location for many years, then this constant block refuge is superior to blended refuges. Lack of compliance by growers with refuge requirements also reduces the expected benefits from block refuges (Pan *et al.*, 2011).

DISCUSSION

This chapter describes a wide variety of cases in which an arthropod species has overcome host-plant resistance (HPR) designed by humans to control crop pests. Resistance by arthropods to HPR can occur to both traditionally bred crops and transgenic insecticidal crops (Gould, 1998; Tabashnik et al., 2008). Much of the focus of HPR is on antibiosis, and more recently, on production of extremely high doses of toxins in transgenic insecticidal crops. IRM will be needed regardless of whether the HPR is caused by antibiosis or by antixenosis.

Mechanisms of antibiosis and the resulting toxicological resistance are much better understood. Major resistance genes have been identified, most notably in insects that are targets of transgenic insecticidal crops. These genes and the resistant insect colonies based on them can be used to study the dynamics of resistance and the efficacy of various management strategies. Examples are *M. destructor* on wheat and *P. xylostella* on broccoli.

The behavior of the arthropod influences how quickly it will evolve resistance to either antibiosis or antixenosis. An HPR variety that deters an herbivore is expected to exert less selection pressure for resistance than one in which the herbivore has no preference between it and the susceptible variety. This does not mean, however, that an herbivore will not evolve behavioral resistance. Indeed, *Tetranychus urticae* still evolved resistance to morphological features of tomatoes and broccoli previously shown to deter mites. *Schizaphis graminum* has evolved resistance to many sorghum and wheat cultivars, most of which have some level of antixenotic effect. Exactly what causes these behavioral changes is yet to be discovered.

Studies of behavior in relation to HPR should be conducted in the appropriate ecological context (Knolhoff and Heckel, 2014). Does a choice test conducted in the laboratory or greenhouse represent a real choice present in nature or a theoretical choice? For insects or insect stages with low mobility, a more appropriate distinction may be acceptance versus rejection. Additionally, temporal effects on starvation (*i.e.*, giving up time) influence the probability of acceptance of a given host plant (McNair, 1982). As Onstad and Carrière (Chapter 10) discuss, larval behavior is very important in evaluating the value and feasibility of mixtures of refuge plants and insecticidal plants.

The history of Gould's modeling work exemplifies the predictions needed to make scientific, preventative, and some would say, "proactive," IRM strategies for HPR. In his first major modeling study, Gould (1984) evaluated the effects of crops expressing one trait for toxicity and one for repellency (Castillo-Chavez et al., 1988). He was the first entomologist to study management of behavioral resistance. He found that behavior modification could be valuable in IRM, but that coordination of strategy implementation would be important because of the off-farm externalities caused by repelling mobile pests to neighboring fields. In later work, he focused on crops with two toxin traits. Gould (1986a) used the model to evaluate sequential deployment of single-gene crops, single-gene cultivar mixtures, and a pyramided two-gene cultivar. He concluded that no single IRM strategy was the best or most durable for all modeled scenarios. Model results indicated that the pyramided crop would be more durable if it is planted along with a plot of the cultivar lacking HPR. Gould (1986b) applied this model to the evaluation of IRM for *Mayetiola destructor* on HPR wheat (discussed in the first section of this chapter along with the alternative model by Cox and Hatchett, 1986). He extended this analysis of pyramided crops containing two toxins by investigating the role of fitness costs in resistance evolution (Gould *et al.*, 2006). He concluded that, when an IRM plan includes a 10 to 20% refuge of cultivars lacking any HPR and the rest of the cultivars are only pyramids of two toxin genes, the risk of resistance is often low when resistant insects experience fitness costs on the susceptible cultivars. However, the IRM plan often fails when the landscape includes a single-gene cultivar (Gould *et al.*, 2006). Again, the need for coordination of strategy implementation is critical for the success of IRM; some agency or stakeholder association must restrict the planting of single-gene cultivars to ensure the durability of pyramided crops. In all of his modeling studies, Gould has emphasized the point that specific IRM plans can only be derived from pest and cropping-system specific models. Additional views concerning IRM models can be found in Chapter 14 of the present volume.

Gould (1988) was one of the first scientists to emphasize the need for IRM plans for transgenic insecticidal crops. By the middle of the 1980s, scientists had successfully transformed plants to express *Bt* toxin, but it was uncertain that anyone in industry, government, or academia had thought of rational approaches to the wise use of such plants in commercial agriculture. Gould was concerned about the intense selection pressure likely to occur when very toxic plants express the toxin in every crop tissue throughout the entire growing season. Chemically inducible Bt toxin expression in a plant is feasible (Bates *et al.*, 2005; Cao *et al.*, 2001, 2006), but commercially grown transgenic insecticidal cotton and corn (*Zea mays*) crops produce Bt toxin continuously. Gould (1988) urged genetic engineers to develop crops that express the toxin only in some tissues and only for a short time. He even challenged these laboratory scientists to have plant expression of the toxin induced only after the pest causes substantial damage (Gould, 1988). Nevertheless, no commercialized transgenic insecticidal cotton or corn plant has been purposefully engineered to limit exposure of a pest to a toxin to short periods and in only a few plant tissues.

Up to now, transgenes have only been used within the nuclear DNA sequence. One approach for the future may be the expression of genes in chloroplasts (Kota et al., 1999; De Cosa et al., 2001). This approach provides tissue specificity and extremely high expression levels (much higher than for nuclear expression).

As with any IRM program, stakeholders should expect and plan for complexity and complications in HPR. Traditional HPR is typically not coordinated in any areawide IPM programs, although effective cultivars may be recognized by many individual farmers and extensively planted across a large region. Management of resistance to traditionally bred or genetically engineered crops will face the complexity of agricultural and natural landscapes. Regions will likely consist of crop cultivars having one, two, or more traits with lethal or sublethal effects on the targeted arthropod. IRM strategies cannot be based on assumptions of single traits with simple effects (Gassmann et al., 2009). Stakeholders should expect expression of insecticidal traits to vary over the growing season because toxins may increase and particularly decrease during certain plant processes and stages. Just as likely, expression of HPR traits can vary in different plant tissues, producing a mosaic of toxin levels within each plant (Chapter 10).

When we manage arthropod resistance to HPR, we should address the following questions. How can we develop or acquire a kind of plant that improves IRM? What kind of toxins should it have? How much of each toxin should the plant produce? In what tissues? Should the toxins be pyramided within a given plant, or should HPR cultivars be planted in rotation or grown sequentially over multiple seasons? To lower the probability of evolution of resistance, how should the plant and its toxin change over a growing season? Of course, solutions to the IRM problem must still allow adequate yield and crop quality in the field where other stresses may be occurring.

Carrière et al. (2001a) describe the efforts by the Arizona Bt Cotton Working Group to develop a rational and scientific IRM strategy for Pectinophora gossypiella. From the beginning, the Working Group knew that they were in the middle of a large-scale field experiment testing the high dose plus refuge strategy and testing their ability to implement and comply with the U.S. Environmental Protection Agency's regulation of Bt cotton. Given the high stakes involved, the Working Group was under pressure to produce educational and scientific results that would not only successfully implement the required IRM strategy in the cotton fields of Arizona but also improve the U.S. Environmental Protection Agency's regulation of Bt cotton. The report by Carrière et al. (2001a) records some of their recommendations to the Agency based on their laboratory and field studies. They also draw some conclusions about coordination of activities. For example, they state that IRM should be a primary objective for grower-funded, commodity-based organizations. The Arizona Bt Cotton Working Group was one way to engage such stakeholder groups.

Under what conditions would government-imposed IRM requirements for traditionally bred HPR crops be beneficial to society? For instance, would wheat growers benefit from a required IRM strategy for *Mayetiola destructor* on HPR wheat? Certainly the mandated coordination of efforts and compliance with the strategy would impose unknown, but substantial, cost on growers and society (Hurley and Mitchell, Chapter 13), but would the benefits in terms of decreased loss of harvested yield from old, but still resistant, cultivars and lower cost of reduced breeding programs exceed these costs over the long term? Part of the question becomes, over what time horizon do we evaluate the economics (Mitchell and Onstad, Chapter 2)?

Two important conclusions can be drawn from field and modeling studies concerning resistance to HPR crops. First, refuges of susceptible cultivars are effective in delaying the evolution of resistance. This point was demonstrated with the field and greenhouse studies of *P. xylostella*, with the field studies of *P. gossypiella*, and with Gould's modeling of *M. destructor*. Second, proper monitoring of resistance and of resistant biotypes or strains requires a significant amount of work (Chapter 15). This is exemplified by the work on *P. gossypiella* and *S. graminum*.

REFERENCES

Agrawal, A.A., 2000. Host-range evolution: adaptation and trade-offs in fitness of mites on alternative hosts. Ecology. 81, 500–508.
Anderson, K.G., Harris, M.O., 2006. Does R gene resistance allow wheat to prevent plant growth effects associated with Hessian fly (Diptera : Cecidomyiidae) attack? J. Econ. Entomol. 99, 1842–1853.

Anderson, K.M., Kang, Q., Reber, J., Harris, M.O., 2011. No fitness cost for wheat's H gene-mediated resistance to Hessian fly (Diptera:Cecidomyiidae). J. Econ. Entomol. 104, 1393−1405.

Anstead, J.A., Burd, J.D., Shufran, K.A., 2002. Mitochondrial DNA sequence divergence among *Schizaphis graminum* (Hemiptera:Aphididae) clones from cultivated and non-cultivated hosts: haplotype and host associations. Bull. Entomol. Res. 92, 17−24.

Anstead, J.A., Burd, J.D., Shufran, K.A., 2003. Over-summering and biotypic diversity of *Schizaphis graminum* (Homoptera:Aphididae) populations on noncultivated grass hosts. Environ. Entomol. 32, 662−667.

Baluch, S.D., Ohm, H.W., Shukle, J.T., Williams, C.E., 2012. Obviation of wheat resistance to the Hessian fly through systemic induced susceptibility. J. Econ. Entomol. 105, 642−650.

Bates, S.L., Cao, J., Zhao, J.-Z., Earle, E.D., Roush, R.T., Shelton, A.M., 2005. Evaluation of a chemically inducible promoter for developing a within-plant refuge for resistance management. J. Econ. Entomol. 98, 2188−2194.

Bates, S.L., Zhao, J.-Z., Roush, R.T., Shelton, A.M., 2005b. Insect resistance management in GM crops: past, present and future. Nat. Biotechnol. 23, 57−62.

Berenbaum, M.R., 2001. Plant-herbivore interactions. In: Fox, C.W., Roff, D.A., Fairbairn, D.J. (Eds.), Evolutionary Ecology: Concepts and Case Studies. Oxford University Press, p. 424, Chapter 23.

Bowling, R., Wilde, G., 1996. Mechanisms of resistance in three sorghum cultivars resistant to greenbug (Homoptera: Aphididae) biotype I. J. Econ. Entomol. 89, 558−561.

Bowling, R., Wilde, G., Margolies, D., 1998. Relative fitness of greenbug (Homoptera: Aphididae) biotypes E and I on sorghum, wheat, rye, and barley. J. Econ. Entomol. 91, 1219−1223.

Buntin, G.D., 1999. Hessian fly (Diptera: Cecidomyiidae) injury and loss of winter wheat grain yield and quality. J. Econ. Entomol. 92, 1190−1197.

Buntin, G.D., Johnson, J.W., Bruckner, P.L., 1990. Hessian fly (Diptera: Cecidomyiidae) management in Georgia by delayed planting of winter wheat. J. Econ. Entomol. 83, 1025−1033.

Buntin, G.D., Ott, S.L., Johnson, J.W., 1992. Integration of plant resistance, insecticides, and planting date for management of the Hessian fly (Diptera: Cecidomyiidae). J. Econ. Entomol. 85, 530−538.

Burd, J.D., Porter, D.R., 2006. Biotypic diversity in greenbug (Hemiptera: Aphididae): characterizing new virulence and host associations. J. Econ. Entomol. 99, 959−965.

Bush, L., Slosser, J.E., Worrall, W.D., Horner, N.V., 1991. Potential of wheat cultivar mixtures for greenbug (Homoptera: Aphididae) management. J. Econ. Entomol. 84, 1619−1624.

Cambron, S.E., Buntin, G.D., Weisz, R., Holland, J.D., Flanders, K.L., Schemerhorn, B. J., et al., 2010. Virulence in Hessian fly (Diptera: Cecidomyiidae) field collections from the southeastern United States to 21 resistance genes in wheat. J. Econ. Entomol. 103, 2229−2235.

Cantelo, W.W., Sanford, L.L., 1984. Insect population response to mixed and uniform plantings of resistant and susceptible plant material. Environ. Entomol. 13, 1443−1445.

Cao, J., Tang, J.D., Strizhov, N., Shelton, A.M., Earle, E.D., 1999. Transgenic broccoli with high levels of *Bacillus thuringiensis* Cry1C protein, control diamondback moth larvae resistant to Cry1A or Cry1C. Mol. Breed. 5, 131−141.

Cao, J., Shelton, A.M., Earle, E.D., 2001. Gene expression and insect resistance in transgenic broccoli containing a *Bacillus thuringiensis* cry1Ab gene with the chemically inducible PR-1a promoter. Mol. Breed. 8, 207−216.

Cao, J., Zhao, J.Z., Tang, J.D., Shelton, A.M., Earle, E.D., 2002. Broccoli plants with pyr-amided cry1Ac and cry1C Bt genes control diamondback moths resistant to Cry1A and Cry1C proteins. Theor. Appl. Genet. 105, 258−264.

Cao, J., Bates, S.L., Zhao, J.-Z., Shelton, A.M., Earle, E., 2006. *Bacillus thuringiensis* protein production, signal transduction, and insect control in chemically inducible *PR-1a/cry1Ab* broccoli plants. Plant Cell Rep. 25, 554−560.

Carrière, Y., Tabashnik, B.E., 2001. Reversing insect adaptation to transgenic insecticidal plants. Proc. Roy. Soc. Lond. B. 268, 1475−1480.

Carrière, Y., Dennehy, T.J., Pedersen, B., Haller, S., Ellers-Kirk, C., Antilla, L., et al., 2001a. Large-scale management of insect resistance to transgenic cotton in Arizona: can transgenic insecticidal crops be sustained? J. Econ. Entomol. 94, 315−325.

Carrière, Y., Ellers-Kirk, C., Liu, Y.-B., Sims, M.A., Patin, A.L., Dennehy, T.J., et al., 2001b. Fitness costs and maternal effects associated with resistance to transgenic cotton in the pink bollworm. J. Econ. Entomol. 94, 1571−1576.

Carrière, Y., Ellers-Kirk, C., Patin, A.L., Sims, M.A., Meyer, S., Liu, Y.-B., et al., 2001c. Overwintering costs associated with resistance to transgenic cotton in the pink boll-worm. J. Econ. Entomol. 94, 935−941.

Carrière, Y., Ellers-Kirk, C., Sisterson, M., Antilla, L., Whitlow, M., Dennehy, T.J., et al., 2003. Long-term regional suppression of pink bollworm by *Bacillus thuringiensis* cot-ton. Proc. Nat'l. Acad. Sci. USA. 100, 1519−1523.

Carrière, Y., Ellers-Kirk, C., Biggs, R., Dennehy, T.J., Tabashnik., B.E., 2004. Effects of gossypol on fitness costs associated with resistance to Bt cotton in pink bollworm. J. Econ. Entomol. 97, 1710−1718.

Carrière, Y., Ellers-Kirk, C., Kumar, K., Heuberger, S., Whitlow, M., Antilla, L., et al., 2005. Long-term evaluation of compliance with refuge requirements for Bt cotton. Pest Manag. Sci. 61, 327−330.

Castillo-Chavez, C., Levin, S.A., Gould, F., 1988. Physiological and behavioral adaptation in varying environments: a mathematical model. Evolution. 42, 986−994.

Chen, B.H., Foster, J.E., Taylor, P.L., Araya, J.E., Kudagamage, C., 1990. Determination of frequency and distribution of Hessian fly (Diptera: Cecidomyiidae) biotypes in the northeastern soft wheat region. Great Lakes Entomol. 23, 217−221.

Chen, M.-S., Echegaray, E., Whitworth, R.J., Wang, H., Sloderbeck, P.E., Knutson, A., et al., 2009. Virulence analysis of Hessian fly populations from Texas, Oklahoma, and Kansas. J. Econ. Entomol. 102, 774−780.

Christou, P., Capell, T., Kohli, A., Gatehouse, J.A., Gatehouse, A.M.R., 2006. Recent developments and future prospects in insect pest control in transgenic crops. Trends Plant Sci. 11, 302−308.

Comins, H.N., 1977. The development of insecticide resistance in the presence of immi-gration. J. Theor. Biol. 64, 177−197.

Cox, T.S., Hatchett, J.H., 1986. Genetic model for wheat/Hessian fly (Diptera: Cecidomyiidae) interaction: strategies for deployment of resistance genes in wheat cultivars. Environ. Entomol. 15, 24−31.

De Cosa, B., Moar, W., Lee, S.-B., Miller, M., Daniell, H., 2001. Overexpression of the Bt cry2Aa2 operon in chloroplasts leads to formation of insecticidal crystals. Nat. Biotechnol. 19, 71−74.

Del Conte, S.C.C., Bosque-Perez, N.A., Schotzko, D.J., Guy, S.O., 2005. Impact of till-age practices on Hessian fly-susceptible and resistant spring wheat cultivars. J. Econ. Entomol. 98, 805−813.

Dhurua, S., Gujar, G.T., 2011. Field-evolved resistance to Bt toxin Cry1Ac in the pink bollworm, *Pectinophora gossypiella* (Saunders) (Lepidoptera: Gelechiidae), from India, Pest Manag Sci., 67. 898-903.

Diehl, S.R., Bush, G.L., 1984. An evolutionary and applied perspective of insect biotypes. Annu. Rev. Entomol. 29, 471–504.

Dixon, A.G.O., Bramel-Cox, P.J., Reese, J.C., Harvey, T.L., 1990. Mechanisms of resistance and their interactions in twelve sources of resistance to biotype E greenbug (Homoptera: Aphididae) in sorghum. J. Econ. Entomol. 83, 234–240.

El Bouhssini, M., Hatchett, J.H., Wilde, G.E., 1999. Hessian fly (Diptera: Cecidomyiidae) larval survival as affected by wheat resistance alleles, temperature, and larval density. J. Agric. Urb. Entomol. 16, 245–254.

El Bouhssini, M.J., Hatchett, J.H., Cox, T.S., Wilde, G.E., 2001. Genotypic interaction between resistance genes in wheat and virulence genes in the Hessian fly *Mayetiola destructor* (Diptera: Cecidomyiidae). Bull. Entomol. Res. 91, 327–331.

Feldman, J., Stone, T., 1997. The development of a comprehensive resistance management plan for potatoes expressing the Cry3A endotoxin. In: Carozzi, N., Koziel, M. (Eds.), Advances in Insect Control: the Role of Transgenic Plants. Taylor & Francis, London, Chapter 14.

Ferro, D.N., 1993. Potential for resistance to *Bacillus thuringiensis*: Colorado potato beetle (Coleoptera: Chrysomelidae)—a model system. Amer. Entomol. 39, 38–44.

Formusoh, E.S., Hatchett, J.H., Black IV, W.C., Stuart, J.J., 1996. Sex-linked inheritance of virulence against wheat resistance gene H9 in the Hessian fly (Diptera: Cecidomyiidae). Ann. Entomol. Soc. Am. 89, 428–434.

Foster, J.E., Ohm, H.W., Patterson, F.L., Taylor, P.L., 1991. Effectiveness of deploying single gene resistances in wheat for controlling damage by the Hessian fly (Diptera: Cecidomyiidae). Environ. Entomol. 20, 964–969.

Fritts, D.A., Michels Jr., G.J., Lazar, M.D., 2000. Greenbug dispersal and colonization on a resistant winter wheat genotype: antixenosis, antibiosis or both? Southwest. Entomol. 25, 113–121.

Frutos, R., Rang, C., Royer, M., 1999. Managing insect resistance to plants producing *Bacillus thuringiensis* toxins. Critical Rev. Biotech. 19, 227–276.

Fry, J.D., 1988. Variation among populations for the twospotted spider mite, *Tetranychus urticae* Koch (Acari: Tetranychidae), in measures of fitness and host-acceptance behavior on tomato. Environ. Entomol. 17, 287–292.

Fry, J.D., 1989. Evolutionary adaptation to host plants in a laboratory population of the phytophagous mite *Tetranychus urticae*. Oecologia. 81, 559–565.

Fry, J.D, 1990. Trade-offs in fitness on different hosts: evidence from a selection experiment with a phytophagous mite. Am. Nat. 136, 569–580.

Fry, J.D., 1992. On the maintenance of genetic variation by disruptive selection among hosts in a phytophagous mite. Evolution. 46, 279–283.

Gallun, R.L., 1977. Genetic basis of Hessian fly epidemics. Ann. N.Y. Acad. Sci. 287, 223–229.

Gallun, R.L., Hatchett, J.H., 1969. Genetic evidence of elimination of chromosomes in the Hessian fly. Ann. Entomol. Soc. Am. 62, 1095–1101.

Gassmann, A.J., Onstad, D.W., Pittendrigh., B.R., 2009. Evolutionary analysis of herbivore adaptation to natural and agricultural systems. Pest Manag. Sci. 65, 1174–1181.

Gassmann, A.J., Petzold-Maxwell, J.L., Keweshan, R.S., Dunbar, M.W., 2011. Field-evolved resistance to Bt maize by western corn rootworm. PLoS One. 6, e22629. Available from: http://dx.doi.org/10.1371/journal.pone.0022629.

Gould, F., 1978a. Predicting the future resistance of crop varieties to pest populations: a case study of mites and cucumbers. Environ. Entomol. 7, 622–626.

Gould, F., 1978b. Resistance of cucumber varieties to *Tetranychus urticae*: genetic and environmental determinants. J. Econ. Entomol. 71, 680–683.

Gould, F., 1979. Rapid host range evolution in a population of the phytophagous mite *Tetranychus urticae* Koch. Evolution. 33, 791–802.

Gould, F., 1983. Genetics of plant-herbivore systems: interactions between applied and basic study. In: Denno, R., McClure, M. (Eds.), Variable Plants and Herbivores in Natural and Managed Systems. Academic, New York, pp. 599—653.

Gould, F., 1984. Role of behavior in the evolution of insect adaptation to insecticides and resistant host plants. Bull. Entomol. Soc. Amer. 30, 34—41.

Gould, F., 1986a. Simulation models for predicting durability of insect-resistant germ plasm: a deterministic diploid, two-locus model. Environ. Entomol. 15, 1—10.

Gould, F., 1986b. Simulation models for predicting durability of insect-resistant germ plasm: Hessian fly (Diptera: Cecidomyiidae)-resistant winter wheat. Environ. Entomol. 15, 11—23.

Gould, F., 1988. Evolutionary biology and genetically engineered crops. BioScience. 38, 26—33.

Gould, F., 1998. Sustainability of transgenic insecticidal cultivars: integrating pest genetics and ecology. Annu. Rev. Entomol. 43, 701—726.

Gould, F., Carroll, C.R., Futuyma, D.J., 1982. Cross-resistance to pesticides and plant defenses: a study of the two-spotted spider mite. Entomol. Exp. Appl. 31, 175—180.

Gould, F., Kennedy, G.G., Johnson, M.T., 1991. Effects of natural enemies on the rate of herbivore adaptation to resistant host plants. Entomol. Exp. Appl. 58, 1—14.

Gould, F., Michael, B., Cohen, J.S.B., George, G.K., Van Duyn, J., 2006. Impact of small fitness costs on pest adaptation to crop varieties with multiple toxins: A heuristic model. J. Econ. Entomol. 99, 2091—2099.

Hare, J.D., 1990. Ecology and management of the Colorado potato beetle. Annu. Rev. Entomol. 35, 81—100.

Harris, M.O., Sandanayaka, M., Griffin, W., 2001. Oviposition preferences of the Hessian fly and their consequences for the survival and reproductive potential of offspring. Ecol. Entomol. 26, 473—486.

Harris, M.O., Stuart, J.J., Mohan, M., Nair, S., Lamb, R.J., Rohfritsch, O., 2003. Grasses and gall midges: plant defense and insect adaptation. Annu. Rev. Entomol. 48, 549—577.

Harris, M.O., Anderson, K.G., Anderson, K.M., Kanno, H., 2006. Proximate cues for reduced oviposition by Hessian fly on wheat plants attacked by conspecific larvae. Environ. Entomol. 35, 83—93.

Hatchett, J.H., Gallun, R.L., 1970. Genetics of the ability of the Hessian fly, *Mayetiola destructor*, to survive on wheats having different genes for resistance. Ann. Entomol. Soc. Am. 63, 1400—1407.

Hawthorne, D., 1998. Predicting insect adaptation to a resistant crop. J. Econ. Entomol. 91, 565—571.

Henneberry, T.J., Naranjo, S.E., 1998. Integrated management approaches for pink boll-worm in the southwest United States. Integr. Pest Manag. Rev. 3, 31—52.

Hoy, C.W., 1999. Colorado potato beetle resistance management strategies for transgenic potatoes. Am. J. Pot. Res. 76, 215—219.

Hoy, C.W., Head, G., 1995. Correlation between behavioral and physiological responses to transgenic potatoes containing *Bacillus thuringiensis* *d*-endotoxin in *Leptinotarsa dedemlineata* Say (Coleoptera: Chrysomelidae). J. Econ. Entomol. 88, 480—486.

Kanno, H., Harris, M.O., 2002. Avoidance of occupied hosts by the Hessian fly: oviposition behaviour and consequences for larval survival. Ecol. Entomol. 27, 177—188.

Kennedy, -G.G., Gould, -F., Deponti, -O.M.B., Stinner, -R.E., 1987. Ecological, agricultural, genetic, and commercial considerations in the deployment of insect-resistant germplasm. Environ. Entomol. 16, 327—338.

Kindler, S.D., Hays, D.B., 1999. Susceptibility of cool-season grasses to greenbug biotypes. J. Agric. Urb. Entomol. 16, 235—243.

Knolhoff, L.M., Heckel, D.G. 2014. Behavioral assays for studies of host plant choice and adaptation in herbivorous insects. Annu. Rev. Entomol. (in press)

Kota, M., Daniell, H., Varma, S., Garczynski, S.F., Gould, F., Moar, W.J., 1999. Overexpression of the Bacillus thuringiensis (Bt) Cry2Aa2 protein in chloroplasts confers resistance to plants against susceptible and Bt resistant insects. Proc. Natl. Acad. Sci. USA. 96, 1840−1845.

Lamb, R.J., Wise, I.L., Smith, M.A.H., McKenzie, R.I.H., Thomas, J., Olfert, O.O., 2002. Oviposition deterrence against *Sitodiplosis mosellana* (Diptera: Cecidomyiidae) in spring wheat (Gramineae). Can. Entomol. 134, 85−96.

Liu, Y.B., Tabashnik, B.E., 1997. Experimental evidence that refuges delay insect adaptation to *Bacillus thuringiensis*. Proc. R. Soc. Lond. B. 264, 605−610.

McNair, J.N., 1982. Optimal giving-up times and the marginal value theorem. Am. Nat. 119, 511−529.

Metz, T.M., Roush, R.T., Tang, J.D., Shelton, A.M., Earle, E.D., 1995. Transgenic broccoli expressing a *Bacillus thuringiensis* insecticidal crystal protein: implications for pest resistance management studies. Mol. Breed. 1, 309−317.

Michels Jr., G.J., 1986. Graminaceous North American host plants of the greenbug with notes on biotypes. Southwest. Entomol. 11, 55−66.

Morton, P.K., Foley, C.J., Schemerhorn, B.J., 2011. Population structure and spatial influence of agricultural variables on Hessian fly populations in the southeastern United States. Environ. Entomol. 40, 1303−1316.

Naranjo, S.E., Ellsworth, P.C., 2010. Fourteen years of Bt cotton advances IPM in Arizona. Southwest. Entomol. 35, 437−444.

Onstad, D.W., Meinke, L.J., 2010. Modeling evolution of *Diabrotica virgifera virgifera* (Coleoptera: Chrysomelidae) to transgenic corn with two insecticidal traits. J. Econ. Entomol. 103, 849−860.

Painter, R.H., 1951. Insect Resistance in Crop Plants. Macmillan, New York, p. 520.

Pan, Z., Onstad, D.W., Nowatzki, T.M., Stanley, B.H., Meinke, L.J., Flexner., J.L., 2011. Western corn rootworm (Coleoptera: Chrysomelidae) dispersal and adaptation to single-toxin transgenic corn. Environ. Entomol. 40, 964−978.

Panda, N., Khush, G.S., 1995. Host Plant Resistance to Insects. CAB International, Wallingford, Oxon, UK, in association with International Rice Research Institute. p. 431.

Porter, D.H., Burd, J.D., Shufran, K.A., Webster, J.A., Teetes, G.L., 1997. Greenbug (Homoptera:Aphididae) biotypes: selected by resistant cultivars or preadapted opportunists? J. Econ. Entomol. 90, 1055−1065.

Porter, D.R., Burd, J.D., Shufran, K.A., Webster, J.A., 2000. Efficacy of pyramiding greenbug (Homoptera: Aphididae) resistance genes in wheat. J. Econ. Entomol. 93, 1315−1318.

Puterka, G.J., Peters, D.C., 1990. Sexual Reproduction and Inheritance of Virulence in the Greenbug, *Schizaphis graminum* (Rondani). In: Campbell, R.K., Eikenbary, R.D. (Eds.), Aphid-Plant Genotype Interactions. Elsevier, Netherlands.

Ratanatham, S., Gallun, R.L., 1986. Resistance to Hessian fly, *Mayetiola destructor*, (Diptera: Cecidomyiidae) as affected by temperature and larval density. Environ. Entomol. 15, 305−310.

Ratcliffe, R.H., 2000. Breeding for Hessian fly resistance in wheat. In: Radcliffe, E.B., Hutchison, W.D. (Eds.), Radcliffe's IPM World Textbook. University of Minnesota, St. Paul, URL: <http://ipmworld.umn.edu>.

Ratcliffe, R.H., Safranski, G.G., Patterson, F.L., Ohm, H.W., Taylor, P.L., 1994. Biotype status of Hessian fly (Diptera: Cecidomyiidae) populations from the eastern United States and their response to 14 Hessian fly resistance genes. J. Econ. Entomol. 87, 1113−1121.

Ratcliffe, R.H., Cambron, S.E., Flanders, K.L., Bosque-Perez, N.A., Clement, S.L., Ohm, H.W., 2000. Biotype composition of Hessian fly (Diptera: Cecidomyiidae) populations from the southeastern, Midwestern, and northwestern United States and virulence to resistance genes in wheat. J. Econ. Entomol. 93, 1319–1328.

Ratcliffe, R.H., Patterson, F.L., Cambron, S.E., Ohm, H.W., 2000. Resistance in durum wheat sources to Hessian fly (Diptera: Cecidomyiidae) populations in Eastern USA. Crop Sci. 42, 1350–1356.

Sambaraju, K.R., Pendleton, B.B., 2005. Fitness of greenbug on wild and cultivated grasses. Southwest. Entomol. 30, 155–160.

Sardesai, N., Nemacheck, J.A., Subramanyam, S., Williams, C.E., 2005. Identification and mapping of *H32*, a new wheat gene conferring resistance to Hessian fly. Theor. Appl. Genet. 111, 1167–1173.

Schuster, D.J., Starks, K.J., 1973. Greenbugs: components of host-plant resistance in sorghum. J. Econ. Entomol. 66, 1131–1134.

Shelton, A.M., 2012. Genetically engineered vegetables expressing proteins from *Bacillus thuringiensis* for insect resistance: successes, disappointments, challenges and ways to move forward. GM Crops and Food: Biotech. Agric. Food Chain. 3, 175–183.

Shelton, A.M., Robertson, J.L., Tang, J.D., Perez, C., Eigenbrode, S.D., Preisler, H.K., et al., 1993. Resistance of diamondback moth (Lepidoptera: Plutellidae) to *Bacillus thuringiensis* subspecies in the field. J. Econ. Entomol. 86, 679–705.

Shelton, A.M., Tang, J.D., Roush, R.T., Metz, T.M., Earle, E.D., 2000. Field tests on managing resistance to Bt-engineered plants. Nat. Biotech. 18, 339–342.

Shufran, K.A., 2011. Host race evolution in *Schizaphis graminum* (Hemiptera: Aphididae): nuclear DNA sequences. Environ. Entomol. 40, 1317–1322.

Shufran, K.A., Burd, J.D., Anstead, J.A., Lushai, G., 2000. Mitochondrial DNA sequence divergence among greenbug (Homoptera: Aphididae) biotypes: evidence for host-adapted races. Insect Mol. Biol. 9, 179–184.

Smith, C.M., 2005. Plant Resistance to Arthropods: Molecular and Conventional Approaches. Springer, Dordrecht, p. 423.

Smith, M.A.H., Lamb, R.J., Wise, I.L., Olfert, O.O., 2004. An interspersed refuge for *Sitodiplosis mosellana* (Diptera: Cecidomyiidae) and a biocontrol agent *Macroglenes penetrans* (Hymenoptera: Pteromalidae) to manage crop resistance in wheat. Bull. Entomol. Res. 94, 179–188.

Sosa Jr., O., 1979. Hessian fly: resistance of wheat as affected by temperature and duration of exposure. Environ. Entomol. 8, 280–281.

Storer, N.P., Kubiszak, M.E., Ed King, J., Thompson, G.D., Santos, A.C., 2012. Status of resistance to Bt maize in *Spodoptera frugiperda*:lessons from Puerto Rico. J. Invertebr Pathol. 110, 294–300.

Stout, M.J., 2012. Reevaluating the conceptual framework for applied research on host-plant resistance. Insect Sci. 20, 263–272.

Stuart, J.J., Hatchett, J.H., 1988a. Cytogenetics of the Hessian fly: I. Mitotic karyotype analysis and polytene chromosome correlations. J. Hered. 79, 184–189.

Stuart, J.J., Hatchett, J.H., 1988b. Cytogenetics of the Hessian fly: II. Inheritance and behavior of somatic and germ-line-limited chromosomes. J. Hered. 79, 190–199.

Stuart, J.J., Chen, M.-S., Shukle, R., Harris, M.O., 2012. Gall midges (Hessian flies) as plant pathogens. Annu. Rev. Phytopath. 50, 339–357.

Tabashnik, B.E., 1994. Evolution of resistance to *Bacillus thuringiensis*. Annu. Rev. Entomol. 39, 47–79.

Tabashnik, B.E., Croft, B.A., 1982. Managing pesticide resistance in crop-arthropod complexes: interactions between biological and operational factors. Environ. Entomol. 11, 1137–1144.

Tabashnik, B.E., Patin, A.L., Dennehy, T.J., Liu, Y.B., Carrière, Y., Sims, M.A., et al., 2000. Frequency of resistance to *Bacillus thuringiensis* in field populations of pink bollworm. Proc. Natl. Acad. Sci. USA. 97, 12,980–12,984.

Tabashnik, B.E., Liu, Y.B., Dennehy, T.J., Sims, M.A., Sisterson, M.S., Biggs, R.W., et al., 2002. Inheritance of resistance to Bt toxin Cry1Ac in a field-derived strain of pink bollworm (Lepidoptera: Gelechiidae). J. Econ. Entomol. 95, 1018–1026.

Tabashnik, B.E., Biggs, R.W., Higginson, D.M., Henderson, S., Unnithan, D.C., Unnithan, G.C., et al., 2005a. Association between resistance to Bt cotton and cadherin genotype in pink bollworm. J. Econ. Entomol. 98, 635–644.

Tabashnik, B.E., Dennehy, T.J., Carrière, Y., 2005b. Delayed resistance to transgenic cotton in pink bollworm. Proc. Natl. Acad. Sci. USA. 102, 15389–15393.

Tabashnik, B.E., Fabrick, J.A., Henderson, S., Biggs, R.W., Yafuso, C.M., Nyboer, M.E., et al., 2006. DNA Screening Reveals Pink Bollworm Resistance to Bt Cotton Remains Rare After a Decade of Exposure. J. Econ. Entomol. 99, 1525–1530.

Tabashnik, B.E., Gassmann, A.J., Crowder, D.W., Carriére, Y., 2008. Insect resistance to Bt crops: evidence versus theory. Nat. Biotechnol. 26, 199–202.

Tabashnik, B.E., Sisterson, M.S., Ellsworth, P.C., Dennehy, T.J., Antilla, L., Liesner, L., 2010. Suppressing resistance to Bt cotton with sterile insect releases. Nat. Biotechnol. 28, 1304–1307.

Tabashnik, B.E., Morin, S., Unnithan, G.C., Yelich, A.J., Ellers-Kirk, C., Harpold, V.S., et al., 2012. Sustained susceptibility of pink bollworm to Bt cotton in the United States. GM Crops and Food: Biotech. Agric. Food Chain. 3, 194–200.

Talekar, N.S., Shelton, A.M., 1993. Biology, ecology, and management of the diamondback moth. Annu. Rev. Entomol. 38, 275–301.

Tang, J.D., Gilboa, S., Roush, R.T., Shelton, A.M., 1997. Inheritance, stability, and lack-of-fitness costs of field-selected resistance to *Bacillus thuringiensis* in diamondback moth (Lepidoptera: Plutellidae) from Florida. J. Econ. Entomol. 90, 732–741.

Tang, J.D., Collins, H.L., Roush, R.T., Metz, T.D., Earle, E.D., Shelton, A.M., 1999. Survival, weight gain, and oviposition of resistant and susceptible *Plutella xylostella* (Lepidoptera: Plutellidae) on broccoli expressing Cry1Ac toxin of *Bacillus thuringiensis*. J. Econ. Entomol. 92, 47–55.

Tang, J.D., Collins, H.L., Metz, T.D., Earle, E.D., Zhao, J.-Z., Roush, R.T., et al., 2001. Greenhouse tests on resistance management of Bt transgenic plants using refuge strategies. J. Econ. Entomol. 94, 240–247.

Taylor, C.E., Georghiou, G.P., 1979. Suppression of insecticide resistance by alteration of dominance and migration. J. Econ. Entomol. 72, 105–109.

Teetes, G.L., Schaefer, C.A., Johnson, J.W., 1974. Resistance in sorghums to the green bug; laboratory determination of mechanisms of resistance. J. Econ. Entomol. 67, 393–396.

Van Rensburg, J.B.J., 2007. First report of field resistance by stem borer, *Busseola fusca* (Fuller) to Bt-transgenic maize. S. Afr. J. Plant Soil. 24, 147–151.

Wan P., Huang Y., Tabashnik B.E., Huang M., Wu K., 2012a. The halo effect: suppression of pink bollworm on non-Bt cotton by Bt cotton in China. PLoS One 7, e42004. Available from: http://dx.doi.org/10.1371/journal.pone.0042004.

Wan, P., Huang, Y., Wu, H., Huang, M., Cong, S., Tabashnik, B.E., et al., 2012b. Increased Frequency of Pink Bollworm Resistance to Bt Toxin Cry1Ac in China. PLoS One. 7, e29975. Available from: http://dx.doi.org/10.1371/journal.pone.0029975.

Weng, Y.Q., Perumal, A., Burd, J.D., Rudd, J.C., 2010. Biotypic diversity in greenbug (Hemiptera: Aphididae): microsatellite-based regional divergence and host-adapted differentiation. J. Econ. Entomol. 103, 1454–1463.

Whalon, M.E., Miller, D.I., Hollingworth, R.M., Grafius, E.J., Miller, J.R., 1993. Selection of a Colorado potato beetle strain resistant to *Bacillus thuringiensis*. J. Econ. Entomol. 86, 226–233.

Wilhoit, L.R., Mittler, T.E., 1991. Biotypes and clonal variation in greenbug (Homoptera: Aphididae) populations from a locality in California. Environ. Entomol. 20, 757–767.

Williams, C.E., Collier, C.C., Sardesai, N., Ohm, H.W., Cambron, S.E., 2003. Phenotypic assessment and mapped markers for *H31*, a new wheat gene conferring resistance to Hessian fly (Diptera: Cecidomyiidae). Theor. Appl. Genet. 107, 1516–1523.

Williams, J.L., Ellers-Kirk, C., Orth, R.G., Gassmann, A.J., Head, G., Tabashnik, B.E., et al., 2011. Fitness cost of resistance to Bt cotton linked with increased gossypol content in pink bollworm larvae. PLoS One.e21863.

Withers, T.M., Harris, M.O., Madie, C., 1997. Dispersal of mated female Hessian flies (Diptera: Cecidomyiidae) in field arrays of host and nonhost plants. Environ. Entomol. 26, 1247–1257.

Zeiss, M.R., Brandenburg, R.L., VanDuyn, J.W., 1993. Suitability of seven grass weeds as Hessian fly (Diptera, Cecidomyiidae) hosts. J. Agric. Entomol. 10, 107–119.

Zhao, J.-Z., Collins, H.L., Tang, J.D., Cao, J., Earle, E.D., Roush, R.T., et al., 2000. Development and characterization of diamondback moth resistance to transgenic broccoli expressing high levels of Cry1C. Appl. Environ. Microbiol. 66, 3784–3789.

Zhao, J.-Z., Li, Y.-X., Collins, H.L., Shelton, A.M., 2002. Examination of the F2 screen for rare resistance alleles to *Bacillus thuringiensis* toxins in the diamondback moth (Lepidoptera: Plutellidae). J. Econ. Entomol. 95, 14–21.

Zhao, J.-Z., Cao, J., Li, Y., Roush, R.T., Earle, E.D., 2003. Transgenic plants expressing two *Bacillus thuringiensis* toxins delay insect resistance evolution. Nat. Biotech. 21, 1493–1497.

Zhao, J.-Z., Cao, J., Collins, H.L., Bates, S.L., Roush, R.T., Earle, E.D., et al., 2005. Concurrent use of transgenic plants expressing a single and two *Bacillus thuringiensis* genes speeds insect adaptation to pyramided plants. Proc. Natl. Acad. Sci. USA. 102, 8426–8430.

> CHAPTER 10

The Role of Landscapes in Insect Resistance Management

David W. Onstad[1] and Yves Carrière[2]
[1]DuPont Pioneer, Wilmington, DE
[2]Department of Entomology, University of Arizona, Tucson, AZ

Chapter Outline

Research concerning insect resistance management (IRM) has primarily emphasized the targeted pest species and the toxin or toxic plant to which it may evolve resistance. Two other components of every system have received less consideration: humans and the environment (Figure 10.1). The roles of stakeholders, policy makers, and pest-management decision makers are discussed by Onstad (Chapter 1), Mitchell and Onstad (Chapter 2), and Hurley and Mitchell (Chapter 13). Onstad *et al.* (Chapter 12) describe the role of natural enemies in IRM. Here we consider impacts of the landscape through its effects on abiotic and biotic conditions, behavior of pests, and selection for resistance.

For efficient development of IRM, the environment needs to be characterized with respect to the spatial and temporal distribution of the pest, the crop or toxin, refuge habitats, and variation within these elements in space and time. We present the following case studies and ideas to promote a greater appreciation for the role of landscapes and to advocate the inclusion of rigorous investigations of the role of landscapes in future IRM research. Without an understanding of temporal and spatial

Insect Resistance Management
DOI: http://dx.doi.org/10.1016/B978-0-12-396955-2.00010-2

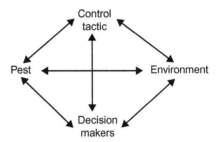

Figure 10.1 Four major components of a pest-management system and their interactions.

variability in natural and human–designed landscapes, IRM plans will be difficult to implement successfully. In the first major section, we highlight how heterogeneity over space influences IRM. In the second major section, we emphasize how variability over time affects IRM.

Spatial Heterogeneity and Management
Landscape Structure and Design: Insights from Simulation Models
Landscapes are typically heterogeneous and dynamic. Although natural variability exists, humans can also design the landscape. In this section, we describe how spatial heterogeneity and the relationships and proximities of various types of habitat patches influence IRM.

Landscape Heterogeneity, Pest Dispersal, and Behavior
A refuge is a pest habitat that is a source of susceptible individuals that can mate with resistant individuals that survive in a habitat where an insecticide is used. When pests are susceptible to a *Bacillus thuringiensis* (Bt) crop or an insecticide, the frequency of resistance alleles is low (Gould, 1998; Carrière *et al.*, 2010). The number of susceptibles emerging from refuges is thus expected to be larger than the number of resistant individuals emerging in treated areas. The general implications of immigration for treated areas from either refuges (two-way gene flow) or a large source of susceptible individuals (gene flow only into treated area) were considered by early IRM modelers (Comins, 1977; Taylor and Georghiou, 1979; Tabashnik and Croft, 1982). This early modeling work affirmed the value of a refuge strategy for IRM.

The refuge strategy, which has been used primarily to delay the evolution of pest resistance to Bt crops, requires the presence of refuges in or near fields where an insecticide is used. For refuges to be effective, the numerous susceptible insects from refuges must mate with the rare resistant pests that survive in treated areas. When resistance is recessive, most hybrid offspring from such matings are killed by the insecticide, which delays the evolution of resistance (Gould, 1998; Carrière et al., 2010). When resistance is not recessive, refuges need to be more abundant to delay resistance evolution effectively (Tabashnik and Gould, 2012; Brévault et al., 2013).

A variety of spatially explicit simulation studies have demonstrated that refuges delay the evolution of resistance to Bt crops. These studies give us an appreciation of the complexities introduced by pest dispersal and other behaviors, as well as population dynamics, within heterogeneous environments. For example, using two models of *Helicoverpa zea* on cotton (*Gossypium hirsutum*), Caprio (2001) studied the relationship between habitat patchiness and pest dispersal, mating, and oviposition. One model was a stochastic, spatially explicit simulation, and the second was a simpler deterministic model. Caprio found that the population density in refuges declined as the rate of movement between Bt crop fields and refuges increased, which accelerated resistance evolution when movement was high. However, resistance also evolved rapidly with low movement, likely because not enough susceptible individuals from refuges mated with resistant individuals surviving on the Bt crop. Caprio (2001) concluded that refuge deployment must account for the demographic characteristics of targeted pests in particular landscapes.

Peck et al. (1999) created a stochastic, spatially explicit model to explore the regional evolution of resistance by *Heliothis virescens* to Bt cotton. They discovered that the spatial scale and temporal pattern of refuge planting significantly affected the evolution of resistance. The time required for resistance to evolve was significantly longer in regions where the same fields were used as refuges year after year, compared with regions where refuge fields were changed randomly from year to year. Their results also suggested that spring and summer dispersal of adults among wild host plants and cotton could significantly affect the evolution of resistance, depending on distances flown.

Storer et al. (2003a,b) used a stochastic, spatially explicit simulation model to examine the role of spatial processes in the evolution of resistance in *H. zea* to Bt cotton and Bt corn (*Zea mays*). They found that selection for resistance was more intense in Bt cotton fields than in Bt corn fields. The results also indicated that local gene frequencies were highly dependent on local abundance of Bt crops despite the high mobility of adults. Storer *et al.* (2003b) found that the proportion of the landscape planted to corn critically affected the evolution of resistance. On a local scale, *H. zea* populations in clusters of fields containing high levels of Bt crops experienced faster evolution of resistance than populations in neighboring clusters with smaller abundances of Bt crops. Storer *et al.* (2003a) concluded that farm-level refuge requirements are important for managing the risk of resistance.

Storer (2003) also created a stochastic, spatially explicit model of *Diabrotica virgifera virgifera* on Bt corn and compared the rate at which resistance evolved under different refuge deployment scenarios. For a given refuge size, the model indicated that planting refuges as a block within each field of Bt corn would delay resistance longer than planting refuges and Bt corn in separate fields. As in Peck *et al.* (1999), Storer found that planting refuges in the same location year after year delayed resistance even further because populations of susceptible individuals reached higher densities in permanent than ephemeral refuges.

Sisterson *et al.* (2004) used a stochastic, spatially explicit model of *Pectinophora gossypiella* on Bt cotton to investigate the influence of population size on resistance evolution. They focused on interactions between carrying capacity for the pest, region size, dispersal, and percentage of fields planted with Bt cotton. The time to resistance decreased as region size increased, because larger regions were most likely to have at least one field in which resistance evolved rapidly and served as a source from which resistance spread throughout the region. They also found that resistance evolution was affected by interactions between carrying capacity, dispersal, and the percentage of fields planted with Bt cotton. Sisterson *et al.* (2005) used the same model to study the influence of relative abundance of refuges and their temporal and spatial distribution on resistance evolution. They found that these three factors significantly affected the evolution of resistance to Bt cotton. Resistance was delayed the longest when refuges had fixed locations over years and were distributed evenly throughout the region to prevent isolation of Bt cotton.

Farmers affect landscape structure by deciding whether or not to fertilize, cultivate, or irrigate crop fields. Onstad and Guse used a two-patch model to investigate the problems of IRM in irrigated fields of corn infested by *Diatraea grandiosella* and *Ostrinia nubilalis* (Guse *et al.*, 2002; Onstad *et al.*, 2002). The two species of stalk-boring Lepidoptera have different adult behaviors, particularly mating, oviposition, and male moth dispersal. Furthermore, these behaviors differ in irrigated and no-irrigated cornfields. Irrigation occurs primarily in southwestern and western corn-growing regions of the United States. Adults prefer to mate and move within moister, growing vegetation than drier vegetation. Guse et al. (2002) discovered that these adult behaviors strongly influenced the evolution of resistance to Bt corn. For example, increasing oviposition in natal refuges delayed the evolution of resistance, both because it increased the source potential of refuges and reduced the intensity of selection. Therefore, Guse et al. (2002) concluded that the interactions of landscape and insect behavior must be understood to properly develop IRM plans.

A common conclusion from the models described above is that pest population dynamics in refuges and resistance evolution are affected by pest dispersal, mating, and oviposition behaviors. Importantly, spatially explicit models reveal that resistance is most likely to emerge in and spread from locations where Bt crop fields are abundant relative to refuges (Peck et al., 1999; Storer et al., 2003a; Sisterson et al., 2004, 2005). This suggests that effective resistance management will be more likely if refuges are in proximity of every field of Bt crops. Furthermore, potential impacts of spatial heterogeneity need to be explicitly considered when developing IRM strategies.

Landscape Heterogeneity and Fitness Costs of Resistance

Novel insecticides are generally effective over the geographical range of target pests, showing that resistance alleles are initially rare across pest populations. This could arise because pests are screened to determine the usefulness of novel compounds. For susceptible pests, however, the consistently low frequency of resistance across populations probably results from a balance between mutation and selection (Crow, 1957). Mutation of alleles conferring susceptibility promotes an increase in resistance frequency, while costs arising from the pleiotropic effects of resistance mutations select against resistance. Refuges where insecticides are not used are thus expected to select against resistance, which is supported by empirical studies that often reveal fitness costs associated with resistance to insecticides (Carrière et al., 1994, Gassmann et al., 2009). Fitness costs have notable implications for the design of agricultural landscapes: When refuges are sufficiently abundant, selection against resistance may counterbalance selection for resistance, and insecticides remain effective for extensive periods.

The mosquito *Culex pipiens* has been controlled with organophosphates and *Bacillus sphaericus* (Bt) toxins for more than 40 years along the coast of southern France. However, *C. pipiens* evolved resistance to both types of insecticides at some locations, and costs associated with resistance were documented (Chevillon *et al.*, 2001; Raymond *et al.*, 2001). Because the number of environmentally acceptable alternatives for controlling this pest is limited, Lenormand and Raymond (1998) explored the possibility that the spread of resistance could be blocked if the areas of treated habitats remained small compared to the areas of untreated habitats. Using a deterministic simulation model and assuming symmetric but limited dispersal among treated and untreated habitats, they calculated the frequency of resistance alleles in treated habitats of different sizes after an equilibrium between selection and migration was attained. Under a broad range of parameters but always assuming nonrecessive costs of resistance (i.e., costs reducing fitness of heterozygote individuals), fixation of resistance alleles did not occur when the treated areas remained sufficiently small compared to the untreated areas. When the equilibrium frequency of resistance was low, most mosquitoes were killed by insecticides in treated areas, but movement form untreated areas increased population density again. Thus, the optimal control strategy consisted in using insecticides in an area small enough to prevent the evolution of resistance, but large enough to dilute the number of mosquitoes moving from untreated areas. For example, when Bt and organophosphate sprays were rotated in two parallel coastal belts approximately 10 km long surrounded by twice as much area left untreated, the equilibrium frequency of resistance alleles was zero and the population density of *C. pipiens* in treated areas was about one quarter of its density in untreated habitats. This modeling approach was extended in spatially explicit simulations of the evolution of pest resistance to a single-toxin Bt crop (Vacher *et al.*, 2003). Still modeling nonrecessive costs, the evolution of resistance did not occur if the proportion of the landscape planted to the Bt crop remained below a certain threshold. These studies indicate that nonrecessive fitness costs combined with gene flow between untreated and treated areas can limit the spread of resistance alleles, if an insecticide is used over a limited area.

Soon after commercialization of Bt cotton in Arizona, the estimated frequency of recessive alleles conferring resistance to Bt cotton was 0.16 (Tabashnik *et al.*, 2000; Morin *et al.*, 2003). Surprisingly, the estimated frequency of resistance alleles declined to 0.007 in the next year and remained low thereafter (Tabashnik *et al.*, 2005, 2010). Carrière and

Tabashnik (2001) used analytical and simulation models to formulate hypotheses that could explain such decline in resistance. Two key model components were fitness costs and incomplete resistance, which was defined as a lower fitness of homozygous resistant individuals on a Bt crop relative to a non-Bt crop (Carrière and Tabashnik, 2001). Incomplete resistance reduces selection favoring resistant individuals over susceptible individuals on Bt crops, while fitness costs increase selection disfavoring resistant individuals in refuges. Furthermore, because most resistance alleles are carried by heterozygous individuals when resistance alleles are rare, even small nonrecessive costs can strongly select for a decline in resistance (Carrière and Tabashnik, 2001). Depending on the abundance of refuges, the dominance of resistance, the extent of incomplete resistance, and the magnitude and dominance of costs, it was shown that resistance could increase, decrease, or remain stable (Carrière and Tabashnik, 2001). A similar modeling approach also demonstrated that fitness costs and incomplete resistance can significantly delay or prevent the evolution of resistance to single-toxin transgenic crops in haplodiploid and parthenogenetic pests, especially when resistance is recessive (Carrière, 2003; Crowder and Carrière, 2009).

Using empirical estimates of critical biological parameters in *P. gossypiella*, Tabashnik *et al.* (2005) further examined how refuge size combines with fitness costs and incomplete resistance to suppress resistance. They showed that thresholds in the abundance of refuges could lead to declines in the frequency of resistance for many realistic combinations of incomplete resistance and cost values. For example, with 50% refuges and incomplete resistance of 0.6 (i.e., the ratio of survival of resistant individuals on Bt cotton relative to non-Bt cotton), fitness costs > 35% resulted in a decline in resistance but costs < 35% caused an increase in resistance. Based on results from models, several nonmutually exclusive hypotheses were proposed to explain the reversal of resistance in *P. gossypiella*. These included among-year variation in climate or types of Bt cultivars that increased the magnitude of fitness costs and incomplete resistance, or among-year reductions in use of insecticides in refuges or increased movement from refuges to Bt crops that promoted declines in resistance (Carrière and Tabashnik, 2001; Carrière *et al.*, 2002; Tabashnik *et al.*, 2005; Cattaneo *et al.*, 2006).

Alphey *et al.* (2008) used related mathematical analyses to more generally investigate the effects of the relative abundance of refuges on the evolution of pest resistance to one-toxin Bt crops. They showed that the trajectory of resistance was determined by the relative abundance of refuges and the relative fitness of genotypes with and without resistance alleles in refuges and on the Bt crop. For a wide range of values for the dominance of resistance, dominance of fitness costs, and incomplete resistance, the change in frequency of resistance varied from positive to negative as the relative abundance of refuges increased. When the average fitness of heterozygotes was higher than fitness of both homozygous genotypes across refuges and Bt crop fields, a stable intermediate frequency of the resistance allele occurred over a range of refuge abundances.

Although the above mathematical models indicate that resistance may not evolve above certain thresholds of refuge abundance in the landscape, such thresholds may often be too large to be practical for resistance management. Manipulation of fitness costs and incomplete resistance have thus been envisaged to increase the capacity of refuges to delay resistance evolution (Crowder and Carrière, 2009; Gassmann et al., 2009; Carrière et al., 2010; Williams et al., 2011). The dominance and magnitude of fitness costs are affected by several environmental factors including natural enemies and host plant, suggesting that use of specific host plants or natural enemies in refuges could contribute to delaying resistance (Gassmann et al., 2009; Williams et al., 2011). Bt crops that produce high concentrations of Bt toxins, two or more Bt toxins, toxins that bypass known resistance mechanisms, or toxins with negative cross-resistance to Bt can also enhance incomplete resistance (Pittendrigh et al., 2004; Crowder and Carrière, 2009; Carrière et al., 2010; Tabashnik et al., 2011). For example, individuals that bear alleles conferring resistance to a single-toxin Bt crop could be hypersusceptible to another toxin, in which case there would be negative cross-resistance between the two toxins (Pittendrigh et al., 2004, Chapter 11). Incomplete resistance would thus be increased on a cultivar that produces both the Bt toxin and the toxin involved in negative cross-resistance. If the effects of the toxin with negative cross-resistance are small, however, a better alternative to delay resistance could be to apply such toxin in refuges, where it would actively select against resistance to Bt (Pittendrigh et al., 2004). So far, most research on the manipulation of costs and incomplete resistance has been done in the laboratory and greenhouse, and no toxin with negative cross-resistance to Bt has been identified. More research is thus needed to enhance our ability to manipulate refuges or Bt cultivars to delay the evolution of resistance.

Transgenic insecticidal crops that express two or more Bt toxins are known as pyramided plants (National Research Council, 2010; Brévault et al., 2013; Tabashnik et al., 2013). Costs are more effective in delaying the evolution of resistance to pyramided plants than to single-toxin plants, implying that refuges of moderate sizes could sometimes block the evolution of resistance. For example, assuming that a two-toxin Bt crop induced high mortality in susceptible individuals, Gould *et al.* (2006) used simulation models to evaluate the effect of relatively small costs on the evolution of resistance. They found that a small recessive cost of 5% to each toxin prevented the evolution of resistance with a 10% refuge, as long as the initial frequency of each resistance allele was below 0.055. Costs were very effective in delaying resistance in this case because only the extremely rare individuals with four resistance alleles were resistant to the Bt crop (resistance to each toxin was recessive), while the most abundant individuals with two resistance alleles at one of the two resistance loci were less fit in the refuge (costs to each toxin were recessive). With nonrecessive or larger costs, the area of refuges blocking the evolution of resistance would be smaller and the initial frequency of resistance alleles required for a positive increase in resistance higher (Gould *et al.*, 2006). For pests highly susceptible to pyramided Bt crops such as *O. nubilalis*, *H. virescens*, and *P. gossypiella*, it would thus seem that preventing the evolution of resistance with practical refuge sizes could be feasible.

A key feature of the refuge strategy is that abundant susceptible insects from refuges mate with the rare homozygote resistant individuals surviving on an insecticidal crop. Unfortunately, the movement of resistant individuals into refuges eventually increases the frequency of resistance alleles in refuges, which triggers an increase in resistance frequency (Comins, 1977; Caprio and Tabashnik, 1992; Caprio, 2001; Sisterson *et al.*, 2004). While the fitness costs of insect resistance select against resistance in refuge and generally contribute to preserving their efficacy, a more drastic procedure has been considered to increase refuge efficacy. The screened refuge concept proposes that movement between refuges and insecticidal crops could be actively restricted by specific procedures, such that resistance alleles would remain rare in refuges but enough susceptible individuals would be supplied to Bt crop fields to delay resistance for long periods (Ringland and George, 2010). For example, a small refuge could be grown in a cage in fields of a Bt crop, and a gap in the cage would encourage movement of susceptible insects from the refuge to the Bt field but limit movement of individuals from the Bt field to the refuge.

Using simulation models, Ringland and George (2010) showed that such a screened refuge could block the evolution of resistance to a single-toxin Bt crop when costs are present, or delay resistance for extensive periods even when costs are absent. However, screened refuges could be costly or difficult to maintain in fields of Bt corn and cotton. Nevertheless, this approach could be appropriate for managing the evolution of resistance to Bt sprays by *Trichoplusia ni* in British Columbia greenhouses (Janmaat and Myers, 2003). In this system, movement of Bt-resistant *T. ni* among greenhouses contributes to the spread of resistance, but long-range migration of Bt-susceptible individuals from regions of California provides a significant source of susceptible individuals (Franklin and Myers, 2008, Franklin *et al.*, 2010). It thus seems that a screened refuge could be envisaged, whereby movement of Bt-resistant individuals from greenhouses to the field would be restricted and entrance of Bt-susceptible migrants in the greenhouses promoted.

Landscape Structure and Design: Insights from Empirical Studies

Spatial and Temporal Variation in Source Potential of Refuges

Spatially explicit simulation models of the regional development of resistance to Bt crops indicate that the abundance and distribution of Bt crop fields and refuges affect pest population dynamics and the evolution of resistance (see above). Such models show that the regional spread of resistance is initiated by an increase in resistance frequency at locations where refuges are rare relative to Bt crops. It is therefore critical for the development of refuge strategies to take into account within-region variation in abundance of Bt crops and refuges.

When many susceptible individuals of a pest are killed by a Bt crop, the net reproductive rate of females that lay eggs on such crops could be reduced enough to cause a decline in the abundance of a pest in refuges. This idea was supported by a simple metapopulation model and a spatially explicit model parameterized according to the life history of the *P. gossypiella*, an ecological specialist of cotton in Arizona (Carrière *et al.*, 2003). Both models showed that pest population density decreased regionally as the abundance of Bt cotton or dispersal between non-Bt and Bt cotton increased, and the net reproductive rate in non-Bt cotton decreased. Model predictions were supported by a ten-year study of 15 regions in Arizona, which compared the abundance of moths before and after deployment of Bt cotton. Moths were trapped in pheromone traps along fields of Bt and non-Bt cotton. While moth density increased or decreased following deployment of Bt cotton in regions where less than about 65% Bt cotton was used, the regional decline in moth density was proportional to the abundance of Bt cotton in regions where more than 65% Bt cotton was used.

Carrière et al. (2004a) studied the association between the local cotton landscape and the abundance of Bt-susceptible *P. gossypiella* moths. Moth captures in pheromone traps at two types of sites were compared: sites with less than 5% non-Bt cotton within a radius of 0.75 km and less than 20% non-Bt cotton within a radius of 1.5 km (such sites did not comply with the refuge strategy mandated for *P. gossypiella* by the U.S. Environmental Protection Agency [EPA]); or sites with more than 5% non-Bt cotton within a radius of 0.75 km or more than 20% non-Bt cotton within 1.5 km (these sites were compliant). During a two-year period, moth captures were significantly lower at the noncompliant than the compliant sites, showing that the source potential of refuges declined at sites where the relative abundance of Bt cotton was high. Although these studies provided a mechanism linking population dynamics in refuges to local and regional declines in abundance of Bt-susceptible individuals, they did not directly demonstrate that the source potential of refuges was reduced by the presence of Bt crops because *P. gossypiella* abundance was not monitored in refuges.

Three studies directly monitored insect abundance in refuges and confirmed that the source potential of refuges is reduced by the presence of Bt crops. In the first study, increased use of Bt cotton in northern China for the control of *Helicoverpa armigera* resulted in regional declines in the density of eggs and larvae in refuges of non-Bt cotton and of corn, peanut (*Arachis hypogaea*), soybean (*Glycine max*) and vegetables (Wu et al., 2008). In northern China, Bt cotton reduces population density of *H. armigera* compared to non-Bt cotton because Bt toxins are more effective than synthetic insecticides for controlling this pest. The regional decline in population density of *H. armigera* occurred even if Bt cotton represented only 10% of the cropland. Cotton was one of the few suitable hosts during part of the growing season, which funneled *H. armigera* populations on cotton and subsequently decreased dispersal to other crops (Wu et al., 2008). A significant long-term decrease in the abundance of *O. nubilalis* larvae in refuges of non-Bt corn was also documented in several states in the north-central United States following the deployment of Bt corn (Hutchison et al., 2010). Finally, a drastic decline in density of *P. gossypiella* eggs and larvae in refuges of non-Bt cotton was observed across six provinces of China after 11 years of Bt cotton use (Wan et al., 2012). Together with several other studies documenting regional population declines of pests occurring after the deployment of Bt crops (Storer *et al.*, 2008; Carrière *et al.*, 2010), these results indicate that the source potential of refuges can be reduced locally or regionally by the presence of Bt crops.

Spatial and Temporal Variation in Scale of Refuge Source Effects

Development of a refuge strategy to delay the evolution of pest resistance to Bt crops requires information on the spatial scale at which refuges can contribute susceptible individuals to Bt crop fields. Direct estimates of dispersal from mark-release-recapture studies, or inferences on dispersal based on variation in pest density between non-Bt and Bt crop fields, can provide a valuable starting point to evaluate the scale at which refuges can provide susceptible insects to Bt crop fields (Tabashnik *et al.*, 1999; Carrière *et al.*, 2001; Spencer *et al.*, 2009). Such estimates may not be sufficient, however, because the source potential of refuges can be affected by characteristics of the pest and landscape (see above).

The scale of refuge source effects can be evaluated in a landscape context with spatially explicit statistical analyses (Carrière *et al.*, 2004a). Such analyses are based on geographic information system mapping of potential refuges and Bt crops in concentric rings around "focal sites" where the abundance of susceptible individuals is measured (Figure 10.2). Because refuges are sources of susceptible individuals, a positive association is expected between the abundance of susceptible insects at the focal sites and the areas of refuges in the rings. However, refuges that are too far from focal sites will have no statistically significant effects on the abundance of susceptible insects (Carrière *et al.*, 2012a). The greatest distance at which the association between the areas of refuge in rings and insect abundance at the focal sites is significant defines the zone of influence of refuges, which is the maximum scale of the source effect of refuges (Carrière *et al.*, 2004a).

To evaluate the zone of influence of refuges of non-Bt cotton for *P. gossypiella* in each of two years, pheromone trap data were used to measure the abundance of susceptible moths at different focal sites (Carrière *et al.*, 2004a). In a first analysis based on data from all sites, the zone of influence of refuges was 0.75 km in the first year and 2.25 km in the second. In a second analysis separating sites that were compliant or not with the refuge strategy mandated by the U.S. EPA (see above), the zone of influence of refuges was still 0.75 km for both types of sites in the first year. In the second year, however, the zone of influence was 2.25 km for

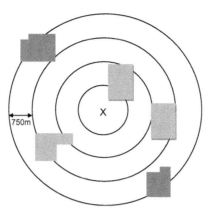

Figure 10.2 Four concentric rings of a width of 0.75 km surrounding a focal site (denoted by a ✕). In the *Pectinophora gossypiella* example discussed, a pheromone trap was located at focal sites to evaluate the abundance of Bt susceptible moths. For each ring, the association between number of moths captured in traps and areas of non-Bt and Bt cotton (respectively, in blue and green) was evaluated. *Reprinted from Carrière et al. (2004a).*

sites where the relative abundance of Bt cotton was low, but less than 0.75 km for sites with higher relative abundance of Bt cotton. These analyses indicated that the scale of the source effects of refuges varied extensively in time and was reduced at sites where the relative abundance of Bt cotton is high (Carrière *et al.*, 2004a).

Fitness costs associated with resistance to Bt select against resistance in refuges and can vary extensively across refuges of different host plants (Gassmann *et al.*, 2009; Carrière *et al.*, 2010). Accordingly, the capacity of refuges to provide susceptible insects may not provide a sufficient basis to fully evaluate how refuges can delay resistance. For example, two types of refuges could produce the same number of susceptible insects and have the same scale of source effects, but could still differ importantly in their capacity to delay resistance if fitness costs were present in one refuge type and absent in the other (Gassmann *et al.*, 2009; Carrière *et al.*, 2010). Ultimately, a full evaluation of the capacity of refuges to delay resistance should be based on analyses of spatial variation in the frequency of resistance alleles.

The spatially explicit statistical approach described above was used to analyze spatial variation in the frequency of resistance to pyriproxyfen in *Bemisia tabaci* (Carrière *et al.*, 2012a). Pyriproxyfen is a juvenile hormone analogue used for *B. tabaci* control in Arizona cotton. Analyses were conducted at distances of up to 3 km from focal sites where resistance to pyriproxyfen was monitored (Figure 10.3). To test the hypothesis that the distribution of refuges and treated cotton affects the evolution of resistance, the effects of four types of habitats were considered in central Arizona: treated cotton and potential refuges of alfalfa (*Medicago sativa*), melons (cucurbits), and untreated cotton. Data from 46 sites sampled from 2002 to 2005 were used to formulate statistical models of the association between resistance to pyriproxyfen and abundance of the four host-plant types, and to determine the maximum spatial scale at which each habitat affected resistance. The areas of alfalfa and melon were not significantly associated with resistance frequency. However, the areas of untreated cotton were negatively associated with resistance, and areas of treated cotton were positively associated with resistance. The zone of influence of cotton refuges was 3.0 km, while treated cotton acted as a source of resistance at a distance of up to 2.75 km.

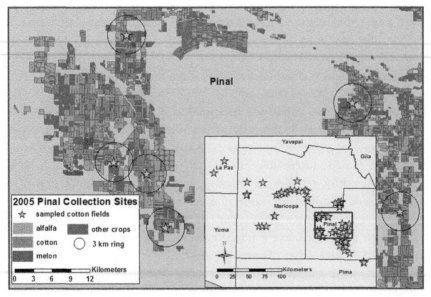

Figure 10.3 Cotton fields sampled for *Bemisia tabaci* resistance to pyriproxyfen between 2002 and 2009. Inset shows egg survival to a discriminating concentration in 84 populations. Larger image shows location of six fields (stars) evaluated for resistance in Pinal County in 2005 with surrounding crops. Rings around sampled cotton fields have a radius of 3 km. *Reprinted from Carrière et al. (2012a).*

The statistical models developed from the 2002–2005 data were then used to make predictions about resistance in 38 populations evaluated for resistance to pyriproxyfen from 2006 to 2009. Two types of models were developed to take into account temporal and regional variation in the frequency of resistance and in the abundance of treated and untreated cotton. Predicted resistance to pyriproxyfen was calculated by substituting the areas of treated and untreated cotton near the 38 sites sampled from 2006 to 2009 into these multiple regression models. The association between predicted and observed resistance to pyriproxyfen was positive and significant in both analyses, confirming that cotton fields treated with pyriproxyfen selected for resistance and untreated cotton fields delayed resistance (Carrière *et al.*, 2012a). Previously, empirical evidence for the refuge strategy had been provided by small-scale experiments and historical comparisons across species and regions (Liu and Tabashnik, 1997; Zhao *et al.*, 2005; Tabashnik *et al.*, 2008, 2009, 2013). The analysis of resistance to pyriproxifen in *B. tabaci* provides more direct support for the effectiveness of the refuge strategy because it is based on large-scale data for a single species in a particular region.

Natural Refuges

Some pests targeted by Bt crops have a broad host range and exploit many cultivated and wild host plants that could serve as "natural" refuges. When natural refuges are sufficiently abundant in a region, it may not be necessary for farmers to plant refuges of a non-Bt crop to delay the evolution of resistance. For example, the requirement that non-Bt cotton refuges be deployed to manage resistance to pyramided Bt cotton in *H. virescens* and *H. zea* was abolished from Texas to the Mid-Atlantic section of the United States because data indicated that weeds and non-Bt crops other than cotton, which can host larvae of both pests, might be abundant enough to delay resistance (U.S. EPA, 2007). However, because the abundance, distribution, and source potential of alternative refuges can vary spatially and temporally (see above), it is critical to take into account such variation when designing a refuge strategy for large regions.

Spatially explicit analyses of biochemical markers of adults trapped in Bt crop fields can be used to evaluate how the source potential of different types of refuges changes across the landscape. Carbon isotope ratios ($^{13}C/^{12}C$) of insect wings indicate whether an individual fed as a larvae on a C3 (e.g., cotton, peanuts, soybeans) or C4 (e.g., corn, *Sorghum bicolor*, grasses) plant (Gould *et al.*, 2002). Gossypol is a phytochemical exclusively produced by the lysigenous glands of cotton. The presence of gossypol residues in moth abdomens shows that the larval host of trapped moths was cotton (Orth *et al.*, 2007).

To determine the source potential of noncorn refuges for fields of Bt corn in the E- and Z-race of *O. nubilalis* in upstate New York, adults of each race were trapped along fields of sweet corn with pheromone traps baited with either the E- or Z-race-specific pheromone blend (O'Rourke *et al.*, 2010). Analyses of carbon isotope ratios revealed that only 4% of Z-race moths developed on C3 (i.e., noncorn) hosts, indicating that refuges of non-Bt corn are important to manage the evolution of resistance in this race. The E-race had a different host-use pattern than the Z-race, as about 18% of E-race moths originated from C3 hosts. However, the proportion of moths from C3 hosts varied from 0 to 40% across five trapping sites. This shows that spatial variation in the distribution of C3 hosts was too high in upstate New York for noncorn refuges to consistently delay the evolution of resistance in the E-race (O'Rourke *et al.*, 2010).

Two-toxin Bt cotton was recently introduced to Burkina Faso and could be used in more regions of the West African cotton belt if commercialization proves successful. It is often assumed that cotton refuges are not required to delay *H. armigera* resistance to Bt cotton in regions where diversified crops and noncultivated hosts are present (Brévault *et al.*, 2012). Because the evolution of resistance in *H. armigera* could cut short the profitability of Bt cotton, three sites in the West African cotton belt of Cameroon were evaluated to determine whether noncotton refuges may consistently delay the evolution of resistance (Brévault *et al.*, 2012). Only non-Bt cotton was grown at these sites, which differed in the abundance of cotton and other cultivated and noncultivated *H. armigera* hosts. Moths were collected with pheromone traps along cotton fields at each site and analyzed for carbon isotope ratios and gossypol residues. When the first generation of *H. armigera* moths emerged from cotton, > 87% of trapped moths were from noncotton hosts at any of the sites. Second-generation moths still commonly originated from noncotton hosts (between 20 and 65%), although fewer moths from the third generation had noncotton larval hosts (between 7.5 and 45%). These data indicated that noncotton refuges were equivalent to ≥ 7.5% non-Bt cotton refuges treated with insecticides throughout the cotton-growing season. However, a simulation model based on data on the source potential of noncotton refuges and key *H. armigera* biological parameters indicated that refuges of non-Bt cotton would likely be needed to significantly delay the evolution of *H. armigera* resistance to two-toxin Bt cotton (Brévault *et al.*, 2012).

Other studies have used biochemical markers to evaluate the source potential of various refuges for Bt corn or Bt cotton, although such studies did not explicitly consider spatial variation within regions (Bontemps *et al.*, 2003, Nagoshi *et al.*, 2007). For example, Gould *et al.* (2002) used carbon isotope analyses of *H. zea* moths trapped in Louisiana and Texas to evaluate the refuge potential of C4 hosts (i.e., noncotton) for Bt cotton fields. Although cultivated and noncultivated C4 hosts were expected to be rare (Louisiana) or absent (Texas) at the end of the cotton-growing season, about 20% of moths trapped at any time during the cotton-growing season were from C4 hosts. Based on these data, Gould *et al.* (2002) proposed that refuges of non-Bt corn were important for managing resistance to Bt cotton in *H. zea*, and that some of these non-Bt corn refuges were probably located in northern locations that contributed moths to cotton fields in Louisiana and Texas via long-range migration.

Building on this work, Head *et al.* (2010) used both carbon isotope and gossypol analyses to evaluate the importance of non-Bt cotton refuges for the management of *H. zea* resistance to Bt cotton. Analyses of carbon isotope ratios were based on moths trapped during two years at four sites in each of five states. Five to six locations were sampled at each of the four sites, resulting in a total of 20–24 sampling locations per state in a given year. After averaging data for each state and year, more than about 15% of moths trapped during the period of moth emergence from cotton originated from C4 hosts, providing some evidence that refuges of non-Bt corn were important for managing *H. zea* resistance to Bt cotton. Gossypol analyses were based on moths trapped using the same design, except that 10–18 sampling locations were used in each of three states. After averaging data for each state and year, about 40% of moths trapped during the period of moth emergence from cotton had detectable gossypol residues. Taken together, data from the carbon isotope and gossypol analyses showed that most moths moving into cotton fields originated from noncotton C3 plants or C4 hosts like corn. Based on these results, Head *et al.* (2010) concluded that cotton refuges would play a minor role in the management of resistance to Bt cotton in *H. zea*, relative to other types of refuges. An important caveat associated with this conclusion, however, is that within-state spatial variation in the source potential of the refuge types was not considered. As shown above, such variation could affect the capacity of refuges to consistently delay the evolution of resistance (O'Rourke *et al.*, 2010; Brévault *et al.*, 2012).

To improve local IRM strategies for Bt cotton in Australia, Sequeira and Playford (2001) examined the suitability of several field crops (*Vigna angularis, Sorghum bicolor, Zea mays, Cajanus cajan,* and *Helianthus annuus*) as refuges for transgenic insecticidal cotton. They based their assessment on the relative production of *Helicoverpa* pupae in each crop. Field assessments showed that *Cajanus cajan* has the greatest potential as a refuge for transgenic cotton. They concluded that postharvest cultivation in cotton fields is largely ineffective for resistance management under Queensland conditions. Sequeira and Playford (2001) proposed an IRM strategy that includes refuge-crop options and the use of late-season trap crops of *Cajanus cajan* as an alternative to postharvest cotton cultivation.

Within-Field Spatial Mosaics

In this section, we describe spatial mosaics that are the consequences of natural phenomena, such as plant-gene expression and pollen dispersal, and typical agricultural practices, such as seed saving. For example, plant-to-plant mosaics can result from variable gene expression within a single field of a Bt crop. Soil moisture, soil nutrients, herbivory, and topography all vary over space and influence the growth of plants and the production of toxin in these plants. Hence, expression of a gene for Bt production may vary among plants from a single cultivar over a crop field, creating a spatial mosaic of toxin doses. Furthermore, many Bt cultivars that express the same toxin gene may produce variable amounts of toxin (Gray *et al.*, 2007; Vaughn *et al.*, 2005; Showalter *et al.*, 2009). As neighboring farmers plant multiple cultivars, even on their own farms, spatial mosaics are created. The consequences of this variability for IRM deserve more attention.

Spatial mosaics also exist in traditional host-plant resistance and in Bt crops that have variable defenses and toxin levels in different plant tissues. In this case, the mosaic does not occur across plants but within them. Many plants have evolved different defenses for reproductive and vegetative organs. Often defenses are induced by injury from various herbivores. Furthermore, plant breeders may breed for protection of only one plant organ that is vulnerable to arthropod damage. Thus, toxin levels in traditionally bred plants are expected to be mosaics of various toxins mixed with nontoxic tissues. In commercialized Bt crops, toxins are constitutively expressed throughout the plant because of the use of particular promoters and transformations in their development. Even in Bt crops, however, within-plant mosaics are common.

Horner *et al.* (2003) observed the feeding behavior of *H. zea* on single-toxin Bt corn ears, which have 25% of their kernels not producing the Bt toxin when kernel production results from cross-pollination between two parents hemizygous for the Bt gene (note that Bt production is dominant; Heuberger *et al.*, 2008a). The larvae seemed to be responding to the variable toxicity of the kernels, and Horner *et al.* (2003) proposed that behavioral resistance could evolve because larvae preferring nontoxic kernels would be favored by natural selection. Burkness and Hutchison (2012) observed that a mosaic of Bt kernels was found throughout the ear of cross-pollinated refuge corn. They found that the ear tip had the highest percentage of cross-pollinated kernels, which may be caused by the synchrony of pollen shed and silking between Bt and non-Bt corn hybrids. Burkness and Hutchison (2012) observed the highest levels of cross-pollination (75%) in the first four rows (3 m) of non-Bt corn adjacent to Bt corn plots. Onstad and Gould (1998a) modeled the scenario in which none of the corn silks and kernels expresses the toxin but stems and leaves did and found that the plant mosaic did not significantly affect the evolution of resistance by *O. nubilalis*. In this case, few larvae fed on the ear, and those that did occurred in only one of two annual generations. Thus, within-plant mosaics must be identified before crop commercialization and IRM strategies have to be adjusted to account for them.

Different types of refuges can be planted to delay resistance. External refuges are relatively large blocks of non-Bt plants grown separately from blocks of Bt plants. Gene flow between Bt and non-Bt plants results in the presence of Bt plants in external refuges and non-Bt plants in fields of Bt crops (Chilcutt and Tabashnik, 2004; Heuberger *et al.*, 2010). Gene flow between Bt and non-Bt plants is mediated by pollen and seeds. Seed-mediated gene flow arises from impurities in purchased seeds, mixing of seeds during planting, or seeds left in the soil in previous seasons (i.e., volunteer plants). Pollen-mediated gene flow is mediated by wind and pollinators. Corn is wind-pollinated: Cross-pollination mainly occurs at distances < 500 m, although isolated events can take place at greater distances (Gealy *et al.*, 2007). Cotton primarily self-pollinates but is also outcrossed by insect pollinators. Pollinator-mediated gene flow typically occurs at distances < 50 m, although honey bees can transport pollen between Bt and non-Bt plants at greater distances (Llewellyn *et al.*, 2007; Heuberger *et al.*,

2010). Pollen-mediated gene flow between Bt and non-Bt plants is expected to decline as a function of distance between fields of these crops. However, the adventitious presence of Bt plants in refuges and non-Bt plants in fields of Bt plants will result in cross-pollination of surrounding plants, which could blur the expected association between distance and gene flow (Heuberger *et al.*, 2008b). The presence of adventitious Bt plants in refuges and non-Bt plants in fields of Bt plants is difficult to prevent completely and in some cases can be high (NRC, 2010; Heuberger *et al.*, 2010).

Pollen dispersal between Bt and non-Bt plants could transform external refuges and fields of Bt plants into seed mixtures for pests feeding on seeds. When larvae move between plants, the presence of Bt plants in refuges are expected to accelerate resistance evolution. In contrast, the presence of non-Bt plants in fields of Bt plants could increase the survival of susceptible, heterozygous, and homozygous resistant insects when larvae are mobile. Depending on whether larval movement increases survival more in susceptible than resistant insects or the reverse, the presence of non-Bt plants in fields of Bt plants could delay or accelerate the evolution of resistance. The potential consequences of pollen dispersal between external refuges and fields of Bt plants for resistance evolution have been compared with models and limited experiments.

P. gossypiella feeds on seeds produced in cotton bolls. Studies conducted in Arizona in 2004 and 2007 showed that most refuges of non-Bt cotton contained < 2.5% of plants producing the Bt toxin Cry1Ac, although the adventitious presence of Bt plants in refuges was as high as 20% (Heuberger *et al.*, 2008b, 2010). Three types of plants producing Cry1Ac occurred in refuges: homozygous Bt plants with 100% Bt seeds, hemizygous Bt plants with about 75% Bt seeds, and non-Bt plants outcrossed with Bt plants with about 15% of Bt seeds (Heuberger *et al.*, 2008b, 2010). To investigate the impact of the adventitious presence of Bt seeds in refuges on the evolution of resistance to Cry1Ac in *P. gossypiella*, Heuberger *et al.* (2008a) measured the survival of Cry1Ac-susceptible and resistant individuals on artificial bolls containing 0, 20, 70, and 100% Bt seeds. Compared to survival in bolls with 100% non-Bt seeds, survival of larvae was progressively reduced in bolls with 20, 70, and 100% Bt seeds. Survival of the three genotypes (homozygous susceptible; heterozygous; homozygous resistant) did not differ in bolls with 100% non-Bt seeds or 20% Bt seeds. However, survival of the homozygous resistant genotype was significantly higher than survival of the other genotypes on bolls with 70 and 100% Bt seeds. The survival of heterozygous and homozygous susceptible larvae did not differ on any of the boll types. Thus, the presence of adventitious Bt seeds in refuges conferred a selective advantage to the rare homozygous resistant individuals and reduced the number of susceptible (and heterozygous) individuals in refuges. Simulation models based on empirical estimates of survival of the genotypes in Bt, non–Bt, and mosaic bolls indicated that typical levels of adventitious presence of Bt seeds in refuges in Arizona would have negligible effects on the evolution of resistance (Heuberger et al., 2008a). Bolls with 100% Bt seeds had greater accelerating effects on resistance than bolls producing a lower proportion of Bt seeds, but the evolution of resistance was only significantly accelerated when > 35% homozygous Bt plants were present in refuges. Other simulations suggested that effects on resistance were small because the presence of Bt seeds in refuges did not increase the fitness of heterozygous larvae compared to susceptible larvae.

Chilcutt and Tabashnik (2004) reported that the Bt toxin Cry1Ab was detected in kernels of non–Bt corn up to 31 m from a plot of Cry1Ab corn. They reported a negative correlation between Cry1Ab concentration in kernels of non–Bt corn and distance between plots of non–Bt and Bt corn. Chilcutt and Tabashnik (2004) estimated that the mean Cry1Ab concentration in kernels of non–Bt corn, located 1 m from a plot of Cry1Ab corn, was 45% of the mean concentration in kernels of Cry1Ab corn. Burkness and Hutchison (2012) observed the highest levels of cross-pollination (75%) in the first four rows (3 m) of a block refuge of non–Bt corn adjacent to Bt corn plots. The configuration of non–Bt and Bt corn blocks influences the proportion of non–Bt ears fertilized by Bt pollen (Burkness et al., 2011).

In a modeling study of O. nubilalis in mixtures of non–Bt and Bt-corn, Kang et al. (2012) showed that Bt-pollen drift has little impact on the evolution of Bt resistance in O. nubilalis. Because adults mate and disperse randomly in the modeled corn landscape and because neonates move only once and move without suffering any harm from starting on Bt corn plants, evolution of resistance was delayed equally well by both seed mixtures and blocks with the same proportion of refuge. Kang et al. (2012) modified and improved the modeling approach used by Onstad and Gould (1998a) to model O. nubilalis in cross-pollinated corn. Kang et al. (2012) determined that 76% of first instars move away from non–Bt plants and 90% move away from Bt plants. The survival rate due to movement was 0.1 (Onstad and Gould, 1998a). A variety of literature sources suggested that a mean of 20% of first instars move to corn ears with a maximum percentage of 40%.

Kang *et al.* (2012) studied a range of proportion of cross-pollinated corn ears in block refuges and assumed 100% cross-pollination in seed mixtures. Burkness *et al.* (2011) estimated that survival of second instars feeding on non-Bt ears fertilized by Cry1Ab pollen and non-Bt ears fertilized by non-Bt pollen are 0.600 ± 0.066 and 1.000 ± 0.0, respectively. Burkness *et al.* (2011) also estimated that survival of second instars feeding on Cry1Ab ears fertilized by Cry1Ab pollen and on Cry1Ab ear fertilized by non-Bt pollen are 0.075 ± 0.053 and 0.029 ± 0.029, respectively, which are not significantly different. Kang *et al.* (2012) used the mean (i.e., 0.052) as the survival of susceptibles on Bt-corn ears, which is much higher than the survival on vegetative tissues of Bt corn. Kang *et al.* (2012) concluded that low-toxin expression in ears of Bt corn can reduce the durability of Bt corn expressing a single toxin. Their results indicated that the survival rate of heterozygous larvae in Bt-corn ears expressing one Bt toxin has more impact on the evolution of resistance in *O. nubilalis* than parameters related to larval movement to Bt ears or survival rate of homozygous susceptible larvae in Bt ears.

In developing countries where farmers grow small fields and often save some of their harvested seeds, the adventitious presence of non-Bt plants in fields of Bt plants and of Bt plants in block refuges could be substantial. Onstad *et al.* (2012) created a model of the pest, *Maruca vitrata*, and cowpea (*Vigna unguiculata*) to study the possible evolution of resistance to Bt cowpea. They focused on population dynamics and genetic factors in a region of West Africa. Onstad *et al.* (2012) assumed that a small proportion (about 0.10) of cowpea fields would be converted to Bt cowpea each year of a five-year period and adoption rates would level off at 50% Bt cowpea. All seed would be saved by cowpea growers at end of year. Therefore, in the next year, only a small proportion of fields would have pure Bt plants. Furthermore, because of 1% cross-pollination each season, fields of pure, traditional cowpea (refuge) would decline each year. The landscape also consisted of wild host plants for *M. vitrata*. The results indicated that as long as a pyramided Bt cowpea could be developed, seed saving by farmers and reliance on natural refuge were not major problems for resistance management. The very small proportion of cross-pollination did not influence model results or conclusions. Onstad *et al.* (2012) concluded that if efforts are made to deploy Bt cowpea only into the regions where *M. vitrata* is not endemic, then there is little to no concern with resistance emerging rapidly in the *M. vitrata* population.

Heuberger *et al.* (2011) examined the effects of gene flow, larval feeding behavior, dominance of resistance, and refuge abundance and type on the evolution of resistance to one-toxin Bt cotton. They modeled three types of larvae: the first type was sedentary and only fed on the natal plant; the second did not discriminate between Bt and non-Bt cotton and moved five times between plants; and the third discriminated between Bt and non-Bt cotton and settled on the first non-Bt plant encountered, although it could still move up to five times between plants. Fitness of the larvae on Bt and non-Bt cotton was based on empirical data for key pests of cotton and reflected incomplete resistance and fitness costs (Tabashnik *et al.*, 2008). Pollen-mediated gene flow between fields of Bt cotton and cotton refuges occurred yearly. Such gene flow resulted in accumulation of off-type plants because seeds were recycled. Percentage of pollen exchanged between Bt fields and cotton refuges was based on empirical data. In simulations with a 5% or 20% refuge, a higher percentage of seeds from non-Bt cotton refuges was fertilized by pollen from Bt cotton fields than the reverse, because the area of Bt cotton was higher than the area of non-Bt cotton. For example, with a 20% refuge, a maximum of 3.25% of seeds in refuges were fertilized by Bt pollen every year, while a maximum of 0.81% of seeds in Bt cotton fields were fertilized by non-Bt pollen. Seed-mediated gene flow occurred only once at the beginning of simulations. However, when seed-mediated gene flow occurred together with pollen-mediated gene flow, plants from seed-mediated gene flow outcrossed other plants, which resulted in the accumulation of off-type plants. Based on empirical data, the maximum rate of seed-mediated gene flow in fields of non-Bt plants was 20% (Heuberger *et al.*, 2010). A similar maximum rate of seed-mediated gene flow was assumed for fields of Bt cotton. Once larvae had completed development, adults mated and oviposited randomly across the entire habitat of Bt and non-Bt plants.

In simulations, effects of gene flow were considered practically significant if gene flow decreased the time to resistance evolution by several years and resistance evolved in less than 20 years (Heuberger *et al.*, 2011). Practically significant effects were more common with a 20% than a 5% refuge, because resistance evolution was faster regardless of gene flow with a 5% refuge. When resistance was recessive or dominant, gene flow had little practical significance because resistance respectively occurred slowly or rapidly. However, both pollen- and seed-mediated gene flow leading to the presence of Bt plants in refuges significantly accelerated resistance in discriminate and indiscriminate larvae when the dominance of resistance was intermediate. Accelerating effects were more important on indiscriminate larvae than on discriminate larvae because the death of susceptible larvae encountering Bt plants was greater in indiscriminate than discriminate larvae. In contrast, pollen-mediated gene flow in fields of Bt plants had little practical effect on resistance evolution, possibly because accumulation of off-type plants was considerably lower in fields of Bt plants than in refuges. Seed-mediated gene flow leading to the presence on non-Bt plants in fields of Bt plants had variable effects: It accelerated resistance evolution in discriminate and indiscriminate larvae, but delayed resistance in sedentary larvae or in discriminate larvae when resistance was recessive. This accelerating effect of seed-mediated gene flow in discriminate and indiscriminate larvae possibly occurred because homozygous resistant and heterozygous larvae were better able to escape mortality imposed by Bt plants than susceptible larvae, when non-Bt plants were present. Additional simulations indicated that larger cotton refuges of 50% significantly mitigated the effects of gene flow only when dominance was intermediate and larvae sedentary. Because noncotton refuges are not affected by gene flow from Bt cotton, increasing the area of noncotton refuges in the landscape was a much better approach to sustain the durability of one-toxin Bt cotton when gene flow was important.

Using a similar approach, Glaum et al. (2011) also examined the impacts of gene flow between one-toxin Bt and non-Bt crops on the evolution of pest resistance. In their simulations, resistance was recessive, and larvae were either sedentary or moved twice indiscriminately between plants. Fitness of larvae on Bt and non-Bt plants, as well as levels of gene flow between Bt and non-Bt plants, was set arbitrarily. As in Heuberger et al. (2011), gene flow from fields of Bt plants to refuges was more important than the reverse, because the area of Bt cotton was higher than the area of non-Bt cotton. However, seed-mediated or pollen-mediated gene flow in fields of non-Bt plants was set at maximum rates higher than in Heuberger et al. (2011). Other differences included density-dependent mortality of larvae in refuges and fields of Bt plants, and restricted movement of adults between refuges and fields of Bt plants.

Comparison of simulations with and without density-dependent mortality revealed that density dependence greatly reduced the number of susceptible adults that moved into fields of Bt plants, which accelerated the rate of resistance evolution Glaum et al. (2011). With density-dependent mortality in refuges, gene flow was more likely to significantly accelerate the evolution of resistance when resistance was recessive, compared to results by Heuberger *et al.* (2011), which did not model density-dependent mortality. The effect of gene flow and larval movement was also affected by the extent of adult movement. High adult movement further reduced the production of susceptible insects in refuges because most eggs laid by females on Bt plants did not survive when resistance alleles were rare. When movement of adults between field types was high and larvae were sedentary, both seed- and pollen-mediated gene flow to fields of Bt plants delayed the evolution of resistance. However, when larvae moved between plants, gene flow did not affect resistance evolution, showing that larval movement significantly accelerated the evolution of resistance when fields of Bt plants contained non-Bt plants. With either seed- or pollen-mediated gene flow to non-Bt fields, gene flow did not significantly affect the evolution of resistance when larvae were sedentary. However, when larvae moved between plants, gene flow did accelerate resistance evolution, showing that larval movement significantly accelerated the evolution of resistance when fields of non-Bt plants contained Bt plants. Thus, with high adult movement and recessive resistance, gene flow in refuges or fields of Bt plants always accelerated the evolution of resistance.

With high adult movement between field types and fields of Bt and non-Bt plants subjected to either seed- or pollen-mediated gene flow at the same time, increased gene flow delayed the evolution of resistance in sedentary larvae but did not affect the evolution of resistance when larvae moved between plants. These trends were most similar to results obtained when gene flow only affected fields of Bt plants. Thus, the presence of non-Bt plants in fields of Bt plants dominated the process of resistance evolution when adult movement was high, contrary to findings by Heuberger *et al.* (2011). Based on these patterns, Glaum *et al.* (2011) concluded that refuge contamination from seed saving in developing countries is unlikely to have major consequences on resistance evolution, when resistance is recessive and movement of adults between fields of Bt and non-Bt plants is high. However, this conclusion likely depends on the level at which population size in refuges was bounded by density-dependent mortality in the simulations. Furthermore, the assumption that strong density-dependent larval mortality occurred in every field of non-Bt and Bt plants is questionable, as many pests are primarily controlled by insecticides in refuges, and spatial variation in use of insecticides can be high. [Onstad *et al.* (Chapter 12) discuss many issues related to density-dependent survival]. In simulations with low adult movement, increased gene flow in both fields of Bt and non-Bt plants no longer affected the evolution of resistance in sedentary larvae, but did accelerate the evolution of resistance when larvae moved between plants. Thus, when adult movement was low, the presence of Bt plants in fields of non-Bt plants seemed to have the greatest influence on resistance evolution, in accord with findings by Heuberger *et al.* (2011).

Additional types of mosaics exist when alternate host plants, particularly weeds, occur in the crop field or in adjacent areas. Transgenic insecticidal crops may be planted in an unplanned mixture with weeds. For example, there is concern that IRM for *D. v. virgifera* on Bt corn can be affected by weeds in corn fields or rotated soybean fields. *D. virgifera virgifera* larvae can survive on almost all grass species for 10 days and develop to at least the final larval stadium on most grass species (Clark and Hibbard, 2004; Oyediran *et al.*, 2004; Wilson and Hibbard, 2004). Thus when target pests can feed on weeds, IRM strategies for Bt crops should consider the influence of weed control.

A special kind of weed is the volunteer or adventitious and unwanted crop plant germinating from seed left in the field. Krupke *et al.* (2009) describe the problems caused by volunteer corn plants that are herbicide-resistant and produce some Bt toxins. They observed high levels of damage to these plants by *D. v. virgifera* in soybean fields. Krupke *et al.* (2009) speculated that intermediate toxins levels could permit more heterozygotes than susceptibles to hatch and survive in soybean fields, leading to selection for resistance to Bt corn.

Seed Mixtures: Designed Mosaics with Nontoxic Plants

A seed mixture or seed blend is an intentionally produced, randomly mixed set of two or more kinds of crop seed. One component is insecticidal seed, such as Bt seed, and the other component is nontoxic seed of the same crop. Thus, for transgenic insecticidal crops, the refuge plants randomly grow within the transgenic crop field. When certain types of larval movement between plants is limited (see below), seed mixtures have value relative to block refuges for five main reasons (Table 10.1; Gould and Tabashnik, 1998; Carrière *et al.*, 2001). First, blending the refuge seeds in bags of Bt seed eliminates the concern about lack of compliance with regulations concerning block-refuge planting (Chapter 13). Second, when a pest species has limited ability to disperse out of large

Table 10.1 Summary of IPM Issues Pertaining to Choice of Deployment of IRM Refuges as Seed Mixtures or Blocks (Onstad et al. 2011)

Issue	Seed Mixture	Block
Pest monitoring	Difficult	Typical
Control of secondary pests	New approaches	Traditional
Biological control	New approaches	Traditional
Insecticide use	Less	More
Quality of refuge relative to Bt corn	Similar	Probably different
Effects on IRM of larval behavior	More risk	Less risk
Effects on IRM of adult behavior	Less risk	More risk
Adoption of PIP technology*	Higher	Lower
Compliance with IRM rules	Higher	Lower

Used by permission of Entomological Society of America
*PIP plant incorporated pesticide

blocks and mate with individuals emerging from a different block, seed mixtures place refuge plants and insects surviving on them closer to Bt plants and the associated survivors (Onstad *et al.*, 2011; Spencer *et al.*, 2013). Third, the non-Bt plants are more likely to be similar in quality to insecticidal plants because they must be planted at the same time and at the same site. Phenological similarity and other qualities are not ensured with the use of block refuges. Fourth, insecticide sprays could be reduced in seed mixtures compared to block refuges where populations of the target pest may increase more rapidly. Fifth, yield may not be reduced by target pests as much in seed mixtures than in comparable areas of block refuges and Bt crop fields.

Wilhoit (1991), Mallet and Porter (1992), and Tabashnik (1994) were the first to evaluate the value of seed mixtures for IRM. The study by Wilhoit is discussed in Chapter 12. Tabashnik (1994) extended the more limited analysis of Mallet and Porter (1992) and concluded that lower durability of seed mixtures compared to separate fields of Bt plants and refuges occurred primarily because mobile susceptible larvae fed on Bt plants in seed mixtures, which reduced the effective size of refuges (Tabashnik, 1994). Seed mixtures were also less durable when heterozygous larvae survived better than susceptible larvae in seed mixtures, which increased the dominance of resistance, as proposed by Mallet and Porter (1992). Although little empirical evidence has supported the idea that movement between Bt and non-Bt plants increases the dominance of resistance (Carrière *et al.*, 2004b; Heuberger *et al.*, 2008a), recent experiments conducted with seed mixtures of non-Bt and Cry1Ac-cotton and

H. zea show that increased dominance of resistance in seed mixtures can substantially accelerate the evolution of resistance compared to situations where blocks of non-Bt and Bt plants are used (Brévault, Tabashnik and Carrière, unpublished results).

General models assessing the evolution of resistance in seed mixtures are based on the assumption that larvae feed on tissues before moving to another plant (Mallet and Porter, 1992; Tabashnik, 1994). Conclusions from such models may not apply for several corn pests that move as neonates before feeding (Kang *et al.*, 2012; Onstad *et al.*, 2011). At least four factors related to larval survival and movement can influence the risk of resistance with seed mixtures: timing, harm, probability of movement, and asymmetry of movement (Onstad *et al.*, 2011). Larvae can have zero, one, two, or more events in which they move from plant to plant. With no larval movement, seed mixtures have no risk and several benefits concerning IRM. If only one event occurs and neonates move prior to feeding, then it is possible that susceptible individuals incur no harm that is not also experienced by heterozygotes during and after movement. Movement as a neonate without feeding is a characteristic of some corn pests (Onstad *et al.*, 2011). Unless the probability of moving away from a non-Bt plant is high, risk is not likely to be high. It is also possible that movement by larvae is asymmetrical with greater movement off Bt plants than non-Bt plants (Onstad *et al.*, 2011). Such asymmetrical movement could result in more larvae developing on non-Bt plants in a mixture than expected based on the relative abundance of Bt and non-Bt plants, which may or may not lower the fitness of individuals developing on non-Bt plants compared to pure refuge blocks. The complexity of the issue makes it difficult to generalize about the consequences of seed mixtures for IRM of all insects and all Bt crops (Table 10.1).

Several models and empirical studies of specific systems have contributed to the debate regarding the value of seed mixtures in IRM (Caprio, 1994; Peck *et al.*, 1999; Arpaia *et al.*, 1998). Shelton *et al.* (2000) and Tang *et al.* (2001) using a model system observed *Plutella xylostella* on transgenic insecticidal broccoli (*Brassica oleracea*) in greenhouse and field studies and concluded that separate block refuges are superior to mixtures of plants for IRM. However, neither of these experiments strongly supports the conclusion that separate refuges delayed resistance more effectively than mixed refuges. In the field experiment involving several generations of *P. xylostella* exposed to seed mixtures or blocks of Bt and non-Bt broccoli, data on susceptibility to Bt indicated no difference between populations exposed to mixed and separate refuges (Shelton et al., 2000). In a similar greenhouse experiment (Tang et al., 2001), data on susceptibility to Bt in larvae exposed to mixed versus separate refuges was not reported at the end of the experiment.

Onstad (2006) created a deterministic IRM model of D. v. virgifera that emphasized processes during the larval stage to evaluate seed mixtures of non-Bt and Bt corn for IRM. The model included many of the same calculations and processes modeled by Onstad and Gould (1998a) and Davis and Onstad (2000) for O. nubilalis. Onstad (2006) concluded that seed mixtures are a reasonable alternative to separate block refuges for D. v. virgifera IRM. Pan et al. (2011) improved the model of Onstad (2006) with better parameter calibration and inclusion of more realistic adult behavior. In the model, the single toxin is expressed at a dose that does not kill all susceptible larvae, and resistance is incompletely recessive. Larvae move from plant to plant but leave insecticidal plants without incurring any harm. Mating is not random in cornfields because females mate where they emerge and males have limited dispersal ability (Spencer et al., 2013). Pan et al. (2011) demonstrated that the seed mixtures produced equal or greater durability than block refuges that were relocated each year. Block refuges maintained at the same location from year to year delayed resistance evolution much more than seed mixtures or relocated blocks. Murphy et al. (2010) observed D. v. virgifera in a two-year study of block and seed mixture refuges. They found that adults emerging from refuge corn emerged more synchronously with those emerging from Bt corn in seed mixtures when compared with block refuges. However, the proportion of adults emerging from non-Bt corn was significantly greater in a block or strip refuge than in a seed mixture. In addition, more adults emerged from Bt corn in the seed mixture.

Onstad and Gould (1998a) concluded that separate block refuges were less risky than seed mixtures because of the uncertainty concerning larval movement and mortality for O. nubilalis in corn. Davis and Onstad (2000) tested the model of Onstad and Gould (1998a) using field data for transgenic insecticidal corn and found that separate block refuges would delay evolution of resistance better than seed mixtures. In a hypothetical analysis based on the Onstad and Gould (1998a) model, Onstad et al. (2011) demonstrated how plant toxicity, larval movement, and larval survival can influence the evolution of resistance with block refuges and seed mixtures. In the model, Z is the probability of a neonate moving from a non-Bt plant, V is the probability of a neonate moving from a Bt plant, and pdts is the survival rate due to predispersal tasting of the toxin (Onstad and Gould, 1998a). No larvae died during movement, and toxin mortality occured after neonates settled on each type of corn. Predispersal tasting survival (pdts) was varied from 0.5–1.0 for homozygous susceptibles (Figure 10.4: Note labels below bars in the figure); pdts was equal to 1 for all other genotypes. Thus, the scenarios presented in Figure 10.4 are the pdts values for susceptible larvae and the movement rates (Z, V) for all genotypes. Mean dominance of the resistance allele was 0.009 in block refuges and for scenarios A and B with seed mixtures. A dominance value of 0 means resistance is recessive. For the other simulated scenarios, the dominance changed as the parameters changed. As usual, the threshold for declaring resistance was a frequency of the resistance allele ≥ 0.5.

Figure 10.4 shows that predictions are sensitive to model parameters concerning larval behavior and survival. When there was no larval movement or movement occurred equally among all plants and no differential

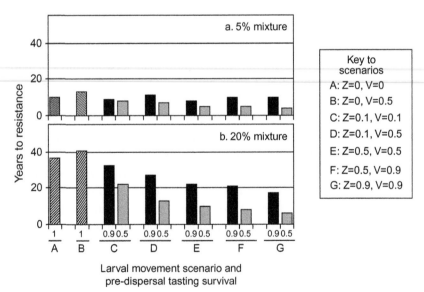

Figure 10.4 Influence of three larval movement and survival parameters on evolution of resistance of *Ostrinia nubilalis* (years to 50% resistance-allele frequency) to single-toxin Bt corn with (a) 5% refuge and (b) 20% refuge seed mixtures. Bar colors represent different predispersal tasting survival rates. Parameters Z and V are probabilities of a neonate moving away from non-Bt corn plant and Bt corn plant, respectively. *Reprinted from Onstad et al. (2011) by permission of Entomological Society of America.*

survival due to tasting occurred among the genotypes (scenario A), 5% and 20% refuges delayed resistance 10 and 36 yr, respectively (Figure 10.4a,b). Block refuge results were the same (not shown). When pdts was equal to 1 and larvae were repelled from Bt corn ($V = 0.5$) but none left non-Bt corn ($Z = 0$), resistance evolved in 13 y for a 5% refuge and 40 y for a 20% seed mixture (scenario B, Figure 10.4a,b). These results show the value of extra larvae being able to find non-Bt plants in a seed mixture, at least when there is no differential selection due to tasting and moving.

In most cases, larval movement will hasten resistance evolution when the extra differential selection exists (pdts <1) and the heterozygotes have a selective advantage over the homozygous susceptibles. Scenario C with pdts $= 0.9$ produced results lower than but similar to the benchmark values in scenario A (Figure 10.4). Almost all other cases in Figure 10.4 have predictions smaller than those for block refuges or seed mixtures without movement (scenario A). The exceptions are scenarios D and F with a 5% seed mixture and with pdts $= 0.9$; evolution takes more than 10 years because the repellency from Bt corn provides the benefit of escape to non-Bt plants and little differential selection due to the high pdts value.

Note that when movement from Bt corn plants (V) is held constant, evolution is faster when movement away from non-Bt corn (Z) increases. This can be seen by comparing scenarios D and E or F and G (Figure 10.4). This is due to larvae leaving non-Bt plants in the seed mixture. In these scenarios, the seed mixture is less effective in producing susceptible insects than is a block refuge because movement of larvae in a block refuge can produce higher numbers of adults (Onstad et al., 2011).

Barclay (1997) created a model of *Pissodes strobi* and its infestations in Sitka spruce (*Picea sitchensis*) to study the role of mixed plantings in managing insect resistance to host-plants. Some Sitka spruce trees are susceptible to insect attacks and others are resistant. Barclay (1997) used the model to evaluate a strategy of interplanting susceptible and resistant trees to delay the evolution of insect resistance to resistant trees. He discovered that if only one gene locus is involved, the development of resistance occurs more quickly than if resistance is governed by two independent loci. Furthermore, Barclay (1997) found that evolution of resistance in the pest is sensitive to processes such as preferential attack of susceptible trees, redistribution of susceptible insects from resistant to susceptible trees, and immigration of susceptible insects in mixed plantings.

TEMPORAL DYNAMICS AND MANAGEMENT

Just as toxin concentration in transgenic insecticidal crops varies over space, toxin concentration is also likely to vary over a single growing season and fluctuate from year to year. As the environment and crop change over time, so will the selection pressure on the targeted pest. The timing of management activities also can determine the outcome of pest management and alter the evolution of resistance. Crop rotation is designed change in the landscape over time. Rotated crops are planted in a field according to a certain schedule. Trap crops may be more or less attractive to pests depending on the planting date and maturation of the trap crop relative to other crops in the landscape (Carrière *et al.*, 2012b).

Crop Rotation

Spencer *et al.* (Chapter 7) provide details about the two species of *Diabrotica* that have evolved resistance to crop rotation; farmers practice rotation of corn crops with other species on over 80% of the agricultural landscape in several regions of the north-central United States. Two mechanisms have contributed to adaptation to the rotation of corn and usually soybean on an annual basis. The larvae cannot survive on soybean roots, so eggs laid in the soil of cornfields during a rotation of crops will die. However, neither species has evolved the ability to feed on soybean. Rather, *Diabrotica barberi* has evolved prolonged diapause in overwintering eggs that allows many eggs to pass through two winters before hatching in a cornfield. In contrast, *D. v. virgifera* has evolved a reduced tendency for laying eggs in the soil of cornfields. Many eggs are oviposited in fields planted to soybeans or other crops, but larvae hatch in cornfields in the next year because soybeans and other crops are frequently rotated with corn.

Onstad *et al.* (2001) used a model to explore several hypotheses concerning the evolution of behavioral resistance to crop rotation by *D. v. virgifera*. Their primary interest was to determine how the landscape design interacts with alternative genetic systems to influence evolution of the pest. In this case, landscape design is determined by cropping practices and the annual switching from corn to soybean and back again. The authors modeled both a 2-allele and a 3-allele genetic system: an X allele for no movement out of corn, a Y allele for the tendency to move to all types of vegetation, and a Z allele for the tendency to move only to the crop most commonly rotated with corn, soybean. The landscape consisted of four patches: continuously planted corn, rotated corn, soybean rotated with corn, and extra vegetation. Landscape diversity was represented by the extra vegetation (not rotated with corn) in the landscape. The rotation level, R, was equal to the sum of the proportions of soybean and rotated corn and tended to be greater than 0.9 in the region where the rotation-resistant phenotype was first detected (Onstad *et al.*, 2001).

Because adult feeding in crops other than corn is associated with egg laying outside of corn, the terms *monophagy* (X-phenotype) and *polyphagy* (Y-phenotype) may help focus attention on the locations where adult insects are present. Wild-type, monophagous individuals move from the natal cornfield and distribute themselves (and their eggs) uniformly across the two patches of corn. Polyphagous individuals move into all patches according to their proportional representation in the region. After emerging in a cornfield, the soybean specialists (Z-phenotype) move only to the soybean patch. If allele expression is additive, the heterozygotes are either polyphagous (XY and YZ) or oligophagous (XZ), with movement only to corn and soybean.

For the X-Y system, the Y-allele frequency only increased at very high levels of rotation. When X or Y was dominant, $R = 0.77$ was the threshold that determined whether Y disappeared or eventually became fixed at 1. For the additive scenario, where the XY-phenotype was expressed, the Y allele disappeared below $R = 0.726$ and became fixed above $R = 0.844$. Between these values of R, stable polymorphisms existed, with the most prevalent genotype switching from XX to XY to YY as R increased. The greater the value of R, the faster the Y-allele frequency increased.

In the X-Z system, only the additive case permitted the Z-allele frequency to increase in the population. When Z and X were additive, Z disappeared for $R < 0.79$. For $R > 0.79$, stable polymorphisms occurred, and the Z-allele frequency increased at a faster rate as R increased. The maximum Z-allele frequency was 0.53 when there was no continuous corn and rotation level, R, was at its highest value. At $R = 0.90$, the Z-allele frequency stabilized at 0.42 after 25 years, and the intermediate oligophagous phenotype (XZ) was the most prevalent. With $R < 0.93$, the frequency of Z did not increase when Z was recessive and it disappeared when Z was dominant.

Given that Z increased only when X and Z were additive, Onstad et al. (2001) evaluated only three scenarios for the 3-allele system:

constant additive X and Z and allele Y with variable expression. When Y was recessive to both X and Z, the results were the same as those described above for the additive X-Z system. When all were additive, Z disappeared from the system, and the later simulated years were similar to results of the 2-allele, additive X-Y system described above, indicating that Y was superior to Z. When Y was dominant to the other alleles and R ≥ 0.80, the polyphagous phenotypes were most prevalent. As R increased, the final simulated frequencies of X and Z both declined to just over 1%.

As the level of rotation increased and the evolution of behavioral resistance to crop rotation occurred, the winners in these simulations were often polyphagous adults. These results match field observations that *D. virgifera virgifera* adults move into a variety of crops (not just soybean) in the areas where resistance to corn-soybean rotation has been reported (Chapter 7). This may mean that the new strains behave most like YY or XY genotypes. The oligophagous XZ-phenotype may be prevalent in reality, but field observations and the model results for the 3-allele system suggest the superiority of the polyphagous insects to the oligophagous or soybean-specialist phenotypes.

Onstad *et al.* (2001) discovered that the diversity of the landscape does not influence the X-Z system, but it does have a significant effect on the evolution of polyphagy in the X-Y system. The Y-allele frequency and the polyphagous phenotypes decreased as the proportions of either extra vegetation or continuous corn increased. Onstad *et al.* (2003) tested this conclusion and its consequences for IRM by using observations from a large region of the north-central part of the United States. Observations of the geographic spread of the variant resistant to crop rotation supported the hypothesis that landscape diversity slows the spread of the variant. Movement by the variant to and egg laying in vegetation that is neither corn nor a crop rotated with corn creates a significant fitness cost for the variant in regions with greater landscape diversity.

Dynamics of Titer Decline and Toxin Concentration

Concentrations of Bt toxins in Bt corn and cotton often decline during the growing season (Dutton *et al.*, 2004; Nguyen and Ja, 2009; Showalter *et al.*, 2009; Kamath *et al.*, 2010). Such declines could hasten resistance evolution by increasing the survival of resistant insects relative to susceptible insects, increasing the survival of heterozygotes relative to susceptible individuals, and invalidating the fundamental assumption of the pyramid strategy: the killing of insects resistant to one toxin by another toxin (Carrière et al., 2010; Brévault et al., 2013).

Onstad and Gould (1998b) studied how corn-plant senescence and resulting reallocation of nitrogen from vegetative tissues to kernels could affect Bt concentration in corn tissues consumed by O. nubilalis. They modeled titer decline and its effects on different genotypes using a function with three parameters. One parameter determined when the decline begins. This parameter is important because it determines how many larvae are impacted and in what life stage. The second parameter described how gradual the decline is—in other words, the time required for mortality to decrease from its usual high level to zero. Again, the length of this period determines how many larvae are affected by this different selection pressure. The third parameter described the time between the reduction in mortality for susceptible homozygotes and the reduction in mortality of heterozygotes. When no time separates the two, no extra differential selection occurs due to titer decline. Onstad and Gould (1998b) found under many conditions that resistance evolved faster because of titer decline favoring heterozygotes in the pest population. However, they noted that the original toxicity and dominance of resistance before titer decline influenced the results. In some cases, resistance evolved more slowly when Bt concentration declined than without titer decline. Because the degree of synchrony between the toxin dynamics and the pest's life stages is critical, the influence of titer decline was affected by the proportion of larvae that entered diapause in the first generation and the timing of this phenomenon relative to corn senescence.

In many pests, the dominance of resistance to Bt toxins increases as toxin concentrations decreases (Tabashnik et al., 2004, Tabashnik and Carrière, 2008). Seasonal declines in Bt toxin concentrations are thus expected to increase the dominance of resistance, especially in pests less susceptible to Bt toxins. This was observed in H. armigera, in which resistance was recessive when larvae fed on Bt cotton in the 5—6 leaf stage, but partially dominant when larvae fed on Bt cotton in the 15 leaf stage (Bird and Akhurst, 2005). Concentration of Cry1Ac was about 75% lower in Bt cotton in the 15 leaf stage than in the 5—6 leaf stage. In Diatraea saccharalis, however, the dominance of resistance did not markedly increase when insects fed on older, reproductive Cry1Ab corn compared to younger vegetative corn, although survival of the genotypes with and without resistance alleles increased on reproductive corn (Wu et al., 2007;

Carrière *et al.*, 2010). In *O. nubilalis*, individuals resistant to Cry1Ab had very low survival on vegetative corn, but resistant individuals and their F1 progeny had significantly higher survival on reproductive corn (Crespo *et al.*, 2009). As plants grew older, new tissues with relatively low Bt toxin concentrations were produced (e.g., silk and pollen). Larvae preferentially fed on such tissues, which could have contributed to the increasing survival of individuals with resistance alleles on reproductive corn (Crespo *et al.*, 2009).

Seasonal declines in Bt toxin concentrations can reduce the success of the pyramid strategy. A central assumption of the pyramid strategy is that insects resistant to one toxin will be killed by the other toxin, which is called redundant killing (Gould, 1986; Roush, 1998). As plants age, toxin concentrations often decline, and resistance to a single toxin could significantly enhance survival on two-toxin Bt crops, if the concentration of the other toxin is not high enough to kill resistant individuals (Carrière *et al.*, 2010). In *H. armigera* selected for resistance to Cry2Ab, survival on cotton producing Cry1Ac and Cry2Ab increased during the growing seasons (Mahon and Olsen, 2009). On field-grown cotton in the pre-square stage, none of the susceptible, F1, and resistant larvae survived. On older cotton in the fruiting stage, however, survival of susceptible, F1, and resistant larvae was, respectively, 1.6%, 1.7%, and 8.5%. Selection for resistance to Cry2Ab did not involve cross-resistance to Cry1Ac (Mahon and Olsen, 2009). Thus, it is likely that increased survival of Cry2Ab-resistant larvae on two-toxin cotton occurred because the concentration of Cry1Ac was not sufficient to kill individuals resistant to Cry2Ab (Mahon and Olsen, 2009). The assumption of redundant killing on cotton producing Cry1Ac and Cry2Ab was also recently tested in *H. zea* (Brévault *et al.*, 2013). After selecting a *H. zea* strain for resistance to Cry1Ac, survival of the unselected and resistant strain was evaluated on field-grown cotton producing Cry1Ac and Cry2Ab. In diet bioassays, resistance to Cry2Ab did not differ significantly between the unselected and selected strain, indicating that cross-resistance between Cry1Ac and Cry2Ab was not extensive. However, survival to adulthood on two-toxin cotton was significantly higher in the resistant than nonselected strain, showing that the concentration of Cry2Ab in pyramided cotton was not sufficient to kill insects resistant to Cry1Ac. These data show that the concentration of Bt toxins in pyramided Bt cotton may not be sufficient to ensure redundant killing in *H. zea* and *H. armigera* during the entire growing season. Thus, management of the evolution of resistance in pests with low susceptibility to

Bt toxins (such as *H. zea* and *H. armigera*) could be enhanced by better knowledge of the effects of seasonal declines in the concentration of Bt toxins (Brévault *et al.*, 2013).

Despite extensive variation among Bt cultivars in the production of Bt toxins during the growing season, seasonal declines in the concentration of some toxins may often be less pronounced than declines in the concentration of others (Greenplate *et al.*, 2003; Showalter *et al.*, 2009; Siebert *et al.*, 2009; Addison and Rogers, 2010). A population genetics model incorporating estimates of key biological parameters affecting *H. armigera* resistance was developed to evaluate the consequence of different seasonal declines in the concentration of Cry1Ac and Cry2Ab in Bt cotton (Brévault *et al.*, 2012). Simulations considered three *H. armigera* generations on cotton from August to October and a decrease in the number of susceptible moths produced by noncotton refuges from the first to last generation (Brévault *et al.*, 2012). More pronounced seasonal declines in Cry1Ac-induced mortality than Cry2Ab-induced mortality was modeled, based on empirical data on the mortality of susceptible *H. armigera* fed Cry1Ac or Cry1Ac and Cry2Ab cotton (Kranthi *et al.*, 2005; Mahon and Olsen, 2009). In most simulations, the initial frequency and dominance of resistance to Cry1Ac had little effect on the evolution of resistance compared to initial frequency and dominance of resistance to Cry2Ab. The toxin Cry1Ab did not kill as many susceptible insects as Cry2Ab at the end of the growing season. Such inefficacy of Cry1Ac facilitated survival of individuals with Cry2Ab resistance alleles, which could explain that alleles for resistance to Cry2Ab dominated the evolutionary process (Roush, 1998; Brévault *et al.*, 2012). These results indicate that seasonal declines in the concentrations of Bt toxin could reduce efficacy of the pyramid strategy to delay the evolution of resistance.

CONCLUSIONS

The term *source-sink dynamics* describes the demographic dynamics of pest populations in Bt crops (sinks) and refuges (sources) (Caprio, 2001; Carrière *et al.*, 2003, 2004a). Because the adoption of Bt crops has not yet caused extinction of target pests (but see Tabashnik *et al.*, 2010), gene flow between source and sink habitats typically allows for repeated colonization of Bt crops and provides long-term opportunities for local adaptation. The local evolution of resistance, in turn, leads to opportunities for the spread of resistance across landscapes. This model of resistance evolution is compatible with the view that source-sink dynamics ultimately govern changes in resistance-allele frequencies across heterogeneous landscapes (Carrière et al., 2012a).

The case studies described in this chapter have clearly demonstrated that the interaction between pest and landscapes is critical for IRM. Processes in the pest population, such as diapause, dispersal, mating, and oviposition, play a critical role. Thus, the behaviors of larvae and adults must be studied in the heterogeneous environments of real systems. Even though most examples presented here describe crop ecosystems, the same conclusions can be drawn for livestock and public health systems (Chapter 6). At the habitat level, selection for and against resistance in Bt crops and refuges play a critical role (e.g., fitness costs and incomplete resistance). The effects of natural enemies on selection and population dynamics are also important (Chapter 12). At the landscape level, temporal and spatial variation in abundance, distribution, and quality of Bt and non-Bt host plants provide the template for resistance evolution. A multi-level approach is critical for understanding and implementing IRM strategies.

Given that the environment also determines how we implement integrated pest management (IPM), there is a connection between IRM and IPM that is more than just coordination of goals and activities (Chapter 1). IPM and IRM will affect each other indirectly through changes in the environment. The two main approaches to IPM are design and control. Design selects the components of the environment and the spatial and temporal patterns that will be used throughout the time horizon. Control selects the amount of inputs to the environment during each decision-making period. Monitoring often affects control decisions. Both design and control influence IRM and make strategies dynamic and complex, as does the natural variability of the environment over time and space.

REFERENCES

Addison, S.J., Rogers, D.J., 2010. Potential impact of differential production of the Cry2Ab and Cry1Ac proteins in transgenic cotton in response to cold stress. J. Econ. Entomol. 103, 1206–1215.

Alphey, N., Coleman, P.G., Bonsall, M.B., Alphey, L., 2008. Proportions of different habitat types are critical to the fate of a resistance allele. Theor. Ecol. 1, 103–115.

Arpaia, S., Chiriatti, K., Giorio, G., 1998. Predicting the adaptation of Colorado potato beetle (Coleoptera:Chrysomelidae) to transgenic eggplants expressing CryIII toxin: the role of gene dominance, migration, and fitness costs. J. Econ. Entomol. 91, 21−29.

Barclay, H.J., 1997. Assessing natural selection in white pine weevils (*Pissodes strobi* Peck) (Coleoptera:Curculionidae) for overcoming resistance in trees: An evolutionary model. Can. Entomol. 129, 1105−1120.

Bird, L.J., Akhurst, R.J., 2005. The fitness of Cry1A-resistant and -susceptible *Helicoverpa armigera* (Lepidoptera, Noctuidae) on transgenic cotton with reduced levels of Cry1Ac. J. Econ. Entomol. 59, 1166−1168.

Bontemps, A., Bourguet, D., Pélozuelo, L., Bethenod, M.T., Ponsard, S., 2003. Managing the evolution of *Bacillus thuringiensis* resistance in natural populations of the european corn borer, *Ostrinia nubilalis*: host plant, host race and pherotype of adult males at aggregation sites. Proc. Roy. Soc. Lond. B. 271, 2179−2185.

Brévault, T., Nibouche, S., Achaleke, J., Carrière, Y., 2012. Assessing the role of non-cotton refuges in delaying *Helicoverpa armigera* resistance to Bt cotton in West Africa. Evol. Appl. 5, 53−65.

Brévault, T., Heuberger, S., Zhang, M., Ellers-Kirk, C., Ni, X., Masson, L., et al., 2013. Potential shortfall of pyramided Bt cotton for resistance management. Proc. Natl. Acad. Sci. USA. 110, 5806−5811.

Burkness, E.C., Hutchison, W.D., 2012. Bt pollen dispersal and Bt kernel mosaics: integrity of non-Bt refugia for lepidopteran resistance management in maize. J. Econ. Entomol. 105, 1773−1780.

Burkness, E.C., O'Rourke, P.K., Hutchison, W.D., 2011. Cross pollination of non-transgenic sweet corn ears with transgenic Bt sweet corn: Efficacy against Lepidopteran pests and implications for resistance management. J. Econ Entomol. 104, 1476−1479.

Caprio, M.A., 1994. *Bacillus thuringiensis* gene deployment and resistance management in single- and multi-tactic environments. Biocontrol Sci. Technol. 4, 487−497.

Caprio, M.A., 2001. Source-sink dynamics between transgenic and nontransgenic habitats and their role in the evolution of resistance. J. Econ. Entomol. 94, 698−705.

Caprio, M.A., Tabashnik, B.E., 1992. Gene flow accelerates local adaptation among finite populations: simulating the evolution of insecticide resistance. J. Econ. Entomol. 85, 611−620.

Carrière, Y., 2003. Haplodiploidy, sex and the evolution of pesticide resistance. J. Econ. Entomol. 96, 1626−1640.

Carrière, Y., Tabashnik, B.E., 2001. Reversing insect adaptation to transgenic insecticidal plants. Proc. Roy. Soc. Lond. B. 268, 1475−1480.

Carrière, Y., Deland, J.P., Roff, D.A, Vincent, C., 1994. Life history costs associated with the evolution of insecticide resistance. Proc. Roy. Soc. Lond. B. 58, 35−45.

Carrière, Y., Dennehy, T.J., Petersen, B., Haller, S., Ellers-Kirk, C., Antilla, L., et al., 2001. Large-scale management of insect resistance to transgenic cotton in Arizona: can transgenic insecticidal crops be sustained? J. Econ. Entomol. 94, 315−325.

Carrière, Y., Dennehy, T.J., Ellers-Kirk, C., Holley, D., Liu, Y.B., Sims, M.A., et al., 2002. Fitness costs, incomplete resistance, and management of resistance to Bt crops. In: Proceedings of the fourth Pacific Rim Conference on the Biotechnology of Bt and its environmental impact. CSIRO. Canberra, Australia, pp. 82−91.

Carrière, Y., Ellers-Kirk, C., Sisterson, M., Antilla, L., Whitlow, M., Dennehy, T.J., et al., 2003. Long-term regional suppression of pink bollworm by *Bacillus thuringiensis* cotton. Proc. Natl. Acad. Sci. USA. 100, 1519−1523.

Carrière, Y., Dutilleul, P., Ellers-Kirk, C., Pedersen, B., Haller, S., Antilla, L., et al., 2004a. Sources, sinks, and zone of influence of refuges for managing insect resistance to Bt crops. Ecol. Appl. 14, 1615−1623.

Carrière, Y., Sisterson, M.S., Tabashnik, B.E., 2004b. Resistance management for sustainable use of *Bacillus thuringiensis* crops. In: Horowitz, A.R., Ishaaya, I. (Eds.), Insect Pest Management: Field and Protected Crops. Springer, New York, pp. 65—95.

Carrière, Y., Crowder, D.W., Tabashnik, B.E., 2010. Evolutionary ecology of adaptation to Bt crops. Evol. Appl. 3, 561—573.

Carrière, Y., Ellers-Kirk, C., Harthfield, K., Larocque, G., Degain, B., Dutilleul, P., et al., 2012a. Large-scale, spatially-explicit test of the refuge strategy for delaying insecticide resistance. Proc. Natl. Acad. Sci. USA. 109, 775—780.

Carrière, Y., Goodell, P.B., Ellers-Kirk, C., Larocque, G., Dutilleul, P., Naranjo, S.E., et al., 2012b. Effects of local and landscape factors on population dynamics of a cotton pest. PLoS One. 7, e39862. Available from: http://dx.doi.org/10.1371/journal.pone.0039862.

Cattaneo, M.G., Yafuso, C., Schmidt, C., Huang, C., Rahman, M., Olson, C., et al., 2006. Farm-scale evaluation of transgenic cotton impacts on biodiversity, pesticide use, and yield. Proc. Natl. Acad. Sci. USA. 103, 7571—7576.

Chevillon, C., Bernard, C., Marquine, M., Pasteur, N., 2001. Resistance to *Bacillus sphaericus* in *Culex pipiens* (Diptera:Culicidae): Interaction between recessive mutants and evolution in Southern France. J. Med. Entomol. 38, 657—664.

Chilcutt, C.F., Tabashnik, B.E., 2004. Contamination of refuges by *Bacillus thuringiensis* toxin genes from transgenic maize. Proc. Natl. Acad. Sci. USA. 101, 7526—7529.

Clark, T.L., Hibbard, B.E., 2004. Comparison of non-maize hosts to support western corn rootworm (Coleoptera: Chrysomelidae) larval biology. Environ. Entomol. 33, 681—689.

Comins, H.N., 1977. The development of insecticide resistance in the presence of immigration. J. Theor. Biol. 64, 177—197.

Crespo, A.L.B., Spencer, T.A., Alves, A.P., Hellmich, R.L., Blankenship, E.E., Magalhães, L.C., et al., 2009. On-plant survival and inheritance of resistance to Cry1Ab toxin from *Bacillus thuringiensis* in a field-derived strain of European corn borer, *Ostrinia nubilalis*. Pest Manag. Sci. 65, 1071—1081.

Crow, J.F., 1957. Genetics of insect resistance to chemicals. Annu. Rev. Entomol. 2, 227—246.

Crowder, D.W., Carrière, Y., 2009. Comparing the refuge strategy for managing the evolution of insect resistance under different reproductive strategies. J. Theor. Biol. 261, 423—430.

Davis, P.M., Onstad, D.W., 2000. Seed mixtures as a resistance management strategy for European corn borers (Lepidoptera:Crambidae) infesting transgenic corn expressing Cry1Ab protein. J. Econ. Entomol. 93, 937—948.

Dutton, A., D'alessandro, M., Romeis, J., Bigler, F., 2004. Assessing expression of Bt-toxin (Cry1Ab) in transgenic maize under different environmental conditions. IOBC/WPRS Bull. 27, 49—55.

Franklin, M.T., Myers, J.H., 2008. Refuges in reverse: The spread of *Bacillus thuringiensis* resistance to unselected greenhouse populations of cabbage loopers *Trichoplusia ni*. Agric. Forest. Entomol. 10, 119—127.

Franklin, M.T., Ritland, C.E., Myers, J.H., 2010. Spatial and temporal changes in genetic structure of greenhouse and field populations of cabbage looper, *Trichoplusia ni*. Mol. Evol. 19, 1122—1133.

Gassmann, A.J., Carrière, Y., Tabashnik, B.E., 2009. Fitness costs of insect resistance to *Bacillus thuringiensis*. Ann. Rev. Entomol. 54, 147—163.

Gealy, D.R., Bradford, K.J., Hall, L., Hellmich, R., Raybould, A., Wolt, J., et al., 2007. Implications of gene flow in the scale-up and commercial use of biotechnology-derived crops: economic and policy considerations. CAST Issue Paper 37, p. 24.

Glaum, P.R., Ives, A.R., Andow, D.A., 2011. Contamination and management of resistance evolution to high-dose transgenic insecticidal crops. Theor. Ecol. Available from: http://dx.doi.org/10.1007/s12080-010-0109-6.

Gould, F., 1986. Simulation models for predicting durability of insect-resitant germ plasm: a deterministic diploid, two-locus model. Environ. Entomol. 15, 1−10.

Gould, F., 1998. Sustainability of transgenic insecticidal cultivars: integrating pest genetics and ecology. Annu. Rev. Entomol. 43, 701−726.

Gould, F., Tabashnik, B.E., 1998. Bt-cotton resistance management. In: Melon, M., Risser, J. (Eds.), Now or Never: Serious Plans to Save a Natural Pest Control. Union of Concerned Scientists, Cambridge, MA, pp. 67−105.

Gould, F., Blair, N., Reid, M., Rennie, T.L., Lopez, J., Micinski, S., 2002. *Bacillus thuringiensis*-toxin resistance management: Stable isotope assessment of alternate host use by *Helicoverpa zea*. Proc. Nat. Acad. Sci. 99, 16581−16586.

Gould, F., Cohen, M.B., Bentur, J.S., Kennedy, G.G., van Duyn, J., 2006. Impact of small fitness costs on pest adaptation to crop varieties with multiple toxins: a heuristic model. J. Econ. Entomol. 99, 2091−2099.

Gray, M.E., Steffey, K.L., Estes, R.E., Schroeder, J.B., 2007. Response of transgenic maize hybrids to variant western corn rootworm larval injury. J. Appl. Entomol. 131, 386−390.

Greenplate, J.T., Mullins, J.W., Penn, S.R., Dahm, A., Reich, B.J., Osborn, J.A., et al., 2003. Partial characterization of cotton plants expressing two toxin proteins from Bacillus thuringiensis: relative toxin contribution, toxin interaction, and resistance management. J. Appl. Entomol.-Zeits Fur Ang. Entomol. 127, 340−347.

Guse, C.A., Onstad, D.W., Buschman, L.L., Porter, P., Higgins, R.A., Sloderbeck, P.E., et al., 2002. Modeling the development of resistance by stalk-boring Lepidoptera (Crambidae) in areas with irrigated, transgenic corn. Environ. Entomol. 31, 676−685.

Head, G., Jackson, R.E., Adamczyk, J., Bradley, J.R., Van Duyn, J., Gore, J., et al., 2010. Spatial and temporal variability in host use by *Helicoverpa zea* as measured by analyses of stable carbon isotope ratios and gossypol residues. J. Appl. Ecol. 47, 583−592.

Heuberger, S., Ellers-Kirk, C., Yafuso, C., Gassmann, A.J., Tabashnik, B.E., Dennehy, T. J., et al., 2008a. Effects of refuge contamination by transgenes on Bt resistance in pink bollworm (Lepidoptera:Gelechiidae). J. Econ. Entomol. 101, 504−514.

Heuberger, S., Yafuso, C., Tabashnik, B.E., Carrière, Y., Dennehy, T.J., 2008b. Outcrossed cottonseed and adventitious Bt plants in Arizona refuges. Environ. Biosafety Res. 7, 87−96.

Heuberger, S., Ellers-Kirk, C., Tabashnik, B.E., Carrière, Y., 2010. Pollen- and Seed-mediated transgene flow in commercial cotton seed production fields. PloS One. 5, e14128. Available from: http://dx.doi.org/10.1371/journal.pone.0014128.

Heuberger, S., Crowder, D.W., Brévault, T., Tabashnik, B.E., Carrière, Y., 2011. Modeling the effects of plant-to-plant gene flow, larval behavior, and refuge size on pest resistance to Bt cotton. Environ. Entomol. 40, 484−495.

Horner, T.A., Dively, G.P., Herbert, D.A., 2003. Development, survival, and fitness performance of *Helicoverpa zea* (Lepidoptera: Noctuidae) in MON810 Bt field corn. J. Econ. Entomol. 96, 914−924.

Hutchison, W.D., Burkness, E.C., Moon, R.D., Leslie, T., Fleischer, S.J, Abrahamson, M., et al., 2010. Areawide suppression of European corn borer with Bt maize reaps savings to non-Bt maize growers. Science. 330, 222−225.

Janmaat, A.F., Myers, J., 2003. Rapid evolution and the cost of resistance to *Bacillus thuringiensis* in greenhouse populations of cabbage loopers, *Trichoplusia ni*. Proc. Roy. Soc. B-Biol. Sci. 270, 2263−2270.

Kamath, S.P., Anuradha, S., Vidya, H.S., Mohan, K.S., Dudin, Y., 2010. Quantification of Bacillus thuringiensis Cry1Ab protein in tissue of Yieldgard (MON810) corn hybrid tested at multiple field locations in India. Crop Prot. 29, 921−926.

Kang, J., Onstad, D.W., Hellmich, R.L., Moser, S.E., Hutchison, W.D., Prasifka, J.R., 2012. Modeling the impact of cross-pollination and low toxin expression in corn kernels on adaptation of European corn borer (Lepidoptera:Crambidae) to transgenic insecticidal corn. Environ. Entomol. 41, 200—211.

Kranthi, K.R., Naidu, S., Dhawad, C.S., Tatwawadi, A., Mate, K., Patil, E., et al., 2005. Temporal and intra-plant variability of Cry1Ac expression in Bt-cotton and its influence on the survival of the cotton bollworm, *Helicoverpa armigera* (Hubner) (Lepidoptera:Noctuidae). Cur. Sci. 89, 291—298.

Krupke, C., Marquardt, P., Johnson, W., Weller, S., Conley, S.P., 2009. Volunteer corn presents new challenges for insect resistance management. Agr. J. 101, 797—799.

Lenormand, T., Raymond, R., 1998. Resistance management: the stable zone strategy. Proc. R. Soc. Lond. B. 265, 1985—1990.

Liu, Y.-B., Tabashnik, B.E., 1997. Experimental evidence that refuges delay insect adaptation to *Bacillus thuringiensis*. Proc R Soc Lond B. 264, 605—610.

Llewellyn, D., Tyson, C., Constable, G., Duggan, B., Beale, S., Steel, P., 2007. Containment of regulated genetically modified cotton in the field. Agr. Ecosyst. Environ. 121, 419—429.

Mahon, R.J., Olsen, K.M., 2009. Limited survival of a Cry2Ab-resistant strain of Helicoverpa armigera (Lepidoptera:Noctuidae) on Bollgard II. J. Econ. Entomol. 102, 708—716.

Mallet, J., Porter, P., 1992. Preventing insect adaptation to insect-resistant crops: are seed mixtures or refugia the best strategy?. Proc. Roy. Soc. Lond. B. 250, 165—169.

Morin, S., Biggs, R.W., Sisterson, M.S., Shriver, L., Ellers-Kirk, C., Higginson, D., et al., 2003. Three cadherin alleles associated with resistance to *Bacillus thuringiensis* in pink bollworm. Proc. Natl. Acad. Sci. USA. 100, 5004—5009.

Murphy, A.F., Ginzel, M.D., Krupke, C.H., 2010. Evaluating western corn rootworm (Coleoptera: Chrysomelidae) emergence and root damage in a seed mix refuge. J. Econ. Entomol. 103, 147—157.

NRC, 2010. The Impact of Genetically Engineered Crops on Farm Sustainability in the United States. The National Academies Press, Washington, DC, p.250.

Nagoshi, R.N., Adamczyk, J.J., Meagher, R.T., Gore, J., Lackson, R., 2007. Using stable isotope analysis to examine fall armyworm (Lepidoptera: Noctuidae) host strains in a cotton habitat. J. Econ. Entomol. 100, 1569—1576.

Nguyen, H.T., Ja, J., 2009. Expression of Cry3Bb1 in transgenic corn MON88017. J. Agric. Food Chem. 57, 9990—9996.

Onstad, D.W., 2006. Modeling larval survival and movement to evaluate seed mixtures of transgenic corn for control of Western corn rootworm (Coleoptera:Chrysomelidae). J. Econ. Entomol. 99, 1407—1414.

Onstad, D.W., Gould, F., 1998a. Modeling the dynamics of adaptation to transgenic maize by European corn borer (Lepidoptera:Pyralidae). J. Econ. Entomol. 91, 585—593.

Onstad, D.W., Gould, F., 1998b. Do dynamics of crop maturation and herbivorous insect life cycle influence the risk of adaptation to toxins in transgenic host plants? Environ. Entomol. 27, 517—522.

Onstad, D.W., Spencer, J.L., Guse, C.A., Isard, S.A., Levine, E., 2001. Modeling evolution of behavioral resistance by an insect to crop rotation. Ent. Expt. Appl. 100, 195—201.

Onstad, D.W., Guse, C.A., Porter, P., Buschman, L.L., Higgins, R.A., Sloderbeck, P.E., et al., 2002. Modeling the development of resistance by stalk-boring Lepidoptera (Crambidae) in areas with transgenic corn and frequent insecticide use. J. Econ. Entomol. 95, 1033—1043.

Onstad, D.W., Crowder, D.W., Isard, S.A., Levine, E., Spencer, J.L., O'Neal, et al., 2003. Does landscape diversity slow the spread of rotation-resistant Western corn rootworm (Coleoptera: Chrysomelidae)?. Environ. Entomol. 32, 992—1001.

Onstad, D.W., Mitchell, P.D., Hurley, T.M., Lundgren, J.G., Porter, R.P., Krupke, C.H., et al., 2011. Seeds of change: corn seed mixtures for resistance management and integrated pest management. J. Econ. Entomol. 104, 343—352.

Onstad, D.W., Kang, J., Ba, N.M., Tamo, M., Jackai, L., Pittendrigh, B.R., 2012. Modeling Evolution of Resistance by *Maruca vitrata* to Transgenic Insecticidal Cowpea in Africa. Environ. Entomol. 41, 1255—1267.

Orth, R.G., Head, G., Mierkowski, M., 2007. Determining larval host plant use by a polyphagous lepidopteran through analysis of adult moths for plant secondary metabolites. J. Chem. Ecol. 33, 1131—1148.

Oyediran, I.O., Hibbard, B.E., Clark, T.L., 2004. Prairie grasses as hosts of the western corn rootworm (Coleoptera: Chrysomelidae). Environ. Entomol. 33, 740—747.

O'Rourke, M.E., Sappington, T.W., Fleischer, S.J., 2010. Managing resistance to Bt crops in a genetically variable insect herbivore, *Ostrinia nubilalis*. Ecol. Appl. 20, 1228—1236.

Pan, Z., Onstad, D.W., Nowatzki, T.M., Stanley, B.H., Meinke, L.M., Flexner, J.L., 2011. Western corn rootworm (Coleoptera: Chrysomelidae) dispersal and adaptation to single-toxin transgenic corn. Environ. Entomol. 40, 964—978.

Peck, S.L., Gould, F., Ellner, S.P., 1999. Spread of resistance in spatially extended regions of transgenic cotton: implications for management of *Heliothis virescens* (Lepidoptera: Noctuidae). J. Econ. Entomol. 92, 1—16.

Pittendrigh, B.R., Gaffney, P.J., Huesing, J., Onstad, D.W., Roush, R.T., Murdock, L.L., 2004. Active refuges can inhibit the evolution of resistance in insects towards transgenic insect-resistant plants. J. Theor. Biol. 231, 461—474.

Raymond, M., Berticat, C., Weill, M., Pasteur, N., Chevillon, C., 2001. Insecticide resistance in the mosquito *Culex pipiens*: what have we learned about adaptation?. Genetica. 112-113, 287—296.

Ringland, J., George, P., 2010. Analysis of sustainable pest control using a pesticide and a screened refuge. Evol. Appl. 4, 459—470.

Roush, R.T., 1998. Two-toxin strategies for management of insecticidal transgenic crops: can pyramiding succeed where pesticide mixtures have not? Philos. Trans. Roy. Soc. B Biol. Sci. 353, 1777—1786.

Sequeira, R.V., Playford, C.L., 2001. Abundance of *Helicoverpa* (Lepidoptera: Noctuidae) pupae under cotton and other crops in central Queensland: implications for resistance management. Aust. J. Entomol. 40, 264—269.

Shelton, A.M., Tang, J.D., Roush, R.T., Metz, T.M., Earle, E.D., 2000. Field tests on managing resistance to *Bt*-engineered plants. Nat. Biotech. 18, 339—342.

Showalter, A.M., Heuberger, S., Tabashnik, B.E., Carrière, Y., 2009. A primer for the use of insecticidal transgenic cotton in developing countries. J. Ins. Sci. 9, 43. Available online: insectscience.org/9.43.

Siebert, M.W., Patterson, T.G., Gilles, G.J., Nolting, S.P., Braxton, L.B., Leonard, B.R., Van Duyn, J.W., Lassiter, R.B., 2009. Quantification of Cry1Ac and Cry1F *Bacillus thuringiensis* insecticidal proteins in selected transgenic cotton plant tissue types. J. Econ. Entomol. 102, 1301—1308.

Sisterson, M.S., Antilla, L., Carrière, Y., Ellers-Kirk, C., Tabashnik, B.E., 2004. Effects of insect population size on evolution of resistance to transgenic crops. J. Econ. Entomol. 97, 1413—1424.

Sisterson, M.S., Carrière, Y., Dennehy, T.J., Tabashnik, B.E., 2005. Evolution of resistance to transgenic crops: interactions between insect movement and field distribution. J. Econ. Entomol. 98, 1751—1762.

Spencer, J., Onstad, D, Krupke, C., Hughson, S., Pan, Z, Stanley, B., et al., 2013. Isolated females and limited males: evolution of insect resistance in structured landscapes. Entomol. Exp. Appl. 146, 38—49.

Spencer, J.L., Hibbard, B.E., Moeser, J., Onstad, D.W., 2009. Behaviour and ecology of the western corn rootworm (*Diabrotica virgifera virgifera* LeConte). Agr. For. Entomol. 11, 9—27.

Storer, N.P., 2003. A spatially explicit model simulating western corn rootworm (Coleoptera:Chrysomelidae) adaptation to insect-resistant maize. J. Econ. Entomol. 96, 1530–1547.

Storer, N.P., Peck, S.L., Gould, F., Van-Duyn, J.W., Kennedy, G.G., 2003a. Spatial processes in the evolution of resistance in *Helicoverpa zea* (Lepidoptera: Noctuidae) to Bt transgenic corn and cotton in a mixed agroecosystem: a biology-rich stochastic simulation model. J. Econ. Entomol. 96, 156–172.

Storer, N.P., Peck, S.L., Gould, F., Van-Duyn, J.W., Kennedy, G.G., 2003b. Sensitivity analysis of a spatially-explicit stochastic simulation model of the evolution of resistance in *Helicoverpa zea* (Lepidoptera:Noctuidae) to Bt transgenic corn and cotton. J. Econ. Entomol. 96, 173–187.

Storer, N.P., Dively, G.P., Herman, R.A., 2008. Landscape effects of insect-resistant genetically modified crops. In: Romeis, J., Shelton, A.M., Kennedy, G.G. (Eds.), Integration of Insect-Resistant Genetically Modified Crops within IPM Programs. Springer, New York, pp. 273–302.

Tabashnik, B.E., 1994. Delaying insect adaptation to transgenic plants: seed mixtures and refugia reconsidered. Proc. Roy. Soc. Lond. B. 255, 7–12.

Tabashnik, B.E., Carrière, Y., 2008. Evolution of insect resistance to transgenic plants. In: Tilmon, K. (Ed.), Specialization, Speciation, and Radiation: the Evolutionary Biology of Herbivorous Insects. University of California Press, Berkeley, pp. 267–279.

Tabashnik, B.E., Croft, B.A., 1982. Managing pesticide resistance in crop-arthropod complexes: interactions between biological and operational factors. Environ. Entomol. 11, 1137–1144.

Tabashnik, B.E., Gould, F., 2012. Delaying corn rootworm resistance to Bt crops. J. Econ. Entomol. 105, 767–776.

Tabashnik, B.E., Patin, A.L., Dennehy, T.J., Liu, Y.B., Miller, E., Staten, R.T., 1999. Dispersal of pink bollworm (Lepidoptera:Gelechiidae) males in transgenic cotton that produces a *Bacillus thuringiensis* toxin. J. Econ. Entomol. 92, 772–780.

Tabashnik, B.E., Patin, A.L., Dennehy, T.J., Liu, Y.B., Carrière, Y., Antilla, L., 2000. Frequency of resistance to *Bacillus thuringiensis* in field populations of pink bollworm. Proc. Natl. Acad. Sci. USA. 21, 12980–12984.

Tabashnik, B.E., Gould, F., Carrière, Y., 2004. Delaying evolution of insect resistance to transgenic crops by decreasing dominance and heritability. J. Evol. Biol. 17, 904–912.

Tabashnik, B.E., Dennehy, T.J., Carrière, Y., 2005. Delayed resistance to transgenic cotton in pink bollworm. Proc. Natl. Acad. Sci. USA. 43, 15389–15393.

Tabashnik, B.E., Gassmann, A.J., Crowder, D.W., Carrière, Y., 2008. Insect resistance to Bt crops: evidence versus theory. Nat. Biotech. 26, 199–202.

Tabashnik, B.E., Van Rensburg, J.B.J., Carrière, Y., 2009. Field-evolved insect resistance to Bt crops: definition, theory, and data. J. Econ. Entomol. 102, 2011–2025.

Tabashnik, B.E., Sisterson, M.S., Ellsworth, P.C., Dennehy, T.J., Antilla, L, Liesner, L., et al., 2010. Suppressing resistance to Bt cotton with sterile insect release. Nat. Biotech. 28, 1304–1308.

Tabashnik, B.E., Huang, F.N., Ghimire, M.N., Leonard, B.R., Siegfried, B.D., Rangasamy, M., et al., 2011. Efficacy of genetically modified Bt toxins against insects with different genetic mechanisms of resistance. Nat. Biotech. 29, 1128–1131.

Tabashnik, B.E., Brévault, T., Carrière, Y., 2013. Insect resistance to Bt crops: Lessons from the first billion acres. Nat. Biotech. 31, 510–521.

Tang, J.D., Collins, H.L., Metz, T.D., Earle, E.D., Zhao, J.Z., Roush, R.T., et al., 2001. Greenhouse tests on resistance management of Bt transgenic plants using refuge strategies. J. Econ. Entomol. 94, 240–247.

Taylor, C.E., Georghiou, G.P., 1979. Suppression of insecticide resistance by alteration of dominance and migration. J. Econ. Entomol. 72, 105–109.

U.S. EPA, 2007. Pesticide News Story: EPA Approves Natural Refuge for Insect Resistance Management in Bollgard II Cotton. U.S. Environmental Protection Agency, Available online at: <http://www.epa.gov/oppfead1/cb/csb_page/updates/2007/bollgard-cotton.htm>.

Vacher, C., Bourguet, D., Rousset, F., Chevillon, C., Hochberg, M.E., 2003. Modelling the spatial configuration of refuges for a sustainable control of pests: a case study of Bt cotton. J. Evol. Ecol. 16, 378–387.

Vaughn, T., Cavato, T., Brar, G., Coombe, T., DeGooyer, T., Ford, S., et al., 2005. A method of controlling corn rootworm feeding using *Bacillus thuringiensis* protein expressed in transgenic maize. Crop. Sci. 45, 931–938.

Wan, P., Huang, Y., Tabashnik, B.E., Huang, M., Wu, K., 2012. The halo effect: suppression of pink bollworm on non-Bt cotton by Bt cotton in China. PLoS One. 7, e42004. Available from: http://dx.doi.org/10.1371/journal.pone.0042004.

Wilhoit, L.R., 1991. Modeling the population-dynamics of different aphid genotypes in plant variety mixtures. Ecol. Model. 55, 257–283.

Williams, J.L., Ellers-Kirk, C., Orth, R.G., Gassmann, A.J., Head, G., Tabashnik, B.E., et al., 2011. Fitness cost of resistance to Bt cotton linked with increased gossypol content in pink bollworm larvae. PLoS One. 6, e21863. Available from: http://dx.doi.org/10.1371/journal.pone.0021863.

Wilson, T.A., Hibbard, B.E., 2004. Host suitability of nonmaize agroecosystem grasses for the western corn rootworm (Coleoptera: Chrysomelidae). Environ. Entomol. 33, 1102–1108.

Wu, K.M., Lu, Y.H., Feng, H.Q., Jiang, Y.Y., Zhao, J.Z., 2008. Suppression of cotton bollworm in multiple crops in China in areas with Bt toxin-containing cotton. Science. 321, 1676–1678.

Wu, X., Huang, F., Leonard, B.R., Moore, S.H., 2007. Evaluation of transgenic *Bacillus thuringiensis* corn hybrids against Cry1Ab-susceptible and −resistant sugarcane borer (Lepidoptera: Crambidae). J. Econ. Entomol. 100, 1880–1886.

Zhao, J.Z., Cao, J., Collins, H.L., Bates, S.L, Roush, R.T., Earle, E.D., et al., 2005. Concurrent use of transgenic plants expressing a single and two *Bacillus thuringiensis* genes speeds insect adaptation to pyramided plants. Proc. Natl. Acad. Sci. USA. 102, 8426–8430.

CHAPTER 11

Negative Cross-Resistance: History, Present Status, and Emerging Opportunities

Barry R. Pittendrigh[1], Joseph Huesing[2], Kent R. Walters, Jr.[1], Brett P. Olds[3], Laura D. Steele[1], Lijie Sun[4], Patrick Gaffney[5] and Aaron J. Gassmann[6]

[1]Department of Entomology, University of Illinois Urbana Champaign, Urbana, IL
[2]USDA-ARS/USAID-BFS, Washington, DC
[3]Department of Animal Biology, University of Illinois Urbana Champaign, Urbana, IL
[4]Synthetic Biology and Bioenergy, J. Craig Venter Institute, San Diego, CA
[5]Department of Statistics, University of Wisconsin-Madison, Madison, WI
[6]Department of Entomology, Iowa State University, Ames, IA

Chapter Outline

INTRODUCTION

Two significant scientific breakthroughs of the twentieth century were the discovery, development, and large-scale use of antibiotics and advanced agricultural technologies, including insecticides and biotechnology. Antibiotics have dramatically reduced human mortality rates resulting from bacterial diseases. Antimicrobial compounds also have been used to reduce

Insect Resistance Management
DOI: http://dx.doi.org/10.1016/B978-0-12-396955-2.00011-4

mortality rates in livestock, thereby allowing for greater levels of production and, in some cases, a reduction in cost for sources of human dietary proteins. The widespread use of advanced agricultural technologies, including the large-scale use of insecticides and genetically modified crops, has increased the quantity and quality of food for the earth's rapidly expanding and urbanizing human population. However, the Achilles' heel of these scientific advances has been the evolution of resistance in microbes, insects, and weeds.

Although efforts have been made to slow the development of resistance to antibiotics and insecticides, the evolution of resistance is likely without the use of carefully developed resistance management plans. Once widespread resistance develops, the chemical, or worse the chemical class, loses much of its utility. Thus, the focus of the academic and industrial research community often has been on identifying and deploying novel antibiotics and insecticides with different modes of action. One alternative to this use-and-discard approach involves negative cross-resistance (NCR) strategies to control organisms containing resistance alleles (Ogita, 1961a, b, c; Chapman and Penman, 1979; Khambay et al., 2001; Cilek et al., 1995; Pittendrigh et al., 2000). Negative cross-resistance occurs when an allele confers resistance to one toxic chemical and hypersusceptibility to another. Thus, in practical terms, a NCR toxin is a compound that can be used to preferentially kill insects that are resistant to another insecticide (Figure 11.1).

The concept of NCR is not new; examples actually date back to the early 1960s (Ogita, 1961a, b, c). NCR also occurs across a wide array of toxins and organisms, including insects (Peiris and Hemingway, 1990; Hemingway et al., 1993), weeds (Oettmeier et al., 1991; Gadamski et al., 2000; Poston et al., 2002), and fungi (Josepovits et al., 1992; Bossche,

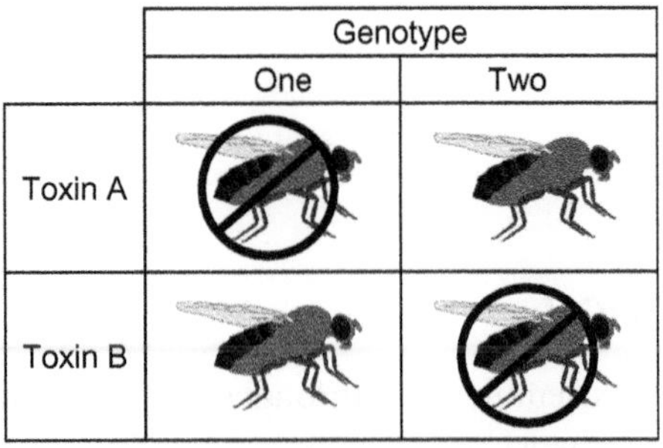

Figure 11.1 Negative cross-resistance refers to a situation in which (i) toxin "A" causes higher mortality rates to a genotype carrying "allele one" than a genotype carrying "allele two" and (ii) toxin "B" causes higher mortality rates to a genotype carrying "allele two" than a genotype carrying "allele one".

1997; Hollomon *et al.*, 1998; Leroux *et al.*, 2000) (Table 11.1). There are, however, few commercial examples of NCR for insect control (Hoy, 1998; Yamamoto *et al.*, 1993; Kamidi and Kamidi, 2005). Why then has NCR played such a limited role in resistance management?

Much of the insecticide discovery effort in industry has focused on two areas: (1) identifying toxins with novel modes of action, and (2) improving the efficacy and spectrum of toxins that have already been discovered (Broadhurst, 1998). Screening for novel toxins typically involves automated systems where tens of thousands of compounds are tested against multiple insect species at once (Broadhurst, 1998). In the agricultural and pharmaceutical industries, high-throughput and ultra-high-throughput screening is used to evaluate in excess of 100,000 compounds a year (Curtis *et al.*, 2004; Kniaz, 2000). In contrast, the systemic investigation of NCR compounds has been restricted to academic laboratories and involved testing of a few dozen compounds (Oettmeier *et al.*, 1991; Palmer *et al.*, 1991; De Prado *et al.*, 1992; Tabashnik *et al.*, 1996; Hedley *et al.*, 1998; Pedra *et al.*, 2004). It is not known to what extent industry has used these systems to develop NCR factors; however, no NCR-based products have been forthcoming. In part, this may be due to business models that necessarily develop new products based on market needs and value capture. In this regard, NCR products would only be developed in response to verified resistance to currently marketed products. However, in principle the same large-scale screening processes used to discover insecticidal compounds could be used for NCR discovery. For example, molecular cloning and expression of peptides in display technologies could allow for the rapid development of NCR products in a reactionary manner.

In this chapter we will (i) explore the current status of NCR in the peer-reviewed literature, (ii) examine discovery strategies in more detail, (iii) discuss how to deploy the resulting NCR compounds and (iv) address potential limitations and possible future opportunities for such an approach in resistance management.

EXISTING EXAMPLES OF NEGATIVE CROSS-RESISTANCE

Negative cross-resistance has been shown to occur in pairs of toxins active against a wide variety of arthropods, including *Plutella xylostella*

Table 11.1 Examples of Organisms where Toxins Pairs or Ecological Factors have been Observed to cause Negative Cross-Resistance

Organism	Toxin Pair(s) or Ecological Factor	References
Insects		
Drosophila melanogaster (Meigen)	DDT and deltamethrin[a]	Pedra *et al.* (2004)
	DDT and phenylthiourea[b]	Ogita (1961a,b,c)
Plodia interpunctella (Hübner)	*Bacillus thuringiensis* (*Bt*) toxins	Van Rie *et al.* (1990)
Helicoverpa zea (Boddie)	*Bt* toxins *Cry1Ac* ad *Cry1F*	Marcus (2005)
Helicoverpa armigera (Hüber)	*Bt* toxins	Liang *et al.* (2000)
Pectinophora gossypiella (Saunders)	*Bt* toxin and gossypol	Carrière *et al.* (2004)
	Bt toxin and *Steinernema riobrave*	Gassmann *et al.* (2006, 2009, 2012) Hannon *et al.* (2010)
	Bt toxin and *Heterorhabditis bacteriophora*	Gassmann *et al.* (2009)
Plutella xylostella (L.)	*Bt* and nucleopolyhedrovirus	Raymond *et al.* (2007)
Musca domestica (L.)	Pyrethroids and N-alkylamides[c]	Khambay *et al.* (2001)
	AaIT and pyrethroids[d]	Elliott *et al.* (1986)
Heliothis virescens (Fabricius)	AaIT and pyrethroids[e]	McCutchen *et al.* (1997)
Haematobia irritans (L.)	Pyrethroids and diazinon[f]	Sheppard and Marchiondo (1987)
Nephotettix cincticeps (Uhler)	N-propylcarbamate and N-methylcarbamate[g]	Yamamoto *et al.* (1993)
Tetranychus urticae (Koch)	Organo-phosphates and synthetic pyrethroids	Chapman and Penman (1979)
Plants		
Conyza Canadensis (L.) Cron. and *Echinochloa crus-galli* (L.) Beauv.	Atrazine and triazine	Gadamski *et al.* (2000)

(*Continued*)

Table 11.1 (Continued)

Organism	Toxin Pair(s) or Ecological Factor	References
Amaranthus hybridus (L.)	Pyrithiobac and imazethapyr	Poston *et al.*, (2002)
Amaranthus hybridus (L.)	triazine and *Disonycha glabrata*	Gassmann and Futuyma (2005)
Fungi		
Ustilago maydis (Persoon) Roussel	Benzimidazoles and diethofencarb	Ziogas and Girgis (1993)
Botrytis cinerea (de Bary) Whetzel, *Venturia nashicola* (S. Tanaka & S. Yamamoto) and *Venturia inaequalis* (Cooke) Wint.	Benzimidazole and N-phenylanilines	Josepovits *et al.* (1992)
Botrytis cinerea (de Bary) Whetzel	Benzimidazoles and phenylcarbamates[h]	Hollomon *et al.* (1998)
Mycosphaerella graminicola (Fuckel)	Triazoles and pyrimidine derivatives/ triflumizole[i]	Leroux *et al.* (2000)

The following alleles, genes, or loci are associated with the respective toxins pairs given above:
[a]*para^{ts1}* (voltage sensitive sodium channel);
[b]*Rst(2)DDT* (differential expression of one or more cytochrome P450s);
[c]*super-kdr (super-knock down resistance)*;
[d,e]*kdr (knock down resistance)*;
[f]cytochrome P450;
[g]AChE(acetylcholinesterases);
[h]β-tubulin (single amino acid changes); and,
[i]P450 sterol 14-demethylase.

(Chen *et al.*, 1993), *Blattella germanica* (Hemingway *et al.*, 1993), *Culex quinquefasciatus* (Peiris and Hemingway, 1990), *Tetranychus urticae* (Hatano *et al.*, 1992), *Cydia pomonella* (Dunley and Welter, 2000), and *Haematobia irritans* (Cilek *et al.*, 1995).

In some cases, NCR has been associated with a single amino acid change in the targeted protein. For example, NCR between the fungicides benzimidazoles and diethofencarb in the fungus *Ustilago maydis* was due to a mutation at a single locus (Ziogas and Girgis, 1993). This single-locus NCR scenario has also been observed in *Drosophila melanogaster*, where Pedra *et al.* (2004) observed that a DDT-resistant strain, known as *para^{ts-1}*, was highly susceptible to deltamethrin.

One of the best studied areas for NCR is pyrethroid resistance associated with the voltage-sensitive sodium channel (VSSC). The VSSC has been well documented as the target site for both DDT and pyrethroid insecticides (Van den Bercken and Vijverberg, 1980; Narahashi and Lund, 1980; Vijverberg et al., 1982; Pittendrigh et al., 1997; Lee et al., 1999). Several mutations can occur in the VSSC, which both result in pyrethroid resistance and confer NCR to other pesticides.

In *D. melanogaster*, the *para*$^{ts-1}$ allele has a mutation (and an alternative splice form) in the α-subunit of the VSSC. The *para*$^{ts-1}$*D. melanogaster* strain is so named because when the fly line is heated to 37°C the flies become paralyzed. When the flies are returned to room temperature, they are no longer paralyzed. The *para*$^{ts-1}$ allele also confers DDT resistance (Pittendrigh et al., 1997). In a small-scale screen of pyrethroids, Pedra et al. (2004) were able to discover a NCR toxin (deltamethrin) for the *para*$^{ts-1}$ allele. Subsequently, population selection experiments were performed to demonstrate that DDT and deltamethrin could be used to increase and decrease, respectively, the frequency of *para*$^{ts-1}$ in a population of *D. melanogaster* containing both the *para*$^{ts-1}$ and wild-type alleles (Figure 11.2).

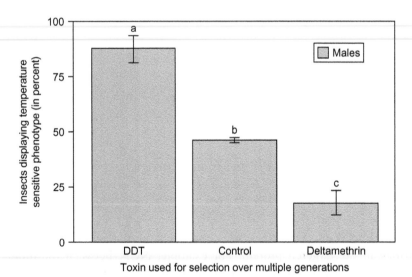

Figure 11.2 Bar graphs showing the number of males displaying the temperature-sensitive phenotype after five generations of selection with: (i) DDT, (ii) no selection, or (iii) deltamethrin. The starting frequency of the wild-type and *para*$^{ts-1}$ alleles were each 50%. In males, the temperature-sensitive phenotype is a direct measure of the allelic frequency of *para*$^{ts-1}$ in the population.

Another example of a possible VSSC NCR scenario involves use of an insect-selective neurotoxic peptide (AaIT), which has been transformed into baculovirus. AaIT is more toxic to house fly *Musca domestica* and tobacco budworm *Heliothis virescens* that are tolerant of pyrethroids (via knockdown resistance, *kdr*) than are pyrethroid susceptible strains (Elliott *et al.*, 1986; McCutchen *et al.*, 1997). The *kdr* phenotype is caused by a single amino acid change in the VSSC and represents a major mechanism of resistance to pyrethroids. Deployment of AaIT against the *kdr* phenotype may be useful in reducing the allelic frequency of the *kdr* alleles in the population (McCutchen *et al.*, 1997). The usefulness of AaIT for minimizing resistance in the field will depend on (i) its selective toxicity to the various alleles of *kdr*-type resistance that may occur in field populations of pest insects and (ii) if scientifically acceptable and economically feasible delivery strategies can be developed.

NCR target site insensitivity has also been observed in *super-knockdown resistance* (*super-kdr*) and pesticide susceptible houseflies (Khambay *et al.*, 2001). The *super-kdr* phenotype is caused by two amino acid changes in the VSSC where one of these amino acid changes is analogous to the *para*$^{ts-1}$ mutation (although they are in different domains of the channel) (Williamson *et al.*, 1996; Pittendrigh *et al.*, 1997). The *super-kdr* houseflies exhibit very high levels of resistance to pyrethroids, but appear to be more sensitive to N-alkylamides than the pyrethroid susceptible strains.

Since several mutations can occur in the VSSC, which result in pyrethroid resistance, it is not known if any one NCR compound would be effective against all these mutations. For example, in *D. melanogaster, para*^{ts} alleles other than *para*$^{ts-1}$ conferred DDT resistance but did not show NCR with deltamethrin (Pittendrigh *et al.*, 1997).

Additionally, *D. melanogaster* has been used as a model system to understand the molecular mechanisms of metabolic pesticide resistance and NCR factors associated with metabolic resistance. Ogita (1961a,b,c) observed that *D. melanogaster* strains metabolically resistant to DDT were more susceptible to phenylthiourea (PTU) in their diets than the DDT-susceptible strains. The DDT-resistant strains are thought to metabolize PTU into the more toxic phenylurea, thus causing greater toxicity in the resistant insects. The DDT susceptible insects are less capable of metabolizing PTU into phenylurea, thus allowing the susceptible strains to better survive on the media containing PTU.

Metabolic NCR also has been observed in several populations of pyrethroid resistant *H. irritans*, which are in turn highly susceptible to diazinon (Sheppard and Marchiondo, 1987; Crosby *et al.*, 1991; Barros *et al.*, 2002). The increased resistance to pyrethroids and susceptibility to diazinon is thought to be due to increased cytochrome P450 activity in the resistant flies (Cilek *et al.*, 1995). Hardstone *et al.* (2007) observed increased pyrethroid resistance in *Culex pipiens quinquefasciatus* that was thought to be the consequence of P450-mediated detoxification, with the resistant strain showing potential NCR to malathion, temephos, and methyl-parathion. Additionally, increased P450 detoxification in combination with reduced acetylcholinesterase sensitivity in *Aphis gossypii* Glover has resulted in what is thought to be NCR to bifenthrin (Shang *et al.*, 2012).

Several practical applications of NCR exist in the literature. For example, *N*-propylcarbamate and *N*-methylcarbamate have been used to control *Nephotettix cincticeps* populations containing mutant and wild-type acetylcholinesterases (Yamamoto *et al.*, 1993). Use of *N*-methylcarbamate on *N. cincticeps* selected for a population that was more susceptible to *N*-propylcarbamate and vice versa. Yamamoto *et al.* (1993) were able to shift the resistance level back and forth by alternating between the two carbamates. Additionally, Chapman and Penman (1979) observed that some mite populations resistant to organophosphates in the field were also hypersusceptible to synthetic pyrethroids.

Kamidi and Kamidi (2005) used what they proposed was an NCR strategy to reduce tick infestation of a Kenyan dairy herd. Using two commercially available acaracides, chlorfenvinphos, and amitraz, they effectively managed resistance and decreased the size of the tick population, thereby reducing the incidence of tick-associated diseases in cattle.

Negative Cross-Resistance versus Just Negatively Correlated Resistance

One critical point to remember is that NCR refers to a scenario in which the locus causing increased levels of resistance to "toxin A" is the same locus causing hypersusceptibility to "toxin B" (Figure 11.3A). Theoretically it is possible that an insect strain displays negatively correlated resistance without actually being negatively cross-resistant. For example, insects may be highly resistant to "toxin A" due to "locus 1" but may be hypersusceptible to "toxin B" due to "locus 2" (Figure 11.3B,C). For negatively correlated resistance, where two separate

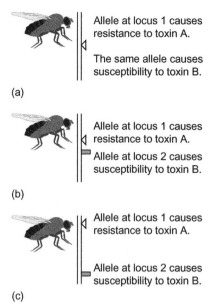

(a)

Allele at locus 1 causes resistance to toxin A.

The same allele causes susceptibility to toxin B.

(b)

Allele at locus 1 causes resistance to toxin A.
Allele at locus 2 causes susceptibility to toxin B.

(c)

Allele at locus 1 causes resistance to toxin A.

Allele at locus 2 causes susceptibility to toxin B.

Figure 11.3 Negative cross-resistance versus strictly negatively correlated resistance. In (A) a given allele (or alleles) from a single given confers resistance to one toxin and hypersusceptibility to a second toxin (negative cross-resistance). In (B) one locus confers resistance to one toxin, and another tightly linked loci confers hypersusceptibility to a second toxin (negatively correlated resistance). In (C) one locus confers resistance to one toxin and another distant locus that confers hypersusceptibility to a second toxin (negatively correlated resistance).

loci are respectively involved in resistance and hypersusceptibility, the linkage between such loci will be important in how effective "toxin A" will be in reducing the allelic frequency of resistance to "toxin B" (and vice versa). The more tightly linked the loci, the more likely they can be selected back and forth in an NCR strategy (Figure 11.3B). However, the less tightly linked these loci are, the less effective such a paired compound strategy will be in managing resistance (Figure 11.3C).

Negative Cross-Resistance in the Context of Population Genetics

For a population with a high frequency of resistance alleles, homozygous resistant individuals are expected to be the most prevalent genotype (Figure 11.4). In such cases, a meaningful reduction in the frequency of resistance alleles can occur for NCR compounds active only against

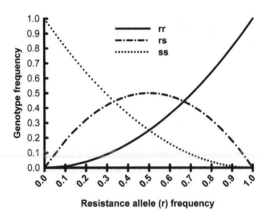

Figure 11.4 Proportion of genotypes as a function of r allele frequency. Graph assumes that the population is at Hardy–Weinberg Equilibrium and is based on a model with one locus and two alleles, where r is the allele for resistance and s is the allele for susceptibility.

homozygous resistant individuals. However, as the frequency of resistance alleles in the population decreases, the proportion of homozygous resistance individuals will decline rapidly, while the frequency of individuals that are heterozygous for resistance will increase rapidly (Figure 11.4). For example, when resistance-allele frequency is 0.9, 81% of the population will be homozygous resistant and 18% will be heterozygous. By contrast when resistance-allele frequency is 0.2, only 4% of the population will be homozygous resistant, but 32% of the population will be heterozygous. Consequently, for NCR compounds to efficiently drive resistance alleles to low levels (e.g., less than 20%), those compounds will need to be active against heterozygous genotypes.

SCREENING AND DEVELOPMENT OF NEGATIVE CROSS-RESISTANCE TOXINS

Based on a lock-and-key understanding of how toxins impact target systems, we can begin to design (or selectively screen) toxins useful in a NCR strategy. Two such examples exist in the literature. First, Oettmeier et al. (1991) were able to demonstrate in plants that specific amino acid changes in the photosystem II D-1 protein confer resistance to a triazinone herbicide. They were also able to determine the position and substituted groups in the herbicide that conferred NCR to the protein coded for by this mutant allele, defining a lock-and-key relationship between the pesticides and the NCR mechanism.

Second, Hedley et al. (1998) were able to selectively screen compounds that provided NCR in insecticide-resistant green peach aphid *Myzus persicae*, based on the knowledge that the mechanism of resistance in this insect was through increased esterase activity. Thus, they tested compounds that were bioactivated by esterase activity, so that aphids with higher esterase activity were more sensitive to their effects than the wild-type insects. Unfortunately, the most potent NCR factor observed, monofluoroacetic acid, would not have practical utility.

Another way of screening for NCR is to use resistant field populations of pest species under selection from pesticide use. By comparing these field populations to susceptible laboratory strains, multiple chemicals can be tested while providing insight into the number of generations it takes to develop resistance in a population. *Spodoptera litura* has been both directly and indirectly exposed to numerous insecticides. When brought into the laboratory and reared for 14 generations of imidacloprid selection, *S. litura* showed 137-fold increase in resistance to imidacloprid, cross-resistance to acetamiprid, and NCR to methomyl (Abbas et al., 2012). Alternatively, Shad et al. (2010) selected *S. litura* with emamectin for only three generations for a tenfold increase of resistance to emamectin. Consequently, selection with emamectin lowered resistance to abamectin, indoxacarb, and acetamiprid (Shad et al., 2010).

The aforementioned studies provide evidence that compounds can be designed or selectively screened based on a priori knowledge of the target site, to provide for NCR to metabolic resistance (Hedley et al., 1998) as well as target site insensitivity (Oettmeier et al., 1991). Structure-based (rational) design of NCR compounds could be employed where we have an in-depth understanding of the molecular mechanisms by which insects have developed resistance to an initial class of pesticides. Industrial laboratories have already used structure-based (rational) design to develop herbicides, fungicides, and insecticides (Walter, 2002). As our understanding of insecticidal molecules, their respective target sites, and the molecular mechanism of resistance in pest species increases (Schnepf et al., 1998), the scientific community will be in a better position to determine the feasibility of developing NCR toxins for specific forms of pesticide resistance.

However, several challenges to the discovery of NCR compounds exist for insect control. If the approach is based on a whole animal screen, separate homozygous susceptible and resistant insect populations must be maintained since it will be critical that NCR compounds are screened against heterozygous insects. Initial screens for NCR compounds will involve bioassays with the homozygous susceptible and resistant insect populations (Pittendrigh and Gaffney, 2001). The NCR compounds discovered from these screens would then have to be tested against heterozygous (crosses between the homozygous susceptible and resistant insects) in order to determine which putative NCR compounds would also be effective in killing heterozygous insects (Pittendrigh and Gaffney, 2001). For example, Nguyen et al. (2007) screened four Drosophila strains against 11 insecticides and found susceptibility to menthofuran and benzothiophene. Rearing resistant strains and performing crosses to maintain them are a cost that will have to be factored into the approach and may prove to be prohibitively expensive. However, live animal screens have a distinct advantage insofar as effects observed in the laboratory give very high confidence of utility in the field.

Screening strategies based on either the specific target site receptor, for example, Bacillus thuringiensis toxins, or on a specific metabolic resistance mechanism, for example, esterases, are very conducive to high-throughput screening approaches routinely used in industrial laboratories. Since the screening and development of NCR compounds are likely to be reactionary given that a priori knowledge of the nature of the resistance is usually lacking, high-throughput approaches will probably be necessary for rapid development of NCR compounds. The development of transgenic insect lines like those developed in D. melanogaster may provide a means of capturing the best attributes of both the live animal and biochemical screening approaches.

Crystal delta-endotoxins (Cry proteins) are a class of insecticidal proteins found in the soil-dwelling bacterium Bacillus thuringiensis (Bt). Cry proteins kill insects through a receptor-mediated process. Cry1Ac is a Bt protein with a high degree of specificity against lepidopteran insects such as Manduca sexta. Drosophila melanogaster lack a midgut receptor for Cry1Ac and so are not negatively affected by it (Figure 11.5A). Gill and Ellar (2002) transformed D. melanogaster with a gene that encodes a Cry1Ac-binding aminopeptidase receptor (APN) obtained from M. sexta. The APN protein was expressed in the digestive system of D. melanogaster, and the flies became susceptible to Cry1Ac (Figure 11.5B).

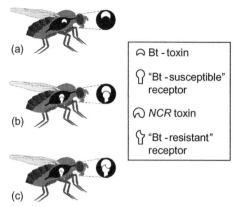

Figure 11.5 Hypothetical example of use of transgenic *Drosophila melanogaster* to discover NCR toxins. In (A) *D. melanogaster* does not have the receptor that would make it susceptible to a given toxins [*e.g.*, nontransgenic *D. melanogaster* and *Bt-Cry1Ac*; Gill and Ellar (2002)]. When in Gill and Ellar (2002) made transgenic *D. mela-nogaster* with an aminopeptidase N (APN) receptor from *Manduca sexta*, the trans-genic flies were now susceptible to *Bt-Cry1Ac*. Transgenic *D. melanogaster* could also be created with alleles that confer resistance to the first toxin, and the resultant strains could be used to discover or test putative NCR toxins useful in pest control.

Other researchers have isolated or identified additional *Bt* receptors from other insects (Flannagan *et al.*, 2005; Griffitts *et al.*, 2005; Morin *et al.*, 2003; Rajagopal *et al.*, 2002; Xie *et al.*, 2004). In cases where resistance is associated with a *Bt* Cry receptor, or a similar receptor-mediated process, this approach could be used to transform *D. melanogaster* with resistance alleles isolated from resistant pest insects, and the resulting transgenic flies could be screened to identify NCR *Bt* Cry variants (or other classes of compounds for NCR).

The utility of combining a biochemical screen with a live animal assay is possible in part because *D. melanogaster* are very easy and inexpensive to rear and maintain (Figure 11.5C). Additionally, separate fly strains could be developed for each novel resistance allele discovered in the field. Transgenic *D. melanogaster* strains can be produced that express both the susceptible and resistant forms of receptor, so that the putative NCR compounds identified in initial screens could subsequently be tested for their potential toxicity against the heterozygous insects. Thus, most or all of the allelic forms of resistance for any given trait could be screened for NCR using a set of transgenic *D. melanogaster* strains.

Future studies will need to be performed to determine the feasibility of such an approach, although all the tools necessary to develop this strategy are currently available for *D. melanogaster*. Regardless of the success of using *D. melanogaster* in such a strategy, the fact that we are continually gaining a better understanding of the molecular mechanisms by which insects develop resistance to pesticides means that in *vivo* or in vitro screening strategies for NCR compounds are now feasible at least for research. Additionally, other emerging technologies hold out possibilities for the discovery and development of peptides or proteins useful in NCR.

An example of a purely biochemical approach to screening is affinity selection using phage display technologies (Sidhu *et al.*, 2000; Pande *et al.*, 2010). Phage display has been employed as a process to rapidly screen very large peptide libraries to select peptides for high-affinity binding to a given target (Huang *et al.*, 2012). Current molecular techniques can easily facilitate production of variant peptides even to the point of saturation mutagenesis of every amino acid residue in a peptide target. The premise is that specific changes in the primary structure that enhance binding affinity will also enhance biological activity. For example, Marzari *et al.* (1997) used phage display to identify Cry1Aa toxin regions implicated in receptor binding. Koiwa *et al.* (1998) used a similar approach to identify plant cysteine proteinase inhibitor variants (cystatins) for use in control of *Callosobruchus maculatus*, a coleopteran pest of cowpeas. Thus, where the target molecule is known, phage display can be used as a strategy to discover molecules that are more effective in killing pest insects (Koiwa *et al.*, 1998, 2001).

Where the pesticide-target protein and NCR allele of the respective gene is known, phage display could be used to identify NCR toxins. The protein that confers resistance could be used as the selective agent in phage display. This approach may be useful in identifying NCR toxin variants or chimeras built from multiple unrelated sources (Figure 11.6). Other approaches, such as combinatorial chemistry, may also be useful, especially for nonpeptide chemistries for development of NCR compounds useful in insect control.

DEPLOYMENT STRATEGIES: THE CASE OF ACTIVE REFUGES AND HIGH-DOSE BT CROPS

A variety of NCR deployment strategies have been suggested, including: (i) rotation of NCR compounds; (ii) periodic pyramiding of the

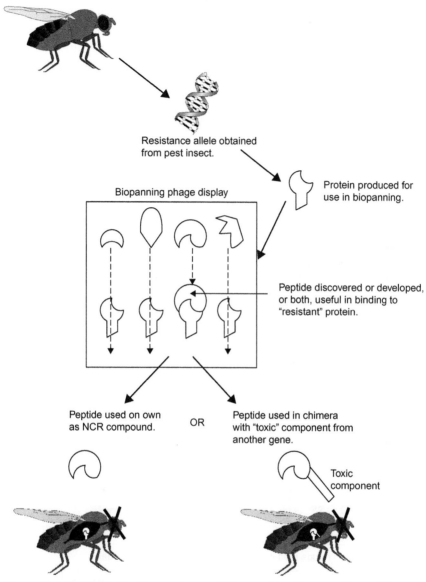

Resistance allele obtained from pest insect.

Biopanning phage display

Protein produced for use in biopanning.

Peptide discovered or developed, or both, useful in binding to "resistant" protein.

Peptide used on own as NCR compound.

OR

Peptide used in chimera with "toxic" component from another gene.

Toxic component

Figure 11.6 Phage display bioscanning could be used to discover polypeptides that selectively interact with proteins coded for by resistance genes.

NCR compounds with a separate group of pesticides (with different modes of action) to both concurrently minimize the pest population and use the NCR compound to minimize the resistance alleles in the insect population; and (iii) continuous pyramiding of the NCR compounds with a separate group of pesticides (Pittendrigh *et al.*, 2000). However, some of these strategies might require constant monitoring of resistance levels in the insect populations, something that may not be economically feasible at present. Additionally, the use of multiple compounds or traits in a cropping system could present significant challenges to industry both in terms of discovery of those compounds as well as in breeding and deployment.

One approach that might prove feasible is the use of an "active refuge" (Pittendrigh *et al.*, 2004). This NCR strategy would take advantage of the high-dose refuge strategy currently employed for the management of pest resistance for transgenic plants expressing insecticidal toxins (Figure 11.7A). The "active refuge" strategy involves pro-active deployment of a NCR factor in the refuge (Pittendrigh *et al.*, 2004), where the NCR toxin acts as a "filter" to remove resistance alleles from the insect population (Figure 11.7B). The active refuge approach is particularly attractive since selection experiments and simulation modeling suggest that the NCR toxins do not have to be particularly effective in killing the homozygous (RR) and heterozygous (RS) resistant insects in order to control resistance in the insect population for many generations (Figure 11.8). Even with a small refuge size (e.g., 4%), an NCR compound deployed in the refuge that killed only about 40% of the heterozygous (RS) insects was highly effective in keeping the resistance allele at a low frequency for many generations beyond the currently used "passive refuge."

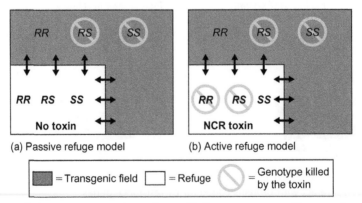

(a) Passive refuge model   (b) Active refuge model

◼ = Transgenic field   ☐ = Refuge   ⊘ = Genotype killed by the toxin

Figure 11.7 Passive versus active refuges. (A) No genotype-specific mortality occurs in the passive refuge. (B) Both the resistant homozygous (*RR*) and heterozygous (*RS*) resistant insects are killed by the negative cross-resistant (NCR) toxin in the active refuge. In both the passive and active refuges, the PPPG selectively kills the *RS* and *SS* individuals. *[reprinted from Pittendrigh et al. (2004), with permission from Journal of Theoretical Biology].*

This fact may prove to be particularly attractive since the discovery of moderately effective compounds or other ecologically based factors (e.g., host-plant resistance or entomopathogens) is far easier than the discovery of "blockbuster" products. Indeed, the focus in large-scale commercial screening is often the discovery of compounds that produce high mortality among heterozygous (RS) insects, or mass-kill toxins. For example, potentially useful NCR compounds might have been deprioritized since they alone would not be highly effective in controlling the size of the insect population and would thus not be commercially viable.

That an NCR approach is feasible with the current array of biotechnology-derived crops is supported by several lines of evidence. One of the first reported cases of NCR with *Bt* toxins was in *Plodia interpunctella* (Van Rie *et al.*, 1990). Negative cross-resistance has also been documented in *Helicoverpa zea* between the *Bt* toxins *Cry1Ac* and *Cry1F* (Marcus, 2005). Additionally, NCR with *Bt* toxins have also been

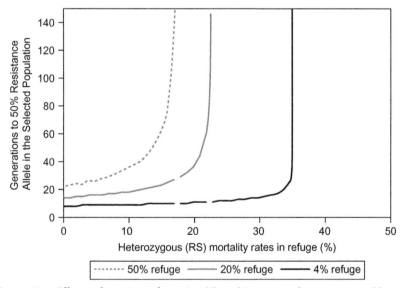

Figure 11.8 Effects of varying refuge size (*G*) and increasing heterozygous, *RS*, mortality rates, in the refuge, on delaying the development of resistance in the insect population when an active refuge is used (i.e., an NCR compound is deployed in the refuge). Details of the assumed conditions are given in Figure 3 of Pittendrigh *et al.* (2004). This figure demonstrates that NCR compounds deployed in the refuge only need to kill a small number of the heterozygous (*RS*) individuals (less than 40%) in order to dramatically delay the time it takes for the resistance allele to become common (50% frequency) in the insect population. Low-toxicity NCR compounds should be easier to discover and develop than high-toxicity NCR compounds.

observed in *Helicoverpa armigera* in China (Liang *et al.*, 2000). These observations hold out the possibility that NCR factors may be discovered or
developed with the potential for use in an active refuge strategy for resistance management. However, the long product development times for
biotech crops as well as the extensive regulatory requirements for the
crops would pose significant hurdles to the use of any "reactive" pest control strategy including NCR.

In addition to future compounds that might be developed, active
refuges may be deployed using currently available crop varieties and
microbial biopesticides (Gassmann *et al.*, 2009a,b). Data indicate that
NCR occurs when *Bt*-resistant insects are exposed to some entomopathogens (i.e., viruses and nematodes). Furthermore, selection of appropriate host-plant varieties and crops for refuges also can impose NCR by
magnifying the fitness costs of *Bt* resistance. As such, active refuges based
on NCR may be deployed in conjunction with development of integrated pest management (IPM) approaches for refuges.

Several studies with *Bt*-resistant strains have found that NCR will
arise on specific host-plant species, with the degree of NCR and its
effects on heterozygous resistant individuals shifting as a function of
which host plant is evaluated. For example, compared with *Bt*-
susceptible conspecifics, Cry1Ac-resistant *H. armigera* displayed reduced
fitness on cotton, pigeon pea, and sorghum; however, the nature of
the effects depended on the life-history trait measured and host plant
(Bird and Akhurst, 2007). While homozygous resistant insects suffered
lower fecundity if they developed on sorghum or pigeon pea, no
effect was present on cotton. By contrast, developmental rate was
delayed for both homozygous resistant and heterozygous insects on
cotton but no effects were detected on pigeon pea (Bird and Akhurst,
2007). Similarly, for *Trichoplusia ni* with resistance to *Bacillus thuringiensis kurstaki*, survival was reduced for homozygous resistant individuals
on cucumber but no NCR was found on pepper or tomato (Janmaat
and Myers, 2005).

In addition to detecting NCR by measuring life-history characters,
selection experiments can provide a highly sensitive method ????
(Gassmann *et al.*, 2009a). Resistant strains are reared in the absence of the
agent to which they are resistant (e.g., *Bt* toxin) but exposed to various
NCR factors. For example, Raymond *et al.* (2007a) found that resistance
to Cry1Ac declined more rapidly when *P. xylostella* was reared on *Brassica
pekinensis* compared to *Brassica oleracea*.

Certain host plants may be able to impose NCR while others do not in part because of differences in host-plant resistance factors such as allelo-chemicals. For example, work by Carrière *et al.* (2004) found that Cry1Ac resistant pink bollworm *Pectinophora gossypiella* suffered NCR when fed diet containing the cotton allelochemical gossypol. When incorporated into artificial diet, gossypol caused greater delays in develop-mental time and reductions in pupal weights in some resistant *P. gossypiella* strains compared to the *Bt*-susceptible strains. Follow-up experiments revealed that resistant larvae had higher concentrations of gossypol, and that loss of fitness depended on the number and type of resistance alleles they carried (Williams *et al.*, 2011).

Both entomopathogenic viruses and nematodes can selectively kill *Bt*-resistant pests in some instances. Several studies have tested the ability of entomopathogenic nematodes to impose NCR on Cry1Ac-resistant *P. gossypiella* (Gassmann *et al.*, 2006, 2008, 2009c, 2012; Hannon *et al.*, 2010). The entomopathogenic nematode *Steinernema riobrave* was found to elicit NCR in most experiments, by causing greater mortality for indi-viduals harboring Bt resistance alleles relative to mortality imposed on homozygous susceptible individuals (Gassmann *et al.*, 2006, 2008, 2009c; Hannon *et al.*, 2010). NCR typically affected only homozygous resistance insects (Gassmann *et al.*, 2006, 2008, 2009c), but data also indicate that NCR for heterozygous individuals does arise in some cases (Hannon *et al.*, 2010).

The capacity of nematodes to impose NCR appears to be species spe-cific and is likely also mediated by other ecological factors. Although some experimental data revealed that *Heterorhabditis bacteriophora* imposed NCR (Gassmann *et al.*, 2009c) other experiments found no evidence for such an effect (Gassmann *et al.*, 2008; Hannon *et al.*, 2010). Two addi-tional species of *Steinernema* and *H. sonorensis* were tested by Hannon *et al.* (2010), but none imposed NCR.

The mechanisms mediating NCR between entomopathogenic nema-todes and Bt-resistant *P. gossypiella* are likely complex, involving the resis-tant insect as well as highly pathogenic bacterial symbionts in the genera *Xenorhabdus* (associated with *Steinernema* spp.) and *Photorhabdus* (associated with *Heterorhabditis* spp.). Previous work has shown that *Photorhabdus* and *Xenorhabdus* produce extremely toxic insecticidal proteins (Forst *et al.*, 1997; Bowen *et al.*, 1998; Brown *et al.*, 2004). Although it is not clear if the nematode or the insecticidal proteins or an interactive combination of the two are responsible for the NCR effect, past research does demonstrate that NCR mechanisms (increased fitness cost for the resistant insects) operate in naturally occurring biological systems.

Entomopathogenic viruses also have the potential to be used as agents of NCR, perhaps as a biopesticide applied to refuges. Raymond *et al.* (2007b) conducted a selection experiment to test the concept of active refuges. Using Cry1Ac-resistant *P. xylostella*, the researchers established population cages containing *Bt*-treated cabbage plants and refuge cabbage plants (not treated with *Bt*); then they monitored changes in resistance over time. Cages in which refuge plants were treated with virus displayed a slower rate of resistance evolution than cages in which an NCR factor was not added to refuges. These data demonstrate the potential utility of entomopathogens, applied either as biopesticides or cultivated as biocontrol agents, to delay resistance to *Bt* toxins.

Computer simulation models have been applied to data from NCR experiments with Cry1Ac-resistant *P. gossypiella* and nematodes, and illustrate the potential benefits and limitations of adopting active refuges (Gassmann *et al.*, 2008, 2009b, 2012; Hannon *et al.*, 2010). All of these models assume that the *Bt* crop is a high-dose event (i.e., only homozygous resistant individuals can survive on the Bt cotton crop) and that resistance is monogenic with two alleles, one for resistance (r) and one for susceptibility (s). When NCR affected only rr individuals, treating all refuge fields with nematodes delayed resistance. However, effects depended on the percentage of the landscape planted to refuges, with increased refuge size bolstering the effectiveness of active refuges to delay resistance (Gassmann *et al.*, 2008). Instances in which NCR affected rs and rr individuals were highly effective at delaying resistance (Gassmann *et al.*, 2009c, Hannon *et al.*, 2010). This is because when resistance first evolves in a population, the vast majority of resistance alleles will be present in heterozygous individuals (Figure 11.4). Computer modeling indicated that when only a small percentage of refuge fields (e.g., 2.5%) were treated with *S. riobrave* that imposed NCR on both rs and rr genotypes, it took twice as long for populations to develop resistance compared with a landscape in which active refuges were absent (Hannon *et al.*, 2010).

Although these results are encouraging, limitations to the benefits of active refuges also exist. Gassmann *et al.* (2012) found that both population bottlenecks and a high initial resistance-allele frequency diminished or eliminated the benefits of active refuges. These results point to the importance of using IPM in conjunction with active refuges. By keeping pests below economic thresholds without the deployment of excessive management practices, the degree to which a pest population experiences bottlenecks should be diminished. Furthermore, using an integrated approach will reduce selection for resistance to any single management tactic, helping to keep the frequency of resistance alleles low, and thus enhancing the capacity of active refuges to delay the onset of resistance.

ADDITIONAL ISSUES
The Third Allele

Regardless of the deployment strategy used, a key question to address will be the likelihood of the development of resistance to a NCR toxin pair. First, let us assume that a screening strategy for NCR compounds reveals a series of putative compounds useful for development of a practical NCR deployment strategy. The next question is, which of these compounds should proceed to commercial development? Many factors will influence this decision, but obviously it would be in a company's best interest to develop an NCR compound that has the longest commercial life. To that end, a factor to consider is the probability that a third allele arises in the insect population that confers resistance to both the first toxin and the NCR toxin. Pittendrigh and Gaffney (2001) outlined a screening strategy to address this issue (Figure 11.9).

Briefly, toxin pairs could be applied to insects that have been muta-genized to screen for alleles capable of surviving both toxins at once. For toxin pairs where dually resistant mutants are not observed, or are observed at a lower frequency than other toxin pairs (NCR compound plus first toxin), then the given NCR compound should be given higher priority for further development (Figure 11.9, left-hand side). Those toxins pairs (the first toxin plus the NCR toxin) in which resistance alleles arise more frequently would be given lower priority for putative development of the NCR compound (Figure 11.9, bottom right-hand side). Such a strategy may help prioritize those NCR compounds for development for practical uses.

Economic Factors

Development and deployment of NCR compounds in the field will ultimately depend on multiple scientific and economic factors. Use of an

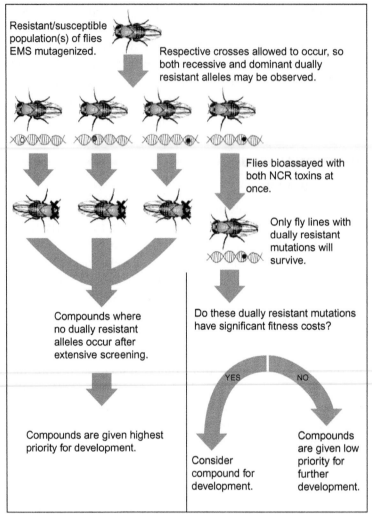

Figure 11.9 A hypothetical screening strategy to test for the existence of a putative third allele (dually resistant allele) that confers resistance to a pair of NCR toxins. *(Flies presented in this figure are reproduced with the permission and copyright of Exploratorium, www.exploratorium.edu).*

NCR strategy may not work in all situations. For example, if multiple forms of resistance occur in a pest species to a particular pesticide (Hemingway *et al.*, 1993), it may be difficult or even impossible to identify a single compound that can provide generalized NCR. In the case of chlorpyrifos and propoxur resistance in cockroaches, of the 14 lines of

B. germanica surveyed for resistance, ten lines showed esterase-like resistance and two lines showed NCR (or negatively correlated resistance) (Hemingway *et al.*, 1993). The development of multiple NCR compounds to deal with this diversity of resistance may be too costly. In contrast, target site-insensitivity to dieldrin is due to amino acid changes that are highly conserved across divergent taxonomic groups (ffrench-Constant *et al.*, 1998) suggesting that commercial development of a single NCR toxin to combat resistance may be feasible.

Economics and competing commercial interests will also influence the decision to develop NCR as an approach to resistance management. Resistance to insecticides that have little commercial value, such as an insecticide useful in a niche market, may not justify the costs of developing an NCR factor. Alternatively, resistance to high-value insecticides may even warrant the development of multiple NCR factors effective against diverse forms of resistance. The ability to rapidly respond to the emergence of resistant alleles via high-throughput screening capabilities as outlined in this chapter greatly enhances the utility of NCR approaches.

CONCLUSIONS

Negative cross-resistance has been observed across a variety of species and chemical classes. However, to date, it has not been used in wide-scale insect resistance management. The lack of forthcoming NCR products may be due to business models that necessitate cost-effective development of new products due to the needs of the marketplace. In this regard, NCR products likely will be developed only in response to verified resistance to currently marketed high-value products, even though the optimal strategy from NCR will often be to apply such compounds proactively rather than reactively. An additional reason for the lack of NCR compounds may be the practical limitations in the methodologies needed for efficient discovery. One solution may be the use of large-scale screening processes modified from those currently used for screening for novel pesticides. Such screening approaches could involve field-resistant insects, or in some specific cases transgenic *D. melanogaster* expressing the resistance trait. Additionally, advances in molecular cloning and expression of peptides in display technologies could allow for the rapid development of NCR products as soon as resistance occurs in insect populations in the field. Rational design of traditional chemistries and proteins is also well established. Thus, field-resistant insects, high-throughput transgenic live insect systems, phage display technologies, rational design approaches, or any combination of these could be used to assess a wide array of receptor/toxin combinations to model a best fit for NCR toxins useful in the field.

Deployment strategies have been presented in order to optimize the usefulness of such NCR toxins in resistance management. In the case of transgenic plants expressing insecticidal toxins, such as *Bt* toxins, an "active refuge model," where the NCR is deployed in the refuge, could delay or prevent resistance. Again, any system developed will have to consider the long development times for the transgenic crops—an active area of research in industry—and the many regulatory hurdles associated with transgenic crops.

The fact that naturally occurring host-plant resistance factors may play a role in NCR (Carrière *et al.*, 2004; Janmatt and Myers ,2005; Bird and Akhurst, 2007; Raymond *et al.*, 2007a) raises an interesting question that has not been sufficiently addressed in the literature. Are there sets of NCR compounds that are found in different classes of plants? If so, how have these compounds shaped food choices and the evolutionary history of different insect species? For example, if an insect evolves the ability to detoxify compounds from one host plant, does this in turn make the insect more susceptible to another group or class of host plants? Suppose an insect population contained a novel allele ("Allele 2") that allowed the population to preferentially survive on host plant "A", but reduced the insect's fitness on its original host plant (host plant "B"). Insects with "Allele 1" would likely remain on their original host plants (host plants "A"), but those insects carrying "Allele 2" would preferentially survive on host plant "B." This dynamic has likely played a role in shaping host-plant specialization by phytophagous insects (Futuyma and Moreno, 1988).

Although NCR has been demonstrated across several classes of synthetic pesticides and for ecological factors, the mechanisms behind entomopathogens and host-plant species causing ecologically based NCR are not well understood. Additional observations and investigations are needed to identify new examples of interacting organisms and their underlying mechanisms. Whether NCR is a widespread ecological phenomenon or only exists in specific cases, its potential applications in IRM is an intriguing possibility worthy of future research.

REFERENCES

Abbas, N., Shad, S.A., Razaq, M., 2012. Fitness cost, cross resistance and realized heritability of resistance to imidacloprid in *Spodoptera litura* (Lepidoptera: Noctuidae). Pestic. Biochem. Phys. 103, 181−188.

Barros, A.T., Gomes, A., Ismael, A.P., Koller, W.W., 2002. Susceptibility to diazinon in populations of the horn fly, *Haematobia irritans* (Diptera: Muscidae), in Central Brazil. Mem. Inst. Oswaldo Cruz. 97, 905−907.

Bird, L.J., Akhurst, R.J., 2007. Effects of host plant species on fitness costs of Bt resistance in *Helicoverpa armigera* (Lepidoptera: Noctuidae). Biol. Control. 40, 196−203.

Bossche, V.H., 1997. Mechanisms of antifungal resistance. Rev. Iberoam. Micol. 14, 44−49.

Bowen, D., Rocheleau, T.A., Blackburn, M., Andreev, O., Golubeva, E., Bhartia, R., et al., 1998. Insecticidal toxins from the bacterium *Photorhabdus luminescens*. Science. 280, 2129−2132.

Broadhurst, M.D., 1998. The influence of the molecular basis of resistance on insecticide discovery. Philos. Trans. R. Soc. London Ser. B. 353, 1723−1728.

Brown, S., Cao, S., Hines, E., Akhurst, R., East, P., 2004. A novel secreted protein toxin from the insect pathogenic bacterium *Xenorhabdus nematophila*. J. Biol. Chem. 279, 14595−14601.

Carrière, Y., Ellers-Kirk, C., Biggs, R., Higginson, D.M., Dennehy, T.J., Tabashnik, B.E., 2004. Effects of gossypol on fitness costs associated with resistance to Bt cotton in pink bollworm. J. Econ. Entomol. 97, 1710−1718.

Chapman, R.B., Penman, D.R., 1979. Negatively correlated cross-resistance to a synthetic pyrethroid in organophosphorous-resistant *Tetranychus urticae*. Nature (London). 218, 298−299.

Chen, Z.H., Liu, C.X., Li, F.L., Han, Z.J., 1993. Development of diamondback moth strains resistant to dimehypo and cartap with reference to the mechanism of resistance. Acta Entomol. Sin. 36, 409−418.

Cilek, J.E., Dahlman, D.L., Knapp, F., 1995. Possible mechanism of diazinon negative cross-resistance in pyrethroid-resistant horn flies (Diptera: Muscidae). J. Econ. Entomol. 88, 520−524.

Crosby, B.L., Byford, R.L., Sparks, T.C., 1991. Bioassay for detecting active site insensitivity in horn fly (Diptera: Muscidae) larvae. J. Econ. Entomol. 84, 367−370.

Curtis, J., Huesing, J., Simpson, R., Elands, J., 2004. Information management for entomology screening. J. Biomol. Screen. 9, 37−43.

De Prado, R., Sanchez, M., Jorrin, J., Dominguez, C., De Prado, R., 1992. Negative cross-resistance to bentazone and pyridate in atrazine-resistant *Amaranthus cruentus* and *Amaranthus hybridus* biotypes. Pestic. Sci. 35, 131−136.

Dunley, J.E., Welter, S.C., 2000. Correlated insecticide cross-resistance in azinphosmethyl resistant codling moth (Lepidoptera: Tortricidae). J. Econ. Entomol. 93, 955−962.

Elliott, M., Farnham, A.W., Janes, N.F., Johnson, D.M., Pulman, D.A., Sawicki, R.M., 1986. Insecticidal amides with selective potency against a resistant (*super-kdr*) strain of houseflies (*Musca domestica* L,). Agric. Biol. Chem. 50, 1347−1349.

Ffrench-Constant, R., Pittendrigh, B., Vaughan, A., Anthony, N., 1998. Why are there so few resistance-associated mutations in insecticide target genes?. Philos. Trans. R. Soc. Lond. B. 353, 1−9.

Flannagan, R., Cao-Guo, Y., Mathis, J., Meyer, T., Xiaomei, S., Siqueira, H., et al., 2005. Identification, cloning and expression of a *CryIAb* cadherin receptor from European corn borer, *Ostrinia nubilalis* (Hübner) (Lepidoptera: Crambidae). Insect. Biochem. Mol. Biol. 35, 33−40.

Forst, S., Dowds, B., Boemare, N., Stackebrandt, E., 1997. *Xenorhabdus* and *Photorhabdus* spp.: bugs that kill bugs. Annu. Rev. Microbiol. 51, 47−72.

Futuyma, D.J., Moreno, G., 1988. The evolution of ecological specialization. Annu. Rev. Ecol. Syst. 19, 207−233.

Gadamski, G., Ciarka, D., Gressel, J., Gawronski, S.W., 2000. Negative cross-resistance in triazine-resistant biotypes of Echinochloa crus-galli and Conyza canadensis. Weed Sci. 48, 176−180.

Gassmann, A.J., Futuyma, D.J., 2005. Consequence of herbivory for the fitness cost of herbicide resistance: photosynthetic variation in the context of plant-herbivore interactions. J. Evol. Biol. 18, 447−454.

Gassmann, A.J., Stock, S.P., Carriere, Y., Tabashnik, B.E., 2006. Effect of entomopathogenic nematodes on the fitness cost of resistance to Bt toxin Cry1Ac in pink bollworm (Lepidoptera: Gelechiidae). J. Econ. Entomol. 99, 920−926.

Gassmann, A.J., Stock, S.P., Sisterson, M.S., Carrière, Y., Tabashnik, B.E., 2008. Synergism between entomopathogenic nematodes and Bacillus thuringiensis crops: integrating biological control and resistance management. J. Appl. Ecol. 45, 957−966.

Gassmann, A.J., Carrière, Y., Tabashnik, B.E., 2009a. Fitness costs of insect resistance to Bacillus thuringiensis. Annu. Rev. Entomol. 54, 147−163.

Gassmann, A.J., Onstad, D.W., Pittendrigh, B.R., 2009b. Evolutionary analysis of herbivorous insects in natural and agricultural environments. Pest. Manag. Sci. 65, 1174−1181.

Gassmann, A.J., Fabrick, J.A., Sisterson, M.S., Hannon, E.R., Stock, S.P., Carrière, Y., et al., 2009c. Effects of pink bollworm resistance to Bacillus thuringiensis on phenoloxidase activity and susceptibility to entomopathogenic nematodes. J. Econ. Entomol. 102, 1224−1232.

Gassmann, A.J., Hannon, E.R., Sisterson, M.S., Stock, S.P., Carrière, Y., Tabashnik, B.E., 2012. Effects of entomopathogenic nematodes on the evolution of pink bollworm resistance to Bt toxin Cry1Ac. J. Econ. Entomol. 105, [In press].

Gill, M., Ellar, E.D., 2002. Transgenic Drosophila reveals a functional in vivo receptor for the Bacillus thuringiensis toxin Cry1Ac1. Insect. Mol. Biol. 11, 619−625.

Griffitts, J., Haslam, S., Yang, T., Garczynski, S., Mulloy, B., Morris, H., et al., 2005. Glycolipids as receptors for Bacillus thuringiensis crystal toxin. Science. 307, 922−925.

Hannon, E.R., Sisterson, M.S., Stock, S.P., Carrière, Y., Tabashnik, B.E., Gassmann, A.J., 2010. Effects of four nematode species on fitness costs of pink bollworm resistance to Bacillus thuringiensis toxin Cry1Ac. J. Econ. Entomol. 103, 1821−1831.

Hardstone, M.C., Leichter, C., Harrington, L.C., Shinji, K., Tomita, T., Scott, J.G., 2007. Cytochrome P450 monooxygenase-mediated permethrin resistance confers limited and larval specific cross-resistance in the southern house mosquito, Culex pipiens quinquefasciatus. Pestic. Biochem. Phys. 89, 175−184.

Hatano, R., Scott, J.G., Dennehy, T.J., 1992. Enhanced activation is the mechanism of negative cross-resistance to chlorpyrifos in the dicofol-IR strain of Tetranychus urticae (Acari: Tetranychidae). J. Econ. Entomol. 85, 1088−1091.

Hedley, D., Khambay, B., Hoopeer, A., Thomas, R., Devonshire, A., 1998. Proinsecticides effective against insecticide-resistant peach-potato aphid (Myzus persicae (Sulzer). Pestic. Sci. 53, 201−208.

Hemingway, J., Small, G.J., Monro, A.G., 1993. Possible mechanisms of organophosphorus and carbamate insecticide resistance in German cockroaches (Dictyoptera: Blattelidae) from different geographical area. J. Econ. Entomol. 86, 1623−1630.

Hollomon, D.W., Butters, J.A., Barker, H., Hall, L., 1998. Fungal beta-tubulin, expressed as a fusion protein, binds benzimidazole and phenylcarbamate fungicides. Antimicrob. Agents Chemother. 42, 2171−2173.

Hoy, M.A., 1998. Myths, models and mitigation of resistance to pesticides. Philos. Trans. Roy. Soc. London B. 353, 1787−1795.

Huang, J., Ru, B., Zhu, P., Nie, F., Yang, J., Wang, X., et al., 2012. MimoDB 2.0: a mimotype database and beyond. Nucleic Acids Res. 40, 271−277.

Janmaat, A.F., Myers, J., 2005. The cost of resistance to *Bacillus thuringiensis* varies with the host plant of *Tricoplusia ni*. Proc. R. Soc. Biol. Sci. Ser. B. 272, 1031−1038.

Josepovits, G., Gasztonyi, M., Mikite, G., 1992. Negative cross-resistance to N -phenylanilines in benzimidazole-resistant strains of *Botrytis cinerea, Venturia nashicola* and *Venturia inaequalis*. Pestic. Sci. 35, 237−242.

Kamidi, R.E., Kamidi, M.K., 2005. Effects of a novel pesticide resistance management strategy on tick control in a smallholding exotic-breed dairy herd in Kenya. Trop. Anim. Health Prod. 37, 469−478.

Khambay, B.P., Denholm, I., Carlson, G.R., Jacobson, R.M., Dhadialla, T.S., 2001. Negative cross-resistance between dihydropyrazole insecticides and pyrethroids in houseflies, *Musca domestica*. Pest. Manag. Sci. 57, 761−763.

Kniaz, D., 2000. Drug discovery adopts factory model. Mod. Drug Discov. 3, 67−72.

Koiwa, H., Shade, R.E., Zhu-Salzman, K., Subramanian, L., Murdock, L.L., Nielsen, S. S., et al., 1998. Phage display selection can differentiate insecticidal activity of soybean cystatins. Plant J. 14, 371−379.

Koiwa, H., D'Urzo, M.P., Assfalg-Machleidt, I., Zhu-Salzman, K., Shade, R.E., An, H., et al., 2001. Phage display selection of hairpin loop soyacystatin variants that mediate high affinity inhibition of a cysteine proteinase. Plant J. 27, 383−391.

Lee, D., Park, Y., Brown, T.M., Adams, M.E., 1999. Altered properties of neuronal sodium channels associated with genetic resistance to pyrethroids. Mol. Pharmocol. 55, 584−593.

Leroux, P.C., Hapeland, F., Arnold, A., Gredt, M., 2000. New cases of negative cross-resistance between fungicides, including sterol biosynthesis inhibitors. J. Gen. Plant Pathol. 66, 75−78.

Liang, G.M., Tan, W.J., Guo, Y.Y., 2000. Study on the resistance screening and cross-resistance of Cotton Bollworm to *Bacillus thuringiensis* (Berliner). Sci. Agric. Sinica (in Chinese). 33, 46−53.

Marcus, M.A., 2005. Fitness studies and cross resistance evaluations of an eastern North Carolina cotton bollworm strain (*Helicoverpa zea*) (Boddie) tolerant to the *Bacillus thuringiensis* delta endotoxin Cry1Ac. Masters thesis. Entomology. NCSU.

Marzari, R., Edomi, P., Bhatnagar, R., Ahmad, S., Selvapandiyan, A., Bradbury, A., 1997. Phage display of *Bacillus thuringiensis* CryIA(a) insecticidal toxin. FEBS Lett. 411, 27−31.

McCutchen, B.F., Hoover, K., Preisler, H.K., Betana, M.D., Herrmann, R., Robertson, J.L., et al., 1997. Interactions of recombinant and wild-type baculviruses with classical insecticides and pyrethroid- resistant tobacco budworm (Lepidoptera: Noctuidae). J. Econ. Entomol. 90, 1170−1180.

Morin, S., Biggs, R., Sisterson, M., Shriver, L., Ellers-Kirk, C., Higginson, D., et al., 2003. Three cadherin alleles associated with resistance to *Bacillus thuringiensis* in pink bollworm. Proc. Natnl. Acad. Sci. 100, 5004−5009.

Narahashi, T., Lund, A.E., 1980. Insect Neurobiology and Pesticide Action. Society of Chemical Industry, London, 497 − 505.

Nguyen, S.N., Song, C., Scharf, M.E., 2007. Toxicity, synergism, and neurological effects of novel volatile insecticides in insecticide-susceptible and−resistant *Drosophila* strains. J. Econ. Entomol. 100 (2), 534−544.

Oettmeier, W., Hilp, U., Draber, W., Fedtke, C., Schmidt, R., 1991. Structure−activity relationships of triazinone herbicides on resistant weeds and resistant *Chlamydomonas reinhardtii*. Pestic. Sci. 33, 399−406.

Ogita, Z., 1961a. An attempt to reduce and increase insecticide-resistance in *D. melanogaster* by selection pressure. Genetical and biochemical studies on negatively correlated cross-resistance in *Drosophila melanogaster* I. Botyu-Kagaku. 26, 7−17.

Ogita, Z., 1961b. Relationship between the structure of compounds and negatively corre-lated activity. Genetical and biochemical studies on negatively correlated cross-resistance in *Drosophila melanogaster* II. Botyu-Kagaku. 26, 18−19.

Ogita, Z., 1961c. Genetical studies on actions of mixed insecticides with negatively corre-lated substances. Genetical and biochemical studies on negatively correlated cross-resistance in *Drosophila melanogaster* III. Botyu-Kagaku. 26, 88−93.

Palmer, C.J., Cole, L.M., Smith, I.H., Moss, M.D., Casida, J.E., 1991. Silylated 1- (4-Ethynylphenyl)- 2,6,7-trioxabicyclo[2.2.2]octanes: structural features and mechanisms of proinsecticidal action and selective toxicity? J. Agric. Food Chem. 39, 1335−1341.

Pande, J., Szewczyk, M.M., Grover, A.K., 2010. Phage display: concept, innovations, applications and future. Biotechnol. Adv. 28 (6), 848−858.

Pedra, J.H.F., Hostetler, A., Gaffney, P.J., Reenan, R.A., Pittendrigh, B.R., 2004. Hyper-susceptibility to deltamethrin in *para*$^{ts-1}$ DDT resistant *Drosophila melanogaster*. Pestic. Biochem. Physiol. 78, 58−66.

Peiris, H.T.R., Hemingway, J., 1990. Temephos resistance and the associated cross-resistance spectrum in a strain of *Culex quinquefasciatus* Say (Diptera: Culicidae) from Peliyagoda, Sri Lanka. Bull. Ent. Res. 80, 49−55.

Pittendrigh, B., Reenan, R., ffrench-Constant, R., Ganetsky, B., 1997. Point mutations in the *Drosophila para* voltage-gated sodium channel gene confer resistance to DDT and pyrethroid insecticides. Mol. Gen. Genet. 256, 602−610.

Pittendrigh, B., Gaffney, P., Murdock, L., 2000. Deterministic modeling of negative-cross resistance for use in transgenic host-plant resistance. J. Theor. Biol. 204, 135−150.

Pittendrigh, B.R., Gaffney, P.J., 2001. Pesticide resistance: can we make it a renewable resource? J. Theor. Biol. 211, 365−375.

Pittendrigh, B.R., Gaffney, P.J., Huesing, J.E., Onstad, D.W., Roush, R.T., Murdock, L. L., 2004. "Active" refuges can inhibit the evolution of resistance in insects towards transgenic insect-resistant plants. J. Theor. Biol. 231, 461−474.

Poston, D.H., Hirata., C.M., Wilson., H.P., 2002. Response of acetolactate synthase from imidazolinone-susceptible and -resistant smooth pigweed populations to various ALS inhibitors. Weed Sci. 50, 306−311.

Rajagopal, R., Sivakumar, S., Agrawal, N., Malhotra, P., Bhatnagar, R., 2002. Silencing of midgut aminopeptidase N of *Spodoptera litura* by double-stranded RNA establishes its role as *Bacillus thuringiensis* toxin receptor. J. Biol. Chem. 277, 46849−46851.

Raymond, B., Sayyed, A.H., Wright, D.J., 2007a. Host plant and population determine the fitness costs of resistance to *Bacillus thuringiensis*. Biol. Lett. 3, 82−85.

Raymond, B., Sayyed, A.H., Hails, R.S., Wright, J.W., 2007b. Exploiting pathogens and their impact on fitness costs to manage the evolution of resistance to *Bacillus thurin-giensis*. J. Appl. Ecol. 44, 768−780.

Schnepf, E., Crickmore, N., Van Rie, J., Lereclus, D., Baum, J., Feitelson, J., et al., 1998. *Bacillus thuringiensis* and its pesticidal crystal protein. Microbiol. Mol. Biol. Rev. 2, 775−806.

Shad, S.A., Sayyed, A.H., Saleem, M.A., 2010. Cross-resistance, mode of inheritance and stability of resistance to emamectin in *Spodoptera litura* (Lepidoptera: Noctuidae). Pest. Manag. Sci. 66, 839−846.

Shang, Q., Pan, Y., Fang, K., Xi, J., Brennan, J.A., 2012. Biochemical characterization of acetylcholinesterase, cytochrome P450 and cross-resistance in an omethoate-resistant strain of *Aphis gossypii* Glover. Crop Prot. 31, 15−20.

Sheppard, D.C., Marchiondo, A.A., 1987. Toxicity of diazinon to pyrethroid resistant and susceptible horn flies, *Haematobia irritans* (L.): laboratory studies and field trials. J. Agric. Entomol. 4, 262−270.

Sidhu, S., Lowman, H., Cunningham, B., Wells, J., 2000. Phage display for selection of novel binding peptides. Methods Enzymol. 328, 333−363.

Tabashnik, B.E., Malvar, T., Liu, Y.B., Finson, N., Borthakur, D., Shin, B.S., et al., 1996. Cross-resistance of the diamondback moth indicates altered interactions with domain II of *Bacillus thuringiensis* toxin. Appl. Environ. Microbiol. 62, 2839−2844.

Van den Bercken, J., Vijverberg, H.P.M., 1980. Insect Neurobiology and Pesticide Action. Society of Chemical Industry, London, 79−85.

Van Rie, J., McGaughey, W.H., Johnson, D.E., Barnett, B.D., Van Mellaert, H., 1990. Mechanism of insect resistance to the microbial insecticide *Bacillus thuringiensis* Science. 247, 72−74.

Vijverberg, H.P., van der Zalm, J.M., van den Bercken, J., 1982. Similar mode of action of pyrethroids and DDT on sodium channel gating in myelinated nerves. Nature. 295, 601−603.

Walter, M.W., 2002. Structure-based design of agrochemicals. Nat. Prod. Rep. 19, 278−291.

Williams, J.L., Ellers-Kirk, C., Orth, R.G., Gassmann, A.J., Head, G., Tabashnik, B.E., et al., 2011. Fitness cost of resistance to Bt cotton linked with increased gossypol content in pink bollworm larvae. PLoS One. 6 (6), e21863. Available from: http://dx. doi.org/10.1371/journal.pone.0021863.

Williamson, M.S., Martinez-Torres, D., Hick, C.A., Devonshire, A.L., 1996. Identification of mutations in the housefly *para*-type sodium channel gene associated with *knockdown resistance* (*kdr*) to pyrethroid insecticides. Mol. Gen. Genet. 252, 51−60.

Xie, R., Zhuang, M., Ross, L., Gomez, I., Oltean, D., Bravo, A., et al., 2004. Single amino acid mutations in the cadherin receptor from *Heliothis virescens* affect its toxin binding ability to Cry1A toxins. J. Biol. Chem. 280, 8416−8425.

Yamamoto, I., Kyomura, N., Takahashi, Y., 1993. Negatively correlated cross resistance: combinations of N-methylcarbamate with N-propylcarbamate or oxidiazolone for green rice leafhopper. Arch. Insect. Biochem. Physiol. 22, 227−288.

Ziogas, B.N., Girgis, S.M., 1993. Cross-resistance relationships between benzimidazole fungicides and diethofencarb in *Botrytis cinerea* and their genetical basis in *Ustilago maydis*. Pestic. Sci. 39, 199−205.

FURTHER READING

Koiwa, H., D'Urzo, M.P., Zhu-Salzman, K., Ibeas, J.I., Shade, R.E., Murdock, L.L., et al., 2000. An in-gel assay of a recombinant western corn rootworm (*Diabrotica virgifera virgifera*) cysteine proteinase expressed in yeast. Anal. Biochem. 282, 153−155.

CHAPTER 12

Insect Resistance, Natural Enemies, and Density-Dependent Processes

David W. Onstad[1], Anthony M. Shelton[2] and J. Lindsey Flexner[3]
[1]DuPont Pioneer, Wilmington, DE
[2]Department of Entomology, Cornell University, Geneva, NY
[3]DuPont Pioneer, Wilmington, DE

Chapter Outline

Entomologists usually focus their attention on mortality caused by an insecticide, insecticidal crop, natural enemy, or cultural control applied against a pest. However, ignoring mortality factors that are either naturally occurring or applied less intensively as part of an integrated pest management (IPM) program significantly influences insect resistance management (IRM). These other mortality factors are familiar to entomologists and include: abiotic and biotic processes, biological control, chemical control, and intraspecific competition. In this chapter, we highlight the variety of effects that these mortality factors can have on IRM and the evolution of resistance.

Models based only on genotypic and allele frequencies would be adequate for studying the evolution of resistance if all individuals within a population experienced the same environment and selection pressures from the same sources. However, for many insect species, genetic and ecological processes are influenced by insect density. Thus, IRM plans, and the models on which they are based, should account for density–dependent processes. Furthermore, if IPM is the foundation of IRM (Chapter 1) and IPM focuses on insect densities and associated economic losses, then insect densities must be considered in damage calculations as well (Chapter 2). A third reason to consider density

Insect Resistance Management
DOI: http://dx.doi.org/10.1016/B978-0-12-396955-2.00012-6

is that stochastic processes, such as genetic drift, may significantly influence system dynamics when populations become small (Caprio and Tabashnik, 1992; Sisterson *et al.*, 2004)

In this chapter, we explore three main subjects related to IRM and insect densities. First, we discuss natural enemies that attack one phenotype more than others. This differential mortality directly imposes additional selection pressure due to differences between susceptible and resistant phenotypes. Table 12.1 summarizes many of the empirical studies on natural enemies and their direct influence on resistant and susceptible individuals. Second, we describe how phenotype-neutral mortality factors, including biological control, can influence resistance evolution when the environment is not homogeneous and pest densities are favored in some areas but not in others.

In the third section, we focus on common intraspecific processes as an important subset of the phenotype-neutral, mortality factors. We examine the importance of density-dependent survival and carrying capacity when refuges for susceptible individuals are deployed. An ecological process that is density-dependent is one in which the response is entirely or partly determined by the density of one or more species. For example, the attack rate of a parasite (number or proportion of hosts attacked) could depend on either the host's density or the parasite's density or both. By density-dependent survival, we mean that the probability of an individual insect surviving is dependent to some extent on that species' density. Carrying capacity is the maximum arthropod density that a specific environment or habitat can support. Food resources often limit arthropod populations. A refuge is habitat for a pest that does not contain a lethal selective agent, such as a toxin.

NATURAL ENEMIES: DIRECT EFFECTS ON SELECTION

Because pests rarely evolve resistance to their natural enemies (but see Chapter 8 for exceptions), the focus of this section is on the influence of natural enemies on the evolution of pest resistance to toxins in insecticides or host plants. With regard to host-plant resistance, Gould *et al.* (1991) took the lead in this subject when they published their conceptual and mathematical models on tritrophic interactions. The most commonly studied tritrophic system consisted of a plant, a herbivore, and a natural enemy.

Table 12.1 Experimental Studies of Biological-Control Effects on IRM

Pest	Selecting Agent	Natural Enemies	Effect on evolution to crop/toxin*	Reference
Helicoverpa armigera	chickpea	pathogen	accelerate	Lawo *et al.* (2008)
Heliothis virescens	tobacco	predators, parasitoids	accelerate	Johnson and Gould (1992)
H. virescens eggs	tobacco/ soybean	predators, parasitoids	no effect	Gould *et al.* (1991)
H. virescens larvae	tobacco/ soybean	predators, parasitoids	accelerate	Gould *et al.* (1991)
H. virescens pupae, adults	tobacco/ soybean	predators, parasitoids	delay/acc.	Gould *et al.* (1991)
H. virescens	tobacco	parasitoid	delay	Johnson *et al.* (1997a)
H. virescens	tobacco	pathogen	accelerate	Johnson *et al.* (1997a,b)
Epilachna varivestis	bean	predators	no effect/acc.	Gould *et al.* (1991)
Leptinotarsa decemlineata	potato	predator	delay	Arpaia *et al.* (1997)
L. decemlineata	potato	predator	delay/acc.	Mallampalli *et al.* (2005)
Myzus persicae	insecticide	parasitoid	delay	Foster *et al.* (2007, 2011)
Pectinophora gossypiella	cotton	nematode	delay	Gassmann *et al.* (2012)
P. gossypiella	cotton	nematode	delay	Hannon *et al.* (2010)
Mayetiola destructor	wheat	parasitoids	no effect	Knutson *et al.* (2002)
Plutella xylostella	*B. thuringiensis*	virus	delay	Raymond *et al.* (2007)

*The natural enemies can delay, accelerate, or have no effect on the evolution of resistance by the pest to the toxin or toxic crop. Note that acc. means accelerate.

Gould *et al.* (1991) realized at the start of their work that hypotheses derived from the deterministic models would be significantly influenced by a variety of interacting ecological, behavioral, and genetic processes acting over single or multiple generations. Their simplest conclusion was that natural enemies that increase differential fitness between susceptible and resistant phenotypes, by attacking more susceptible individuals than resistant individuals, will accelerate the evolution of resistance by the herbivore to the host plant. The opposite effect on resistance evolution is expected when the natural enemy attacks more resistant individuals than susceptible ones. Thus, the early focus of research was on the phenotypic fitness costs imposed on the pest by natural enemies (Gassmann et al., 2009a).

To evaluate the hypotheses postulated by Gould et al. (1991), Gould and his colleagues performed a series of experiments using transgenic insecticidal tobacco (Nicotiana tobacum) and potato (Solanum tuberosum), both expressing insecticidal proteins from the bacterium, Bacillus thuringiensis (Bt). Johnson and Gould (1992) conducted field experiments to examine interactions of Heliothis virescens, its natural enemies, and Bt tobacco plants considered partially resistant to H. virescens. They then calibrated a model to study the influence of natural enemies on the evolution of resistance to Bt tobacco. Simulation results indicated that biological control could accelerate evolution to resistant plants. Johnson et al. (1997a,b) carried out controlled studies with a parasitoid species and a pathogenic fungus that attack H. virescens on tobacco. They concluded that the parasitoid would likely delay the evolution of resistance to Bt tobacco, while the pathogen would likely promote the evolution of resistance. Arpaia et al. (1997) investigated predation of Leptinotarsa decemlineata on Bt potato plants in greenhouse and field studies. They included predation rates in a mathematical model to simulate the impact of natural enemies on the evolution of resistance by L. decemlineata to Bt potato. Simulations also included refuges of non-Bt potato plants. Results showed that predation could decrease the rate of evolution. Mallampalli et al. (2005) performed field studies to calibrate a simulation model of L. decemlineata on Bt potatoes to determine the influence of predation on IRM. They discovered that different prey species for a generalist predator that also eats L. decemlineata have different effects on the evolution of resistance to Bt potato: One prey species may delay the evolution of resistance, while the other could potentially accelerate it.

Gassmann et al. (2012) performed a meta-analysis on previous experiments, a new experiment, and simulation modeling to explore the interaction of entomopathogenic nematodes and Bt cotton (Gossypium hirsutum) on the evolution of resistance by Pectinophora gossypiella to Bt cotton. Their work extended the findings of Hannon et al. (2010) and demonstrated the effectiveness of entomopathogenic nematodes for reducing the relative fitness of the Bt-resistant pest on cotton. The nematodes attacked the larvae and reduced the fitness of Bt-resistant moths more than susceptible moths. Fitness was the same without nematodes. Simulation modeling demonstrated that an initial resistance-allele frequency > 0.015 and population bottlenecks can diminish or eliminate the resistance-delaying effects of fitness costs. Hannon *et al.* (2010) and Gassmann *et al.* (2012) concluded that some species of nematodes could delay resistance by *P. gossypiella* to Bt cotton under some conditions.

Raymond *et al.* (2007) studied *Plutella xylostella*, a microbial insecticide containing Bt, and a pathogenic nucleopolyhedrovirus. They found that the virus increased the fitness costs for Bt-resistant *P. xylostella*. Raymond *et al.* (2007) then used a model to investigate how the virus can be used to delay the evolution of resistance to Bt. One option that they advocated is the application of the virus only to refuges not sprayed with Bt. They did not model simultaneous evolution of resistance to both Bt and the virus.

Resistant *Myzus persicae* exhibit a fitness trade-off between resistance to insecticides and avoidance of parasitism through defensive behavior (Foster *et al.*, 2007). Foster *et al.* (2007) observed a variety of genotypes during periods of exposure to the parasitoid, *Diaeretiella rapae*, in the presence and absence of measured amounts of alarm pheromone. Wild-type, insecticide-susceptible individuals responded to alarm pheromone in ways that reduced parasitism. Insecticide-resistant *M. persicae* incurred significantly higher levels of parasitism. Foster *et al.* (2011) studied the reduced response to alarm pheromone in resistant *M. persicae* at three spatial scales: a single leaf of *Brassica napus* var *chinensis* with a single parasitoid, one plant of *B. napus* with one parasitoid, and eight plants with five parasitoids. At all scales, fewer insecticide-susceptible individuals became parasitized compared to insecticide-resistant ones. At the largest spatial scale, more susceptible individuals than resistant ones moved from their inoculation leaves to other leaves on the same plant after exposure to parasitoids. Given the fitness cost of insecticide resistance, evolution of resistance would likely be delayed by parasitism.

Knutson *et al.* (2002) evaluated parasitism by several parasitoids of *Mayetiola destructor* infesting five wheat (*Triticum aestivum*) cultivars with various levels of host-plant resistance. Parasitism in field cages and in open fields did not vary among wheat cultivars, and parasitism rates were independent of host density. Knutson *et al.* (2002) concluded that parasitism of *M. destructor* is compatible with host-plant resistance and may extend the usefulness of wheat cultivars by slowing the increase of resistant/virulent pest populations.

Lawo *et al.* (2008) experimentally determined that *Helicoverpa armigera* larvae resistant to insecticidal chickpea (*Cicer arietinum*) were more tolerant of infection by the entomopathogenic fungus *Metarhizium anisopliae* than were susceptible larvae. They concluded that resistance to the insecticidal crop did not influence fitness costs relative to this natural enemy. Lawo *et al.* (2008) also measured larval movement by the phenotypes on both conventional and insecticidal chickpea plants. Movement did not differ, so neither phenotype would be exposed to the fungus more than the other. Although the authors concluded that biological control would be compatible with this biotechnology, it is possible, as has been previously documented, that resistance to the insecticidal crop could evolve faster if the fungus infects more susceptible individuals than resistant ones. When attack by natural enemies is greater on resistant individuals, this effect can be related to negative cross-resistance (Pittendrigh *et al.*, Chapter 11; Gassmann *et al.*, 2009b).

NATURAL ENEMIES: DENSITY-INDEPENDENT AND DENSITY-DEPENDENT EFFECTS

Heimpel *et al.* (2005) were among the first to point out that density-dependent mortality will limit the relative effectiveness of refuges during the early stages of evolution because the refuge population (mostly susceptible individuals) will be limited by the habitat's carrying capacity or the population's density-dependent mortality. (Pest populations being extirpated should be considered an entirely different case.) Even before population densities reach high levels, natural survival rates for rare resistant individuals in insecticidal fields are likely to be higher than those in refuges where the populations are more dense. Furthermore, because density-dependent effects may occur in most systems deploying a refuge for IRM, any density-independent mortality factors could potentially mitigate the density-dependent effects in the refuge, as many of the case studies below indicate.

Heimpel *et al.* (2005) modeled various levels and forms of pest egg mortality: density independence, positive density dependence, and inverse density dependence. Heimpel *et al.* (2005) found that both the magnitude and form of egg mortality can influence the rate of resistance evolution to a hypothetical insecticidal crop. They demonstrated that high-density-independent or density-dependent egg mortality (independent of genotype) delays the evolution of resistance. Furthermore, they concluded that for genotype-independent mortality to influence evolution in a landscape consisting of refuge and toxic habitats, it must be followed by density-dependent mortality in a later life stage. Because densities tend to be higher in refuges and because susceptible individuals have higher densities in refuges, indirect selection can occur by equalizing mortality that otherwise would favor the resistant phenotypes in the insecticidal crop. Thus, Heimpel *et al.* (2005) demonstrated that natural enemies can influence the evolution of pest resistance even when the attacks on pests are neutral with respect to genotype.

Chilcutt and Tabashnik (1999) simulated a model of the interactions of foliar sprays of Bt and a parasitoid in the control of *Plutella xylostella*. They also modeled the population genetics of *P. xylostella* and its evolution of resistance to Bt. They concluded that the use of parasitoids could slow the evolution of resistance to Bt by decreasing the number of generations in which insecticide treatments would be required.

In a series of experiments at Cornell University's experiment station in Geneva, New York, USA, Xiaoxia Liu, Mao Chen, and Anthony Shelton performed a multigeneration study of a greenhouse system consisting of Bt broccoli (*Brassica olereacea*), a population of *P. xylostella* carrying a low percentage of alleles resistant to the Bt protein in broccoli, foliar insecticides, and one natural enemy of *P. xylostella*. The natural enemy was either the predator, *Coleomegilla maculate*, or the parasitoid, *Diadegma insulare*. Liu *et al.* (2011, 2012) determined that the predator had no preference for either the resistant or susceptible phenotype of *P. xylostella*. Liu *et al.* (2011) found no effects of *P. xylostella* genotypes on parasitism. Liu *et al.* (2012) observed one-third as much parasitism on broccoli treated with the insecticide lambda-cyhalothrin compared to nontreated plants, but they observed no effect when broccoli was treated with spinosad.

Onstad *et al.* (2013) used information about the insecticidal broccoli and *P. xylostella* system to create a model to study the influence of the parasitoid on the long-term pest management and evolution of resistance by *P. xylostella*. The model included density-dependent mortality of *P. xylostella* caused by both intraspecific competition and parasitism. Parasitism rate depended on both host and parasitoid density on broccoli. They evaluated the evolution of resistance to Bt broccoli and the two types of foliar insecticides. Simulations demonstrated that density–dependent parasitism provided the most reliable, long–term control of *P. xylostella* populations. Density-dependent parasitism always delays the evolution of resistance to insecticidal broccoli, especially when the refuge size is large. Parasitism also maintains the pest population at the lowest densities over the long run compared to all other treatments, including Bt broccoli by itself. Onstad *et al.* (2013) also included rainfall, which is an abiotic, pest-density-independent mortality factor, in the model. Results indicated that resistance evolution is delayed with significant rainfall mortality of pest eggs and neonates. These results with density-independent rainfall mortality support the conclusions about egg mortality and evolution drawn by Heimpel *et al.* (2005). Onstad *et al.* (2013) also demonstrated that density-dependent but genotype-independent mortality caused by natural enemies can delay evolution in patchy landscapes in the same way.

INTRASPECIFIC, DENSITY-DEPENDENT FACTORS

Comins (1977a,b) and Georghiou and Taylor (1977a,b) were the first to prepare a model for IRM. Their models included density-dependent population growth. However, Tabashnik and Croft (1982) were the first to use a density-dependent survival function for larvae in an IRM model. Alstad and Andow (1995) and Pittendrigh *et al.* (2004) both used density-dependent survival functions to represent intraspecific competition in their models of resistance evolution in landscapes with refuge and insecticidal crop fields. However, none of these authors discussed the significance of density-dependence in resistance evolution.

May and Dobson (1986) described the influence of the density-dependent survival of adults on the evolution of resistance to an insecticide in landscapes without refuges. They classified density-dependent survival as either overcompensating or undercompensating according to the way that the population returns to its long-run equilibrium after a perturbation. May and Dobson (1986) claimed that most insect species exhibit under-compensating density dependence because they recover gradually and monotonically after a disturbance. May and Dobson (1986) concluded that species with undercompensating density dependence evolve resistance more slowly than species with overcompensating density dependence.

Mitchell and Onstad (2005) used a model of *Diabrotica barberi* to study the influence of density-dependent larval survival on the evolution of resistance to Bt corn (*Zea mays*) planted with refuge. They determined that this pest exhibits undercompensating density-dependent survival. Model results indicated that increasing the maximum survival rate (or fecundity) reduced the undercompensating property of the model (the population recovers more quickly from perturbations), and so resistance evolves more quickly.

Sisterson *et al.* (2004) used a stochastic model and discovered that with high adoption of a transgenic insecticidal crop, evolution of resistance occurred faster when carrying capacities for the pest in the crop were reduced.

Glaum *et al.* (2012) used the same density-dependent survival function and parameters as Alstad and Andow (1995). In their Figure 6, they showed that survival of larvae in refuge declines relative to survival in a field of transgenic insecticidal crop as refuge proportion increases. Thus, they concluded that density-dependent survival always produces faster evolution of resistance in a heterogeneous landscape relative to a scenario without density-dependent survival.

Density-dependent survival due to intraspecific competition significantly influences the evolution of resistance of *Diabrotica virgifera virgifera* to Bt corn (Storer, 2003; Onstad *et al.*, 2003; Crowder and Onstad, 2005; Crowder *et al.*, 2005b; Onstad, 2006). In these models, Bt corn expresses a low dose, and oviposition is not uniform across the corn landscape (Pan *et al.*, 2011). Thus, planting the refuge in the same location year after year can delay resistance evolution by permitting susceptible populations to grow in refuges without major disturbance (Pan *et al.*, 2011). Additionally, in the models, as density-dependent, intraspecific competition reduces the number of susceptible beetles in the refuge, it has little impact on the resistant individuals in the insecticidal cornfield and resistance evolves more quickly. For example, omitting density-dependent survival from the model delays evolution of resistance to Bt corn, whereas increasing the maximum survival experienced by the larvae at the lowest densities (typically in Bt cornfields) caused resistance to evolve more quickly (Onstad *et al.*, 2003; Crowder *et al.*, 2005b; Crowder and Onstad, 2005). The type of density-dependent survival was determined by several field studies (Onstad *et al.*, 2001; Onstad *et al.*, 2006,;Hibbard *et al.*, 2010) and was included in the models of Onstad *et al.* (2001), Crowder and Onstad (2005), Crowder *et al.*, 2005b, Onstad and Meinke (2010), and Pan *et al.* (2011).

Onstad (2006) included two kinds of density-dependent survival functions in a model of *D. virgifera virgifera*. One was for intraspecific competition (described above), and the other represented a decrease in mortality caused by the Bt toxin in plant tissues as larval density increases. Onstad postulated that the wounds made by the initial feeding larvae permit other larvae to access less toxic root tissue. Results were sensitive to the intraspecific competition with resistance evolution delayed when density-dependence was eliminated. Onstad (2006) found little difference in the results produced by the density-dependent and typical density-independent functions for toxin mortality.

After Onstad (1988) developed a density-dependent survival function for *O. nubilalis* larvae, most of Onstad's subsequent papers related to IRM for *O. nubilalis* used that function or a carrying capacity of 22 larvae/plant (Chapter 14). Kang *et al.* (2012) created a different density-dependent survival function for *O. nubilalis* larvae based on an unpublished dataset. Because high-dose, insecticidal corn can extirpate the populations when 20% or less of the corn landscape consists of refuge (Onstad and Guse, 1999; Bell *et al.*, 2012; Hutchison *et al.*, 2010), model results indicated that the intraspecific competition in this species is not important for IRM.

In a valuable demonstration of modeling based on alternative representations of nature, Wilhoit (1991) used two hypothetical models to demonstrate how seed mixtures of toxic and nontoxic plants in combination with biological control could delay or prevent the evolution of resistance by aphids. Wilhoit (1991) created a simple deterministic model and a complex stochastic model. The deterministic nonlinear model simulated intraspecific competition between two asexual phenotypes in a field of resistant and susceptible plants. It included mating, sexual reproduction during only one generation at the end of the season. The stochastic simulation model included many nonlinear equations for plant growth, aphid behavior, and predation. The model simulated plant-to-plant movement by both predators and aphids. In both models, the rate of immigration into the field from overwintering sites could differ between the two aphid phenotypes.

Wilhoit (1991) determined that the seed mixture reduced the probability of the resistant (superior) phenotype dominating the aphid population. He also discovered that the resistant aphid could be excluded by the susceptible phenotype because of delayed arrival time by the former into the field. Wilhoit stated that this late-arrival disadvantage is likely to happen when aphids reach the field by random immigration and when resistant aphids are initially less numerous. The effect also depends on mortality due to predation increasing as density increases. In both models, the end-of-season sexual activity and genetics had little effect on the results compared to the competition during the season between asexual aphids.

Cannibalism is an extreme form of intraspecific competition that has been included in IRM models for two species of insects. For both *Diatraea grandiosella* (Guse *et al.*, 2002; Onstad *et al.*, 2002) and *Helicoverpa zea* (Storer *et al.*, 2003a,b) at most one larva typically survives on a corn plant per insect generation. Storer *et al.* (2003a,b) assumed that resistant larvae would mature faster and therefore more likely win cannibalistic encounters with susceptible larvae. Cannibalism therefore acts to increase the fitness differential between resistant and susceptible phenotypes on insecticidal corn. Storer *et al.* (2003a) stated that the role played by *H. zea* cannibalism in resistance evolution is both density dependent and resistance-allele frequency dependent. Model results indicated that as cannibalism in *H. zea* populations intensified, the rate of evolution of resistance to insecticidal corn increased (Storer *et al.*, 2003b).

Horner and Dively (2003) found that *H. zea* feeding on sublethal levels of insecticidal corn reduced the frequency of cannibalistic behaviors when old and young larvae were paired together. Exposure to the insecticidal corn had no significant effect on the timing or the level of mortality due to cannibalism. However, Horner and Dively (2003) postulated that cannibalistic encounters could result in partially resistant larvae feeding on nontoxic younger larvae, thus temporarily providing an escape from exposure to the plant toxin and increase the selective differential between susceptible and resistant individuals. Chilcutt (2006) observed slightly more cannibalism by *H. zea* larvae reared on insecticidal corn than on conventional corn. He concluded that the negative effects of insecticidal corn on larvae were compensated by increased cannibalism but higher survival of winners on Bt corn in comparison with larvae reared on conventional corn. However, Chilcutt *et al.* (2007) drew different conclusions when more younger larvae and fewer older larvae were observed on Bt corn.

CONCLUSIONS

For IPM, we typically try to maintain pests far below the carrying capacity of the environment. For example, Peck and Ellner (1997) modeled a system in which the pest is maintained below an economic threshold for an insecticide to which the pest evolves resistance. In this case, greater population growth and higher pest densities cause more frequent selection for resistance. Because of the traditional emphasis on maintaining pests at very low densities, it is not surprising that the density-dependent survival of many important pests has not been measured. However, entomologists often investigate this phenomenon because of an interest in intraspecific competition or biological control. Given the cases and conclusions summarized in this chapter, it seems prudent for density-dependent survival to be measured and included in future models, particularly because the use of refuge is so prevalent in IRM for transgenic insecticidal crops. Furthermore, all IRM models should include all major mortality factors so that credible and accurate predictions can be made. This recommendation matches the main theme of the book: that IRM must be considered part of IPM and that IRM will be most effective with good IPM. Of course, all of these recommendations become harder to implement when a pest infests multiple crops or when the pest inhabits crops and wild host plants in the landscape. Too often we have little knowledge of the pest's biology and ecology on its alternative hosts in an agroecosystem, although behavior and demography in the entire landscape can have significant consequences for resistance evolution (see Onstad and Carrière, Chapter 10).

It is not easy to measure density-dependent mortality or even the carrying capacity for a pest population in a given environment. It is likely more difficult to measure density-dependent fecundity and dispersal. Nevertheless, almost all IRM models depend on simplifying assumptions about fecundity, larval movement, and adult dispersal, and these processes could determine the effectiveness of a refuge as much as density-dependent larval survival. In addition, any density-independent processes that cause individuals emerging in a refuge to remain and mate and lay most of their eggs in the refuge will promote population growth closer to the higher densities and limits. Onstad and Meinke (2010) found that reductions in density-independent fecundity had the same effect on simulated resistance evolution as increased egg mortality modeled by Heimpel et al. (2005) and described above. Onstad and Meinke (2010) also demonstrated that even genotype-specific fecundity reductions that seem to favor resistant phenotypes can actually delay resistance evolution because of density-dependent survival in the refuge.

In this chapter, we have demonstrated that biological control can significantly affect IRM (Riddick et al., 2000; Lundgren et al., 2009). Interactions of natural enemies with other control tactics imposing selection pressure on pests can complicate IPM or lead to simple solutions. For example, two modeling studies have demonstrated how natural enemies and geographically distributed pest populations interact over time (but neither accounted for the evolution of resistance) in landscapes with conventional and insecticidal crops. Bell *et al.* (2012) demonstrated how a microsporidian pathogen and Bt corn interacted to drive populations of *O. nubilalis* to historically low levels in the central United States. Sisterson and Tabashnik (2005) modeled a pest and its parasitoid in a 9000-ha region with 900 fields, each planted with either an insecticidal crop or a refuge. They concluded that risk of regional parasitoid loss can be assessed from its life-history traits and reduced by increasing the percentage of refuge fields, fixing refuge locations, and reducing insecticide applications in refuges.

With the introduction of Bt plants into agricultural systems, it has now become increasingly possible to dramatically reduce the use of broad-spectrum insecticides while still effectively controlling key target pests (Qaim *et al.*, 2008; Romeis *et al.*, 2008; Shelton, 2012). There is well-documented evidence that the currently deployed Bt plants, with their narrow spectrum of activity, have no direct negative effect on natural enemies (Romeis *et al.*, 2006; Naranjo, 2009; Tian *et al.*, 2012). This is in stark contrast to the use of traditional, broader spectrum insecticides that decrease natural enemy abundance and the biological control function they exert (Wolfenbarger *et al.*, 2008; Naranjo 2009; Chen *et al.*, 2008). Furthermore, studies have shown that the pest insects can rapidly evolve resistance to traditional broad-spectrum insecticides, while important natural enemies that could regulate the pest populations do not have such an ability (Xu *et al.*, 2001). This phenomenon often results in rapid pest outbreaks and crop losses. The evidence is clear that Bt plants can contribute to natural enemy conservation and help maintain pest populations at lower levels. It is becoming increasingly evident that the combination of Bt plants and biological control agents can also delay the evolution of resistance by the pest species to Bt plants (Onstad *et al.*, 2013) and be a useful combination for IPM.

As stakeholders attempt to improve IPM by taking advantage of both biological control and host-plant resistance, interactions between natural enemies and pests targeted by insecticidal crops will become even more important in the future (Onstad *et al.*, 2011). The extensive and interesting history of the genetic modification and use of toxin-resistant natural enemies could not be incorporated into this chapter. Good reviews can be found in publications by Hoy (1990, 2003).

REFERENCES

Arpaia, -S., Gould, -F., Kennedy, -G., 1997. Potential impact of *Coleomegilla maculata* predation on adaptation of *Leptinotarsa decemlineata* to Bt-transgenic potatoes. Entomol. Exp. Appl. 82, 91–100.

Alstad, D.N., Andow, D.A., 1995. Managing the evolution of insect resistance to transgenic plants. Science. 268, 1894–1896.

Bell, J.R., Burkness, E.C., Milne, A.E., Onstad, D.W., Abrahamson, M., Hamilton, K.L., et al., 2012. Putting the brakes on a cycle: bottom-up effects damp periodicity. Ecol. Lett. 15, 310–318.

Caprio, M.A., Tabashnik, B.E., 1992. Gene flow accelerates local adaptation among finite populations: simulation the evolution of insecticide resistance. J. Econ. Entomol. 85, 611–620.

Chen, M., Zhao, J.-Z., Collins, H.L., Earle, E.D., Cao, J., Shelton, A.M., 2008. A critical assessment of the effects of Bt transgenic plants on parasitoids. PLoS One. 3, e2284. Available from: http://dx.doi.org/10.1371/journal.pone.0002284.

Chilcutt, -C.F., Tabashnik, -B.E., 1999. Simulation of integration of *Bacillus thuringiensis* and the parasitoid *Cotesia plutellae* (Hymenoptera: Braconidae) for control of susceptible and resistant diamondback moth (Lepidoptera: Plutellidae), 1999. Environ. Entomol. 28, 505–512.

Chilcutt, C.F., 2006. Cannibalism of *Helicoverpa zea* (Lepidoptera: Noctuidae) from *Bacillus thuringiensis* (Bt) transgenic corn versus non-Bt corn. J. Econ. Entomol. 99, 728–732.

Chilcutt, C.F., Odvody, G.N., Correa, J.C., Remmers, J., 2007. Effects of *Bacillus thuringiensis* transgenic corn on corn earworm and fall armyworm (Lepidoptera: Noctuidae) densities. J. Econ. Entomol. 100, 327–334.

Comins, H.N., 1977a. The development of insecticide resistance in the presence of migration. J. Theo. Biol. 64, 177–197.

Comins, H.N., 1977b. The management of pesticide resistance. J. Theo. Biol. 65, 399–420.

Crowder, D.W., Onstad, D.W., 2005. Using a generational time-step model to simulate the dynamics of adaptation to transgenic corn and crop rotation by western corn rootworm (Coleoptera: Chrysomelidae). J. Econ. Entomol. 98, 518–533.

Crowder, D.W., Onstad, D.W., Gray, M.E., Pierce, C.M.F., Hager, A.G., Ratcliffe, S.T., et al., 2005b. Analysis of the dynamics of adaptation to transgenic corn and crop rotation by western corn rootworm (Coleoptera: Chrysomelidae) using a daily time-step model. J. Econ. Entomol. 98, 534–551.

Foster, S.P., Tomiczek, M., Thompson, R., Denholm, I., Poppy, G., Kraaijeveld, A.R., et al., 2007. Behavioural side-effects of insecticide resistance in aphids increase their vulnerability to parasitoid attack. Anim. Behav. 74, 621–632.

Foster, S.P., Denholm, I., Poppy, G.M., Thompson, R., Powell, W., 2011. Fitness trade-off in peach-potato aphids (*Myzus persicae*) between insecticide resistance and vulnerability to parasitoid attack at several spatial scales. Bull. Entomol. Res. 101, 659–666.

Gassmann, A.J., Carriere, Y., Tabashnik, B.E., 2009a. Fitness costs of insect resistance to *Bacillus thuringiensis*. Annu. Rev. Entomol. 54, 147–163.

Gassmann, A.J., Onstad, D.W., Pittendrigh, B.R., 2009b. Evolutionary analysis of herbivore adaptation to natural and agricultural systems. Pest Manag. Sci. 65, 1174–1181.

Gassmann, A.J., Hannon, E.R., Sisterson, M.S., Stock, S.P., Carriere, Y., Tabashnik, B.E., 2012. Effects of entomopathogenic nematodes on evolution of pink bollworm resistance to *Bacillus thuringiensis* toxin Cry1Ac. J. Econ. Entomol. 105, 994–1005.

Georghiou, G.P., Taylor, C.E., 1977a. Genetic and biological influences in the evolution of insecticide resistance. J. Econ. Entomol. 70, 319–323.

Georghiou, G.P., Taylor, C.E., 1977b. Operational influences in the evolution of insecticide resistance. J. Econ. Entomol. 70, 653–658.

Glaum, P.R., Ives, A.R., Andow, D.A., 2012. Contamination and management of resistance evolution to high-dose transgenic insecticidal crops. Theor. Ecol. 5, 195–209.

Gould, -F., Kennedy, -G.G., Johnson, -M.T., 1991. Effects of natural enemies on the rate of herbivore adaptation to resistant host plants. Entomol. Exp. Appl. 58, 1–14.

Guse, C.A., Onstad, D.W., Buschman, L.L., Porter, P., Higgins, R.A., Sloderbeck, P.E., et al., 2002. Modeling the development of resistance by stalk-boring Lepidoptera (Crambidae) in areas with irrigated, transgenic corn. Environ. Entomol. 31, 676–685.

Hannon, E.R., Sisterson, M.S., Stock, S.P., Carriere, Y., Tabashnik, B.E., Gassmann, A.J., 2010. Effects of four nematode species on fitness costs of pink bollworm resistance to *Bacillus thuringiensis* toxin Cry1Ac. J. Econ. Entomol. 103, 1821–1831.

Heimpel, G.E., Neuhauser, -C., Andow, D.A., 2005. Natural enemies and the evolution of resistance to transgenic insecticidal crops by pest insects: the role of egg mortality. Environ. Entomol. 34, 512–526.

Hibbard, B.E., Meihls, L.N., Ellersieck, M.R., Onstad, D.W., 2010. Density-dependent and density-independent mortality of the western corn rootworm: impact on dose calculations of rootworm-resistant Bt corn. J. Econ. Entomol. 103, 77–84.

Horner, T.A., Dively, G.P., 2003. Effect of MON810 Bt field corn on *Helicoverpa zea* (Lepidoptera: Noctuidae) cannibalism and its implications to resistance development. J. Econ. Entomol. 96 (931-934).

Hoy, M.A., 1990. Pesticide resistance in arthropod natural enemies: variability and selection responses. In: Roush, R.T., Tabashnik, B.E. (Eds.), Pesticide Resistance in Arthropods. Chapman and Hall, New York, Chapter 8.

Hoy, M.A., 2003. Transgenic pest and beneficial insects for pest management programs, Insect Molecular Genetics: An Introduction to Principles and Applications. second ed. M. A. Hoy. Academic Press, Chapter 14.

Hutchison, W.D., Burkness, E.C., Mitchell, P.D., Moon, R.D., Leslie, T.W., Fleischer, S. J., et al., 2010. Areawide suppression of European corn borer with Bt maize reaps savings to non-Bt maize growers. Science. 330, 222–225.

Johnson, M.T., Gould, F., 1992. Interaction of genetically engineered host plant resistance and natural enemies of *Heliothis virescens* (Lepidoptera: Noctuidae) in tobacco. Environ. Entomol. 21, 586–597.

Johnson, M.T., Gould, -F., Kennedy, -G.G., 1997a. Effects of natural enemies on relative fitness of *Heliothis virescens* genotypes adapted and not adapted to resistant host plants. Entomol. Exp. Appl. 82, 219–230.

Johnson, M.T., Gould, -F., Kennedy, -G.G., 1997b. Effect of an entomopathogen on adaptation of *Heliothis virescens* populations to transgenic host plants. Entomol. Exp. Appl. 83, 121–135.

Kang, J., Onstad, D.W., Hellmich, R.L., Moser, S.E., Hutchison, W.D., Prasifka, J.R., 2012. Modeling the impact of cross-pollination and low toxin expression in corn kernels on adaptation of European corn borer (Lepidoptera: Crambidae) to transgenic insecticidal corn. Environ. Entomol. 41, 200–211.

Knutson, A.E., Rojas, E.A., Marshal, D., Gilstrap, F.E., 2002. Interaction of parasitoids and resistant cultivars of wheat on Hessian fly, *Mayetiola destructor* Say (Cecidomyiidae). Southwest. Entomol. 27, 1–10.

Lawo, N.C., Mahon, R.J., Milner, R.J., Sarmah, B.K., Higgins, T.J.V., Romeis, J., 2008. Effectiveness of *Bacillus thuringiensis*-transgenic chickpeas and the entomopathogenic fungus *Metarhizium anisopliae* in controlling *Helicoverpa armigera* (Lepidoptera: Noctuidae). Appl. Environ. Microbiol. 74, 4381–4389.

Liu, X., Chen, M., Onstad, D., Roush, R., Shelton, A.M., 2011. Effect of Bt broccoli and resistant genotype of *Plutella xylostella* (Lepidoptera: Plutellidae) on development and host acceptance of the parasitoid *Diadegma insulare* (Hymenoptera: Ichneumonidae). Transgenic. Res. 20, 887−897.

Liu, X., Chen, M., Collins, H., Onstad, D., Roush, R., Zhang, Q., et al., 2012. Effect of insecticides and *Plutella xylostella* genotype on a predator and parasitoid and implications for the evolution of insecticide resistance. J. Econ. Entomol. 105, 354−362.

Lundgren, J.G., Gassmann, A.J., Bernal, J.S., Duan, J.J., Ruberson, J.R., 2009. Ecological compatibility of GM crops and biological control. Crop Prot. 28, 1017−1030.

Mallampalli, N., Gould, F., Barbosa, P., 2005. Predation of Colorado potato beetle eggs by a polyphagous ladybeetle in the presence of alternate prey: potential impact on resistance evolution. Entomol. Exp. Appl. 114, 47−54.

May, R.M., Dobson, A.P., 1986. Population dynamics and the rate of evolution of pesticide resistance. National Research Council, Committee on Strategies for the Management of Pesticide Resistance Pest Populations, "Pesticide Resistance: Strategies and Tactics for Management.". National Acad. Press, Washington, DC.

Mitchell, P.D., Onstad, D.W., 2005. Effect of extended diapause on the evolution of resistance to transgenic *Bacillus thuringiensis* corn by northern corn rootworm (Coleoptera: Chrysomelidae). J. Econ. Entomol. 98, 2220−2234.

Naranjo, S.E., 2009. Impacts of Bt crops on non-target organisms and insecticide use patterns. *CAB Reviews: Perspect. Agric., Vet. Sci., Nutrit. Nat. Resour.* 4 (No.011), Available from: http://dx.doi.org/10.1079/PAVSNNR20094011.

Onstad, D.W., 1988. Simulation model of the population dynamics of *Ostrinia nubilalis* (Lepidoptera: Pyralidae) in maize.. Environ. Entomol. 17, 969−976.

Onstad, D.W., 2006. Modeling larval survival and movement to evaluate seed mixtures of transgenic corn for control of western corn rootworm (Coleoptera: Chrysomelidae). J. Econ. Entomol. 99, 1407−1414.

Onstad, D.W., Guse, C.A., 1999. Economic analysis of the use of transgenic crops and non-transgenic refuges for management of European corn borer (Lepidoptera:Pyralidae). J. Econ. Entomol. 92, 1256−1265.

Onstad, D.W., Meinke, L.J., 2010. Modeling evolution of *Diabrotica virgifera virgifera* (Coleoptera: Chrysomelidae) to transgenic corn with two insecticidal traits. J. Econ. Entomol. 103, 849−860.

Onstad, D.W., Crowder, D.W., Mitchell, P.D., Guse, C.A., Spencer, J.L., Levine, E., et al., 2003. Economics versus alleles: balancing IPM and IRM for rotation-resistant western corn rootworm (Coleoptera: Chrysomelidae). J. Econ. Entomol. 96, 1872−1885.

Onstad, D.W., Hibbard, B.E., Clark, T.L., Crowder, D.W., Carter, K.G., 2006. Analysis of density-dependent survival of *Diabrotica* (Coleoptera: Chrysomelidae) in cornfields. Environ. Entomol. 35, 1272−1278.

Onstad, D.W., Mitchell, P.D., Hurley, T.M., Lundgren, J.G., Porter, R.P., Krupke, C.H., et al., 2011. Seeds of change: corn seed mixtures for resistance management and integrated pest management. J. Econ. Entomol. 104, 343−352.

Onstad, D.W., Liu, X., Chen, M., Roush, R., Shelton, A.M., 2013. Modeling the integration of parasitism, insecticide, and transgenic insecticidal crop for the long-term control of an insect pest. J. Econ. Entomol.

Pan, Z., Onstad, D.W., Nowatzki, T.M., Stanley, B.H., Meinke, L.J., Flexner, J.L., 2011. Western corn rootworm (Coleoptera: Chrysomelidae) dispersal and adaptation to single-toxin transgenic corn. Environ. Entomol. 40, 964−978.

Peck, S.L., Ellner, S.P., 1997. The effect of economic thresholds and life-history parameters on the evolution of pesticide resistance in a regional setting. Am. Nat. 149, 43−63.

Pittendrigh, B.R., Gaffney, P.J., Huesing, J., Onstad, D.W., Roush, R.T., Murdock, L.L., 2004. Active refuges can inhibit the evolution of resistance in insects towards transgenic insect-resistant plants. J. Theor. Biol. 231, 461−474.

Qaim, M., Pray, C.E., Zilberman, D., 2008. Economic and social considerations in the adoption of Bt crops. In: Romeis, J., Shelton, A.M., Kennedy, G.G. (Eds.), Integration of Insect-Resistant Genetically Modified Crops Within IPM Programs Springer. Springer, Dordrecht, The Netherlands, pp. 329−356.

Raymond, B., Sayyed, A.H., Hails, R.S., Wright, D.J., 2007. Exploiting pathogens and their impact on fitness costs to manage the evolution of resistance to *Bacillus thuringiensis*. J. Appl. Ecol. 44, 768−780.

Riddick, E.W., Dively, G., Barbosa, P., 2000. Season-long abundance of generalist predators in transgenic versus nontransgenic potato fields. J. Entomol. Sci. 35, 349−359.

Romeis, J., Meissle, M., Bigler, F., 2006. Transgenic crops expressing *Bacillus thuringiensis* toxins and biological control. Nat. Biotechnol. 24, 63−71.

Romeis, J., Shelton, A.M., Kennedy, G.G., 2008. Integration of Insect-Resistant Genetically Modified Crops Within IPM Programs. Springer, Dordrecht, The Netherlands, p. 441.

Shelton, A.M., 2012. Genetically engineered vegetables expressing proteins from *Bacillus thuringiensis* for insect resistance: Successes, disappointments, challenges and ways to move forward. GM Crops & Food. 3, 175−183.

Sisterson, M.S., Tabashnik, B.E., 2005. Simulated effects of transgenic bt crops on specialist parasitoids of target pests. Environ. Entomol. 34, 733−742.

Sisterson, M.S., Antilla, L., Carrière, Y., Ellers-Kirk, C., Tabashnik, B.E., 2004. Effects of insect population size on evolution of resistance to transgenic crops. J. Econ. Entomol. 97, 1413−1424.

Storer, N.P., 2003. A spatially explicit model simulating western corn rootworm (Coleoptera: Chrysomelidae) adaptation to insect-resistant maize. J. Econ. Entomol. 96, 1530−1547.

Storer, -N.P., Peck, -S.L., Gould, -F., Van-Duyn, -J.W., Kennedy, -G.G., 2003a. Spatial processes in the evolution of resistance in *Helicoverpa zea* (Lepidoptera: Noctuidae) to Bt transgenic corn and cotton in a mixed agroecosystem: a biology-rich stochastic simulation model. J. Econ. Entomol. 96, 156−172.

Storer, -N.P., Peck, -S.L., Gould, -F., Van-Duyn, -J.W., Kennedy, -G.G., 2003b. Sensitivity analysis of a spatially-explicit stochastic simulation model of the evolution of resistance in *Helicoverpa zea* (Lepidoptera: Noctuidae) to Bt transgenic corn and cotton. J. Econ. Entomol. 96, 173−187.

Tabashnik, B.E., Croft, B.A., 1982. Managing pesticide resistance in crop-arthropod complexes: interactions between biological and operational factors. Environ. Entomol. 11, 1137−1144.

Tian, J., Collins, H.L., Romeis, J., Naranjo, S.E., Hellmich, R.L., Shelton, A.M., 2012. Using field-evolved resistance to Cry1F maize in a lepidopteran pest to demonstrate no adverse effects of Cry1F on one of its major predators. Transgenic. Res. 21, 1303−1310.

Wilhoit, L.R., 1991. Modeling the population-dynamics of different aphid genotypes in plant variety mixtures. Ecol. Modell. 55, 257−283.

Wolfenbarger, L.L., Naranjo, S.E., Lundgren, J.G., Bitzer, R.J., Watrud, L.S., 2008. Bt Crop effects on functional guilds of non-target arthropods: A meta-analysis. PLoS One. 3, e2118. Available from: http://dx.doi.org/10.1371/journal.pone.0002118.

Xu, J., Shelton, A.M., Cheng, X., 2001. Variation in susceptibility of *Diadegma insulare* to permethrin. J. Econ. Entomol. 94, 541−546.

> CHAPTER 13

Insect Resistance Management: Adoption and Compliance

Terrance M. Hurley[1] and Paul D. Mitchell[2]
[1]Department of Applied Economics, University of Minnesota, St. Paul, MN
[2]Department of Agricultural and Applied Economics, University of Wisconsin, Madison, WI

Chapter Outline

Because effective pest management exploits biological adaptations commonly observed in the targeted species, it is easy to think of insect resistance to management as a purely evolutionary process driven solely by insect population dynamics and genetics. But it is the act of pest management that drives the evolution of insect resistance, and pest management is fundamentally a human activity in agricultural production. Therefore, thinking of insect resistance as purely driven by insect biology ignores human behavior and other important aspects of the problem.

This chapter discusses aspects of human behavior that affect the evolution of insect resistance to management and shows how a better understanding of this behavior can be used to improve insect resistance management (IRM). While IRM can be thought of in terms of individual farmers, Clark and Carlson (1990) find that individual farmers treat insect resistance as a common property problem, which means they do not have the incentive to manage it appropriately from a societal perspective (see Mitchell and Onstad, Chapter 2). Therefore, this chapter focuses on the problem from a public policy perspective. From this perspective, government regulators like the U.S. Environmental Protection Agency (EPA) or stakeholder groups like the Arizona Cotton Growers Association (ACGA)

Insect Resistance Management
DOI: http://dx.doi.org/10.1016/B978-0-12-396955-2.00013-8

421

are interested in formulating and implementing IRM policies in order to promote pest-management practices that provide a greater benefit to society or association members. Since pest-management decisions are ultimately made by farmers, the regulator or stakeholder group can only influence IRM indirectly. This creates what is referred to as a principal-agent problem (Laffont and Martimort, 2001). The principal (e.g., the EPA or ACGA) would like the agent (e.g., farmers) to use prescribed management strategies that may not be wholly in the interest of the agent. Therefore, the agent's response to the principal's prescription plays an important role in the principal's ability to achieve his or her objectives. This principal-agent problem can be further complicated by the fact that farmer decisions are influenced by the decisions of seed, chemical, and other farm-input suppliers through which regulators may choose to act.

EPA regulation of transgenic insecticidal crops such as Bt corn and Bt cotton offers a useful illustration of the nature of the problem. Transgenic insecticidal crops are genetically engineered to express insecticidal proteins. The EPA has determined that these insecticidal proteins are safer for human health and the environment, so it would like to promote sustainable use through insect resistance management (IRM) (U.S. EPA, 1998; Berwald et al., 2006). EPA IRM requirements obligate farmers to plant conventional crop varieties (i.e., refuge) along with transgenic insecticidal varieties and also dictate where refuge varieties should be planted in relation to transgenic insecticidal varieties (i.e., refuge configuration). The EPA implements these requirements by dictating the types of contractual arrangements transgenic insecticidal seed providers must enter into with farmers. Farmers who plant the prescribed refuge forgo the production benefits of transgenic insecticidal crops on refuge acreage, which can be costly. They must also devote costly management and planting time to meeting EPA configuration requirements. These additional production costs give farmers an incentive to ignore EPA IRM requirements. If some farmers do not comply with EPA requirements, there may not be enough refuge planted in a suitable configuration to meet IRM objectives for all or part of the pest population.

This example illustrates how one aspect of human behavior (i.e., regulatory compliance behavior) can influence the success of IRM. Another aspect of human behavior that this chapter will explore is technology adoption behavior by farmers. The evolution of resistance to pest management depends crucially on the extent to which an insect population is exposed to a particular management strategy. The extent to which an insect population is exposed to management depends on how many farmers choose to adopt the strategy and how intensively they use it.

The remainder of the chapter outlines a basic conceptual model for framing the effects of farmer adoption and compliance behavior on IRM. We then discuss factors that have been identified as influencing adoption and compliance behavior. Simulations of the basic model are used to illustrate how adoption and compliance behavior constrain IRM policies, while alternative approaches used to characterize adoption and compliance behavior are discussed and opportunities for future research are proposed. The conclusions reiterate key implications.

CONCEPTUAL FRAMEWORK

A basic model of the evolution of resistance is useful for framing the discussion of how farmer behavior affects IRM. Consider a simplified agricultural production region with a single pest. A proportion of the pest population in the region is managed. Let $1 \geq \tau_t \geq 0$ be the proportion of managed pests in period t where the period reflects discrete pest generations. Following standard Hardy–Weinberg assumptions, suppose a randomly mating pest population with nonoverlapping generations and resistance conferred by a single, non–sex-linked gene (Hartl, 1988). There are two types of alleles: resistant and susceptible. The proportion of resistant alleles in period t is $1 \geq p_t \geq 0$. The Hardy–Weinberg model implies that the proportion of each genotype can be represented by $[p_t^2, 2p_t(1 - p_t), (1 - p_t)^2]$, with elements corresponding to the proportion of resistant homozygous, heterozygous, and susceptible homozygous pests.

The Hardy–Weinberg model assumes no selection pressure—survival rates are the same for all genotypes. Pest management selects for resistant pests by eliminating susceptible pests. Let $[1, 1 - hs, 1 - s]$ be the genotypic survival rates for managed pests relative to unmanaged pests, with elements corresponding to resistant homozygous, heterozygous, and susceptible homozygous pests, where $1 \geq h \geq 0$ is a dominance coefficient for resistance to management and $1 \geq s \geq 0$ is a selection coefficient. For $h = 0$, the survival rates of resistant homozygous and heterozygous pests are the same, which occurs when resistance is a completely dominant trait. For $h = 1$, the survival rates of susceptible homozygous and heterozygous pests are the same, which occurs when resistance is a completely recessive trait. Values between 0 and 1 represent varying degrees of incomplete dominance with values closer to zero representing a higher degree of dominance. Given the proportion of the pest population managed in period t, the genotypic survival rates are $[1, 1 - \tau_t hs, 1 - \tau_t s]$.

Since each surviving pest contributes two alleles, resistant homozygous pests contribute two resistant alleles, and heterozygous pests contribute one resistant allele, the proportion of resistant alleles in the subsequent period is

$$p_{t+1} = \frac{p_t^2 + (1 - \tau_t hs)p_t(1 - p_t)}{p_t^2 + 2(1 - \tau_t hs)p_t(1 - p_t) + (1 - \tau_t s)(1 - p_t)^2} \tag{13.1}$$

Equation 13.1 with parameters h and s and initial condition p_0 describes a dynamic biological system where the evolution of resistance, p_t, is controlled by the proportion of managed pests, τ_t for all t. While this chapter focuses on managing resistance by controlling the proportion of the pest population exposed to management (i.e., the notion of refuge), it is also possible to think of managing resistance by influencing selection (s) and dominance (h). For example, a regulator may restrict pesticide application rates to influence s or decline to register pesticides where resistance exhibits a relatively high degree of dominance to influence h.

From a naive perspective, insect resistance in this system can be effectively managed by choosing τ_t for all t of interest to meet desired objectives. From a more practical perspective, the choice of τ_t is constrained by human behavior, which is influenced by economic, environmental, sociological, and psychological factors, as well as insect biology and behavior.

HUMAN BEHAVIOR

Turning again to the EPA example of the regulation of transgenic insecticidal crops, many of the models developed to help guide IRM policy have assumed $\tau_t = \tau$ for all t where $1 - \tau$ represents the refuge requirement set by the EPA. These models often focus on describing how quickly resistance evolves as τ is varied for different assumptions regarding important biological parameters (e.g., h and s). Although such assumptions are convenient in terms of studying how biological factors affect insect resistance, they do not withstand empirical scrutiny. In the context of transgenic insecticidal crops, assuming $\tau_t = \tau$ for all t means that all farmers fully adopt the pest-management strategy and that they all fully comply with EPA IRM requirements.

There is substantial evidence to refute the assumption that farmers will fully adopt transgenic insecticidal crops for pest management. For example, Figure 13.1 shows the proportion of U.S. corn and cotton acreage planted to transgenic insecticidal corn and cotton from 2000 to 2012 (USDA-NASS, 2001−2012). What is more important to recognize from this figure is that even after 17 years of availability, farmers still have not fully adopted these crops, though adoption has trended upward. For the crop years 2008 to 2011, crop insurance premium reductions were approved in some states for farmers planting at least 75% for their corn with approved "triple-stack" corn hybrids that contained Bt traits for control of both lepidopteran and coelopteran pests and herbicide tolerance (USDA RMA, 2011). Figure 13.2 shows the geographic variability in farmer participation in this program in 2010. The highest participation rates are evident in southwestern and south-central Minnesota and in central Illinois, with the maximum participation rate of 70.7% in Murray

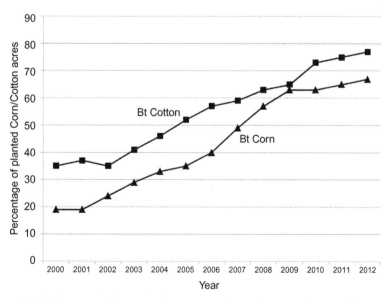

Figure 13.1 Bt corn and cotton adoption trends in the United States from 2000 to 2012.

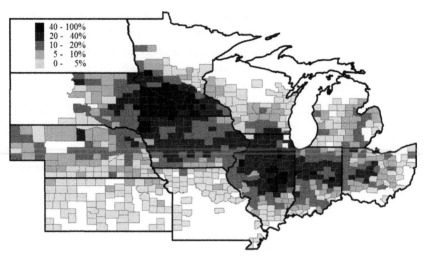

Figure 13.2 Percentage of planted corn acres enrolled in the biotech endorsement in 2010.

County, Minnesota, meaning that almost 71% of the corn acres planted in this county were at least 75% triple-stack Bt corn. Figure 1 in Hutchison *et al.* (2010) shows a similar geographic heterogeneity in farmer adoption of Bt corn for control of European corn borer in 2006. These data show that full adoption fails to characterize either individual or aggregate farmer behavior, even in regions with high adoption.

There is also substantial evidence to refute the assumption that farmers who adopt transgenic insecticidal crops will fully comply with EPA IRM regulations (Jaffe (2003a,b, 2009); ABSTC, 2005; Merrill *et al.*, 2005; Martinez and Reynolds, 2009). Table 13.1 summarizes third-party compliance survey results for 2002 to 2008 reported by Martinez and Reynolds (2009). These results show compliance rates ranging from 63 to 96% for EPA refuge size and distance requirements. They also show a precipitous decline in compliance in 2007 and 2008, though the survey methodology used to collect the compliance data changed in 2007, which confounds the explanation for this decline.

The adoption pattern witnessed with the release of transgenic insecticidal crops is not idiosyncratic. Neither is the compliance pattern witnessed with the EPA IRM requirements. There is a substantial literature that documents and explains (1) systematic trends in the adoption of new production technologies, including new pest-management strategies; (2) why farmers who choose to adopt a particular production practice

Table 13.1 Refuge Size and Distance Compliance Rates for *Ostrinia nubilalis* (ECB), *Diabrotica* Corn Rootworm (CRW), and Stacked ECB & CRW Bt Corn

	Size			Distance		
Year	ECB	CRW	ECB & CRW	ECB	CRW	ECB & CRW
2002	86%	NA	NA	89%	NA	NA
2003	92%	NA	NA	93%	NA	NA
2004	91%	NA	NA	96%	NA	NA
2005	92%	93%	NA	96%	87%	NA
2006	89%	89%	78%	96%	82%	92%
2007	80%	80%	70%	88%	79%	66%
2008	78%	74%	72%	88%	66%	66%

NA: Product was not available.
Source: Martinez and Reynolds (2009).

may not fully utilize it; and (3) regulatory compliance. It is instructive to briefly review this literature to better understand the likely implications of its observations and explanations for IRM.

Adoption Behavior

The two important facets of adoption behavior are known as extensive and intensive adoption (Feder *et al.*, 1985). Extensive adoption refers to a farmer's decision to initially adopt a particular production practice. Intensive adoption refers to the degree to which a farmer utilizes a particular production practice once it has been adopted. To initially adopt a new production practice, a farmer must take time to learn about the new practice and how to implement it on his farm. He may also have to invest in new equipment. This time and investment are costly and must be borne regardless of the degree to which a farmer utilizes the new practice. Once a farmer has borne these costs, there are additional costs that depend on the degree to which the farmer utilizes the practice. For example, the cost of managing pests using a chemical pesticide increases as the amount of treated area increases. Costs that do not vary with the intensity of utilization are referred as fixed costs, whereas costs that vary with the intensity of utilization are called variable costs. Only variable costs are important in terms of intensive adoption. For extensive adoption, fixed and variable costs are important.

Adoption of new production practices exhibits a predictable sigmoidal time trend (e.g., Griliches, 1957). The rate of adoption is initially slow, but increasing, eventually reaches a peak and begins to decline ultimately becoming zero or even negative when new and superior practices are introduced. A variety of economic and sociological theories explain this trend and identify influential factors.

Sociological explanations of this trend describe adoption as a sequential process (Rogers, 1983). First, farmers become aware of a new technology. Awareness leads them to seek information. The decision to adopt the new technology then follows from an evaluation of this information. Five important characteristics are identified as playing a crucial role in the adoption decision: (i) the perception that the new technology is better due to either economic or social factors, (ii) the compatibility of the technology with tradition and past experience, (iii) the complexity of the technology, (iv) the feasibility of experimenting with the technology, and (v) the visibility of the results of the technology. Ultimately, sigmoidal adoption patterns follow from individual differences among farmers in terms of these characteristics, with farmers often cast into categories such as innovators, early adopters, early majority, late majority, and laggards, depending on how rapidly they adopt new production practices.

Economic explanations of the trend are generally consistent with sociological explanations, though greater emphasis is placed on factors such as profitability and risk. All else being equal, economists argue that farmers will choose the most profitable production practices. Of particular importance in terms of profitability for extensive adoption decisions are the fixed costs of adoption (e.g., Just et al., 1980; Feder and O'Marra, 1981). However, when considering a new practice, there is considerable uncertainty regarding profitability due to a lack of experience, which introduces the notion of risk (e.g., Heibert, 1974; Linder et al., 1979). Economists think of risk in terms of risk preferences and risk perceptions. Risk preferences characterize the degree to which an individual does not like variability or does not like to gamble. Individuals who do not like variability are referred to as risk averse. Economists typically assume an individual's degree of risk aversion is stable overtime. However, different individuals exhibit differing degrees of risk aversion, which can drive differences in risky behavior such as the adoption of a new pest-management strategy. Risk perceptions characterize the likelihood of unknown outcomes; for example, the probability that a cotton farmer will experience a severe pest infestation in the coming year. Different individuals may have different risk perceptions because they have different personal experiences. Furthermore, risk perceptions can evolve over time as individuals have new experiences. Therefore, individual differences in risky behavior can also be explained by different risk perceptions. Sigmoidal adoption trends follow from inherent differences in profitability, risk preferences, and risk perceptions.

Profitability and risk can also explain intensive adoption patterns (e.g., Feder, 1979; Horowitz and Lichtenberg, 1993, 1994; Hurley et al., 2004), though there are important differences in perspective. Profitability decisions for intensive adoption ignore fixed costs, which is not the case for extensive adoption. Intensive adoption decisions offer a mechanism for managing risk through diversification, while extensive decisions are seen as characterized by additional risk due to a lack of experience with and uncertainty about the profitability of the new practice.

Several additional specific factors are commonly identified as important determinants of farmer adoption behavior: credit constraints, farm structure or size, human capital, labor supply, and physical environment (Feder et al., 1985). Credit constraints limit a farmer's ability to invest in the skills and equipment needed to adopt new production practices, especially when there are high fixed costs (e.g., El-Osta and Morehart, 1999). In terms of farm size or structure, when the fixed costs of adoption are large, profitability will tend to be higher for larger farms because they can spread these fixed costs over higher levels of production (Just et al., 1980; Feder and O'Marra, 1981). Human capital refers to the skills and experience a farmer has acquired, often measured by a farmer's years of formal education and farming. Higher levels of human capital are often associated with increased profitability. Additionally, farmers with more human capital tend to be better at collecting and critically evaluating new information, which means they are able to identify and adopt profitable new production practices more rapidly (e.g., Schultz, 1964, 1981). Labor supply refers to a farmer's own labor, as well as unpaid family labor and hired labor. For labor-intensive production practices such as integrated pest management (IPM), a lack of adequate labor can deter adoption (McNamara et al., 1991). Alternatively, labor-saving production practices such as planting herbicide tolerant crops are more likely to be adopted when labor resources are scarce. Heterogeneity in the physical environment influences profitability and ultimately the adoption of new production practices (e.g., Green et al., 1996; Thrikawala et al., 1999). For example, farmers with inherently more productive soils often find pest management is more profitable because the increased yield potential also results in increased loss potential from pest damage.

Recent research on factors affecting the adoption of new pest-management practices focuses on IPM and transgenic herbicide tolerant and insecticidal crops. In terms of IPM, Fernandez-Cornejo et al. (1994) found that for vegetable growers, adopters tend to be less risk averse and use more managerial time. Farm size is positively related to adoption, as is the availability of unpaid farm labor. In terms of transgenic herbicide tolerant and insecticidal crops, Carpenter and Gianessi (1999) found that the adoption of herbicide tolerant soybean depends more on simplicity, flexibility, and the fit with existing production practices than on profitability. Alexander et al. (2002) found that more risk-averse farmers were less likely to adopt transgenic insecticidal corn, while risk aversion was not related to the likelihood of adoption of herbicide tolerant soybean. Hubbell et al. (2000) found that adoption of transgenic insecticidal cotton was negatively influenced by the price of the cotton seed and a farmer's share of income derived from cotton production, while more education, total cotton area, and experience with insect resistance positively influenced adoption. Fernandez-Cornejo and McBride (2002) found that farmers with larger farms, production and marketing contracts, more education, and more severe pest problems were more likely to adopt transgenic insecticidal corn active against stalk-boring lepidopteran pests such as the European corn borer (*Ostrinia nubilalis*). Payne et al. (2003) found that, for transgenic insecticidal corn active against pests such as the western and northern corn rootworm (*Diabrotica virgifera virgifera* and *Diabrotica barberi*), the likelihood of adoption increased with a farmer's age and farm size up to about 49 years of age and 1175 hectares, but then decreased. Farmers who specialized in corn production, had experienced more severe corn rootworm (*Diabrotica* sp.) problems, and had used insecticides to control these pest species were also more likely to adopt. Farmers who performed more off-farm work or lived in regions where more corn is exported were less likely to adopt.

Compliance Behavior

Extensive and intensive decisions are as important for compliance behavior as they are for adoption behavior. Extensive compliance refers to the choice of whether or not to comply with a regulation, whereas intensive compliance refers to the degree of compliance with a regulation. For example, the EPA dictates refuge size and configuration requirements for transgenic insecticidal crops. Goldberger et al. (2005) found that while some Minnesota and Wisconsin corn farmers chose to plant sufficient refuge in 2003, they did not always meet configuration requirements. Alternatively, some farmers did not plant enough refuge, but met configuration requirements with what they did plant. As with adoption, intensive compliance decisions depend on variable costs, while extensive decisions depend on variable and fixed costs. Economists, sociologists, and psychologists have all sought explanations for observed patterns in regulatory compliance.

Dominant psychological perspectives on compliance behavior include cognitive and social learning theories. Cognitive theory emphasizes the internal characteristics of an individual that influence compliance behavior, such as individual morality and moral development (e.g., Kohlberg, 1969, 1984; Levine and Tapp, 1977; Tapp and Kohlberg, 1977). Individuals with a higher degree of morality or moral development are more likely to comply with regulations. Social learning theory places less emphasis on individuals and more emphasis on the social interactions that shape behavior through forces such as peer suasion (e.g., Aronfreed, 1968, 1969; Bandura, 1969; Mischel and Mischel, 1976; Akers *et al.*, 1979; Akers, 1985). Individuals are more likely to comply with a regulation when they perceive that others are also complying.

Dominant sociological perspectives include normative and instrumental theories (Tyler, 1990). Normative theory explains compliance behavior in terms of social justice and morality such that an individual's perceptions of the appropriateness and fairness of a regulation and the legitimacy of the regulator are key factors. Individuals are more likely to comply with a regulation that they perceive is appropriate and fair. They are also more likely to comply if they accept the regulator's authority. Instrumental theory stresses individual self-interest in terms of the rewards an individual receives from ignoring a regulation and the likelihood and severity of sanctions for noncompliant behavior. Individuals are more likely to comply when the cost of compliance is relatively low or the likelihood and severity of sanctions are high.

Like instrumental sociologists, economists have tended to emphasize individual self-interest to explain compliance behavior (Becker, 1968). Specifically, economists have focused on the cost of complying with a regulation and the likelihood and severity of sanctions for noncompliant behavior. Risk is also an important facet of compliance behavior because whether or not an individual faces sanctions for noncompliant behavior is usually uncertain.

Empirically, two clear results emerge from the literature. First, compliance behavior is influenced by economic factors such as compliance costs, and the likelihood and severity of sanctions for noncompliant behavior. Second, these economic factors are not sufficient to fully explain compliance behavior. Generally, regulatory compliance is higher than what economic and instrumental sociological theories predict. Therefore, recent efforts to improve theories of compliance behavior integrate models developed by economists, sociologists, and psychologists. For example, Sutinen and Kuperan (1999) propose a theory of compliance behavior that includes compliance costs; the likelihood and severity of sanctions for noncompliant behavior; individual morality; moral suasion; and legitimacy in terms of the fairness and efficiency of the regulatory process and of regulatory outcomes.

Recent studies of compliance with EPA IRM regulations for transgenic insecticidal crops include Jaffe (2003a,b, 2009), ABSTC (2005), Merrill et al. (2005), Carriere et al. (2005), and Martinez and Reynolds (2009). Data for these studies come from a variety of sources, including farmer surveys (mail, personal, and telephone) and field surveys. All assess whether farmers plant enough refuge. Several also consider farmer compliance with EPA configuration requirements. The results of these studies indicate that larger farms tend to have higher compliance rates. Refuge size requirements are not always binding (i.e., some farmers plant more refuge than required), which can be explained by typical patterns in intensive adoption behavior. Initially, compliance rates trended upward, though recent evidence suggests this may no longer be the case (Martinez and Reynolds, 2009). Compliance rates for transgenic insecticidal corn in the Corn Belt of the United States have been higher than for transgenic insecticidal corn in the Cotton Belt where the EPA refuge size requirement has been higher. Compliance with refuge size requirements is positively correlated with refuge configuration requirements, such that farmers who do not meet size requirements are more likely to fail to meet configuration requirements.

Implications of Human Behavior

The common adoption and compliance patterns identified in the literature ultimately constrain the ability of regulators to implement an IRM policy that increases the societal benefits of pest management. However, it is not immediately clear whether these constraints will mitigate or exacerbate the evolution of insect resistance. To understand why, several rudimentary modifications to the conceptual model described by Equation 13.1 are explored. Next, alternative methods that have been used to provide a richer characterization of adoption and compliance behavior are reviewed and opportunities for future research are proposed.

Two distinct types of IRM policies are explored within the framework of Equation 13.1. The first type is referred to as a refuge policy, and the second is referred to as an IPM policy. The refuge policy is analogous to the EPA's strategy for managing insect resistance to transgenic insecticidal crops where farmers are required to leave a proportion of the pest population unmanaged by planting refuge. The IPM policy is analogous to IPM strategies in which control tactics are used only when pest abundance exceeds some economic threshold. The primary distinction between these alternative policies in the context of Equation 13.1 is the temporal pattern of the proportion of managed pests. For the refuge policy, the regulator tries to ensure that only a proportion of the pest population is managed every generation. For the IPM policy, the regulator tries to ensure that the pest population is only managed in generations when pests are relatively abundant and likely to cause significant crop damage.

Refuge Policy

Consider the basic model in Equation 13.1 where

$$\tau_t = \tau \tag{13.2}$$

for all t such that $1.0 \geq 1 - \tau \geq 0.0$ is the mandated proportion of refuge. This version of the model is referred to as Full Adoption and Compliance because it assumes that farmers fully and immediately adopt the new pest-management strategy and fully comply with the mandated refuge requirement.

Alternatively, suppose

$$\tau_t = a(t)\tau \tag{13.2a}$$

where $a(t)$ represents the proportion of extensive and intensive farmer adoption of a new pest-management strategy in period t. Assume

$$a(t) = \frac{1}{1 + e^{\alpha_a + \beta_a t}} \tag{13.3}$$

where α_a and β_a are parameters that describe how rapidly farmers adopt and utilize this new strategy over time. Note that the parameters are subscripted with a to indicate that they are related to adoption. For the transgenic insecticidal corn adoption trend observed in Figure 13.1, ordinary least-squares estimates assuming two generations of pest per year are $\alpha_a = 2.34$ and $\beta_a = -0.101$. Figure 13.3 shows the predicted adoption rates for 50 periods given these estimates. These predicted adoption rates follow the typical pattern. The adoption rate initially increases over time at an increasing rate (up to about 50%). The adoption rate then continues to increase (above 50%) but at a decreasing rate. This version of the model is referred to as Partial Adoption and Full Compliance because it assumes that not all farmers immediately adopt the new pest-management strategy, but those that do fully comply with the mandated refuge requirement.

Now, let $c(\tau)$ be the proportion of pests managed by farmers that comply with the mandated refuge requirement such that $c(\tau)\tau$ reflects the proportion of the pest population exposed to management by compliant farmers and $1 - c(\tau)$ reflects the proportion of the population exposed to management by noncompliant farmers. Adding these two proportions yields the total proportion of the population exposed to management:

$$\tau_t = c(\tau)\tau + 1 - c(\tau) \tag{13.2b}$$

For this specification, noncompliant farmers are assumed to manage all their pests (i.e., they plant no refuge), while compliant farmers abide by the mandated refuge requirement.

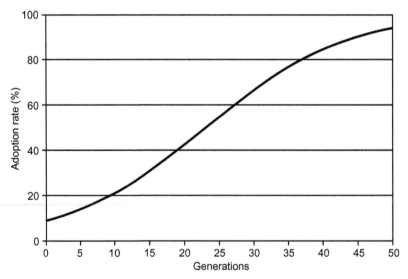

Figure 13.3 Estimated adoption function.

The proportion of compliant farmers is assumed to be decreasing in the mandated proportion of refuge, $1 - \tau$, because compliance costs are increasing as the proportion of refuge increases. They are increasing because compliant farmers must forgo the benefits of pest management on more land as the proportion of refuge increases. This assumption is also consistent with results reported by ABSTC (2005) for transgenic insecticidal corn and cotton. Assume

$$c(\tau) = \frac{1}{1 + e^{\alpha_c + \beta_c(1-\tau)}} \tag{13.4}$$

where α_c and β_c are parameters that describe how sensitive farmer compliance is to the mandated refuge requirement. Note that the parameters are subscripted with c to indicate that they are related to compliance. In 2001, compliance rates for transgenic insecticidal corn refuge size requirements in the Corn and Cotton Belts of the United States were 87 and 77% (Hurley, 2005). Refuge requirements for these regions were 20 and 50%. Assuming these differences in compliance rates can be explained solely by the difference in refuge requirements implies $\alpha_c = -2.36$ and $\beta_c = 2.31$. Figure 13.4 shows the predicted compliance rates as the refuge requirement increases from 0.0 to 1.0 given these estimates. In this figure, the compliance rate falls from about 90 to 50% as the refuge requirement increases from 0 to 100%. This version of the model is referred to as Full Adoption and Partial Compliance because it assumes that all farmers

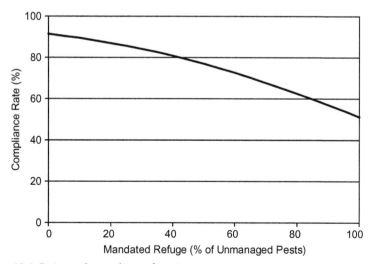

Figure 13.4 Estimated compliance function.

immediately adopt the new pest-management strategy, but only a portion of farmers fully comply with the mandated refuge requirement.

Finally, suppose

$$\tau_t = a(t)[c(\tau)\tau + 1 - c(\tau)] \tag{13.2c}$$

where $a(t)$ and $c(\tau)$ are as previously defined. This version of the model is referred to as Partial Adoption and Compliance because it assumes that only a portion of farmers adopt the new pest-management strategy over time and only a portion of farmers adopt and fully comply with the mandated refuge requirement.

Figure 13.5 shows the minimum feasible refuge requirement needed to maintain the proportion of resistant alleles below 0.5 for 50 periods for each version of the model as the selection (s) and dominance (h) coefficients vary between 0 and 1 and initial resistance is $p_0 = 0.001$. Note that variations in initial resistance do not qualitatively change the reported results. In some cases, it is not possible to maintain the proportion of resistant alleles below 0.5 for 50 periods even when the refuge requirement is 100% due to noncompliance. Therefore, Figure 13.6 shows the

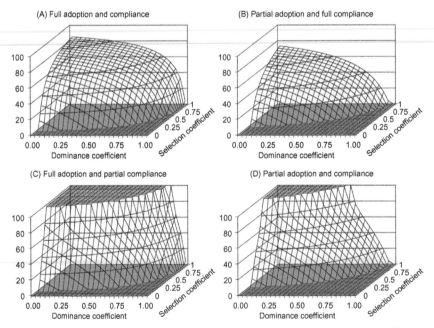

Figure 13.5 Minimum feasible refuge required to maintain the proportion of resistant alleles below 0.5 for 50 periods.

proportion of resistant alleles after 50 periods given the refuge requirement in Figure 13.5. Comparing the results of these four versions of the model provides insight into the potential implications of ignoring farmer behavior when trying to identify a refuge policy that maintains the proportion of resistant alleles below 0.5 for 50 periods.

Figure 13.5 and 13.6 reveal a consistent pattern in the refuge requirement and resistance for all four versions of the model. When selection or degree of dominance is low (s is close to zero or h is close to 1, resistance is recessive), resistance is of little concern and there is no need for refuge. Alternatively, as selection and the degree of dominance increase (s approaches 1 and h approaches 0, resistance is dominant), increasing the refuge requirement helps slow the evolution of insect resistance. From a practical standpoint, it is important to recognize that a pest-management strategy with a low selection coefficient will not be attractive to farmers because it will not be efficacious.

Comparing alternative versions of the model provides insight into how ignoring the implications of farmer adoption and compliance behavior can bias policy prescriptions. Comparing Figures 13.5 and 13.6(A) Full Adoption

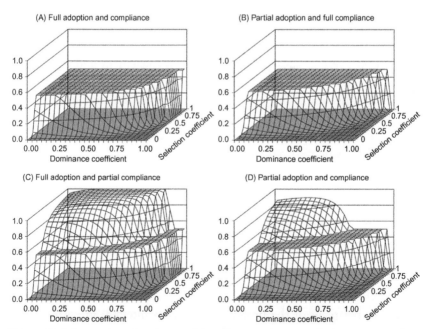

Figure 13.6 Proportion of resistant alleles after 50 periods with the minimum feasible refuge requirement.

and Compliance to Figures 13.5 and 13.6(B) Partial Adoption and Full Compliance shows that not accounting for adoption typically results in overestimates of the refuge requirement and resistance. Alternatively, comparing Figures 13.5 and 13.6(A) Full Adoption and Compliance to Figures 13.5 and 13.6(C) Full Adoption and Partial Compliance shows that not accounting for noncompliance typically leads to underestimates of the refuge requirement and resistance. Finally, comparing Figures 13.5 and 13.6(A) Full Adoption and Compliance to Figures 13.5 and 13.6(D) Partial Adoption and Compliance shows that not accounting for adoption and noncompliance can result in over- or underestimates of the refuge requirement and resistance depending on the degree of selection and dominance. It typically leads to overestimates for relatively low selection or dominance, while it typically leads to underestimates for relatively high selection and dominance. For example, when $s = 0.95$ and $h = 0.8$, the minimum feasible percentage of refuge required to maintain resistance below 0.5 for 50 periods is 60 assuming Full Adoption and Compliance, but only 49 assuming Partial Adoption and Compliance. Alternatively, when $s = 1.0$ and $h = 0.75$, the minimum feasible percentage of refuge required to maintain resistance below 0.5 for 50 periods is 66 assuming Full Adoption and Compliance, but 68 assuming Partial Adoption and Compliance.

The results of this analysis are intuitive. Since farmers tend to gradually rather than fully adopt new production practices over time, assuming full adoption overestimates pest exposure to management and the evolution of resistance. Since not all farmers will comply with mandated refuge requirements, assuming full compliance underestimates pest exposure to management and the evolution of resistance. Given these two opposing effects, the net bias relative to ignoring these effects is generally indeterminate and must be resolved empirically.

IPM Policy

Now consider the basic model outlined above where

$$\tau_t = \begin{cases} 1, & \text{for } N > \tau \\ 0, & \text{otherwise} \end{cases} \tag{13.2d}$$

for all t, where $N_t \geq 0$ reflects pest abundance (e.g., the normalized number of pests per plant or the proportion of plants infested) in period t and $\tau \geq 0.0$ is now a management threshold mandated by the regulator. That is, farmers are allowed to manage the pest only if pest abundance exceeds the mandated threshold. For this version of Full Adoption and Compliance, the regulator manages insect resistance by restricting pest management to periods when pest abundance is relatively high by setting the threshold τ. For the Partial Adoption and Full Compliance version of the IPM policy model, assume

$$\tau_t = \begin{cases} a(t), & \text{for } N_t > \tau \\ 0, & \text{otherwise} \end{cases} \tag{13.2e}$$

where $a(t)$ is the proportion of extensive and intensive farmer adoption of a new pest-management strategy in period t. For the Full Adoption and Partial Compliance version, assume

$$\tau_t = \begin{cases} 1, & \text{for } N_t > \tau \\ 1 - c(t), & \text{otherwise} \end{cases} \tag{13.2f}$$

where $c(\tau)$ is the proportion of pests managed by farmers that comply with the mandated threshold requirement. Finally, for the Partial Adoption and Compliance version, assume

$$\tau_t = \begin{cases} a(t), & \text{for } N_t > \tau \\ a(t)(1 - c(t)), & \text{otherwise} \end{cases} \tag{13.2g}$$

where $a(t)$ and $c(\tau)$ are as previously defined. For the partial adoption versions of the model, the adoption rate determines the proportion of pests managed by farmers when pest abundance exceeds the threshold requirement. The remaining proportion of the pest population is unmanaged because some farmers choose not to adopt the new management strategy. For the partial compliance versions of the model, the compliance rate determines the proportion of pests managed by farmers even when pest abundance does not exceed the threshold requirement. There is also a proportion of unmanaged pests in some generations attributable to the portion of farmers who comply with the threshold requirement.

To operationalize the IPM policy versions of the model, the evolution of pest abundance must be described in addition to the population genetics. Consider the common growth function

$$N_{t+1} = r\left(1 - \frac{N_t}{K}\right)N_t + N_t \tag{13.5}$$

where N_t is the size of the pest population in generation t (e.g., average number of pests per plant), r is the population's intrinsic rate of growth, and K is the population's carrying capacity. Equation 13.5 describes the population dynamics assuming no pest management. To account for the effect of pest management on this dynamic, N_t must be adjusted to reflect the proportion of pests removed from the population due to management, so define

$$\omega_t = p_t^2 + 2p_t(1 - p_t)(1 - \tau_t hs) + (1 - p_t)^2(1 - \tau_t s) \qquad (13.6)$$

and rewrite Equation 13.5 as

$$N_{t+1} = r\left(1 - \frac{\omega_t N_t}{K}\right)\omega_t N_t + \omega_t N_t \qquad (13.7)$$

Figure 13.7 shows the minimum feasible threshold requirement $(K \geq \tau \geq 0.0)$ needed to maintain the proportion of resistant alleles below 0.5 for 50 periods for each version of the model as the selection (s) and dominance (h) coefficients vary between 0 and 1, initial resistance is $p_0 = 0.001$, intrinsic rate of growth is $r = 1$, carrying capacity is $K = 1$, and initial pest pressure equals the carrying capacity ($N_0 = K$). Although the general pattern of results and comparisons of different versions of the

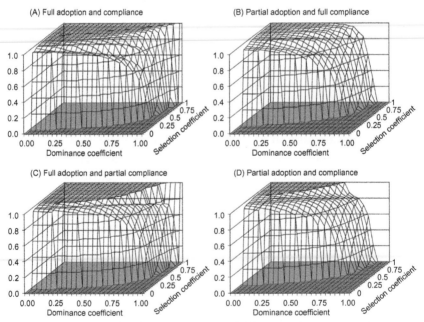

Figure 13.7 Minimum feasible treatment threshold required to maintain the proportion of resistant alleles below 0.5 for 50 periods.

models is similar to Figure 13.5, there is a notable difference. With the IPM policy, comparing Figure 13.7(A) Full Adoption and Compliance to Figure 13.7(C) Full Adoption and Partial Compliance reveals that not accounting for compliance behavior can result in over- or underestimating the minimum threshold requirement. Assuming full compliance tends to lead to overestimates when selection is relatively high, dominance is relatively low, and the selection and dominance coefficients are nearly equal. The implication of this result is that, in some cases, noncompliance can actually slow the evolution of resistance such that regulatory policy does not need to be as restrictive, which is contrary to the results obtained for the effect of noncompliance on a refuge policy.

Figure 13.8 helps explain why noncompliance can slow the evolution of resistance with an IPM-based insect resistance management policy. The top of the figure shows the evolution of pest abundance over time for the Full Adoption and Compliance and Full Adoption and Partial

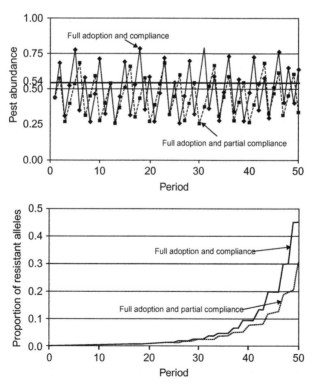

Figure 13.8 Evolution of pest abundance and proportion of resistant alleles over 50 periods.

Compliance versions of the model when the threshold requirement is $\tau = 0.54$, selection coefficient is $s = 0.75$, dominance coefficient is $h = 0.90$, initial resistance is $p_0 = 0.001$, intrinsic rate of growth is $r = 1$, carrying capacity is $K = 1$ and initial pest pressure equals the carrying capacity. The bottom of the figure shows the evolution of the proportion of resistant alleles.

First note that, in Figure 13.8, the proportion of resistant alleles after 50 periods is lower for the Full Adoption and Partial Compliance version of the model. Therefore, ignoring compliance behavior could result in mandating an IPM threshold that is more restrictive than necessary to maintain resistance below 0.5 for 50 periods. This result emerges because when pest abundance is relatively low, no pest management occurs under Full Adoption and Compliance, but a modicum of management occurs under Full Adoption and Partial Compliance due to noncompliant farmers. This modicum of management has the effect of modestly increasing resistance. But it also has the effect of slowing population growth, which reduces pest abundance and the likelihood of all farmers employing pest management in future periods. Indeed, with Full Adoption and Compliance pest management is triggered in 22 periods, while it is only triggered in 16 periods with Full Adoption and Partial Compliance. By delaying pest management by all farmers in the future, the evolution of resistance is slowed. In Figure 13.8, the effect of increased resistance due to noncompliance is outweighed by the effect of decreased resistance due to delayed pest management by all farmers over 50 periods.

Characterizations of Adoption and Compliance: A Review

Exploring the effectiveness of refuge and IPM-based IRM policies in the context of this basic model highlights the importance of having a better understanding of human behavior. For the rudimentary adoption and compliance models, farmer adoption and compliance behavior can either mitigate or exacerbate the evolution of resistance depending on important biological factors. Therefore, it is not possible to make general recommendations regarding how regulatory policy should be adjusted to account for this behavior. Instead, reasonable policy recommendations require a better understanding of the specific factors affecting farmer adoption and compliance decisions for a given pest-management strategy and regulatory policy—a point emphasized by Onstad et al. (2011).

Two general approaches have been employed to provide a richer description of farmer adoption and compliance behavior. One approach captures farmer adoption or compliance behavior using simulation models based on fundamental economic principles (Peck and Ellner, 1997; Hurley *et al.*, 2001, 2002; Mitchell *et al.*, 2002; Livingston *et al.*, 2004; Hurley, 2005; Mitchell and Hurley, 2006). Many, though not all, of these models have then been incorporated into biological models of the evolution of resistance in order to evaluate alternative IRM policies. The other approach uses farmer survey data to econometrically estimate farmer adoption functions that are conditional on important sociological and regulatory factors in addition to important economic factors (Hubbell *et al.*, 2001; Hurley *et al.*, 2006). These estimated functions could also be incorporated into biological models of the evolution of resistance, but typically they have not been.

Simulation Approaches

The simplest models used to describe farmer adoption behavior have relied on concepts such as economic thresholds (e.g., Peck and Ellner, 1997; Hurley *et al.*, 2001, 2002; Crowder *et al.*, 2006). In these models, farmers are assumed to adopt pest management only when pest abundance exceeds some threshold. The threshold in these models is a fixed parameter and does not take into account losses in pest-management efficacy due to the evolution of resistance, which would typically change the optimal threshold over time.

More complex models have dynamically adjusted adoption over time based on pest abundance and a loss in pest-management efficacy due to the evolution of resistance (e.g., Livingston *et al.*, 2004). These models define the returns to pest management as the difference in crop revenues and management costs. Crop revenues are the product of the price and yield, with yield nonincreasing in pest survival. Pest survival is nondecreasing in the initial pest abundance and resistance, and nonincreasing in the level of pest management. Management costs are increasing in the level of pest management. The model then assumes that farmers choose the level of pest management that maximizes returns, where the level of pest management may be constrained by regulatory policy like a refuge requirement.

Hurley (2005) extends the basic framework of Livingston *et al.* (2004) in two important directions. First, Hurley's model takes into account the inherent variability of pest abundance and the risk it entails by assuming a random pest population that is conditional on pest survival in the previous generation. Second, it accounts for both compliance and adoption behavior. To account for differences in risk attitudes across individual farmers, it assumes that adoption is increasing and compliance is decreasing in the difference between expected returns with and without pest management. The model permits adoption even when the returns to pest management are negative, which can result when pest management reduces risk and farmers are risk averse. Farmer perceptions of expected returns in the current period are captured by averaging returns from previous periods, implying that farmers use past experiences to form expectations about future returns to pest management.

Mitchell *et al.* (2002) provides a richer description of compliance behavior that includes regulatory policy instruments for encouraging compliance: compulsory refuge insurance, voluntary refuge insurance with subsidized premiums, and enforcement with monitoring and sanctions for noncompliant behavior. Their model accounts for risk more explicitly than Hurley (2005) using traditional economic models of risk preferences and perceptions, but does not consider adoption behavior. Mitchell and Hurley (2006) extend Mitchell *et al.* (2002) to include adoption behavior and also explicitly account for the effect of this behavior on a monopolistic seed company's optimal pricing and IRM enforcement decisions. While both of these models were used to evaluate the feasibility of implementing policies to ensure regulatory compliance, they were not integrated with a biological model of the evolution of resistance to evaluate the effect of policy levers, like the frequency and severity of sanctions for noncompliant behavior, on the efficacy of IRM. However, Mitchell and Onstad (2005) combine a more complete biological model with a simplified adoption model, to examine how adoption behaviors such as insecticide use on refuge and crop rotation patterns interact with extended diapause, Bt toxin dose, and several population genetics parameters to affect the evolution of resistance.

Survey-Based Econometric Approaches

Hubbell *et al.* (2000) and Hurley *et al.* (2006) use farmer survey data to estimate adoption functions. The survey used by Hubbell *et al.* (2000) asked farmers whether they adopted transgenic insecticidal cotton at the current price and if so, how much they planted. If farmers had not adopted transgenic insecticidal cotton at the current price, the survey asked hypothetically if they would have adopted it at a lower price, and if so, how much they would have planted. Econometric methods are used with the survey data to estimate both extensive and intensive adoption functions that depend on the price of transgenic insecticidal cotton seed as well as other important socioeconomic factors. Hubbell *et al.* (2000) used the estimated adoption function to evaluate how government subsidies for transgenic insecticidal cotton could be used to increase adoption while reducing the use of other insecticides. The results were not used to evaluate the effect of adoption behavior on IRM policy.

There are a number of important differences between Hurley *et al.* (2006) and Hubbell *et al.* (2000). Hurley *et al.* (2006) focuses on farmer adoption of transgenic insecticidal corn active against western and northern corn rootworm (*D. virgifera virgifera* and *D. barberi*) and stalk-boring lepidopteran species such as *Ostrinia nubilalis*. Also, the survey data used by Hurley *et al.* (2006) were collected before, not after, the commercial release of the product. Therefore, the survey questions were purely hypothetical, which provided an opportunity to focus on the effect of regulatory policy on adoption. Specifically, the survey described a new transgenic insecticidal corn variety that provided corn rootworm control or corn rootworm and *O. nubilalis* control; indicated whether the transgenic insecticidal corn crop would be approved for export; and described the refuge requirements the farmer would be obligated to follow if they planted the new variety. The refuge requirements varied across surveys in terms of size, configuration, and whether supplemental refuge insecticide treatments were permitted in years of heavy infestation. Finally, the survey told farmers what the additional seed costs would be, which also varied across surveys. Farmers were then asked if they would adopt the new variety if it were available. Econometric methods were used to estimate an adoption function that depended on the price of Bt corn, the spectrum of control, export market approval, and regulatory policy. The results indicated that adoption would increase if the corn hybrid controlled both European corn borer and corn rootworm and was approved for export, while it would decrease as the size of refuge increased and if supplemental refuge insecticide treatments were not permitted. The adoption function was used to estimate compliance costs for alternative regulatory policies. The results were not used to evaluate the effect of adoption behavior on IRM policy.

The simulation and survey-based econometric approaches to characterizing farmer adoption and compliance behavior each has its advantages and disadvantages. Applications of the simulation approach have done a good job of integrating important economic aspects of the problem such as profitability with important biological aspects of the problem such as pest population dynamics, behavior, and genetics, resulting in rich and dynamic descriptions of farmer adoption and compliance behavior. These approaches have not done as well integrating less tangible sociological and psychological factors. Applications of the survey-based econometric approach have done a good job integrating important economic and sociological factors, but have not done as well in integrating biological factors. Furthermore, econometric applications have yet to take advantage of the benefits of longitudinal surveys, which would offer a richer description of temporal variability in farmer adoption and compliance behavior.

Opportunities for Future Research

The development and commercial release of the first transgenic insecticidal crops reinvigorated IRM research, resulting in important advances in understanding how the complex interactions between pest biology and behavior, and farmer adoption and compliance behavior can affect IRM. Still, there are many opportunities to further advance this understanding.

Some of the richer simulation models that have been developed still remain to be integrated with biological models in order to more fully understand important IRM policy trade-offs. Simulation approaches to modeling adoption and compliance behavior could also be further developed to more accurately characterize extensive versus intensive patterns of adoption and compliance. Two possible ways to accomplish this objective are to add fixed costs and farmer experience to the problem.

Survey-based econometric approaches could benefit from employing longitudinal survey data that includes more extensive information on farmer perceptions of the frequency and severity of pests. With longitudinal data, it becomes possible to directly measure the effect of farmer experience on adoption behavior. With more extensive information on farmer perceptions of the frequency and severity of pests, it becomes possible to estimate adoption equations that can be integrated with pest population dynamics and genetics. Survey-based econometric approaches based on well-designed surveys might also be employed to obtain a better understanding of the sensitivity of farmer compliance to alternative IRM policies.

The majority of work on the effect of human behavior on IRM has focused on farmer adoption and compliance without much regard for other important actors such as chemical and seed companies. Alix and Zilberman (2003) and Noonan (2003) note that new pesticides are often patented by chemical companies, which gives them exclusive control over distribution for a specified time period. This exclusive control provides an incentive to charge higher prices in order to achieve higher returns from pesticide sales.

These higher prices slow adoption by increasing the cost of pest management to farmers, thereby slowing the evolution of resistance. Incorporating this observation into an IRM model, Alix-Garcia and Zilberman (2005) argue that regulatory policy may in fact be unnecessary.

For most transgenic insecticidal crops, however, several companies compete for a share of the crop acreage planted with them. These companies can try to increase market share by lowering prices, but they can also try to increase it by differentiating the transgenic varieties they offer farmers. For example, some companies have introduced stacked crop varieties that include multiple toxins for the targeted pests (Bt-pyramids) and herbicide tolerant traits, while others rely less on stacking traits. The introduction of these stacked varieties has resulted in less restrictive IRM requirements that make these varieties more attractive to farmers, such as seed mix refuges. As a result, these varieties create a trade-off between adoption and compliance—seed mixtures may increase adoption and ensure compliance, while hybrids requiring refuge-based IRM will have lower adoption as well as lower compliance (Onstad *et al.*, 2011). Some companies have also been successful in securing crop insurance endorsements that lower the premiums farmers pay to use their varieties, but only for some regions for a limited period of time.

To date, little published research exists on the consequences of these trends for farmer adoption and compliance behavior and the effect of this farmer behavior on the efficacy of current IRM policies. However, a recent example is Carroll *et al.* (2012), who use a simulation approach to examine the trade-off between noncompliance and seed mix refuges as a way to slow the evolution of resistance with Bt-pyramids. Critical research needs include developing strategies for managing and mitigating insect and weed resistance in light of trends for stacking multiple traits and for releasing new traits, particularly with the rapid emergence of herbicide resistant weeds and the recent documentation of field-evolved corn rootworm resistance to Cry3Bb1 Bt corn (Gassman *et al.*, 2011; Green, 2012).

CONCLUSIONS

Pest management is a fundamental human activity in agricultural production. This activity is inherently constrained by the evolution of insect resistance. Insect resistance management (IRM) has the potential to increase societal benefits achieved through pest management. Farmers are unlikely to effectively manage insect resistance to the benefit of society on their own due to the common property nature of the problem (see Mitchell and Onstad, Chapter 2). Therefore, public policy can play an important role in helping manage insect resistance to the benefit of society. An understanding of insect biology and behavior is not enough to design effective IRM policies because predictable farmer behavior also plays an important role in the evolution of resistance. Furthermore, this predictable farmer behavior depends in complex ways on important economic, sociological, and psychological factors, as well as pest behavior and biology. These complex interactions limit the possibility of using sweeping generalities to help guide IRM policy, which ensures the continued importance of both practical and innovative IRM research by biological and social scientists.

REFERENCES

Agricultural Biotechnology Stewardship Technical Committee (ABSTC), 2005. Insect resistance management grower survey for corn borer-resistant Bt field corn: 2004 growing season. Available on the World Wide Web: <http://www.pioneer.com/biotech/irm/survey.pdf>.

Akers, R.L., 1985. Deviant Behavior: A Social Learning Approach. third ed. Wadsworth, Belmont, CA.

Akers, R.L., Krohn, M.D., Lonza-Kaduce, L., Radosevitch, M., 1979. Social learning and deviant behavior: a specific test of a general theory. Am. Sociol. Rev. 44, 635–655.

Alexander, C., Fernandez-Cornejo, J., Goodhue, R.E., 2002. Determinants of GMO use: a survey of Iowa Maize-Soybean farmers' acreage allocation. In: Santaniello., V., Evenson, R.E., Zilberman, D. (Eds.), Market Development for Genetically Modified Foods. CABI, Wallingford, U.K. and New York, pp. 127–139.

Alix, J., Zilberman, D., 2003. Industrial organization and institutional considerations in agricultural pest resistance management. In: Laxminarayan, R. (Ed.), Battling Resistance to Antibiotics and Pesticides. Resources for the Future, Washington, DC.

Alix-Garcia, J., Zilberman, D., 2005. The Effect of Market Structure on Pest Resistance Buildup. Working Paper, Department of Economics, University of Montana.

Aronfreed, J., 1968. Conduct and Conscience. Academic Press, New York.

Aronfreed, J., 1969. The Problem of imitation. In: Lipsutt, L.P., Reese, H.W. (Eds.), Advances in Child Development and Behavior, vol. 4. Academic Press, New York.

Bandura, A., 1969. Social-learning theory of identificatory processes. In: Goslin, D.A. (Ed.), Handbook of Socialization Theory and Research. Rand McNally, Chicago, IL.

Becker, G., 1968. Crime and punishment: an economic approach. J. Polit. Econ. 76 (2), 169–217.

Berwald, D., Matten, S., Widawsky, D., 2006. Economic analysis and regulating pesticide biotechnology at the U. S. environmental protection agency. In: Just, R.E., Alston,

J.M., Zilberman, D. (Eds.), Regulating Agricultural Biotechnology: Economics and Policy. Springer, New York, pp. 21–36.

Carpenter, J., Gianessi, L., 1999. Herbicide tolerant soybeans: why growers are adopting roundup ready varieties. AgBioforum. 2, 65–72.

Carriere, Y., Ellers-Kirk, C., Kumar, K., Heuberger, S., Whitlow, M., Antilla, L., et al., 2005. Rapid report long-term evaluation of compliance with refuge requirements for Bt cotton. Pest. Manag. Sci. 61, 327–330.

Carroll, M.W., Head, G., Caprio, M., 2012. When and where a seed mix refuge makes sense for managing insect resistance to Bt plants. Crop Prot. 38, 74–79.

Clark, J.S., Carlson, G.A., 1990. Testing for common versus private property: the case of pesticide resistance. J. Environ. Econ. Manage. 19, 45–60.

Crowder, D.W., Onstad, D.W., Gray, M.E., 2006. Planting transgenic insecticidal crops based on economic thresholds: consequences for integrated pest management and insect resistance management. J. Econ. Entomol. 99, 899–907.

El-Osta, H.S., Morehart, M.J., 1999. Technology adoption decisions in dairy production and the role of herd expansion. Agric. Resour. Econ. Rev. 28, 84–95.

Feder, G., 1979. Pesticides, information, and pest management under uncertainty. Am. J. Agric. Econ. 61, 97–103.

Feder, G., O'Marra, G.T., 1981. Farm size and the adoption of green revolution technology. Econ. Dev. Cult. Change. 30, 59–76.

Feder, G., Just, R.J., Zilberman, D., 1985. Adoption of agricultural innovations in developing countries: a survey. Econ. Dev. Cult. Change. 33, 255–298.

Fernandez-Cornejo, J., McBride, W.D., 2002. Adoption of Bioengineered Crops. Economic Research Service, US Department of Agriculture, Washington. Agricultural Economic Report no. 810. <http://www.ers.usda.gov/publications/aer810/aer810.pdf>.

Fernandez-Cornejo, J., Beach, E.D., Huang, W., 1994. The adoption of IPM By vegetable growers in Florida, Michigan, and Texas. J. Agric. Appl. Econ. 26 (1), 158–172.

Gassman, A.J., Petzold-Maxwell, J.L., Keweshan, R.S., Dunbar, M.W., 2011. Field-evolved resistance to Bt Maize by Western corn rootworm. PLoS One. 6 (7), e22629. Available from: http://dx.doi.org/10.1371/journal.pone.0022629.

Goldberger, J., Merrill, J., Hurley, T.M., 2005. Bt corn farmer compliance with insect resistance management requirements in minnesota and wisconsin. AgBioForum. 8, 151–160.

Green, G., Sunding, D., Zilberman, D., Parker, D., 1996. Explaining irrigation technology choices: a microparameter approach. Am. J. Agric. Econ. 78, 1064–1072.

Green, J.M., 2012. The benefits of herbicide-resistant crops. Pest. Manage. Sci. Forthcoming. Available from: http://dx.doi.org/10.1002/ps.3374.

Griliches, Z., 1957. Hybrid corn: an exploration in the economics of technological change. Econometrica. 25, 501–522.

Hartl, D.L., 1988. A Primer of Population Genetics. second ed. Sinauer and Associates, Inc., Sunderland, MA.

Heibert, D., 1974. Risk, learning and the adoption of fertilizer responsive seed varieties. Am. J. Agric. Econ. 56, 764–768.

Horowitz, J.K., Lichtenberg, E., 1993. Insurance, moral hazard, and chemical use in agriculture. Am. J. Agric. Econ. 75, 926–935.

Horowitz, J.K., Lichtenberg, E., 1994. Risk-reducing and risk increasing effects of pesticides. J. Agric. Econ. 45, 82–89.

Hubbell, B.J., Marra, M.C., Carlson, G.A., 2000. Estimating the demand for a new technology: Bt cotton and insecticide policies. Am. J. Agric. Econ. 82 (1), 118–132.

Hurley, T.M., 2005. Bt resistance management: experiences from the U.S.. In: Wessler, J. (Ed.), Environmental Costs and Benefits of Transgenic Crops in Europe, vol. 7. Springer, Dordrecht, pp. 81–93. Wageningen UR Frontis Series.

Hurley, T.M., Babcock, B.A., Hellmich, R.L., 2001. Bt corn and insect resistance: an economic assessment of refuges. J. Agric. Resour. Econ. 26 (1), 176–194.

Hurley, T.M., Secchi, S., Babcock, B.A., Hellmich, R.L., 2002. Managing the risk of European corn borer resistance to Bt corn. Environ. Resour. Econ. 22, 537–558.

Hurley, T.M., Mitchell, P.D., Rice, M.E., 2004. Risk and the value of Bt corn. Am. J. Agric. Econ. 86 (2), 345–358.

Hurley, T.M., Langrock, I., Ostlie, K., 2006. Estimating the benefits of Bt corn and cost of insect resistance management Ex Ante. J. Agric. Resour. Econ. 31 (2), 355–375.

Hutchison, W.D., Burkness, E.C., Mitchell, P.D., Moon, R.D., Leslie, T.W., Fleischer, S. J., et al., 2010. Areawide suppression of European corn borer with Bt maize reaps savings to non-Bt maize growers. Science. 330, 222–225.

Jaffe, G., 2003a. Planting trouble: are farmers squandering Bt corn technology? Center for Science in the Public Interest (CSPI). 1875 Connecticut Avenue, NW, Suite 300, Washington, DC.

Jaffe, G., 2003b. Planting trouble update. Center for Science in the Public Interest (CSPI). 1875 Connecticut Avenue, NW, Suite 300, Washington, DC.

Jaffe, G., 2009. Complacency on the Farm. Center for Science in the Public Interest (CSPI). 1875 Connecticut Avenue, NW, Suite 300, Washington, DC.

Just, R.E., Zilberman, D., Rauser, G.C., 1980. A putty clay approach to the distributed effects of new technology under risk. In: Yaron, D., Tapiero, C. (Eds.), Operations Research in Agriculture and Water Resources. North-Holland Publishing Co., Amsterdam.

Kohlberg, L., 1969. Stage and sequence: the cognitive-development approach to socialization. In: Goslin, D.A. (Ed.), Handbook of Socialization Theory and Research. Rand McNally, New York.

Kohlberg, L., 1984. Essays on Moral Development, vol. II. Harper & Row, San Francisco, CA.

Laffont, J.J., Martimort, D., 2001. The Theory of Incentives: The Principal-Agent Model. Princeton University Press, Princeton, NJ.

Levine, F.J., Tapp, J.L., 1977. The dialectic of legal socialization in community and school. In: Tapp, J.L., Levine, F.J. (Eds.), Law, Justice and the Individual in Society: Psychological and Legal Issues. Holt, Rinehart & Winston, New York.

Linder, R.K., Fisher, A.J., Pardey, P., 1979. The time to adoption. Econ. Lett. 2, 187–190.

Livingston, M.J., Carlson, G.A., Fackler, P.L., 2004. Managing resistance evolution in two pests to two toxins with refugia. Am. J. Agric. Econ. 86 (1), 1–13.

Martinez, M., Reynolds, A., 2009. EPA's Perspective on Compliance of Stewardship for Bt Crops. Entomological Society of America Annual Meeting, December 16, 2009.

McNamara, K.T., Wetzstein, M.E., Douce, G.K., 1991. Factors affecting peanut producer adoption of integrated pest management. Rev. Agric. Econ. 13, 129–139.

Merrill, J., Goldberger, J., Hurley, T., 2005. Bt corn farmer compliance with insect resistance management requirements in Minnesota and Wisconsin. AgBioForum. 8 (2&3), 151–160.

Mischel, W., Mischel, H.N., 1976. A cognitive social-learning approach to morality and selfregulation. In: Lickona, T. (Ed.), Moral Development and Behavior. Holt, Rinehart & Winston, New York.

Mitchell, P.D., Hurley, T.M., 2006. Adverse selection, moral hazard, and grower compliance with Bt corn refuge. In: Just, R.E., Alston, J.M., Zilberman, D. (Eds.), Regulating Agricultural Biotechnology: Economics and Policy. Springer, New York, pp. 599–624.

Mitchell, P.D., Onstad, D.W., 2005. Effect of extended diapause on evolution of resistance to transgenic *Bacillus thuringiensis* corn by Northern corn rootworm (Coleoptera: Chrysomelidae). J. Econ. Entomol. 98, 2220–2234.

Mitchell, P.D., Hurley, T.M., Babcock, B.A., Hellmich, R.L., 2002. Insuring the stewardship of Bt corn: a carrot versus a stick. J. Agric. Resour. Econ. 27, 390–405.

Noonan, D., 2003. An economic model of a genetic resistance commons: effects of market structure applied to biotechnology in agriculture. In: Laxminarayan, R. (Ed.), Battling Resistance to Antibiotics and Pesticides. Resources for the Future, Washington, DC.

Onstad, D.W., Mitchell, P.D., Hurley, T.M., Lundgren, J.G., Porter, R.P., Krupke, C.H., et al., 2011. Seeds of change: corn seed mixtures for resistance management and IPM. J. Econ. Entomol. 104, 343–352.

Payne, J., Fernandez-Cornejo, J., Daberkow, S., 2003. Factors affecting the likelihood of corn rootworm Bt seed adoption. AgBioForum. 6, 79–86.

Peck, S.L., Ellner, S.P., 1997. The effect of economic thresholds and life-history parameters on the evolution of pesticide resistance in a regional setting. Am. Nat. 149 (1), 43–63.

Rogers, E.M., 1983. Diffusion of Innovations. third ed. Free Press, New York.

Schultz, T.W., 1964. Transforming Traditional Agriculture. Yale University Press, New Haven, CT.

Schultz, T.W., 1981. Investing in People: The Economics of Population Quality. University of California Press, Berkeley, CA.

Sutinen, J.G., Kuperan, K., 1999. A socio-economic theory of regulatory compliance. Int. J. Soc. Econ. 26, 174–192.

Tapp, J.L., Kohlberg, L., 1977. Developing senses of law and legal justice. In: Tapp, J.L., Levine, F.J. (Eds.), Law, Justice and the Individual in Society: Psychological and Legal Issues. Holt, Rinehart & Winston, New York.

Thrikawala, S., Weersink, A., Kachanoski, G., Fox, G., 1999. Economic feasibility of variable-rate technology for nitrogen on corn. Am. J. Agric. Econ. 81, 914–927.

Tyler, T., 1990. Why People Obey the Law. Yale University Press, New Haven, CT.

U.S. Environmental Protection Agency, 1998. The Environmental Protection Agency's White Paper on Bt Plant-Pesticide Resistance Management. Biopesticides and Pollution Prevention Division, Office of Pesticide Programs, Office of the Assistant Administrator for Prevention, Pesticides and Toxic Substances, U.S. Environmental Protection Agency, Washington, DC.

USDA National Agricultural Statistics Service, 2001–12. Acreage. Washington, DC. Available on the World Wide Web: <http://usda.mannlib.cornell.edu/MannUsda/viewDocumentInfo.do?documentID=1000>.

USDA Risk Management Agency, 2011. Biotechnology Endorsement. Kansas City, MO. Available on the World Wide Web: <http://www.rma.usda.gov/policies/bye.html>.

CHAPTER 14

Modeling for Prediction and Management

David W. Onstad
DuPont Pioneer, Wilmington, DE

Chapter Outline

Modeling plays a critical role in understanding and managing resistance. In the study of complex ecological systems, modeling can be used to identify important gaps in knowledge, predict consequences of management, assess risks, and perform virtual experiments that are impossible to perform in reality because of cost, logistics, or ethics. To make predictions about evolution, insect population dynamics, and pest management, scientists need to use the best possible information and logic available and integrate them using mathematics. Once the model is made, scientists can study the system dynamics over time and space. The wide variety of models used in insect resistance management (IRM) is similar to the range of models used in population ecology and integrated pest management (IPM).

Insect Resistance Management
DOI: http://dx.doi.org/10.1016/B978-0-12-396955-2.00014-X

Modeling is no longer simply an academic exercise. Stakeholders concerned about public health and environmental protection are interested in IRM modeling. For instance, governments and developers of transgenic insecticidal crops have focused on modeling since transgenic crops began to be regulated (Glaser and Matten, 2003). This chapter describes the process of modeling so that models can be clearly understood and evaluated.

MODEL DEVELOPMENT AND EVALUATION
What is Modeling?

Modeling is the process of creating a conceptual, diagrammatic, algorithmic, or mathematical representation of reality. We all use conceptual models when we try to mentally predict what will happen in situations that we experience every day. We try to understand causes and imagine the effects. Computer programs and mathematical models describe population processes that cause changes to occur in the quality and quantity of populations. We calculate or work through the model to answer questions, test hypotheses, or make decisions. Much of model analysis involves understanding your ecological system and your goals well enough to follow leads provided by early calculations.

The model is given credibility when it is supported by theory and/or data. Greater credibility is achieved by testing the model against independent data and finding an adequate match between model results and observations. To create a model that is credible, one must use logic and the best data available during its calibration. The best models are made by the most logical, critical, and careful scientists with access to the best data. Logic helps eliminate mistakes.

Analyzing some models is like performing a traditional scientific experiment (Royama, 1971). The typical laboratory or field experiment involves conditions that are held constant, other conditions that are allowed to vary over time or space, and treatments that are evaluated. Replication and multiple trials are performed to account for the variation that cannot be controlled. With a model, particularly a deterministic one, the scientist has complete control over all conditions. Stochastic models require replication because of the variability of results. Thus, a model is an experimental design, and the calculation of the model is the experimental trial. Hypotheses are derived from the modeler's interpretation and generalization of the results. The entire process of modeling is summarized in Table 14.1.

Table 14.1 The Process of Modeling
  I. Select subject and purpose of model
      a. Determine time horizon
      b. Determine maximum spatial boundaries
  II. Review existing models and literature about experiments
      a. Find quantitative information; note the scales of the data, especially time
      b. Identify relevant theories
      c. Take advantage of techniques used in existing models
  III. Create mathematical functions from logic and data
      a. Convert data to appropriate units
      b. Interpolate—where does change occur?
      c. Extrapolate—what are logical limits?
      d. Compromises may be required because of interactions between functions
         and processes occurring at different time scales
  IV. Verification
      a. Check logic of entire model
      b. Check conversion of math into computer code
  V. Validation
      a. Relate to goal of model
      b. Use independent data
  VI. Analysis and experimentation
      a. Sensitivity analyses
      b. Assess risk
      c. Evaluate economics

The Role of the Goal

There are an infinite number of representations of any given ecological system. All will be simplifications. Our goal or the one given to us by stakeholders or supervisors determines which subset of the possible models should be considered. Of course, deadlines and budgets also determine which models are attempted. Do we have enough time and money to make a credible model given the goal? The goal not only directs us during the creation of the model, it also helps others understand and evaluate the model and its results. Therefore, it is important to clearly express the goal and purpose for making the model(s).

Every model must be based on a goal and judged according to that goal.

Given the goal, the preferable approach would be to create at least two different models of the system. If they both provide the same results with regard to your most important questions, then you have more confidence in your results. The answer and solution may be robust to changes in your model influenced by budget, deadlines, and personal bias.

Kinds of Models

Models used for IRM are usually labeled according to several major characteristics that are easily identified. Some models can be solved without a computer while others require a computer. Each model is based on a particular style of mathematics. Characteristics that currently seem important are discussed below.

One conceptual division of models is based on whether a model incorporates variability in processes or conditions. This variability is formally represented by probability distributions in stochastic models that account for variability within a population or variability in the environment. Deterministic models do not contain any probability distributions with random variables. They are often justified by focusing on mean values for large populations. Stochastic models contain one or more functions that are based on a random variable with a random number generator providing a value each time the function is calculated. The underlying probability distributions are either based on data or assumed to be a particular type.

Sometimes choices are made regarding the use of linear and nonlinear functions in models. Linear functions consist of multipliers of a variable such as insect density that do not change as density changes. For example, if oviposition is linear with respect to female density, then 1 female will lay 10 eggs and 100 females will lay 1000 eggs. A nonlinear function causes density, for instance, to produce different effects: 1 female lays 10 eggs but 100 females only lay 500 eggs. Well-known nonlinear functions are (a) dose response for toxin mortality in each genotype, (b) density-dependent, competition-based survival, (c) mating, (d) density-dependent dispersal by larvae or adults, and (e) oviposition. Modelers typically explain why they include or ignore density-dependent and other nonlinear functions.

To aggregate or disaggregate, that is the question, at least in many situations and models. Aggregation combines variables into one or a few variables or reduces space from many units into one or a few patches. With aggregation we omit age or stage structure. Disaggregation enables the modeling of each life stage, each sex, or each genotype. The chosen level of simplification should be based on the purpose of the model and the availability of data to support the disaggregation.

Choices must be made concerning the representation of time and space. Modelers can choose to make time discrete or continuous. Typical time steps for calculation in IRM models are a day or a pest generation. Several choices can be made concerning the representation of space. Often space is discretized into explicit patches, but it can also be continuous or essentially one location without dimensions. Modelers must choose to make space either homogeneous or heterogeneous. For heterogeneous landscapes, models usually consist of many units; each unit representing a plant, a plot, a field, or some other rational area.

Simplicity, Generality, Realism, and Precision

The goal helps us focus on the real ecological system and sharpens the focus on the major components of interest. However, you must always remember that

> Every model of the same system has the same total number of implicit and explicit assumptions.

One model of a pest and its environment may explicitly include only one natural enemy and only temperature as a climatic influence. Another model of the same pest may include two natural enemies and no weather. A third may not include natural enemies, but it may focus on the variety of host plants, some with partial resistance. Of course, real systems include most, if not all, of these factors. And each model rests on implicit assumptions about the influence of factors not explicitly described with mathematical functions.

The recognition of implicit and explicit assumptions is particularly important for evaluating the representation of space in IRM models. Every model that does not explicitly represent space and its heterogeneity implicitly considers space to be either uniform or random depending on the explicit functions for the modeled components.

In general, if another model can add a function (explicit assumption) for a new process or component, then another model without that function can be considered to have one more implicit assumption regarding the same ecological system. In other words, when we state that a modeler or model ignores some aspect of the system, then we are also implying that an implicit assumption has been made about that factor.

Why is this viewpoint important? Some modelers who make simple models do not mention the many implicit assumptions that they had to make about the real system. This is a serious concern when they claim that the simple model is generally applicable to many ecological systems. These same modelers claim that complex models have too many explicit functions. Yet the viewpoint expressed above sees both simple and complex models having the same number of assumptions. Thus, both kinds are complex. One kind is just more easily solved mathematically rather than numerically on a computer.

Onstad (1988) addressed the issues of simplicity, generality, realism, and precision. He concluded that generality is not a property of a model that can be identified, nor did he believe it could be proclaimed at the time of a model's creation. A model earns a designation of being general after it has been tested against many systems. A model's accuracy and precision can be evaluated when it is tested against independent data. Complex and simple models can produce the same degree of accuracy and precision. Modeling and experimentation in basic science are similar because each is a simplification of reality. Both can be faulted at times for being too simple or unrealistic.

Since all models are simplifications of reality, can one model be more real than another? Certainly! A model that permits all individuals to reproduce without accounting for the differences of males and females and immatures and adults cannot be considered as real as a model that has more realistic reproduction. For example, exponential growth of a population may seem to work fine at some gross scales of time, but the same function cannot be used when only immatures are alive over a given period.

Models should be made as realistic and as simple as possible to achieve the goal.

The realism of a model can usually be determined at its creation. Its generality and precision must be determined as it is tested. The value of a model and the question of whether it should be more or less simple are determined by how well it helped the modeler achieve the goal and satisfy the stakeholders.

Validation

As mentioned above, validation helps modelers and stakeholders determine the accuracy, precision, and credibility of a model relative to its purpose. If a model is tested against data from a variety of systems, then its generality can also be determined. Individual processes, such as larval survival or interfield dispersal of males, can be tested, or the predictions and observations of major variables representing the dynamics of an entire system over time and space (e.g., allele frequency, pest density) can be compared. The comparison of predictions to observations may simply involve evaluations of qualitative patterns that can be presented in plots of model results and field data. Or the validation may involve statistical analysis of the quantitative data and model output. Tabashnik (1990) provides a good review of efforts to validate or test IRM models.

Two issues are important in validation of IRM models. First, it is often difficult to collect the allele-frequency data over enough pest generations (years) to test predictions in the field. For species that can be reared under laboratory conditions, there is greater potential for extensive model validation. For other species, partial validation through testing of population dynamics and individual processes is usually the most that can be expected. Second, models can only be tested against independent data collected in the field, greenhouse, or laboratory. Data are independent when they have not contributed to the calibration or construction of the model. Early validation studies for specific systems were performed by Taylor *et al.* (1983), Tabashnik and Croft (1985), Tabashnik (1986), Denholm *et al.* (1987), and Mason *et al.* (1989).

More recently, several modelers have demonstrated proper and informative validation techniques. Storer (2003) created a spatially explicit stochastic IRM model primarily to study the evolution of resistance by *Diabrotica virgifera virgifera* to transgenic insecticidal corn. He was able to test several ecological processes against field data of population dynamics. Furthermore, he tested the entire model against historical field data concerning the pest's evolution of resistance to specific insecticides. Carriere and Roff (1995) tested a simple model simulating the simultaneous evolution of diapause propensity and insecticide resistance in *Choristoneura rosaceana*. Differences between the optimal solution to the model and the field data allowed them to explain the population genetics of the system. The IRM model created by Zhao *et al.* (2005) failed to match the population dynamics of *Plutella xylostella* in greenhouse studies, but the predicted changes in allele frequencies were generally similar to those observed over 26 generations. Boivin *et al.* (2005) created a phenological model of *Cydia pomonella* that represented populations of a susceptible and two insecticide-resistant homozygous genotypes. Model simulations for each genotype were compared with pheromone trap catches recorded in field populations over an eight-year period.

Sensitivity Analysis

A sensitivity analysis permits the modeler and stakeholders to determine how sensitive the model results are to small changes in model parameters. The less sensitive the model results are to these changes, the more confidence we have in the decisions that are made based on the model. Model results and recommendations are considered robust when they are not sensitive to changes in the model.

Usually, modelers focus on the parameters that they are uncertain about. When model results are sensitive to small changes in a parameter, such as fecundity or survival, and the values are not known with great certainty, the sensitivity analysis can be used to guide future data collection and experimentation to reduce the uncertainty.

The sensitivity analysis will likely be influenced by the modeler's goal. Thus, the goal and the procedures for every analysis should be clearly presented in reports. Sensitivity analyses can also investigate synergistic effects due to changes to multiple parameters and the substitution of mathematical functions representing complete ecological or genetic processes.

Risk Assessment for IRM

Risk assessment is important in IRM because stakeholders need to understand the consequences of implementing alternative IRM strategies in an uncertain future (Andow and Zwahlen, 2006). Modeling can contribute to these assessments. What is the risk that a population will evolve resistance to a management tactic? Associated with this risk are the economic risks that farmers face when a pest evolves resistance (Chapters 2 and 13) and the risk that substituted tactics will harm the environment. The three basic steps of a risk assessment—problem formulation, analysis, and risk characterization—are essentially the same as those used in an IRM modeling project (Jensen and Bourgeron, 2001). They both start with stakeholders selecting a goal and accepting system boundaries for the assessment. The time horizon and the spatial scale must be clear (Chapters 2 and 5). Ecological risk assessment identifies the management tactic (the stressor) as the threat to the pest population (the receptor). In other risk-assessment jargon, the endpoint consists of the entity, which is the targeted population; and the characteristic of the population that is measured is the allele frequency or some other measure of resistance.

The IRM model calculates the population's exposure to the tactic and the level of selection pressure (the effect on the stressor). With deterministic models, the emphasis is on the expected or mean threat, exposure, and selection pressure, whereas, with stochastic models, the modeler can use the probabilities of a range of threats, exposures, and selection events to calculate the probabilities of change to the population. Of course, determining the probabilities of the threats, exposures, and selection events can be difficult.

STOCHASTIC MODELS AND UNCERTAINTY ANALYSIS

Stochastic models have been used for predicting the consequences of IRM for several decades (Caprio, 1994; Caprio and Tabashnik, 1992; Caprio and Hoy, 1994; Caprio et al., 2006, 2008; Edwards et al., 2013; Peck et al., 1999; Sisterson et al., 2004, 2005; Storer, 2003; Storer et al., 2003a,b). A stochastic model has parameter values that change randomly either within each simulation (over time and/or space) or from one simulation to the next while keeping the parameter constant within each simulation. An example of the former is a mortality rate that changes each time step (e.g., day); an example of the latter is a mortality rate that varies between simulations but does not change in each time step.

Caprio et al. (2008) produced an excellent overview of issues, examples, and challenges concerning the use of stochastic models of resistance evolution. Uncertainty is inherent in all risk assessments, and these authors provided an excellent guide to the use of uncertainty analysis. According to Caprio et al. (2008), to be useful for risk assessments, models need to produce probabilistic outputs represented by distributions for each endpoint of interest.

An uncertainty analysis concerning model parameters in a deterministic model is similar to the second type of stochastic model. The goal of an uncertainty analysis is not just to obtain a prediction or an explanation, but to explore the influence of the uncertainty in many parameters on the reliability and credibility of model results. Uncertainty analyses for stochastic models require that the fundamental parameters for each probability distribution be drawn from another probability distribution representing the uncertainty. For instance, an uncertainty analysis pertaining to stochastic fecundity changing over time according to a normal distribution would require that the mean and variance each be selected from other distributions at the start of each simulation.

Three major factors contribute to uncertainty in model results. Once the design or set of biological processes have been chosen for the model, uncertainty exists in two factors: choice of mathematical function and parameter value. Nevertheless, model design, the third factor, cannot be selected with certainty because all models are simplifications of reality. A committee of the National Research Council (NRC, 2007) concluded that model-design uncertainty might have a much greater impact on output uncertainty than the other two types. The choice of mathematical function to represent a process involves less uncertainty because the choice is likely constrained by other functions in the model (overall logic), but the uncertainty is still very large.

Some scientists define two major contributions to parameter uncertainty: variability and incertitude. Variability in data is caused by variation in reality; precision of our estimates is based on our sampling this variability. Incertitude is the rest of the uncertainty due to what we do not know and have not sampled. Parameter uncertainty is constrained the most by logic of the mathematical function containing the parameter and by the potential to make observations. For a few conditions, we can measure variability in data. But we still may be uncertain about the parameter value for other conditions that may be simulated with the overall model. A major limitation of variability analysis can be the lack of representativeness of the data (U.S. Environmental Protection Agency [EPA], 1997).

Much uncertainty is due to modelers and decision makers attempting to use a model for many situations. Some situations never existed before, while others exist in different locations. One example is extrapolation of a model beyond the range of data that was used to calibrate the model. Extrapolation is common and often necessary because it is expensive and possibly infeasible to measure components of a system under all conditions. Prediction over time also requires use of a model to represent periods not originally studied. It is possible that parameter values measured at one time will not be constant in the future.

The following principle regarding modeling implies that one can easily and routinely overwhelm and dominate a modeling effort with uncertainty.

Given that a system can be modeled with almost infinite detail, the possible number of parameters and mathematical functions are also close to infinite.

If one knows the mean and variance for the result of a process, then uncertainty in subprocesses is constrained by the overall knowledge of variability in that process. However, if a process is known to occur but we have no information about it, then it contributes to uncertainty in the model result. Does the uncertainty in a model result increase if we explicitly model an increasing number of subprocesses? For example, we know that mating occurs, but we do not know how or when or where. We can model it as simply as possible with many implicit assumptions and two explicit parameters, or we can model it with 20 explicit parameters in three explicit functions. The presence or absence of moderate to strong correlations or dependencies between the input variables must be discussed and accounted for in the uncertainty analysis (EPA, 1997).

Every mathematical modeling approach has limitations. Therefore, we must determine the logical and conceptual limitations of any approach used for uncertainty analysis. Limitations involve extrapolation beyond the range of existing data, difficulties using expert opinion, and other issues. The value of probabilistic models, as with any model, depends on their credibility and the benefit that model results provide to the decision-making process. They can provide more information to decision makers, but how valuable is this extra set of numbers? What is the quality of results? In the end, we need an evaluation process for these models that is transparent and useful for determining credibility and utility.

In 2007, a committee formed by the NRC studied the use and evaluation of models used by the EPA for regulatory decision making (NRC, 2007) and made the following statement: "A wide range of possibilities is available for performing model uncertainty analysis. At one extreme, all model uncertainties could be represented probabilistically, and the probability distribution of any model outcome of interest could be calculated. However, in assessing environmental regulatory issues, these analyses generally would be quite complicated to carry out convincingly, especially when some of the uncertainties in critical parameters have broad ranges or when the parameter uncertainties are difficult to quantify. Thus, although probabilistic uncertainty is an important tool, requiring EPA to do complete probabilistic regulatory analyses on a routine basis would probably result in superficial treatments of many sources of uncertainty. The practical problems of performing a complete probabilistic analysis stem from models that have large numbers of parameters whose uncertainties must be estimated in a cursory fashion" (pp. 163–164).

An Example of Uncertainty Analysis

This section provides an example of what the NRC committee was concerned about and demonstrates that, when the distributions are not based on data, it can be difficult to know which and how many distributions to use. In the scenarios presented below, either the model design or the mathematical functions, or both, is different. For each parameter, the decision is made to make it a constant or allow it to vary according to a probability distribution represented by a mode, mean, shape, and extremes. Interactions among random variables are considered.

Model description

The following model is the simplest possible model of evolution for a single di-allelic gene (one major gene for resistance) in a landscape of toxic habitat and refuge (Tabashnik *et al.*, 2004, Chapter 5). Mating is random, and eggs are deposited uniformly across the crop landscape. The fitnesses for the three genotypes are WTss, WTrs, and WTrr in the toxic crop habitat with proportional area T, and WRss, WRrs, and WRrr in the refuge having proportional area $R = (1-T)$. In all scenarios and simulations, WRss equals 1. The values of WTss, WTrr, and WRrr are presented in Table 14.2 and explained in the Scenarios and Parameters section below. The fitnesses of the heterozygotes are calculated with

$$WRrs = hR \times WRrr + (1 - hR) \times WRss \qquad (14.1)$$

$$WTrs = hT \times WTrr + (1 - hT) \times WTss \qquad (14.2)$$

where hR and hT are the dominance parameters for the resistance allele in each habitat. A value of $h = 0$ means that the allele is recessive. The total fitness for each genotype is calculated with

$$Wg = T \times WTg + R \times WRg \qquad (14.3)$$

for each genotype g. The mean population fitness is

$$\Omega = p^2 \times Wss + 2pq \times Wrs + q^2 \times Wrr \qquad (14.4)$$

where q is the proportion (frequency) of the resistance allele in the population and $p = (1-q)$. The change in q each generation is

$$\Delta q = pq[q(Wrr - Wrs) + p(Wrs - Wss)]/\Omega \qquad (14.5)$$

The initial allele frequency is q0. Only q, p, and Ω change over generations.

Table 14.2 Scenarios, Functions, and Parameters

Scenario	q0	WTrr	hT	WRrr	hR
			Parameter		
1	$0.0001-0.01^{a}$ 0.001^{b}	1	$0-1$ 0.05	1	0
2	$0.0001-0.01$ 0.001	1	$0-1$ 0.05	$0.7-1$ 0.85	$0-1$ 0.05
3	$0.0001-0.01$ 0.001	$0.4-1$ 0.7	$0-1$ 0.05	1	0
4	$0.0001-0.01$ 0.001	$0.4-1$ 0.7	$0-1$ 0.05	$0.7 + (1\text{-WTrr})/2$	$0-1$ 0.05
5	$0.0001-0.01$ 0.001	$0.4-1$ 0.7	$0-1$ 0.05	$0.7 + (1\text{-WTrr})/2$	0.05
6	$0.0001-0.01$ 0.001	$0.4-1$ 0.7	$0-1$ 0.05	$0.7-1$ $0.7 + (1\text{-WTrr})/2$	$0-1$ 0.05
7	$0.0001 + (1\text{-WTrr})$ x0.0165	$0.4-1$ 0.7	$0-1$ 0.05	1	0
8	$0.0001-0.01$ $0.0001 + (1\text{-WTrr})$ x0.0165	$0.4-1$ 0.7	$0-1$ 0.05	1	0
9	$0.0001-0.001$ 0.001	1	$0-0.5$ 0.05	1	0
10	$0.0001-0.01$ $0.0001 + (1\text{-WTrr})$ x0.0165	$0.4-1$ 0.7	$0-1$ 0.05	$0.7-1$ $0.7 + (1\text{-WTrr})/2$	$0-1$ 0.05

[a]Range of values in probability distribution.
[b]Most likely value in PERT distribution.
All scenarios and simulations use WRss = 1 and a PERT distribution for WTss with $0-0.1$ and 0.001 as most likely, including those that use uniform distribution for other parameters.

Probability distributions

I use the PERT and uniform distributions to represent the possible values of parameters. The uniform distribution is based on a minimum and maximum value with the mean exactly in the middle. The PERT distribution is a Beta distribution with three parameters: minimum, maximum, and most likely (Vose, 2000). The mode of the PERT distribution matches the most likely value. The mean is calculated from (min + 4 x most likely + max)/6, which, as Vose notes, causes the mean to be four times more sensitive to the most likely value than to the minimum or maximum. Furthermore, the standard deviation of the PERT distribution is also less sensitive to the estimates of the extremes.

Scenarios and parameters

There are ten scenarios (Table 14.2). Scenario 1 represents the typical set of functions and parameters used in the most basic IRM models. There is no fitness cost due to resistance, and it is assumed that WTrr = 1. These are conservative, but very common, assumptions that likely cause the model to calculate fast evolution of resistance. The most likely values for WTss and hT represent high-dose scenarios with mostly recessive resistance, moderate but less conservative assumptions. Scenario 2 adds a probabilistic fitness cost that is mostly recessive (in PERT simulations) for resistant pests in the refuge. Scenario 3 omits the fitness cost but makes WTrr a probabilistic function with values as low as 0.4, representing cases of incomplete resistance. Scenario 4 extends Scenario 3 by bringing fitness cost back into the model as a deterministic function of WTrr.

$$\text{WRrr} = 0.7 + (1 - \text{WTrr})/2 \qquad (14.6)$$

Several authors believe that the greater the effect on resistance, the greater the fitness cost (lower WRrr) of the resistance allele (Lande, 1983; Groeters and Tabashnik, 2000). Scenario 5 is a comparison for Scenario 4: Dominance for fitness cost becomes a constant instead of a probabilistic function. Scenario 6 converts the deterministic function for fitness cost (Scenario 4) into a probabilistic function that uses that same function of WTrr to calculate the most likely value in the PERT simulations.

Scenario 7 replaces the probabilistic function for initial allele frequency with a deterministic function of WTrr (Table 14.2).

$$q0 = 0.0001 + (1 - \text{WTrr}) \times 0.0165 \qquad (14.7)$$

Several authors believe that the greater the effect on resistance, the rarer the resistance allele (Lande, 1983; Groeters and Tabashnik, 2000). Scenario 8 converts the deterministic function for q0 (Scenario 7) into a probabilistic function that uses that same function of WTrr to calculate the most likely value in the PERT simulations. Scenario 9 represents a system with a history of transgenic insecticidal crop use with no observation of resistance. The lack of observed field resistance permits us to assume that the maximum values of q0 and hT should be lower than those used in the other scenarios. (More common resistance alleles and higher dominance would cause resistance to evolve more quickly.) Scenario 10 combines the two functions of WTrr into one model to calculate the most likely values of q0 and WRrr in PERT simulations (Table 14.2).

Simulations of the first nine scenarios were repeated with a uniform distribution representing all probabilistic functions, except for the one for WTss. In many cases, scientists will not know the most likely value for WTrr, hT, WRrr, hR, and q0. However, we may have some confidence in our choices of minima and maxima. This type of situation justifies the use of uniform distributions for parameter values. For example, selection of standard values in IRM models is based more on tradition and conservative bias than on any theoretical or empirical justification. Generally, we know much more about control mortality for wild-type, susceptible pests than we do for any other population genetic parameter. Before commercializing an insecticide or insecticidal crop, industry and public scientists experimentally determine the control mortality. Therefore, I continued to use the PERT distribution for WTss in this second set of simulations.

The proportion of the landscape that is transgenic insecticidal crop is 0.8. Thus, 20% of the local landscape is refuge.

Simulation approach

Model results were calculated using Monte Carlo simulations for each scenario. Each simulation lasted 100 generations. I recorded the value of q at the end of the fifteenth generation and the generation during which q first exceeds 0.5. I calculated the proportion of simulations that had q first exceed 0.5 after the end of the fourteenth generation.

A preliminary analysis of the distribution of generations required for $q > 0.5$ in the first scenario permitted me to determine how many Monte Carlo simulations to calculate per scenario. With five replicates of 1000 simulations for the PERT distribution, I obtained values of 0.381, 0.345, 0.360, 0.326, and 0.346, with a mean of 0.35. With five replicates of 10,000 simulations for the PERT distribution, I obtained values of 0.366, 0.364, 0.359, 0.363, and 0.361 with a mean of 0.36. Based on the latter results, I decided to run one replicate of 10,000 simulations per scenario. Similar variability was found in simulations for Scenario 2 and for the uniform distribution in both scenarios. In other projects, more replicates based on unique seeds for the random number generator may be needed.

In the sensitivity analysis, I evaluated scenarios 1, 2, 3, and 9. I varied each parameter by 10% of its original value (i.e., 0.1 becomes 0.09 or 0.11) and determined its effect on the proportion of simulations in which q exceeded 0.5 after generation 14. For values of 0 for q0, hT, and hR, I simply increased the minimum to 0.01. For WTss, I increased the 0 to 0.0001 to study the sensitivity of the minimum. The sensitivity analysis was repeated, with the model using the uniform distribution for most parameters.

Results

There are two modes in the distributions of q for generation 15. This pattern was repeated for all scenarios and for both distributions. For the PERT-based analysis of Scenario 1, more than 20% of simulations had $q < 0.1$ and circa 60% had $q > 0.9$ by generation 15. For the analysis of Scenario 1 based on a uniform distribution, fewer had any change in resistance-allele frequency (5% with $q < 0.05$), while many more simulations had $q > 0.9$ (about 90%). In all scenarios, the cumulative density functions for PERT simulations were much different from those for uniform distributions in terms of the first generation in which q exceeded 0.5.

The results in Table 14.3 highlight the differences between the two types of distributions. The higher proportions in Table 14.3 signify slower evolution of resistance. Scenario 1 actually is a relatively conservative parameter set. Resistance evolved fastest with uniform distributions and second fastest with PERT distributions in Scenario 1 (Table 14.3). [With $q0 = 0.0001$, resistance evolves at one of the slowest rates; 67% and 24% of simulations resulted in $q > 0.5$ after generation 14 for the PERT and uniform distributions, respectively.] Adding a fitness cost to Scenario 2 with most likely values WRrr = 0.85 and hR = 0.05 did not change the results very much. As expected, allowing incomplete resistance in Scenarios 3−6 slowed evolution of resistance (Table 14.3). The adjustments related to fitness cost in Scenarios 4−6 had little effect. The change to calculation of q0 in Scenarios 7−8 had a greater effect on results when uniform distributions were used. The uniform-distribution based simulations produced slower evolution relative to the basic scenario compared to those simulated with PERT distributions.

Scenario 9 slowed the evolution of resistance as expected (Table 14.3), but the magnitude of the differences with Scenario 1 were not anticipated. More than a third and two-thirds of the simulations produced evolution of resistance after generation 14 when uniform and PERT distributions were used. The maximum initial allele frequency and the maximum dominance of resistance were reduced significantly. Thus, a lack of historical resistance evolution may allow us to use parameter values that suggest slow evolution in the future (assuming conditions remain the same).

Table 14.3 Proportion of Simulations that Result in the Frequency of Resistance
Allele q Exceeding 0.5 After 14 Generations

| Scenario | Distribution | | | |
	PERT	Rank	Uniform	Rank
1	0.36	8	0.09	8
2	0.38	7	0.11	7
3	0.57	3	0.16	5
4	0.59	2	0.19	3
5	0.57	3	0.17	4
6	0.59	2	0.20	2
7	0.42	6	0.15	6
8	0.43	5	0.16	5
9	0.68	1	0.36	1
10	0.47	4	0.20	2[a]

[a]Scenarios 6 and 10 are equivalent for the uniform distribution.

Evolution occurred more quickly in Scenario 10 than in Scenario 1 when a PERT distribution was employed (Table 14.3). However, because Scenarios 6 and 10 are equivalent with respect to the simulations using the uniform distributions, evolution of resistance is not increased because of the extremes of the distributions. Changes to WTrr and fitness costs should have slowed evolution, but the recessive nature of the fitness cost and the fivefold higher, most likely value of $q0$ counteracted these effects. This is also the only difference between Scenarios 6 and 10 with respect to the PERT distributions. I omitted the probability distribution for $q0$ and set $q0 = 0.005$ as a constant. This adjustment to Scenario 10 produced 47% of the simulations with $q > 0.5$ after generation 14. Clearly, the expected initial allele frequency is the primary factor influencing the results in Scenarios 6 and 10.

In the sensitivity analysis, few of the 10% changes in the parameter values caused more than a 0.01 change in the proportion of simulations in which q exceeded 0.5 after generation 14. Even fewer 10% changes in parameters caused more than a 10% change in the result. In Scenario 1, a 10% reduction in WTrr from 1 to 0.9 caused the proportions to change from 0.36 to 0.41 with the PERT distributions and from 0.09 to 0.10 with the uniform distributions. When the maximum of $q0$ was increased by 10% from 0.01 to 0.011, the proportion in the output increased to 0.38 with the PERT distributions. The third significant effect seen under the first scenario occurred when maximum hT was reduced from 1 to 0.9.

This raised the proportions to 0.40 and 0.10 for the PERT and uniform cases, respectively. For Scenarios 2, 3, and 9 I focused on parameters that were unique to those scenarios. In Scenario 2, a 10% reduction in the most likely value of WRrr changed the result from 0.38 to 0.40 with the PERT distributions. In Scenario 3, the change in the most likely value of WTrr decreased the result to 0.52 or increased it to 0.61. In Scenario 9, 10% reductions in maximum hT and most likely hT caused the results to change from 0.68 to 0.73 and 0.71, respectively, with the PERT distributions. Simulations with PERT distributions were generally more sensitive than those with uniform distributions.

How can probabilistic modeling help IRM experts and regulators make better decisions? If the models are based on good knowledge of the pest system, then we can expect them to at least contribute as much as deterministic models. In general, the better the data, the better the models. Unfortunately, when IRM decisions must be made at the beginning of commercialization, we usually do not have satisfactory knowledge and data. Therefore, we tend to rely on expert opinion about worst-case scenarios and most likely values. Sometimes professional and experiential biases lead scientists to promote most likely values for parameters that are either conservative or personal favorites. Ultimately, many of these suggestions are guesses, particularly those about heterozygous and homozygous resistant insects. Fortunately, we can usually take advantage of data and even empirical distributions for the survival of susceptibles exposed to the toxin, WTss, at doses encountered in the field.

I purposefully chose the simplest possible model to explore because I did not want the messages to be hidden by the details of more complicated models. Caprio *et al.* (2008) used a similar model to evaluate two scenarios to demonstrate that model results are very sensitive to choice of parameter distribution. Certainly, more complicated models would likely require even more parameters and additional evaluation. Fortunately, many demographic parameters relevant to population dynamics are known with much more certainty or at least have empirical distributions that can support stochastic models. Thus, another approach to uncertainty analysis could involve the development of probabilistic functions for demographic parameters while keeping the guesses as constants.

The results of this simple modeling exercise clearly show that the benefits of probabilistic modeling may be obscured or negated by the difficulty in choosing probability distributions, selecting parameters, and maintaining the scientific logic that may not have been an issue when constants and functions were used in deterministic models. The probability of resistance evolving after 15 generations varied from 9 to 68% across the ten scenarios (Table 14.3), indicating that model design and the choice of mathematical function are critically important . Thus, even an expert should consider more than one version of nature in a probabilistic model. Most of the ten scenarios evaluated in this study are simple yet profound adjustments in the ecological and genetic logic commonly utilized by IRM experts and modelers. Some adjustments caused major changes in the distributions of results that would significantly influence the development of an IRM strategy.

There were serious differences in results for simulations based on PERT and uniform distributions (Table 14.3). Caprio *et al.* (2008) repeatedly state that more than one probability distribution should be considered to represent distributions of parameters. In fact, they list a variety of well-known mathematical forms. Of course, models can be hybrids of the two, or they could use one of the many well-known distributions. When we have more confidence in the most likely value as the most common value in nature, and not as a biased standard, then the PERT distribution is a reasonable choice. However, the PERT distribution should be used with caution if the most likely value is a complete guess (Vose, 2000).

Caprio *et al.* (2008) caution the reader to be aware of possible correlations and interactions between parameters. They warn modelers that risk estimates can be highly sensitive to the type of input distribution and that there is considerable inherent uncertainty. They warn decision makers that uncertainty distributions will often contain elements of subjectivity. Despite their own warnings about subjectivity and inherent uncertainty in the chosen probability distributions, Caprio *et al.* (2008) repeatedly declare that stochastic and probabilistic modeling are the only useful approaches and are necessary for risk assessment. These statements put them at odds with the NRC committee that wrote the book on evaluation of models for environmental regulatory decision making. The NRC committee concluded that deterministic scenario analysis (described at end of this chapter) has a valuable role in decision making (NRC, 2007, p. 164).

Uncertainty analysis can be a powerful tool, but it can be used foolishly. Probabilistic modeling is more work for the modeler, or at least it should be if done properly. More must be added to a model, more must be explained, and more must be verified and evaluated. Modelers are taught not to include functions and parameters that are not related to the goal of the project or to the hypothesis to be tested. Does the doubling or tripling of parameters in uncertainty analyses or stochastic models contribute to goals and add clarity, or does it simply triple the confusion? Uncertainty analysis based on probabilities can be useful if it is focused and narrowly applied to cases in which the results will identify which processes require greater understanding and will suggest experiments that should be performed with greater precision and effort to improve our understanding. Otherwise, a simple sensitivity analysis can be performed to evaluate the importance of parameters and functions that cannot be feasibly measured or determined before a decision is made. The EPA (1997) concluded that probabilistic assessments should be restricted to analyses of significant pathways and parameters identified in sensitivity analyses. This was supported by the NRC committee (2007, p. 136): "it may not be appropriate for analysts to attach probability distributions to critical quantities that are highly uncertain, especially if the uncertainty is itself difficult to assess."

IRM MODELS

The historical development of modeling for IRM coincides with the development of scientific IRM strategies (Taylor, 1983; Tabashnik, 1990). If models did not provide the foundation of IRM strategies, their results at least supported them from the beginning. Sometimes abstract models have been used to study the evolution of resistance and the consequences of management practices (Taylor, 1983; Tabashnik, 1990). Several abstract models are discussed in other chapters.

Various authors have advocated the creation of species and system-specific models to develop the most credible results for applied IRM (Roush and McKenzie, 1987; Kennedy et al., 1987; McKenzie and Batterham, 1994; McKenzie, 1996). They generally support the points made about realistic models discussed above. The following two topics demonstrate how models can be used to explore issues not normally considered in traditional population-genetics models.

Effects of Pest Phenology

Models can be used to investigate how pest phenology influences the efficacy of IRM tactics. Follett et al. (1993) developed a simulation model to predict the rate of evolution in *Leptinotarsa decemlineata* on potato (*Solanum tuberosum*). DeSouza *et al.* (1995) created a model for *Helicoverpa armigera* on cotton (*Gossypium hirsutum*). Both studies showed that the timing of insecticide applications relative to the period in which the pest is in diapause has a significant effect on the evolution of resistance to insecticides. For some pests in some cropping systems, diapause can provide a temporal refuge. DeSouza *et al.* (1995) discovered that the effect of diapause was different in Australia and India; diapause conserved resistance in Australia but conserved susceptibility in India.

Ferro (1993) modeled the effects of a low-dose strategy for transgenic insecticidal potatoes that could delay maturation of *Leptinotarsa decemlineata*. He concluded that a low-dose potato crop could reduce the number of generations of *L. decemlineata* per year, but warned that nonrandom mating among susceptible and resistant beetles could occur because of the delays in maturation. Carrière *et al.* (2001) used a simulation model to demonstrate how the cultural-control tactics and cotton-planting date could be changed to delay evolution of resistance by *Pectinophora gossypiella* to transgenic insecticidal cotton. In essence, Carrière *et al.* (2001) showed that adult emergence in the spring should be the focus of both IPM and IRM.

Mitchell and Onstad (2005) developed a model for *Diabrotica barberi* to examine the effect of extended egg diapause on the evolution of resistance to transgenic insecticidal corn (*Zea mays*). They attempted to mimic conditions found in the center of the extended-diapause problem along the Minnesota–South Dakota–Iowa borders (Spencer, Hughson, and Levine, Chapter 7). Sensitivity analysis indicates that toxin dose and farmer management practices (e.g., insecticide use on refuge corn, rotation pattern) generally have a larger impact on the evolution of resistance than many parameters concerning population dynamics and genetics. In the region where extended diapause already exists, increasing extended diapause (increasing hatch rates after two and/or three winters while holding total hatch constant), tends to increase resistance to transgenic insecticidal corn. However, this is not always the case, since combinations of rotation pattern, toxin dose, and soil insecticide use exist for which the net effect of extended diapause decreases resistance to insecticidal corn (Mitchell and Onstad, 2005).

Some models even explore the simultaneous evolution of pest phenology and resistance to a toxin. Carrière and Roff (1995) used two models to study the simultaneous evolution of diapause propensity and insecticide resistance in *Choristoneura rosaceana*. One model was based on ecological optimality theory, and the other was a quantitative-genetic model simulated with two threshold traits. The models predicted that more of the population would enter diapause early in the summer because of insecticide applications. Refuges occupied during diapause may permit larvae to escape insecticide. The evolution of the timing of events in a pest's life cycle will likely be an important research subject in the future.

Complex Biological Models with Simple Economic Analyses

Models that include economic factors, management processes, and even human behavior are described by Mitchell and Onstad (Chapter 2) and Hurley and Mitchell (Chapter 13). The following models combine complex biological models with relatively simple economic functions to calculate the benefits of certain strategies for IRM.

Gutierrez *et al.* (1979) were perhaps the first to combine complex biological processes and economics into a model for IRM. They modeled the population dynamics of the pest, the dynamics of crop growth, the population genetics of insecticide resistance, and the economics of harvested crop yield and insecticide use. Gutierrez *et al.* (1979) evaluated a scenario in which the insecticide is applied according to a predetermined schedule and a second scenario in which insecticide applications are adjusted each generation based on observations of pest density and resistance-allele frequency. They concluded that sampling (second scenario) delayed the evolution of resistance but not significantly.

Since the mid-1990s, Onstad has created a variety of deterministic simulation models to evaluate IRM for pests of corn. These data-based models tend to have intermediate complexity, with age structure and behavior explicitly incorporated. Some models have daily time steps and more details that permit hypotheses to be addressed concerning intragenerational issues. Alternative models for the same species have life stages and processes aggregated to permit the use of a time step equal to the period of the insect's generation. The purpose of these models was to study the evolution of the pest, to evaluate the alternative strategies and tactics for IRM, and to demonstrate how some IRM strategies economically compare to other IPM practices. In addition to the models described below, other modeling efforts by this group are described in Chapters 7 and 10.

Onstad's original focus was on IRM for *Ostrinia nubilalis* in transgenic insecticidal cornfields (Onstad and Gould, 1998a). In these deterministic models with a daily time step, mating is random in the landscape and eggs are oviposited uniformly across the cornfields. Onstad and Gould (1998a) evaluated block, row-strip, and corn-ear refuges, seed mixtures, and sequential planting of transgenic insecticidal and nontransgenic corn as IRM tactics. Onstad and Gould (1998a,b) created a model that simulated toxin concentration decline in plants during senescence. The daily time-step permitted the simulation of various synchronies and asynchronies between toxin dose and larval maturation. Seasonal dynamics of toxin concentrations significantly increased the risk of evolution of resistance. Onstad and Guse (1999) extended the model to include density-dependent larval mortality and the economics of crop damage. Onstad and Guse economically evaluated the block and row-strip refuges and concluded that a 20% refuge along with a constant high dose of toxin was a robust strategy for European corn borer IRM. Both studies emphasized the need for research on seasonal decline in toxin concentration in transgenic insecticidal crops before commercialization (Onstad and Gould, 1998b; Onstad and Guse, 1999).

Onstad *et al.* (2002) created a different model for *Ostrinia nubilalis* IRM that had a time-step of one generation. They used the model to study the effects of insecticide use in refuge corn on the evolution of resistance to both transgenic insecticidal corn and the insecticide. Figure 14.1 clearly demonstrates the importance of limiting insecticide use in the refuge. Evolution to transgenic insecticidal corn occurs faster as either the refuge size decreases or as the frequency of insecticide use in the refuge increases. For example, with a 20% refuge, the number of years to resistance decreases from 30 to 15 when the insecticide use changes form none (0.0) to every fourth generation (0.25), which is equivalent to one application every other year. A consistently sprayed refuge, consisting of less than 40% of the cornfields, was an inadequate IRM strategy for *Ostrinia nubilalis* even when a low-efficacy insecticide (70% mortality) was used.

Onstad *et al.* (2001a) investigated IRM for *Diabrotica virgifera virgifera* in cropping systems with transgenic insecticidal corn. They used a deterministic model that included two nonlinear functions: (1) density-dependent larval mortality due to competition, predation, or parasitism and (2) a density-dependent function for calculating the proportion of female beetles that mate each day in each field. Results indicated that toxin concentrations at intermediate levels in the transgenic plants would be worse for IRM than either higher or lower doses. Sensitivity analyses demonstrated that the function for density-dependent survival was important, but the nonlinear mating function had little effect on the results.

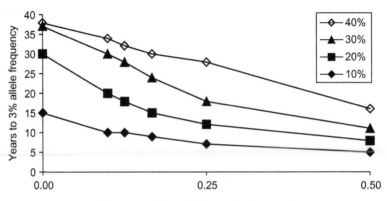

Frequency of insecticide application over time.

Figure 14.1 Number of years required for *O. nubilalis* to evolve resistance to trans-genic insecticidal corn (3% allele frequency) as a function of the proportion of gen-erations treated in the refuge for four refuge sizes. Initial allele frequency is 0.001, with a practical high-dose crop, partially recessive expression of the transgenic resis-tance allele, and 90% insecticide efficacy with no evolution of insecticide resistance. *Reprinted from Onstad* et al. *(2002) by permission of Entomological Society of America.*

Onstad *et al.* (2001b, 2003) modeled the evolution of behavioral resis-tance in *D. virgifera virgifera*. One of the most interesting results of the modeling by Onstad *et al.* (2001b) was the discovery that landscape diver-sity, represented by the proportional area of vegetation in the environ-ment that is not part of the annual rotation of corn and soybean (*Glycine max*), reduces the rate of evolution of rotation resistance. Essentially, in this very simple deterministic model, the oviposition on lands not rotated with cornfields is a fitness cost for resistant beetles. Onstad *et al.* (2001b) also used the 3-allele, single-locus model to determine that a soybean spe-cialist would be less fit than a polyphagous phenotype that oviposits on lands covered by a wide variety of vegetation.

Onstad *et al.* (2003) used a model to analyze the economics of IRM for rotation-resistant *D. virgifera virgifera*. They explored the economic consequences for six alternative IRM scenarios and compared them to the standard two-year rotation of corn and soybean. They concluded that IRM utilizing transgenic insecticidal corn in the two-year rotation is a robust, efficient strategy.

Crowder and Onstad investigated the simultaneous evolution of resis-tance by *D. virgifera virgifera* to crop rotation and transgenic insecticidal corn (Crowder and Onstad, 2005; Crowder *et al.*, 2005a,b, 2006). At the same time, they evaluated how management of the cropping system (design and control of what is planted, where, and when) influences IRM for both problems. Thus, the deterministic models, which were extensions of the models created by Onstad *et al.* (2001a, 2003), considered a two-gene system: one for behavioral resistance and one for toxicological resistance. These models included mating by females before dispersal to oviposit, mating by males before and after dispersal from the field of emergence, differential, density-independent, field-to-field dispersal based on gender and phenotype, and density-dependent larval survival. Crowder and Onstad determined that a robust strategy for delaying resistance to transgenic corn in areas where rotation resistance is a problem is to plant transgenic corn in rotated cornfields. This also helps delay resistance to crop rotation.

Crowder *et al.* (2005a, 2006) investigated the economics of IRM strategies for *D. virgifera virgifera* in areas with crop rotation and transgenic insecticidal corn. In areas with and without populations adapted to a two-year rotation of corn and soybean (rotation-resistant), the standard management strategy is the planting of 80% of a cornfield (rotated and continuous) to a transgenic corn hybrid each year. In each area, Crowder *et al.* (2005a) also studied dynamic management strategies where the adoption of transgenic corn increased over time in a region. They also analyzed management strategies for a single field that is the first to adopt transgenic corn within a larger unmanaged region. In all areas, increasing the expression of the toxin in the plant increased economic returns for farmers. In areas without rotation resistance, planting 80% transgenic insecticidal corn in the continuous cornfield each year generated the greatest returns with an intermediate or greater toxin dose. In areas with alleles for rotation resistance at low initial levels, a two-year rotation of nontransgenic corn and soybean may be the most economical strategy if resistance to crop rotation is recessive. If resistance to crop rotation is additive or dominant, planting transgenic insecticidal corn in the rotated cornfield was the most effective strategy. In areas where rotation resistance is already a severe problem, planting transgenic insecticidal corn in the rotated cornfield each year was always the most economical strategy. In some cases, the strategies that increased the proportion of transgenic insecticidal corn in the region over time increased returns compared with the standard strategies. With these strategies, the evolution of resistance to crop rotation occurred more rapidly, but resistance to transgenic insecticidal corn was delayed compared with the standard management strategy.

In areas not managed by a regional norm, increasing the proportion of transgenic corn and increasing toxin dose in the managed field generally increased returns.

Crowder *et al.* (2006) explored the use of sampling and economic thresholds to improve IPM and IRM when transgenic insecticidal corn is used for *D. virgifera virgifera* management. In the model, transgenic insecticidal corn was planted only when adult densities in the previous year were above a threshold calculated from single-season economics for IPM. The use of economic thresholds slightly slowed the evolution of resistance to transgenic insecticidal corn. In areas with or without rotation-resistant phenotypes, the use of sampling and economic thresholds generated similar returns compared with strategies of planting transgenic corn every season. Because many transgenic insecticidal crops are extremely effective, farmers may be inclined to plant transgenic crops every season rather than implementing costly and time-consuming sampling protocols.

CONCLUSIONS

Some IRM models will emphasize evolution of resistance, while others will focus on economic outcomes. For management of insect resistance, we need both types of models. In a single chapter, it is impossible to adequately describe or even represent all the important models that have been created to study and improve IRM; more models are described in other chapters.

So what can model analysis do for prediction and decision making, given all the uncertainty described above? Modeling allows the subject matter experts and the decision makers to synthesize limited information into a set of alternative views of the future. The modeling process encourages developers to incorporate important factors that may not always be considered, justify their inclusion, and justify the exclusion of other possibly important factors. Modeling should also identify important factors that are uncertain. Modeling forces everyone to confront the alternative realities simulated with the model. Decision making usually focuses on choosing from a small number of alternative scenarios. Therefore, we need to know whether the ranking of choices/decisions changes with consideration of uncertainty.

The EPA (1997) concluded that one approach to uncertainty analysis uses expert judgment to define and separately analyze alternate, but plausible, scenarios. Qualitative evaluations of uncertainty, including relative ranking of the sources of uncertainty, may be acceptable, especially when objective measures are not available (EPA, 1997). Scenario-planning techniques are fundamental to determining strategy under conditions of uncertainty, but scenario-based planning rarely includes the probabilities of future scenarios. Schwartz (1991) stated that modelers should avoid assigning probabilities to different scenarios. The NRC committee (NRC, 2007) proposed that a small number of key parameters having large, poorly characterized uncertainty be fixed at various plausible levels and then that probabilities be used to describe all other sources of uncertainty. This avoids the extremes of assigning probabilities to all factors or ignoring probabilities altogether. The committee stated that use of this scenario analysis may be worthwhile in some circumstances (NRC, 2007). Each plausible scenario would be simulated deterministically to produce a single result. The committee of the NRC (2007) stated, "For example, one might consider scenarios with such names as highly optimistic, optimistic, neutral, pessimistic, or highly pessimistic. This approach makes no formal use of probability theory and can be simpler to present to stakeholders who are not fully versed in probability theory and practice. One advantage of the scenario approach is that many of those involved in modeling activities, including members of stakeholder groups and the public, may attach their own risk preference (such as risk seeking, risk averse, or risk neutral) to such scenario descriptions. However, even using multiple scenarios ranging from highly optimistic to highly pessimistic will not necessarily ensure that such scenarios will bracket the true value."

One of the most important challenges for modelers is effectively communicating results from an uncertainty analysis (EPA, 1997). One of the main concerns of the NRC committee that studied modeling for environmental regulatory decision making was that (NRC, 2007, p. 136) "there are substantial problems in reducing the results of a large-scale study with many sources of uncertainty to a single number or even a single probability distribution. We contend that such an approach draws the line between the role of analysts and the role of policy makers in decision making at the wrong place." They reiterated this point on pages 164–165, "Effective decision making will require providing policy makers with more than a single probability distribution for a model result.

Such summaries obscure the sensitivities of the outcome to individual sources of uncertainty." The NRC committee (NRC, 2007, p. 165) warned decision makers and modelers that "Probabilistic uncertainty analysis should not be viewed as a means to turn uncertain model outputs into policy recommendations that can be made with certitude." The committee (NRC, 2007) concluded that "the notion that reducing the results of a large-scale modeling analysis to a single number or distribution is at odds with one of the main themes that began this chapter, that models are tools for helping make decisions and are not meant as vehicles for producing decisions. In sounding a cautionary note about the difficulties of both carrying out and communicating the results of probabilistic uncertainty analyses, we are trying to avoid the outcome of having models (and a probabilistic uncertainty analysis is the output of a model) make decisions."

Modeling can be a valuable tool in predicting the consequences of IRM strategies deployed under a given set of environmental, economic, and social conditions. This chapter was written to help scientists and stakeholders to understand how models are created, tested, and analyzed. Everyone should realize that models and associated analytical procedures must be considered in the same manner as reports about experiments. The reader must be able to duplicate the model and its calculation, judge its credibility, and interpret the analysis.

REFERENCES

Andow, D.A., Zwahlen, C., 2006. Assessing environmental risk of transgenic plants. Ecol. Lett. 9, 196−214.

Boivin, T., Chadoeuf, J., Bouvier, J.C., Beslay, D., Sauphanor, B., 2005. Modelling the interactions between phenology and insecticide resistance genes in the codling moth *Cydia pomonella*. Pest Manag. Sci. 61, 53−67.

Caprio, M.A., 1994. *Bacillus thuringiensis* gene deployment and resistance management in single and multi-tactic environments. Biocontrol Sci. Tech. 4, 487−497.

Caprio, M.A., Hoy, M.A., 1994. Metapopulation dynamics affect resistance development in a predatory mite. J. Econ. Entomol. 87, 525−534.

Caprio, M.A., Tabashnik, B.E., 1992. Gene flow accelerates local adaptation among finite populations: simulating the evolution of insecticide resistance. J. Econ. Entomol. 85, 611−620.

Caprio, M.A., Nowatzki, T., Siegfried, B., Meinke, L.J., Wright, R.J., Chandler, L.D., 2006. Assessing risk of resistance to aerial applications of methyl-parathion in western corn rootworm (Coleoptera: Chrysomelidae). J. Econ. Entomol. 99, 483−493.

Caprio, M.A., Storer, N.P., Sisterson, M.S., Peck, S.L., Maia, A.H.N., 2008. Assessing the risk of the evolution of resistance to pesticides using spatially complex simulation models. In: Whalon, M.E., Mota-Sanchez, D., Hollingworth, R.M. (Eds.), Global Pesticide Resistance in Arthropods. CABI International, Cambridge, MA.

Carrière, Y., Roff, D.A., 1995. The joint evolution of diapause and insecticide resistance: a test of an optimality model. Ecology. 76, 1497—1505.

Carrière, Y., Ellers-Kirk, C., Pederson, B., Haller, S., Antilla, L., 2001. Predicting spring moth emergence in the pink bollworm: Implications for managing resistance to transgenic cotton. J. Econ. Entomol. 94, 1012—1021.

Crowder, D.W., Onstad, D.W., 2005. Using a generational time-step model to simulate the dynamics of adaptation to transgenic corn and crop rotation by western corn rootworm (Coleoptera: Chrysomelidae). J. Econ. Entomol. 98, 518—533.

Crowder, D.W., Onstad, D.W., Gray, M.E., Mitchell, P.D., Spencer, J.L., Brazee, R.J., 2005a. Economic analysis of dynamic management strategies utilizing transgenic corn for control of western corn rootworm (Coleoptera:Chysomelidae). J. Econ. Entomol. 98, 961—975.

Crowder, D.W., Onstad, D.W., Gray, M.E., Pierce, C.M.F., Hager, A.G., Ratcliffe, S.T., et al., 2005b. Analysis of the dynamics of adaptation to transgenic corn and crop rotation by western corn rootworm (Coleoptera:Chrysomelidae) using a daily time-step model. J. Econ. Entomol. 98, 534—551.

Crowder, D.W., Onstad, D.W., Gray, M.E., 2006. Planting transgenic insecticidal crops based on economic thresholds: consequences for integrated pest management and insect resistance management. J. Econ. Entomol. 99, 899—907.

DeSouza, K., Holt, J., Colvin, J., 1995. Diapause, migration and pyrethroid-resistance dynamics in the cotton bollworm, *Helicoverpa armigera* (Lepidoptera:Noctuidae). Ecol. Entomol. 20, 333—342.

Denholm, I., Sawicki, R.M., Farnham, A.W., 1987. Laboratory simulation of selection for resistance. In: Ford, M.G., Holloman, D.W., Khambay, B.P.S., Sawicki, R.M. (Eds.), Combating Resistance to Xenobiotics: Biological and Chemical Approaches. Horwood, Chichester, England, pp. 138—149.

Edwards, K.T., Caprio, M.A., Allen, K.C., Musser, F.R., 2013. Risk assessment for *Helicoverpa zea* (Lepidoptera: Noctuidae) resistance on dual-gene versus single-gene corn. J. Econ. Entomol. 106, 382—392.

Environmental Protection Agency, 1997. Guiding principles for Monte Carlo analysis.

Ferro, D.N., 1993. Potential for resistance to *Bacillus thuringiensis*: Colorado potato beetle (Coleoptera:Chrysomelidae)—a model system. Amer. Entomol. 39, 38—44.

Follett, P.A., Kennedy, G.G., Gould, G., 1993. REPO: a simulation model that explores Colorado potato beetle (Coleoptera:Chrysomelidae) adaptation to insecticides. Environ. Entomol. 22, 283—296.

Glaser, J.A., Matten, S.R., 2003. Sustainability of insect resistance management strategies for transgenic Bt corn. Biotech. Adv. 22, 45—69.

Groeters, F.R., Tabashnik, B.E., 2000. Roles of selection intensity, major genes, and minor genes in evolution of insecticide resistance. J. Econ. Entomol. 93, 1580—1587.

Gutierrez, A.P., Regev, R., Shalit, H., 1979. An economic optimization model of pesticide resistance: alfalfa and Egyptian alfalfa weevil — an example. Environ. Entomol. 8, 101—107.

Jensen, M.E., Bourgeron, P.S., 2001. A Guidebook for Integrated Ecological Assessments. Springer-Verlag, New York.

Kennedy, G.G., Gould, F., Deponti, O.M.B., Stinner, R.E., 1987. Ecological, agricultural, genetic, and commercial considerations in the deployment of insect-resistant germplasm. Environ. Entomol. 16, 327—338.

Lande, R., 1983. The response to selection on major and minor mutations affecting a metrical trait. Heredity. 50, 47—65.

Mason, G.A., Tabashnik, B.E., Johnson, M.W., 1989. Effects of biological and operational factors on evolution of insecticide resistance in *Liriomyza* (Diptera: Agromyzidae). J. Econ. Entomol. 82, 369—373.

McKenzie, J.A., 1996. Ecological and Evolutionary Aspects of Insecticide Resistance. Academic Press, Austin.

McKenzie, J.A., Batterham, P., 1994. The genetic, molecular and phenotypic consequences of selection for insecticide resistance. Trends Ecol. Evol. 9, 166–169.

Mitchell, P.D., Onstad., D.W., 2005. Effect of extended diapause on the evolution of resistance to transgenic *Bacillus thuringiensis* corn by northern corn rootworm (Coleoptera: Chrysomelidae). J. Econ. Entomol. 98, 2220–2234.

National Research Council, 2007. Models in Environmental Regulatory Decision Making. National Academies Press, Washington, DC.

Onstad, D.W., 1988. Population-dynamics theory: the roles of analytical, simulation and supercomputer models. Ecol. Modell. 43, 111–124.

Onstad, D.W., Gould, F., 1998a. Modeling the dynamics of adaptation to transgenic maize by European corn borer (Lepidoptera:Pyralidae). J. Econ. Entomol. 91, 585–593.

Onstad, D.W., Gould, F., 1998b. Do dynamics of crop maturation and herbivorous insect life cycle influence the risk of adaptation to toxins in transgenic host plants? Environ. Entomol. 27, 517–522.

Onstad, D.W., Guse, C.A., 1999. Economic analysis of the use of transgenic crops and nontransgenic refuges for management of European corn borer (Lepidoptera: Pyralidae). J. Econ. Entomol. 92, 1256–1265.

Onstad, D.W., Guse, C.A., Spencer, J.L., Levine, E., Gray, M., 2001a. Modeling the adaptation to transgenic corn by western corn rootworm (Coleoptera: Chrysomelidae). J. Econ. Entomol. 94, 529–540.

Onstad, D.W., Spencer, J.L., Guse, C.A., Isard, S.A., Levine, E., 2001b. Modeling evolution of behavioral resistance by an insect to crop rotation. Ent. Expt. Appl. 100, 195–201.

Onstad, D.W., Guse, C.A., Porter, P., Buschman, L.L., Higgins, R.A., Sloderbeck, P.E., et al., 2002. Modeling the development of resistance by stalk-boring Lepidoptera (Crambidae) in areas with transgenic corn and frequent insecticide use. J. Econ. Entomol. 95, 1033–1043.

Onstad, D.W., Crowder, D.W., Mitchell, P.D., Guse, C.A., Spencer, J.L., Levine, E., et al., 2003. Economics versus alleles: balancing IPM and IRM for rotation-resistant western corn rootworm (Coleoptera: Chrysomelidae). J. Econ. Entomol. 96, 1872–1885.

Peck, S.L., Gould, F., Ellner, S.P., 1999. Spread of resistance in spatially extended regions of transgenic cotton: implications for management of *Heliothis virescens* (Lepidoptera: Noctuidae). J. Econ. Entoml. 92, 1–16.

Roush, R.T., McKenzie, J.A., 1987. Ecological genetics of insecticide and acaricide resistance. Annu. Rev. Entomol. 32, 361–380.

Royama, T., 1971. A comparative study of models for predation and parasitism. Res. Popul. Ecol. 13 (1), 1–91.

Schwartz, P., 1991. The Art of the Long View. Doubleday, New York.

Sisterson, M.S., Antilla, L., Ellers-Kirk, C., Carrière, Y., Tabashnik, B.E., 2004. Effects of insect population size on evolution of resistance to transgenic crops. J. Econ. Entomol. 97, 1413–1424.

Sisterson, M.S., Carrière, Y., Dennehy, T.J., Tabashnik, B.E., 2005. Evolution of resistance to transgenic Bt crops: interactions between movement and field distribution. J. Econ. Entomol. 98, 1751–1762.

Storer, N.P., 2003. A spatially explicit model simulating western corn rootworm (Coleoptera: Chrysomelidae) adaptation to insect-resistant maize. J. Econ. Entomol. 96, 1530–1547.

Storer, N.P., Peck, S.L., Gould, F., Van Duyn, J.W., Kennedy, G.G., 2003a. Spatial processes in the evolution of resistance in *Helicoverpa zea* (Lepidoptera: Noctuidae) to Bt transgenic corn and cotton in a mixed agroecosystem: a biology-rich stochastic simulation model. J. Econ. Entomol. 96, 156–172.

Storer, N.P., Peck, S.L., Gould, F., Van Duyn, J.W., Kennedy, G.G., 2003b. Sensitivity analysis of a spatially-explicit stochastic simulation model of the evolution of resistance in *Helicoverpa zea* (Lepidoptera:Noctuidae) to Bt transgenic corn and cotton. J. Econ. Entomol. 96, 173—187.

Tabashnik, B.E., 1986. A model for managing resisatnce to fenvalerate in diamondback moth (Lepidoptera: Plutellidae). J. Econ. Entomol. 79, 1447—1451.

Tabashnik, B.E., 1990. Modeling and evaluation of resistance management tactics. In: Roush, R.T., Tabashnik, B.E. (Eds.), Pesticide Resistance in Arthropods. Chapman and Hall, New York, Chapter 6.

Tabashnik, B.E., Croft, B.A., 1985. Evolution of of pesticide resistance in apple pests and their natural enemies. Entomophaga. 30, 37—49.

Tabashnik, B.E., Gould, F., Carriere, Y., 2004. Delaying evolution of insect resistance to transgenic crops by decreasing dominance and heritability. J. Evol. Biol. 17, 904—912.

Taylor, C.E., 1983. Evolution of resistance to insecticides: the role of mathematical models and computer simulations. In: Georghiou, G.P., Saito, T. (Eds.), Pest Resistance to Pesticides. Plenum, New York, pp. 163—173.

Taylor, C.E., Quaglia, F., Georghiou, G.P., 1983. Evolution of resistance to insecticides: a cage study on the influence of migration and insecticide decay rates. J. Econ. Entomol. 76, 704—707.

Vose, D., 2000. Risk Analysis: A Quantitative Guide. second ed. John Wiley & Sons, Chichester, pp. 418.

Zhao, J.Z., Cao, J., Collins, H.L., Bates, S.L., Roush, R.T., Earle, E.D., et al., 2005. Concurrent use of transgenic plants expressing a single and two *Bacillus thuringiensis* genes speeds insect adaptation to pyramided plants. Proc. Nat. Acad. Sci. USA. 102, 8426—8430.

CHAPTER 15

Monitoring Resistance

Bruce H. Stanley
DuPont Agricultural Biotechnology, Lodi, NY

Chapter Outline

SUSCEPTIBILITY AND TOLERANCE

One of the goals of an insect resistance management (IRM) program is to keep the proportion of susceptible individuals in a population as large as possible. Ideally, all individuals in the population are susceptible to the dosage of compound, presence of the crop trait, or cultural practice to which they are exposed, although this may not be the case even for a new practice, trait, or compound. In this chapter, the emphasis of discussion is on assessing the susceptibility of agricultural pests to toxins, the area of the author's expertise, and little focus will be placed on other

Insect Resistance Management
DOI: http://dx.doi.org/10.1016/B978-0-12-396955-2.00015-1

organisms or systems. Other measures of compound or trait susceptibility, both direct and indirect (van Kretschmar *et al.*, 2013), may prove to be the most expedient given the constraints of the system under study. However, the concepts presented here can be generalized to most other systems.

The tolerance of an individual arthropod to a substance, such as a pesticide, is governed by the phenotypic expression of its genotype (Andow and Ives, 2002). This has led researchers to measure the heritability of the presumed resistance-conferring gene in artificially selected cultures (Ban *et al.*, 2012; Devos *et al.*, 2012). Yue *et al.* (2008) was able to estimate the resistance allele frequency using an F_1 screen for *Diatraea saccharalis*. The set of phenotypes within a population of arthropods is continually changing through time (Martinson *et al.*, 1991; Kranti *et al.*, 2002), during increases in age and weight (Robertson *et al.*, 1984; Reissig *et al.*, 1986; Reyes and Sauphanor, 2008) and over space as a result of the insect's response to the environment in terms of age, gender, density, mating, and reproduction (Bouvier *et al.*, 2002; Sparks *et al.*, 1990). This has led some researchers to treat their test subjects under field conditions (Schouest and Miller, 1991).

As noted in the subsection ahead, The Importance of Positive and Negative Controls for the Monitoring Method, appropriate positive and negative controls are needed to ensure the trustworthiness of the techniques used to measure the status of a population.

The Concept of a Distribution of Tolerances

When dealing with a population of arthropods, each individual is likely to vary in some fashion that affects intoxication or expression of a response. This is particularly important when there is great spatial and temporal variability in host use, as demonstrated by Head *et al.* (2010) for *Helicoverpa zea* in the United States. Accordingly, the result will be a statistical distribution of response probabilities that is a function of the stimulus or level of toxin. Because a monitoring program will most likely not measure the entire population, the responses in the part of the population sampled will follow a statistical sampling distribution. Even if the response to a nondestructive stimulus is measured on a single individual through time, the response will most likely follow a distribution because the response will be affected by the attributes of the individual that are sensitive to the ever-changing environmental conditions that it experiences.

This is one reason why it is so important in a resistance-monitoring program to conduct the evaluation under a standard environmental regime.

The typical pattern of susceptibility of organisms to a harmful or burdensome stimulus is that first all are observed to be susceptible, and then eventually through evolution by natural selection, they all are observed to be tolerant. This results in a shift of the tolerance distribution to higher rates as shown in Figure 15.1. Note that the variance in the population tends to be small both before and after selection.

Tolerance is usually governed by a small set of genes and in many cases by one primary gene. This results in a characteristic shape of response when plotted against dose if a single mode of action dominates, and a smoother response when many factors are involved in defining susceptibility to an insecticide. Pittendrigh et al. (Chapter 3) and Nelson and Alves (Chapter 4) describe the importance of classifying toxins according to mode of action. For a monogenic trait, one might expect a chair-shaped tolerance distribution representing a mixture of the susceptible and tolerant individuals (i.e., two distributions of susceptibility). The height of the "seat" of the "chair" is the proportion of susceptible

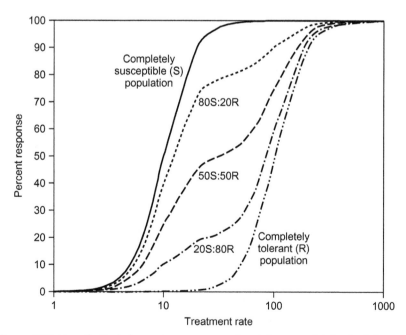

Figure 15.1 Hypothetical dose-response curve (i.e., tolerance distribution) for mixtures of susceptible and tolerant individuals.

individuals. Because a population is usually assayed over a finite set of doses, a graph of the percent response against treatment rate or stimulus intensity will look like a shallow sigmoid curve, the typical dose-response curve with a shallow slope. When many genes are involved in determining tolerance, there will not be a clear separation among populations, and one would expect a wider range of susceptibilities. This, too, will result in a shallow, sigmoidal tolerance distribution.

Care must be used when interpreting a plot of percent response versus treatment rate or stimulus intensity. Chilcutt and Tabashnik (1995) conducted an extensive review of the literature and did not observe a relationship between potency (e.g., the LD_{50}) and the slope of dose-response curves. So, although theoretically plausible, the empirical evidence does not support a relationship between potency and the slope of the dose-response curve. Regardless, the researcher needs to maximize his or her understanding of population susceptibility from the data he or she generates, and needs to check it for consistency with reasonable models.

Typically, the distribution of tolerances tends to be approximately lognormally or loglogistically distributed. The lognormal distribution tends to be very similar to the loglogistic distribution. Both are approximately normal on a logarithmic scale (any base), but the loglogistic distribution tends to have slightly thicker "tails." Details of both the lognormal and loglogistic distribution can be found in Finney (1971) and McCullagh and Nelder (1989). Because the loglogistic distribution tends to be mathematically simpler, I will use the loglogistic distribution in the following discussion.

Timing of Monitoring and Treatment

Care must be taken when a treatment, such as a pesticide, is applied before a population has reproduced. The survivors are from the more tolerant individuals of the pest population, and their tolerance distribution will appear to have shifted to the right or "more resistant." The individuals making up this population may still be "genetically susceptible," and upon reproducing may generate a normal susceptible population. If the genes involved now confer some tolerance to the offspring, then one might observe a slow progression toward a stable tolerant population. This illustrates the importance of resistance monitoring. It allows a researcher to understand how the susceptibility of a population may be shifting. This apparent shift in the tolerance of a population previously

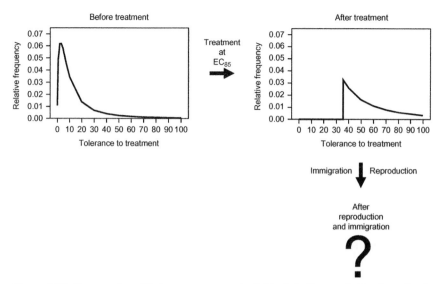

Figure 15.2 Frequencies of tolerances among individuals before and after hypothetical selection at the EC_{85} before recruitment.

treated at rate R units is illustrated in Figure 15.2. Assuming a loglogistic ("logit") tolerance distribution, we find that the density of individual tolerances, f_{DR}, and the tolerance distribution for the population, F_{DR}, of a treatment at dosage D units would be

$$f_{D|R} = \begin{cases} 0 & \text{if} \quad D \leq R \\ \dfrac{b \cdot F_D \cdot (1 - F_D)}{D^{b+1} \cdot F_R} & \text{if} \quad D > R \end{cases}$$

and

$$F_{D|R} = \begin{cases} 0 & \text{if} \quad D \leq R \\ \dfrac{F_D}{F_R} & \text{if} \quad D > R \end{cases}$$

respectively, where

$$F_X = \frac{X^b}{X^b + (Dose_{50})^b}$$

is the tolerance distribution at rate X (X = D or R), D is the any dose, R is the dose of the treatment, $LD_{50} = Dose_{50}$, is the median lethal dose, and $LD_{90} = r \cdot Dose_{50}$ units is the dose eliciting a 90% effect, r is the ratio of the LD_{90} to LD_{50}, and b is the slope of the logit versus log dose line. For example, parameter b is calculated as

$$b = \frac{\log(r)}{\log\left(\frac{P}{1-P}\right)} = \frac{\log(5)}{\log(9)} = 0.73$$

for an LD_{50} of 10 and LD_{90} of 50 units.

Again, this is shown in Figure 15.2 for an LD_{50} of 10 units and an LD_{90} of 50 units and a previous treatment at the LD_{85} of 35.6 units. As can be seen in Figure 15.2, the theoretical shape of the tolerances of the residual population is very similar to that of the unexposed population. It is unlikely that a researcher would be able to discriminate the two tolerance distribution shapes using conventional methods. Accordingly, one could erroneously assume that a significant shift in tolerance has occurred. This is one reason that researchers only get concerned when the LD_{50} shifts by a factor of some larger number, such as 10. However, in a monitoring program any shift should be investigated using follow-up testing.

Visualizing Tolerance

Monitoring programs will generate a large amount of information. This information should be summarized in a way that permits easy communication of the results and allows changes in tolerance to be readily recognized. One method that lends itself well to visualizing changes in the distribution of tolerances is the probability plot. Probability plots are easy to generate and communicate graphically.

The following is a recipe for creating an empirical probability plot of the distribution of population tolerances in a resistance monitoring program:

1. Rank the dose-response quantiles (e.g., LD_{50s}) from low to high.
2. Calculate the sample-size adjusted percentile for each quantile [e.g., $\% = 100 \times (r + 0.5)/(n + 1)$].
3. Plot the adjusted percentiles against the ranked quantiles.

This method is illustrated later in this chapter in the section Monitoring as Part of the Resistance Monitoring Program. In the case study presented there, probability plotting was used to assess the geographic variability of *Helicoverpa armigera* susceptibility to cypermethrin in six West African countries.

The Importance of Positive and Negative Controls for the Monitoring Method

It is critical that relevant positive and negative controls are used concurrently as part of the monitoring program. The positive control demonstrates that the method is working as intended when there is a positive detection. The negative control verifies that the monitoring method does not erroneously give a false detection. These controls are extremely important in assessing the validity of the monitoring program. They reduce the possibility that erroneous actions and communications are made that might undermine the credibility of those managing the monitoring program, prevent the possibility of large expenditures in unneeded corrective action, and have large impacts on sales, if a commercial product is involved. The incorporation of positive and negative controls into a monitoring program does not have to be costly or complicated as will be shown in the West African cypermethrin example.

QUANTIFYING TOLERANCE

A population can be very heterogeneous in its response to a stimulus, and the population will vary in response relative to its environment. Also, the individuals in a population are often exposed to a mixture of stimuli that can, to varying degrees, elicit similar responses. However, the treatment or crop trait of interest usually elicits a significant, distinct response. Accordingly, researchers need to balance the resources spent measuring a response with the likelihood that it is stable. The goal is to get a representative measurement at a reasonable cost. Researchers must ask a series of questions to ensure that the tolerances are measured logically. Is the susceptibility stable? If so, is the population already tolerant? Does the frequency of tolerance observed in the laboratory reflect the frequency in the field? Researchers should maximize the effectiveness of their monitoring program relative to all of these questions.

Thus, an effective monitoring assay should do the following:

Remove the natural response or mortality.

Quantify the level of tolerance.

Allow ranking of sites in order of tolerance.

Allow tracking of changes in tolerance through time.

Test hypotheses and explore sources of variability in tolerance.

If these requirements are met in a monitoring assay, the researcher will be able to readily assess changes in susceptibility and respond quickly with the development or implementation of a management strategy.

Single, Discriminating Dose Approach

The discriminating or diagnostic dose assay has been by far the most widely used method for monitoring susceptibility in the field. It is easy and relatively resource efficient, and it supplies a clear "answer." A local population is either susceptible or tolerant. The goal of the discriminating dose assay is to determine whether or not the status of the population's susceptibility has changed, ideally, in time for remedial action to be taken.

The three important considerations for designing a single, discriminating dose monitoring program are

1. Establishing the "diagnostic dose" to separate individuals with susceptible phenotypes from the resistant phenotypes.
2. Determining the sample size to be collected at each location.
3. Determining the appropriate response to a survivor of the discriminating dose.

Extensive work has been done on the diagnostic-dose monitoring approach. A very notable example is the *Heliothis* susceptibility monitoring program (Staetz, 1985; Plapp *et al.*, 1990a,b). Knight *et al.* (1990b), Marcon *et al.* (2000), and Magalhaes *et al.* (2012) discuss the development of diagnostic doses for other arthropods. Venette *et al.* (2002) assess the discriminating dose approach thoroughly. They point out that the discriminating-dose approach is most likely used to detect resistance determined by a dominant allele. Venette *et al.* (2002) conclude that an average of 1400 individuals would need to be randomly sampled to detect a 1% rate of the phenotypic expression of resistance.

Dose-Response Approach

The most precise method to assess the susceptibility of a population to a compound or crop trait is the classical dose-response bioassay (Busvine, 1971; Finney, 1971). This is a regression of responses, usually employing the probit or logit transformation, against the logarithm of the dosage. It is an oversimplification to assume that all populations follow either a logit (i.e., loglogistic) or probit (i.e., lognormal) tolerance distribution. As was previously illustrated, the population surviving after exposure to a trait or chemical will be a truncated tolerance distribution that may or may not recover to the pretreatment distribution. The tolerance distribution of a population is ever changing in response to a dynamic environment. Accordingly, the researcher must be flexible in assuming and fitting a tolerance distribution to the data. Often the LC_{50} estimated from the dose-response curve is used to define the baseline susceptibility (Miller et al., 2010). The LC_{50} is a logical choice for the benchmark because it is the response quantile that can be measured with the greatest precision in dose-response bioassays (Finney, 1971).

There is a substantial literature, both in the biological and statistical disciplines, describing the virtues and limitations of the probit and logit (Finney, 1971; McCullagh and Nelder, 1989). In essence, the two methods ensure that there are only responses for nonzero dosages and that the shape of the tolerance curve is sigmoid with a long positive tail. Both the logit and probit curves have a clear theoretical genesis, but one is hard pressed to consider one to be logically superior to the other. I tend to favor the logit (i.e., loglogistic tolerance distribution) simply because it lends itself to easier mathematical manipulation.

A tolerance model can be fit to a set of data in a number of ways. These include linear regression on transformed data, minimum chi-square regression (Berksen, 1955; Smith et al., 1984), maximum likelihood (Finney, 1971; McCullagh and Nelder, 1989), and maximum quasi-likelihood (Wedderburn, 1974; McCullagh and Nelder, 1989), of which maximum likelihood is a special case. All of the methods have their strengths. One should select the appropriate method for one's needs. If in doubt, it is highly recommended that the researcher work with a biostatistician. Because of the time and resources that it takes to conduct a bioassay, one should try to extract as much information with the least possible bias from the data.

Background or natural response should be accounted for in any bioassay. If it can be assumed that the factors affecting background response are independent of those affecting tolerance to the compound or crop trait, then the background response is easily removed from the effect of the chemical or trait using Abbott's formula (Abbott, 1925) or the equivalent Schneider-Orelli formula (Schneider-Orelli, 1947). This again reinforces the importance of a negative control in the monitoring program.

Background immunity (Sakuma, 2013) limits the maximum response that may be observed. Similarly to the background response, it defines a threshold for the maximum response. The traditional method for modeling the dose-response curve when background response and immunity are present is to model the dose-response curve as a probit or logit curve bounded by the two background responses. Background immunity can arise when the complete upper dosage was not delivered; for instance, the toxin did not go into solution. Background immunity can also arise when the test subjects are from a mixture of populations, with one population being much more tolerant than the other; in this case, the upper dosages define the "seat" of the "chair" of the tolerance distribution.

The design of multiple-dose bioassays is well described (Finney, 1971; Smith and Robertson, 1984; Robertson et al., 1984; Smith et al., 1984). The DOSESCREEN (Smith and Robertson, 1984) approach is particularly useful. A helpful modification is to design to the relative length of the confidence interval of the EC_x value rather than the absolute length.

The Two-Dose Approach

A simplification of the multiple-dose monitoring approach that has proved very informative and cost effective in susceptibility monitoring has been the "two-dose" approach. In reality, it requires three treatments—an untreated control group to assess background mortality; a lower discriminating dose that allows a shift in susceptibility to be detected; and a higher discriminating dose that allows the researcher to determine what proportion of the population is tolerant to the treatment. This method is simple and requires only a modest number of test subjects. The two-dose approach is only slightly more expensive than a single discriminating-dose monitoring program because only one more treatment is added. Most of the expense of a bioassay is in finding the sample site and collecting the samples. The dose-response curve may be fit using traditional maximum likelihood methods. However, reasonable estimates may be obtained using the technique described below when computing equipment is not available. The recipe for conducting the simplified two-dose bioassay is given in Table 15.1.

The point estimate for the LD_{50} is estimated as a perfect fit through the two responses adjusted for background mortality using Abbott's formula (Abbott, 1925). The formula for the LD_{50} estimate adjusted for background response is given as

$$LD_{50} = 10 \left[\log_{10}[L] + \frac{\log_{10}\left[\frac{S_L}{S_C - S_L}\right]}{\log_{10}\left[\frac{S_L(S_C - S_H)}{S_H(S_C - S_L)}\right]} (\log_{10}[H] - \log_{10}[L]) \right]$$

Table 15.1 Recipe for Conducting a Two-Dose Bioassay

1. Work with number alive as the endpoint to reduce misclassification error.
2. Define four categories of local results:
   a. Highly Susceptible—Survival at the low dose is zero.
   b. Intermediate—Survival at the low dose is less than the control.
   c. Highly Tolerant—Survival at the high dose is equal to the control.
   d. Inconsistent—Survival at the high dose is less than the low dose or greater than the control.
3. Estimate the LD_{50} for the tests in the intermediate category using logit analysis techniques that adjust for background mortality using Abbott's formula (Abbott, 1925).
4. Use probability plot techniques and nonparametric, rank-based analysis methods to understand the changes in the distribution of susceptibility over space and time or among management approaches.

where C is the zero dose (control), L is the lower discriminating dose (5 mg), and H is the higher discriminating dose (30 mg). The reduced bias estimate of the percentage of larvae surviving exposure to dose D, $S_D = 100x(a_D + 0.5)/(n_D + 1)$, where a_D is the number of larvae alive after exposure to dose D, and n_D is the number of larvae exposed to dose D (D = C, L, or H).

Since the responses at the two dosages are binomial proportions, confidence intervals can be created around the responses. If the confidence intervals around the two dosages are connected, approximate confidence bands around the tolerance curve, and thus any quantile such as the LD_{50}, can be approximated. The approach is to find the dose where the lower and upper confidence interval of a binomial proportion with mean equal to the fitted curve crosses the percentile of interest, such as the LD_{50}, is estimated from 50% survival. This is illustrated in Figure 15.3.

The parameter estimates from a two-dose bioassay are used just like those generated by a multiple-dose bioassay. They generally will not be as precise as those estimated from a multiple-dose assay and will often be in the highly susceptible category in Table 15.1. However, the two-dose bioassay is very cost efficient and lends itself well to a monitoring program. It is recommended that additional follow-up testing using multiple-dose bioassays be conducted for populations falling into the intermediate category (Table 15.1) to understand the real tolerance distribution of a population.

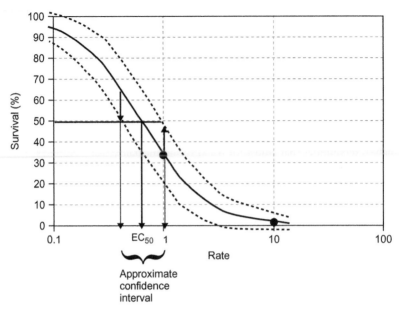

Figure 15.3 The two-dose bioassay.

Other Methods for Quantifying Tolerance

As the biochemical mechanisms for resistance are discovered, scientists can use this knowledge to create techniques for identifying biochemical markers in the tissues of resistant arthropods (ffrench-Constant and Roush, 1990; Tabashnik *et al.*, 2006). The marker may be an allele for a resistance gene, or it may be a chemical created by the resistance gene. Edwards and Hoy (1995) used a DNA marker to monitor insecticide-resistant parasitoids in orchards. Van Kretschmar *et al.* (2011, 2013) describe a colormetric feeding disruption test for detecting the presence of resistant individuals. An overview of some of the methods with applications to resistance monitoring can be found in Kranthi (2005). Because of the greater certainty in these procedures, lower sample sizes will likely be required to obtain the same precision as for methods described above (Venette *et al.*, 2002). However, unless a resistance mechanism is known a priori for a new insecticide, multiple biochemical assays may be needed to discover tolerance in the field (ffrench-Constant and Roush, 1990).

Venette *et al.* (2002) summarize the feral screen and F_2 screen for measuring tolerance in wild populations of pests. Andow and Ives (2002) describe the statistics and costs for these methods for *Ostrinia nubilalis* on transgenic insecticidal corn (*Zea mays*). They state that the F_2 screen is the most labor-intensive monitoring method. Zhao *et al.* (2002) used an F_2 screen for *Plutella xylostella* on transgenic insecticidal broccoli (*Brassica oleracea*). They concluded that using transgenic plants expressing a high level of toxin may underestimate the frequency of resistance alleles with high false negatives or fail to detect true resistance alleles. Zhao *et al.* (2002) urged scientists to carefully validate the screening method for each insect-crop system before use of an F_2 screen.

An important new technique is the in-field screen (Venette *et al.*, 2000; Tabashnik *et al.*, 2000; Venette *et al.*, 2002; Andow and Ives, 2002). This approach uses the toxic crop field as the diagnostic or discriminatory dose to distinguish resistant from susceptible individuals. In essence, the survivors in the toxic field indicate the level of tolerance relative to the survivors in a field of nontoxic plants. Because actual crop fields are used to screen insects, the costs are lower than those for other methods (Andow and Ives, 2002).

MONITORING TO DETECT THE EARLY DEVELOPMENT OF RESISTANCE

Geographic scale and complexity add to the challenges of resistance monitoring for effective and timely action plans for the maintenance of the effectiveness of a chemical or crop trait for pest control (Blanco *et al.*, 2007; Head *et al.*, 2010). Many biological factors influence the evolution of resistance. Resistance will likely become established in a locally small region. This brings up questions of (1) whether it is sustainable within the larger population and (2) whether it brings with it deleterious fitness costs.

The most important question concerning the detection of resistance in its early stages of development is, "what is the likelihood of resistance given that it was not detected?" Agrentesi and Cavalloro (1983) developed a technique for estimating the confidence interval of the number of individuals present given that none were detected. This method allows one to bound the frequency of resistant individuals given that none were detected.

MONITORING AS PART OF RESISTANCE MANAGEMENT PROGRAM

To be effective, resistance monitoring must be calibrated and broadly coordinated with clear and efficient communication. All of the information needs to flow into a decision-making body that will take action based on the analysis and distribution of measurements. Before the process starts, the decision-making body needs to understand the costs and ensure that the stakeholders can afford the monitoring program. If not, the decision makers need to reevaluate the goals to determine if their needs can be met with a subset of the program or a different approach. At the end of each cycle, a prudent manager should evaluate whether the measurements can be obtained more efficiently, and adjust the program before the next cycle. In all adjustments, the manager should modify the scheme to preserve as much value as possible from any earlier work. In essence, this is the "adaptive management" with "adaptive management interventions" approach set forth by Andow and Ives (2002). Sometimes, the decision makers may decide that no resistance management is cost effective and will let the progression of resistance unfold. They may choose to continue monitoring to track the status of the system.

Accordingly, the manager leading a resistance-monitoring program should:

1. Ensure that a method (bioassay) is available to detect tolerance.
2. Make an assessment that the monitoring program is logistically and cost feasible.
3. Ensure that the detection method can be used correctly by everyone involved.
4. Ensure that everyone communicates accurately and promptly.
5. Ensure all of the information is synthesized to yield an accurate picture from which decisions can be made.
6. Have an action plan in place to respond in some way to the results. (Monitoring is usually not a goal unto itself.)
7. Review, to assess whether the next iteration might be done more efficiently.

Energy and resources focused on the logistics of the monitoring program will ensure a successful program. Timing is a key issue. Ensuring that information flows accurately and promptly will greatly enhance the feasibility and usefulness of any resistance monitoring program.

Logistics, Costs, and Practical Issues

Monitoring and its associated logistics have a cost, which can be substantial. Alacalde *et al.* (2007) advocated that the monitoring of transgenic insecticidal crops be conducted by an industry working group. This model has been followed by the Insecticide Resistance Action Committee (http://www.irac-online.org/) and the Agricultural Biotechnology Stewardship Technical Committee (http://www.excellencethroughstewardship.org/). As discussed by Andow and Ives (2002), the manager needs to assess the potential cost of a monitoring program against the benefit of information generated by it to determine if it is worthwhile. This suggests that the manager should map out the entire program and his actions for each of the possible outcomes. This could be an iterative process where the manager tailors his monitoring program to fit comfortably within a budget. Usually, the manager can optimize the program within the cost constraints to achieve a very cost-effective program. The manager should also explore novel ways to conduct the monitoring, perhaps with new bioassays or new contracting opportunities, to maximize the chances of a successful program. If the manager decides that monitoring is not worth the cost, then other strategies may be needed to determine when to implement change or how to accomplish stakeholders' goals. It must be remembered that monitoring may show that there is no problem, and all that is needed is future vigilance.

Rust *et al.* (2011) demonstrated that there has been no decrease in control by imidacloprid of *Ctenocephalides felis* despite widespread use of the insecticide as a topical treatment for pets. Rust *et al.* (2011) sampled 1437 fleas collected by veterinary clinics in Australia, Germany, France, the United Kingdom, and 29 states in the United States from 2002 to 2009. Seventy-one percent of the collections were from cats. Of the 1437 collections received, 1064 contained adequate numbers of eggs for testing, but about 22% of untreated eggs did not hatch and were not considered in final bioassays. Bioassays confirmed sustained susceptibility of *C. felis* to imidacloprid.

There are two considerations associated with any set of measurements. The first is the central tendency and intrinsic variability among the items being measured. This is the true magnitude and heterogeneity among the items being measured. The second is any variability and bias added to the actual values by the measurement technique. This is the error associated with the measurement system. The goal of a monitoring program is to minimize the error and bias, so that the monitoring team gets an accurate understanding of the magnitude and variability of the system. The manager must balance the resources spent on reducing the bias and error and the cost. This will yield an optimum monitoring program relative to the goals of the project.

A philosophy that lends itself well to resistance monitoring is the adequate-precision approach. Under this approach, the manager assesses early in the program how much precision there must be in the measurement of the effectiveness of a treatment or crop trait. The program is then designed to measure to this level of precision. However, because unforeseen needs often arise over the course of a program that require a bit higher precision, the manager should design the program to be slightly more precise than needed to ensure that the measurement will be sufficient for all needs. Of course, the program needs to be below the budgeted costs, so that it is feasible over the time horizon defined by stakeholders' goals.

A key resource in any monitoring program is time. The manager should design a program that allows him to assess the state of the system, decide upon a course of action, and implement program adjustments in time to influence the population genetics to preserve the effectiveness of the treatment or crop trait. This can be achieved by clearly communicating program goals and response options, and automating reporting electronically to promote near real-time population status assessments. Again, a cost-benefit analysis is appropriate so that the manager is confident that the effectiveness of the communication effort is worth the cost.

All of the preceding points regarding standardization, logistics, and precision are achievable through planning. The manager should budget some time and resources for planning before initiating the monitoring program. One of the simplest methods to promote feasibility and responsiveness is to standardize all aspects of the program and communication. Also, attention to the logistics of the program will pay great dividends toward feasibility. The key is to have a program that is cost effective and meets all of the goals of the program. Although good planning can be the key to a successful resistance monitoring program, the manager should not expend unreasonable time or resource in planning. Planning should not be an end unto itself, and it should be in balance with all of the other activities in the monitoring program.

The importance of training and communication to the success of a resistance monitoring program cannot be overemphasized. Because the natural variation in response is usually already high, there is great payback in reducing procedural deviations among workers and errors due to poor communication. All of these problems mask an already variable and, hopefully, small change in response. If the change in susceptibility is large, then monitoring adds only minimal value. The manager is already confronting the need for a major change in the management of the pest and may need to adapt the monitoring approach to a new set of goals and responses.

Effective and timely communication is important, particularly in a large geographical area, because decisions need to be implemented quickly to manage susceptibility. Anything that promotes communication approaching real time has value. There are many logistic considerations involved in changing a management program, so quick, effective communication allows time for critical changes to be made quickly and correctly. This maximizes the chances that an IRM program will be successful.

It is important to allocate time and resources in a monitoring program to benchmark the status of the program. The manager needs to verify that the monitoring program is not drifting and that all parties involved are communicating well and are ready to respond to any changes in susceptibility. Inevitably, the manager will want to fine-tune the program while it is in progress. This can best be done by developing and testing possible changes in a research environment outside of an existing monitoring program. Proposed changes can then be incorporated simultaneously in all locations. This ensures that the monitoring results are broadly comparable and have an expected and well-delineated shift in measured response. This avoids a drift in response over a wide geographic range that could easily be misinterpreted as a change in population susceptibility.

A nice example of a large-scale resistance monitoring program with an action plan was the European project to monitor and manage pyrethroid resistance in several species of pests in the genus *Meligethes* (Slater *et al.*, 2011). It provides a good standard for the development and implementation of a large-scale resistance monitoring program.

Major Limitations: Sampling and Surveying

Determining how to quantify tolerance in a population requires significant scientific work, but the primary costs of monitoring are the sampling and handling of all the insects or their tissues. Thousands of sample units must be checked for a chemical and a gene, or tested in a bioassay. This section focuses on the feasibility of collecting the tens of thousands of samples needed for precise estimation of tolerance in a population over time and space.

The goal of the project determines the region over which resistance will be monitored, the time horizon during which the population will be sampled (Chapter 2), and the resolution of sampling in space and time. The larger the region and the longer the time horizon, the greater the effort devoted to and the cost of the monitoring program. The resolution indicates how small the areas and time intervals will be for sampling. For example, will samples be collected monthly or yearly? Will samples be taken from every field, farm, or county? Obviously, as the resolution becomes finer with smaller units of time and space sampled, the greater the cost.

A good general book on insect or animal surveys should be consulted for a description of the various types of sampling approaches that may be valuable in IRM. The collections may be obtained from simple random sampling or from stratified sampling. Geographic stratification can be based on farm, county, watershed, community, or any other rational or political division of the world. There are now reliable sampling methods for almost every pest species and crop posted on the Internet.

The likelihood that a monitoring program will detect the first tolerant individual is very low. This issue has been discussed in detail by Roush and Miller (1986), Andow and Ives (2002), and Venette et al. (2002). Roush and Miller (1986) describe the formula for calculating the randomly selected sample size necessary to detect one tolerant individual in a population of 1,000,000 individuals with probability 0.95.

$$n = \frac{\log[1 - P(x \geq 1)]}{\log[1 - f]} = \frac{\log[1 - 0.95]}{\log[1 - 0.000001]} = 2,995,731 \; individuals$$

where x is the number of resistant individuals in the sample, and f is the frequency of the resistant phenotype. It is very unlikely that any monitoring program will have the capability to make this feasible. Calculation of the sample size needed to detect the U.S. Environmental Protection Agency's hypothetical resistant-allele frequency of 0.001 (U.S. EPA, 2010) is 2995 individuals. Andow and Ives (2002) extended this analysis using more complicated models. They determined that one must detect recessive resistant alleles at a frequency less than 5×10^{-3} to provide enough time to implement an adaptive management program. If the resistance allele is dominant, then it must be detected at a frequency less than 1×10^{-7}.

As can be surmised from the above analyses of gene frequencies, it will be difficult to detect resistance in the field unless a monitoring program collects large numbers of individuals using methods such as pheromone-baited traps (Riedl *et al.*, 1985; Haynes *et al.*, 1987; Shearer *et al.*, 1994) and attractant baits (Siegfried *et al.*, 2004). Even with this possibility, it is very unlikely that the manager is going to be able to sample over the entire range of the population. Accordingly, for organisms with large reproductive power, such as arthropods, it is likely that the tolerant gene will become quickly fixed in the population, at least locally. Eradication of the tolerant gene will then become challenging and costly.

Turning Measurement into Action

Unless the goal of one's effort is solely to assess the geographic and temporal susceptibility of a population to a stimulus, the decision makers should have a plan of action developed that is tied to the observed susceptibility. This may range from slight changes in use patterns to completely removing the product or crop trait from a region. For example, Huang *et al.* (2006) proposed insecticide use patterns to prolong product life while preserving the level of crop protection. The decision makers should continually refine contingencies as they learn more about the system. In practice, the most challenging aspect of a resistance management program is responding in a time frame that allows changes to purposefully and deliberately affect the monitored system. Having a plan in place allows one to react quickly and in a coordinated manner to the observed results from a monitoring program. The uninterrupted continuation of the monitoring program should also allow an assessment of whether remediation actions have improved the susceptibility of a population to the product or crop trait.

As described in Chapter 14, models are often used to study the evolution of resistance and the management of resistance. Government regulators use modeling results to develop regulations and evaluate IRM plans provided by industry. Although the testing of model results against field observations is not usually possible before products are registered by regulators, researchers can use time series of data to test or validate model results after products and crop traits have been commercialized for several years.

EXAMPLES OF MONITORING PROJECTS

There have been many insect resistance monitoring programs in the past. Each had its specific goals and was constrained by resources, time, and commitment. Their reports and methodology provide useful guides to designing effective monitoring programs. Some have examined the variation in susceptibility over large spatial scales (Carriere *et al.*, 2001; Knight *et al.*, 1990a; Knight and Hull, 1990). Farinos *et al.* (2011) monitored for three years to assess the level of erosion of Cry1Ab to control the *Sesemia nonagriodes* in corn. Many different taxonomic groups have been studied, particularly heliothines (Bailey *et al.*, 2001; Kanga *et al.*, 1995; Riley,1990; Rogers *et al.*, 1990;Wu *et al.* ,2006) and other Lepidoptera (Leeper, 1984; Zhao *et al.*, 2006; and Osorio *et al.*, 2008). Croft *et al.* (1989), Grafton-Cardwell and Vehrs (1995), Grafton-Cardwell *et al.* (2004), Hockland *et al.* (1992), Nauen and Elbert (2003), Sanderson and Rousch (1992), and Ouyang *et al.* (2010) examined homopterans. Dennehy *et al.* (1990) and Reissig *et al.* (1987) examined mites. Coleman and Hemingway (2007) discussed the importance of resistance monitoring for effective disease vector (mosquito) management. Peterson (2005) gave an extension service perspective on monitoring resistance. Slater *et al.* (2012) assessed the susceptibility of European populations of *Myzus persicae* to neonicotinoids and recommended a coordinated IRM strategy employing products encompassing multiple modes of action to preserve neonicotinoid susceptibility. Emery *et al.* (2011) recommended further research and fumigation strategies to manage already high levels of resistance in pests of grain stored in Australian grain bins. Spencer *et al.* (Chapter 7) discuss monitoring resistance to transgenic insecticidal corn by *Diabrotica virgifera virgifera*.

Once pyrethroid resistance was suspected in the region, a group of entomologists in the southeastern United States investigated the problem with a monitoring project (Plapp *et al.*, 1990a,b). They used pheromone traps to collect large numbers of adult male *Heliothis virescens*. Over three years and five states, they collected over 55,000 moths: no more than 13,000 in one state in one year. Plapp *et al.* (1990b) believed that their samples were large enough to base valid conclusions concerning the occurrence of resistance. This is certainly the case given that resistance allowed at least 10 to 20% of the male moths to survive higher doses of pyrethroid. Even though the standard deviations for their estimates of survival were large, with these high levels of survival it was impossible to not discover resistance in the population. Clearly, Plapp *et al.* (1990a,b) were not trying to find the earliest indications of resistance (Roush and Miller, 1986).

In studies described by Plapp *et al.* (1990a), field experiments demonstrated that estimates of resistance based on larval sampling yielded more accurate results than those based on the sampling of adults. However, higher accuracy at one site based on more difficult larval sampling must be balanced with the greater precision provided by the huge numbers of easily captured male moths. Hopkins and Pietrantonio (2010) looked at the differential efficacy of three commonly used pyrethroids to *Helicoverpa zea* and developed an optimal resistance management strategy based on the results.

Steven N. Irving led a very efficient and innovative resistance monitoring study in Africa in 1998 as part of an initiative by Insecticide Resistance Action Committee International. The concern was that field control failures of the pyrethroids against larval *Helicoverpa armigera* in Western Africa were due to resistance. The goal of the monitoring study was to survey cypermethrin efficacy against larval *H. armigera* in a number of countries in Western Africa to determine whether the control failures were due to the buildup of tolerance in *H. armigera* to the pyrethroids. There was a relatively small budget, and the monitoring had to be implemented in rural farming regions where a number of languages are spoken. Prior to initiating the study, Irving organized a small team of bioassay and statistics experts to design the monitoring plan. They decided to use the two-dose (plus control) method to monitor susceptibility. Irving did not observe a relationship between the estimated slopes and EC_{50}s. To minimize variability, he prepared all of the test vials from a common source of cypermethrin in France and shipped the prepared, color-coded vials to his investigators in Africa. He also sent instructions for use of the vials in the form of simple, plastic-coated cards with a pictorial representation of bioassay procedures (Figure 15.4). He then trained the local investigators to use the vials in accordance with the instructions, and he had the raw data sheets sent back to France for analysis. A graphical summary of the results is presented in Figure 15.5. Irving demonstrated that cypermethrin was still efficacious. Efficacy was greatest in Senegal and Chad and weakest in Burkino Faso and Benin. This highly coordinated effort allowed data from very rural, multicultural regions to be generated uniformly, summarized expertly, and reported in the same field season within budget.

The Insecticide Resistance Action Committee conducted a broad-scale monitoring program on the susceptibility of *Meligethes* spp. to pyrethroids in Europe. Slater *et al.* (2011) presented the results of the first three years of the monitoring program and recommended that a resistance management plan be put into place. Meligethes beetles are a major pest

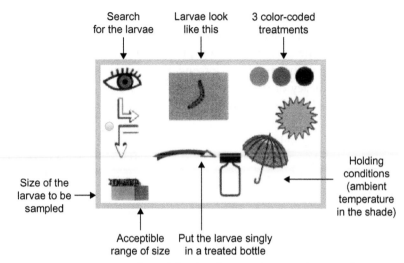

Search for the larvae — Larvae look like this — 3 color-coded treatments

Size of the larvae to be sampled

Holding conditions (ambient temperature in the shade)

Acceptible range of size — Put the larvae singly in a treated bottle

Figure 15.4 Instruction card used by Dr. Steven N. Irving (Insecticide Resistance Action Committee) to monitor cypermethrin susceptibility in West Africa.

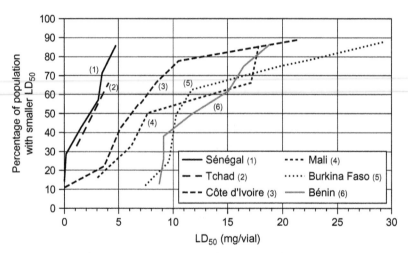

Figure 15.5 Probability plots representing the ranges of susceptibility of different field populations from six West African countries in a monitoring study sponsored by Insecticide Resistance Action Committee and managed by Dr. Steven N. Irving.

of European crops of *Brassica napus*. Resistance to pyrethroid insecticides and control failures have been recorded in Europe since 1999. Since 2007, beetle samples have been collected from 20 countries. Slater *et al.* (2011) observed a trend of increasing frequency and spread of resistance.

Pyrethroid-resistant beetles dominate in western and central Europe and are becoming established in the northern and eastern parts of the continent.

CONCLUSION

Monitoring involves two major steps. First, the arthropod population must be sampled or an indirect estimate of resistance must be measured (e.g., yield loss, control failure). Second, the sample must then be tested or interpreted based on criteria designed to make a judgment about resistance. Often a bioassay is performed. Some newer approaches, such as the in-field and F_2 screens, try to combine the two steps. In addition to providing an overview of the kinds of assays and statistics that can be used in monitoring programs, this chapter has highlighted three issues that are critical for the economic evaluation of all plans: goals, precision, and cost. These issues pertain to monitoring all forms of resistance described in this book.

Monitoring can be an important aspect of a resistance management program. It reveals the current state of the system and allows the magnitude and direction of changes to be tracked. Monitoring may be valuable for assessing population status, understanding potential risks, determining whether a resistance management program is stabilizing the efficacy of a compound, crop trait, or program, and projecting future trends.

Coordination among workers and among stakeholders is critical for successful resistance monitoring (Brent, 1986). Whether only one agency or corporation chooses to monitor on its own or whether multiple agencies, corporations, public health organizations, farmers, or ranchers all contribute to the monitoring effort, coordination is and communication must be maintained. Thus, coordination for monitoring is no different than the overall coordination required for mplementation of an effective IRM plan.

Since most monitoring programs will involve some type of biological assay, details of a number of techniques were presented in this chapter to help managers maximize the value of the information generated. In practice, I have found the two-dose approach very useful. It requires only slightly more resources than a single-dose assay and produces much of the information given by multiple dose assays. It has been very useful in practice.

Resistance monitoring has been widely and successfully used in a number of pest management systems. Monitoring should be a tool for a large decision-making strategy. The strategy should be understood by all participants, and attention should be given to effective communication so that the whole process can nimbly respond to changes in the status of the system under study.

A final note is that monitoring the evolution and management of resistance allows us to study evolution in near real time. We are able to see indirectly the changes in gene frequency over geographic space during relatively short time frames. This allows in-field observation and experimentation around gene flow and the consequences of large-scale selection by pesticides and transgenic insecticidal crops, patchy environments, and local extinctions. Resistance monitoring will be an important dimension of studies into any of thesse phenomena.

REFERENCES

Abbott, W.S., 1925. A method of computing the effectiveness of an insecticide. J. Econ. Entomol. 18, 265−267.

Agrentesi, F., Cavalloro, R., 1983. A new theoretical approach to the control and management of fruit-fly systems. In: Cavalloro, R. (Ed.), Fruit Files of Economic Importance. A. A. Balkema, Rotterdam, pp. 297−306, p. 642.

Alacalde, E., Amijee, F., Blache, G., Bremer, C., Fernandez, S., Garcia-Alonso, M., et al., 2007. Insect resistance monitoring for Bt maize cultivation in the EU: proposal from the industry IRM Working Group. J. Verbrauch. LebenSM. 2, 47−49.

Andow, D.A., Ives, A.R., 2002. Monitoring and adaptive resistance management. Ecolog. Appl. 12, 1378−1390.

Bailey, W.D., Brownie, C., Bacheler, J.S., Gould, F., Kennedy, G.G., Sorenson, C.E., et al., 2001. Species diagnosis and Bacillus thuringiensis resistance monitoring of Heliothis virescens and Helicoverpa zea (Lepidoptera:Noctuidae) field strains from the southern United States using feeding disruption bioassays. J. Econ. Entomol. 94, 76−85.

Ban, L., Zhang, S., Huang, Z., He, Y., Peng, Y., Gao, C., 2012. Resistance monitoring and assessment of resistance risk to pymetrozine in Laodelphax striatellus (Hemiptera: Delphacidae). J. Econ. Entomo. 105, 2129−2135.

Berksen, J., 1955. Maximum likelihood and minimum χ^2 estimates of the logistic function. J. Am. Statist. Assoc. 50, 130−162.

Blanco, C.A., Perera, O.P., Boykin, D., Abel, C., Gore, J., Matten, S.R., et al., 2007. Monitoring Bacillus thuringiensis-susceptibility in insect pests that occur in large geographies: How to get the best information when two countries are involved. J. Invertebr. Pathol. 95, 201−207.

Bouvier, J.C., Boivin, T., Beslay, D., Sauphanor, B., 2002. Age-dependent response to insecticides and enzymatic variation in susceptible and resistant codling moth larvae. Arch. Ins. Biochem. Physiol. 51, 55−66.

Brent, K.J., 1986. Detection and monitoring of resistant forms: an overview. Pesticide Resistance: Strategies and Tactics for Management. National Research Council. National Academy Press, Washington, DC, 298-312.

Busvine, J.R., 1971. A Critical Review of the Techniques for Testing Insecticides. Commonwealth Agricultural Bureaux, London, p. 345

Carriere, Y., Dennehy, T.J., Pedersen, B., Haller, S., Ellers-Kirk, C., Antilla, L., et al., 2001. Large-scale management of insect resistance to transgenic cotton on Arizona: can transgenic insecticidal crops be sustained? J. Econ. Entomol. 94, 315–325.

Chilcutt, C.F., Tabashnik, B.E., 1995. Evolution of pesticide resistance and slope of the concentration-mortality line: Are they related?. J. Econ. Entomol. 88, 11–20.

Coleman, M., Hemingway, J., 2007. Insecticide resistance monitoring and evaluation in disease transmitting mosquitoes. J. Pestic. Sci. 32, 69–76.

Croft, B.A., Burts, E.C., van de Baan, H.E., Westigard, P.H., Riedl, H., 1989. Local and regional resistance to fenvalerate in Psylla pyricola Foerster (Homoptera:Psyllidae) in western North America. Can. Entomol. 121, 121–129.

Dennehy, T.J., Nyrop, J.P., Martinson, T.E., 1990. Characterization and exploitation of instability of spider mite resistant to acaricides. ACS-Symp. Ser. 421, 77–91.

Devos, Y., Meihls, L.N., Kiss, J., Hibbard, B.E., 2012. Resistance evolution to the first generation of genetically modified Diabrotica-active Bt-maize events by western corn rootworm: management and monitoring considerations. Transgenic. Res.1–31 [Article in Press].

Edwards, O.R., Hoy, M.A., 1995. Random amplified polymorphic DNA markers to monitor laboratory-selected, pesticide resistant Trioxys pallidus (Hymenoptera: Aphidiidae) after release into three California walnut orchards. Environ. Entomol. 24, 487–496.

Emery, R.N., Nayak, M.K., Holloway, J.C., 2011. Lessons learned from phosphine resistance monitoring in Australia. Stewart Postharvest Rev. 7, article 8.

Farinos, G.P., Andreadis, S.S., de la Poza, M., Mironidis, G.K., Ortego, F., Savopoulou-Soultani, M., et al., 2011. Comparative assessment of the field-susceptibility of Sesamia nonagrioides to the Cry1Ab toxin in areas with different adoption rates of Bt maize and in Bt-free areas. Crop Prot. 30, 902–906.

ffrench-Constant, R.H., Roush, R.T., 1990. Resistance detection and documentation: the relative roles of pesticidal and biochemical assays. In: Roush, R.T., Tabashnik, B. E. (Eds.), Pesticide Resistance in Arthropods. Chapman and Hall, New York, Chapter 2.

Finney, D.J., 1971. Probit Analysis. third ed. Cambridge Univ. Press, Cambridge, p. 333

Grafton-Cardwell, E.E., Vehrs, S.L.C., 1995. Monitoring for organophosphate- and carbamate-resistant armored scale (Homoptera:Diaspididae). San Joaquin Valley citrus. 88, 495–504.

Grafton-Cardwell, E.E., Ouyang, Y., Striggow, R.A., Christiansen, J.A., Black, C.S., 2004. Role of esterase enzymes in monitoring for resistance of California red scale, Aonidiella aurantii (Homoptera: Diaspididae), to organophosphate and carbamate insecticides. J. Econ. Entomol. 97, 606–613.

Haynes, K.F., Miller, T.A., Staten, R.T., Li, W.G., Baker, T.C., 1987. Pheromone trap for monitoring insecticide resistance in the pink bollworm moth (Lepidoptera: Gelechiidae): new tool for resistance management. Environ. Entomo. 16, 84–89.

Head, G., Jackson, R.E., Adamczyk, J., Bradley, J.R., Duyn, J.V., Gore, J., et al., 2010. Spatial and temporal variability in host use by Helicoverpa zea as measured by analyses of stable carbon isotope ratios and gossypol residues. J. Appl. Ecol. 47, 583–592.

Hockland, S., Devonshire, A.L., Devine, G.J., Dewar, A.M., Read, L.A., 1992. Monitoring insecticide resistance of aphids in sugar beet and potatoes in England and Wales 1982–1991. Aspects of Appl. Biol. 1992, 81–88.

Hopkins, B.W., Pietrantonio, P.V., 2010. Differential efficacy of three commonly used pyrethroids against laboratory and field-collected larvae and adults of Helicoverpa zea (Lepidoptera: Noctuidae) and significance for pyrethroid resistance management. Pest Manag. Sci. 66, 147–154.

Huang, S., Xu, J., Han, Z., 2006. Baseline toxicity data of insecticides against the common cutworm Spodoptera litura (Fabricius) and a comparison of resistance monitoring methods. Int. J. Pest Manag. 52, 209–213.

Kanga, L.H.B., Plapp Jr., R.W., Elzen, G.W., Hall, M.L., Lopez Jr., J.D., 1995. Monitoring for resistance to organophosphorus, carbamate, and cyclodiene insecticides in tobacco budworm adults (Lepidoptera: Noctuidae). J. Econ. Entomol. 88, 1144–1149.

Knight, A., Hull, L., Rajotte, E., Hogmire, H., Horton, D., Polk, D., et al., 1990a. Monitoring azinphosmethyl resistance in adult male Platynota idaeusalis (Lepidoptera: Tortricidae) in apple from Georgia to New York. J. Econ. Entomol. 83, 329–334.

Knight, A.L., Hull, L.A., 1990. Areawide patterns of azinphosmethyl resistance in adult male Platynota idaeusalis (Lepidoptera:Tortricidae) in southcentral Pennsylvania. J. Econ. Entomol. 83, 1194–1200.

Knight, A.L., Beers, E.H., Hoyt, S.C., Riedl, H., 1990b. Acaracide bioassays with spider mites (Acari: Tetranychidae) on pome fruits: evaluation of methods and selection of discriminating concentrations for resistance monitoring. J. Econ. Entomol. 83, 1752–1760.

Kranthi, K.R., 2005. Insecticide Resistance -Monitoring, Mechanisms and Management Manual. Published by CICR, Nagpur, India and ICAC, Washington DC.

Kranthi, K.R., Russell, D., Wanjari, R., Kherde, M., Munje, M., Lavhe, N., et al., 2002. In-season changes in resistance to insecticides in Helicoverpa armigera (Lepidoptera: Noctuidae) in India. J. Econ. Entomol. 95, 134–142.

Leeper, J.R., 1984. Monitoring fall armyworm for susceptibility. Fla. Entomol. 67, 339–342.

Magalhaes, L.C., Van Kretschmar, J.B., Barlow, V.M., Roe, R.M., Walgenbach, J.F., 2012. Development of a rapid resistance monitoring bioassay for codling moth larvae. Pest Manag. Sci. 68, 883–888.

Marcon, P.C.R.G., Siegfried, B.D., Spencer, T., Hutchison, W.D., 2000. Development of diagnostic concentrations for monitoring Bacillus thuringiensis resistance in European corn borer (Lepidoptera: Crambidae). J. Econ. Entomol. 93, 925–930.

Martinson, T.E., Nyrop, J.P., Dennehy, T.J., Reissig, W.H., 1991. Temporal variability in repeated bioassays of field populations of European red mite (Acari:Tetranychidae): implications for resistance monitoring. J. Econ. Entomol. 84, 1119–1127.

McCullagh, P., Nelder, J.A., 1989. Generalized Linear Models. second ed. Chapman & Hall, London, p. 511.

Miller, A.L.E., Tindall, K., Leonard, B.R., 2010. Bioassays for monitoring insecticide resistance. J. Vis. Exp.46, art. no. e2129.

Nauen, R., Elbert, A., 2003. European monitoring of resistance to insecticides in Myzus persicae and Aphis gossypii (Hemiptera:Aphididae) with special reference to imidacloprid. Bull. Entomol. Res. 93, 47–54.

Osorio, A., Martinez, A.M., Schneider, M.I., Diaz, O., Corrales, J.L., Aviles, M.C., et al., 2008. Monitoring of beet armyworm resistance to spinosad and methoxyfenozide in Mexico. Pest Manag. Sci. 64, 1001–1007.

Ouyand, Y., Chueca, P., Scott, S.J., Montez, G.H., Grafton-ardwell, E.E., 2010. Chlorpyrifos bioassay and resistance monitoring of San Joaquin Valley California citricola scale populations. J. Econ. Entomol. 103, 1400–1404.

Peterson, J.L., 2005. An extension perspective on monitoring pesticide resistance. J. Ext. 43, 1077.

Plapp Jr., F.W., Campanhola, C., Bagwell, R.D., McCutchen, B.F., 1990a. Management of pyrthroid-resistant tobacco budworms on cotton in the United States. In: Roush, R.T., Tabashnik, B.E. (Eds.), Pesticide Resistance in Arthropods. Chapman and Hall, New York, Chapter 9.

Plapp, F.W., Jackman, J.A., Campanhola, C., Frisbie, R.E., Graves, J.B., Luttrell, R.G., et al., 1990b. Monitoring and management of pyrethroid resistance in tobacco budworm (Lepidoptera: Noctuidae) in Texas, Mississippi, Louisiana, Arkansas, and Oklahoma. J. Econ. Entomol. 83, 335−341.

Reissig, H., Welty, C., Weires, R., Dennehy, T., 1987. Monitoring and management of resistance of the European red mite to plictran in New York apple orchards. Proceedings of the Annual Meeting of Mass. Fruit Growers' Association. 93, pp. 97−107.

Reissig, W.H., Stanley, B.H., Hebding, H.E., 1986. Azinphosmethyl resistance and weight-related response of obliquebanded leafroller (Lepidoptera: Tortricidae) larvae to insecticides. J. Econ. Entomol. 79, 329−333.

Reyes, M., Sauphanor, B., 2008. Resistance monitoring in codling moth: a need for standardization. Pest Manag. Sci. 64, 945−953.

Riedl, H., Seaman, A., Henrie, F., 1985. Monitoring susceptibility to azinphosmethyl in field populations of the codling moth (Lepidoptera: Tortricidae) with pheromone traps. J. Econ. Entomol. 78, 692−699.

Riley, S.L., 1990. Pyrethroid resistance in Heliothis spp.: Current monitoring and management programs. ACS Symp. Ser. 421, 134−148.

Robertson, J.L., Smith, K.C., Savin, N.E., Lavigne, R.J., 1984. Effects of dose selection and sample size on the precision of lethal dose estimates in dose-mortality regression. J. Econ. Entom. 77, 833−837.

Rogers, B., Fleece, H.D., Riley, S.L., Staetz, C.A., Whitehead, J., Mullins, W., 1990. Heliothis resistance monitoring in the cotton belt: PEG US update. Proc. Beltwide Cotton Res. Conf. 1990, 163−165.

Roush, R.T., Miller, G.L., 1986. Considerations for design on insecticide resistance monitoring programs. J. Econ. Entomol. 79, 293−298.

Rust, M.K., Denholm, I., Dryden, M.W., Payne, P., Blagburn, B.L., Jacobs, D.E., et al., 2011. Large-scale monitoring of imidacloprid susceptibility in the cat flea, Ctenocephalides felis. Med. Vet. Entomol. 25, 1−6.

Sakuma, M., 2013. The Instructions for PriProbitver. 1.63. Kyoto University, Japan. Available for free from at: <http://www.ars.usda.gov/Services/docs.htm?docid=11284>.

Sanderson, J.P., Rousch, R.T., 1992. Monitoring insecticide resistance in greenhouse whitefly (Homoptera: Aleyrodidae) with yellow sticky cards. J. Econ. Entomol. 85, 634−641.

Schneider-Orelli, O., 1947. Entomologisches Praktikum. Aarau, 2. Auflage.

Schouest Jr., L.P., Miller, T.A., 1991. Field incubation of insects for insecticide toxicity assays in resistance monitoring programs. Environ. Entomol. 20, 1526−1530.

Shearer, P.W., Varela, L.G., Riedl, H., Welter, S.C., Jones, V.P., 1994. Topical pheromone trap assay for monitoring insecticide resistance of Phyllonorycter elmaella (Lepidoptera: Gracillariidae). J. Econ. Entomol. 87, 1441−1449.

Siegfried, B.D., Meinke, L.J., Parimi, S., Scharf, M.E., Nowatzki, T.J., Zhou, X., et al., 2004. Monitoring western corn rootworm (Coleoptera: Chrysomelidae) susceptibility to carbaryl and cucurbitacin baits in the areawide management pilot program. J. Econ. Entomol. 97, 1726−1733.

Slater, R., Ellis, S., Genay, J.P., Heimbach, U., Huart, G., Sarazin, M., et al., 2011. Pyrethroid resistance monitoring in European populations of pollen beetle (*Meligethes* spp.): A coordinated approach through the Insecticide Resistance Action Committee (IRAC). Pest Manag. Sci. 67, 633−638.

Slater, R., Paul, V.L., Andrews, M., Garbay, M., Camblin, P., 2012. Identifying the presence of neonicotinoid resistant peach-potato aphid (*Myzus persicae*) in the peach-growing regions of southern France and northern Spain. Pest Manag. Sci. 68, 634−638.

Smith, K.C., Robertson, J.L., 1984. DOSESCREEN: a computer program to aid dose placement. Gen. Tech. Rep. PSW-78. Pacific SW Forest and Range Expt Station, Forest Service, USDA, Berkeley, CA. p. 12.

Smith, K.C., Savin, N.E., Robertson, J.L., 1984. A Monte Carlo comparison of maximum likelihood and minimum chi square sampling distributions in logit analysis. Biometrics. 40, 471−482.

Sparks Jr., A.N., Norman Jr., J.N., Raulston, J.R., Wolfenbarger, D.A., 1990. *Heliothis* resistance monitoring: effect of age and gender of moths on cypermethrin. Proc. Beltwide Cotton Res. Conf. 1990, 163−165.

Staetz, C.A., 1985. Susceptibility of *Heliothis virescens* (F.) (Lepidoptera: Noctuidae) to permethrin from across the cotton belt: a five-year study. J. Econ. Entomol. 78, 505−510.

Tabashnik, B.E., Patin, A.L., Dennehy, T.J., Liu, Y.B., Carrière, Y., Sims, M.A., et al., 2000. Frequency of resistance to *Bacillus thuringiensis* in field populations of pink bollworm. Proc. Natl. Acad. Sci. USA. 97, 12,980−12,984.

Tabashnik, B.E., Fabrick, J.A., Henderson, S., Biggs, R.W., Yafuso, C.M., Nyboer, M.E., et al., 2006. DNA screening reveals pink bollworm resistance to *Bt* cotton remains rare after a decade of exposure. J. Econ. Entomol. 99, 1525−1530.

United States Environmental Protection Agency (US EPA), 2010. Biopesticides Registration Document: Bacillus thuringiensis Cry34Ab1 and Cry35Ab1 Proteins and the Genetic Material Necessary for Their Production (PHP17662 T-DNA) in Event DAS-59122-7 Corn (OECD Unique Identifier: DAS-59122-7). Office of Pesticide Programs, Biopesticides and Pollution Prevention Division.PC Code: 006490 (link: <http://www.epa.gov/oppbppd1/biopesticides/pips/2010%20Cry3435Ab1%20BRAD.pdf>).

van Kretschmar, J.B., Bailey, W.D., Arellano, C., Thompson, G.D., Sutula, C.L., Roe, R.M., 2011. Feeding disruption tests for monitoring the frequency of larval lepidopteran resistance to Cry1Ac, Cry1F and Cry1Ab. Crop. Prot. 30, 863−870.

van Kretschmar, J.B., Cabrera, A.R., Bradley, J.R., Roe, R.M., 2013. Novel adult feeding disruption test (FDT) to detect insecticide resistance of lepidopteran pests in cotton. Pest Manag. Sci. 69, 652−660.

Venette, R.C., Hutchison, W.D., Andow, D.A., 2000. An in-field screen for early detection of insect resistance in transgenic crops: practical and statistical considerations. J. Econ. Entomol. 93, 1055−1064.

Venette, R.C., Moon, R.D., Hutchison, W.D., 2002. Strategies and statistics of sampling for rare individuals. Annu. Rev. Entomol. 47, 143−174.

Wedderburn, R.W.M., 1974. Quasi-likelihood functions, generalized linear models, and the Guass-Newton method. Biometrika. 61, 439−447.

Wu, K., Guo, Y., Hea, G., 2006. Resistance monitoring of *Helicoverpa armigera* (Lepidoptera: Noctuidae) to Bt insecticidal protein during 2001−2004 in China. J. Econ. Entomol. 99, 893−898.

Yue, B., Huang, F., Leonard, B.R., Moore, S., Parker, R., Andow, D.A., et al., 2008. Verifying an F1 screen for identification and quantification of rare *Bacillus thuringiensis* resistance alleles in field populations of the sugarcane borer, *Diatraea saccharalis*. Entomol. Exp. Applic. 129, 172−180.

Zhao, J.Z., Li, Y.X., Collins, H.L., Shelton, A.M., 2002. Examination of the F_2 screen for rare resistance alleles to *Bacillus thuringiensis* toxins in the diamondback moth (Lepidoptera: Plutellidae). J. Econ. Entomol. 95, 14—21.

Zhao, J.Z., Collins, H.L., Li, Y.X., Mau, R.F.L., Thompson, G.D., Hertlein, M., et al., 2006. Monitoring of diamond back moth (Lepidoptera: Plutellidae) resistance to spinosad, indoxacard, and emamectin benzoate. J. Econ. Entomol. 99, 176—181.

> **CHAPTER 16**

IPM and Insect Resistance Management

David W. Onstad
DuPont Pioneer, Wilmington, DE

Chapter Outline

The final chapter of the first edition predicted that insect resistance management (IRM) would change quickly and significantly over just a few years. The changes and additions to this second edition support that prediction. However, most IRM specialists could not have predicted all of the many problems and concepts that have developed over the past six years. The events of the past several years have also demonstrated the need to emphasize integrated pest management (IPM) to protect crops, livestock, humans, and the environment.

Insect Resistance Management
DOI: http://dx.doi.org/10.1016/B978-0-12-396955-2.00016-3

In this chapter, I present several case studies concerning synthetic chemical insecticides that provide additional lessons for IRM scholars. Then I describe recommendations for IRM practitioners that can guide IRM over the next several years.

CASE STUDIES

Leptinotarsa decemlineata

One of the most serious pests of potato (*Solanum tuberosum*) around the world is *Leptinotarsa decemlineata*. Roush and Tingey (1992) described the biological and management factors that contribute to the evolution of resistance by *L. decemlineata* and proposed an IRM strategy for insecticides. Three important biological characteristics of this species include its 40-fold population growth during each generation, the tendency for the populations to inhabit only one or a few treated crops, and the existence of a single mechanism capable of causing broad cross-resistance. The use of soil insecticides and the treatment of all active stages of the pest are the management practices that contribute to the evolution of the pest. Based on this knowledge of the pest and the agro-ecosystem, Roush and Tingey (1992) recommended that insect resistance be managed by the replacement of soil insecticides with foliar applications and the rotation of insecticides each generation. In addition, monitoring pest densities and applying insecticides only when needed along field edges reduces insecticide use (Roush and Tingey, 1992). They also recommended the use of annual rotation of potatoes with nonhost plants to reduce the pest population in alternating years. Crop rotation also supports the reduction in insecticide use, thereby lowering the selection pressure. Thus, the IRM plan reduces population growth, the proportion of each population treated during a generation, and the frequency of insecticide use. Alyokhin *et al.* (2013) also advocate IPM for this pest, but caution that *L. decemlineata* has the potential to evolve resistance to crop rotation through extended diapause.

Bemisia tabaci

Bemisia tabaci is a serious cosmopolitan pest of many crops, including cotton (*Gossypium hirsutum*) and vegetables. It causes three kinds of damage: (1) direct damage through piercing and sucking sap from plant foliage,

(2) indirect damage caused by the accumulation of honeydew produced by *B. tabaci* which leads to mold growth on foliage, and (3) transmission of viruses. *B. tabaci* has evolved resistance to many types of insecticides (see Table 1.1 in Chapter 1). Recently, insecticides with new modes of action, such as the neonicotinoid insecticides, insect growth regulators that inhibit chitin biosynthesis, and a juvenile hormone mimic, have been introduced.

The Arizona cotton IPM program consists of a multilevel, multicomponent approach that focuses *B. tabaci* management on three subjects: sampling, effective chemical use, and avoidance (see Figure 16.1) (Ellsworth and Martinez-Carrillo, 2001; Palumbo *et al.*, 2003). Naranjo and Ellsworth (2009) described the development, evolution, validation, and implementation of the IPM program for *B. tabaci*. Research-based economic thresholds were integrated with rapid sampling plans into validated decision-making tools that were widely adopted by consultants and growers. The *B. tabaci* economic threshold helps the Arizona cotton grower or consultant decide when to make insecticide applications that will minimize crop damage. Naranjo and Ellsworth (2009) described how extensive research that measured the interactions among pest populations,

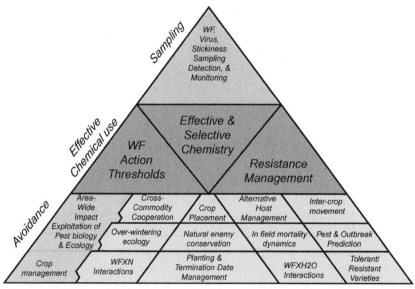

Figure 16.1 Successful management of *B. tabaci* depends on a range of tactics that include effective chemical use and IRM, and requires coordination across crops. *Reprinted from Ellsworth and Martinez-Carrillo (2001) with permission from* Crop Protection *and Elsevier.*

indigenous natural enemies, and selective insecticides (described above) demonstrated the complementary roles played by chemical and biological control in cotton IPM. Selective insecticides preserve natural enemies while providing effective *B. tabaci* control. The improved cotton IPM program produced a 70% reduction in foliar insecticides, more than $200 million worth of reduced control costs, and yield-loss over 14 years (Naranjo and Ellsworth, 2009).

Successful adoption of the "pyramid" approach has depended on organized education efforts through the University of Arizona Extension Service, including grower and/or crop consultant training and literature. The program has also benefited from simultaneous improvements in IPM for other cotton pests. Since 1996, transgenic insecticidal cotton has been widely adopted to control *Pectinophora gossypiella* and other lepidopteran pests in Arizona, replacing more broad-spectrum insecticides (Onstad and Knolhoff, Chapter 9) and thereby conserving biological control agents.

IPM and IRM programs for *B. tabaci* often must focus on several crops in a landscape because of the broad host range of this pest. For example, the University of Arizona, in cooperation with the U.S. National Cotton Council, developed a multiple commodity *B. tabaci* management program with the objective of managing *B. tabaci* and harmonizing insecticide use in multicrop systems (Palumbo *et al.*, 2003; Figure 16.1). The goal of this program is to maximize the efficacy and longevity of all current insecticides for all crops in Arizona attacked by *B. tabaci*. The program is based on a set of guidelines for insecticide use that differ depending on the crop and the cropping system. Recommendations are provided for insecticide use on cotton, melons, and vegetables that vary depending on whether the cropping system is cotton-intensive, cotton and melons, or a multicrop system. Table 16.1 indicates the maximum number of neonicotinoid applications recommended for cotton, melons, and vegetables in these different cropping systems. For instance, in a cotton/melon system, one neonicotinoid application can be made on both cotton and melons, whereas in a multicrop system, one application can be made on both melons and vegetables but no neonicotinoid applications should be made on cotton. These guidelines are intended to avoid sequential exposure of multiple generations of *B. tabaci* to insecticides with the same mode of action, with particular emphasis on limiting the total number of neonicotinoid applications (i.e., no more than two neonicotinoid applications should be made per season in any cropping system).

Table 16.1 Recommended Maximum Number of Neonicotinoid Insecticide Applications per Crop Season in Three Different Sorts of Cropping Systems in Arizona (Palumbo *et al.*, 2003)

Cropping System	Cotton	Melons	Vegetables
Multicrop	0	1[a]	1[b]
Cotton/melon	1	1[a]	—
Intensive cotton	1	—	—

[a]Soil applications only.
[b]Soil or foliar applications.

Crowder *et al.* (2006, 2008) evaluated IRM strategies for *B. tabaci* and pyriproxyfen, a juvenile hormone analogue, using mathematical models. They discovered that in some situations, strategies that delayed resistance were effective from an IPM perspective. However, some strategies that delayed resistance resulted in higher population densities. Results suggested that cotton growers could adjust several operational and environmental factors to delay the evolution of resistance to pyriproxyfen. Crowder *et al.* determined that refuges for susceptible *B. tabaci* could delay resistance to pyriproxyfen. Carrière *et al.* (2012) tested the refuge strategy with eight years of data on refuges and resistance to pyriproxyfen in 84 populations of *B. tabaci* from cotton fields in central Arizona. They found that spatial variation in resistance within each year was not affected by refuges in other crops near cotton fields. However, resistance was negatively associated with the area of cotton refuges and positively related to the area of cotton treated with pyriproxyfen. Carrière *et al.* (2012) used a statistical model (based on the first four years of data) to adequately predict the spatial variation in resistance observed in the last four years of the study. These results confirmed that cotton refuges delayed resistance and treated cotton fields accelerated resistance.

Cydia pomonella

Chemical insecticides, pheromones, and viral insecticides are used to control *Cydia pomonella*, a major orchard pest found in most temperate regions of the world. In a series of studies, Boivin *et al.* (2003, 2004, 2005) investigated the evolution of changes in maturation in *Cydia pomonella* in France. Boivin *et al.* (2003) investigated the pleiotropic effects of resistance on maturation. Resistant homozygotes had significantly shorter developmental times relative to susceptible homozygotes.

The resistant moths had much higher propensity for diapause than susceptible phenotypes. Boivin *et al.* (2004) investigated whether genetic variation associated with selection for insecticide resistance may be a source of divergence in the photoperiodic timing of diapause through pleiotropic interactions. Boivin *et al.* (2004) observed higher critical photoperiods for diapause induction in resistant homozygotes than in other genotypes. Boivin *et al.* (2005) then compared model simulations for each genotype with pheromone-trap catches recorded in insecticide-treated orchards over an eight-year period. They found a significant delay in adult emergence relative to the prediction of the model for susceptible homozygotes. The delay was positively correlated with increasing frequencies of insecticide resistance in the sampled field population. The predicted emergence for resistant homozygotes matched those recorded in the field. Boivin *et al.* (2005) suggested that phenological modeling can be used as a forecasting tool for IRM.

Tetranychus urticae

Tetranychus urticae is the most common pest of orchards and a frequent target of pesticide applications. This mite has a long history of evolving resistance to acaricides. Flexner *et al.* (1989) determined that fitness costs and immigration of susceptibles could cause reversion of acaricide resistance when selection pressure is relaxed. Flexner *et al.* (1995) also concluded that the immigration of susceptible mites into pear orchards (*Pyrus* sp.) could be important for IRM. They studied the dynamics of resistance in *T. urticae* in pear orchards for seven years. They compared five treatments involving two acaricides: (1 and 2) consecutive use of one acaricide (two applications per year), (3) alternation of both within a single year, (4) rotation of both on a yearly basis, and (5) a combination at half rates of both acaricides. Flexner *et al.* (1995) concluded that the field durability of the acaricides was not extended by rotations or half-rate combinations compared with consecutive uses. Alternate, consecutive uses may give greater than 33% longer control compared with control for other programs. However, this advantage depends on which acaricide is used first, because one acaricide conferred cross-resistance to the other. Flexner *et al.* (1995) concluded that better IPM, including the use of economic thresholds and biological control, could reduce the number of applications and delay resistance.

Musca domestica

Few scientists have performed field experiments in which the population dynamics and behavior of a pest are modified. These studies are even more difficult when immigration of susceptible insects is manipulated. Imai (1987) performed field experiments to determine how immigration of susceptibles could reduce insecticide resistance in a population of *Musca domestica*. Imai (1987) released 163,000 susceptible flies to mate with resistant flies at a waste disposal site in late 1977. Five months after the releases, susceptibility increased in the field population. For a second experiment, between 31,000 and 46,000 susceptible pupae were used in each of five releases during late 1980. Genetic markers were used to permit morphological identification of susceptible flies. The field population became more susceptible within six months after the second series of susceptible-fly releases.

GUIDELINES FOR MANAGING INSECT RESISTANCE

The following section highlights rules that emphasize management. Ecology, molecular biology, and population genetics are important foundations for IRM, but success in the real world depends on using knowledge to accomplish important goals. These rules can be used to keep our attention on the management side of IRM. They are not listed in order of importance or in the sequence that they might be followed. They are all important and should be considered simultaneously.

Always Consider IRM within an IPM Framework

As explained in Chapter 1, IRM will be more effective and more valuable when it is incorporated into integrated pest management (IPM). The best IRM will take advantage of the best IPM, including all design and control opportunities. Both IPM and IRM must deal with problems related to pest phenology and pest behavior, and sometimes these problems can be turned into opportunities to improve management. IRM may complicate IPM because of the use of refuges for susceptibles and tactics involving negative cross-resistance, but these complications will be less than those caused by the evolution of resistance. As with IPM, IRM cannot simply rely on and reuse old tactics. New IRM strategies must be proposed, evaluated, implemented, and evaluated again to improve management of insect resistance.

The relationship between IPM and IRM may not always be easy, however, particularly if stakeholders and scientists continue to view the goals of IPM and IRM as separate issues. For example, Onstad *et al.* (2003) warned about the conflicts that will continue to exist, if IRM is kept separate from and perhaps even elevated above IPM. They stated:

> *Management recommendations and decisions by growers must be made even when we do not know how evolution will develop and when new phenotypes will invade our landscapes. IPM must address management concerns without being overwhelmed by these uncertainties and being overridden by IRM. In its best sense, IPM has accounted for both the ecological and economic factors of a problem. Too often in the past, evolutionary changes in pest populations have caused scientists to emphasize IRM over IPM, as if one gene and one risky tactic was more important than basic IPM. Perhaps this narrow focus on IRM makes sense when only one tactic is available. But some believe that IPM cannot be true IPM if it permits evolution of resistance. From this perspective the extensive use of crop rotation (for insect control), though very effective for many years, was not truly IPM. However, resistance by insects to highly effective IPM strategies is inevitable, but long-term success for IPM is not inevitable. Thus, we must constantly strive for better approaches that combine landscape design, host plant resistance, biological control, and other feasible tactics for pest management (p. 1884).*

Always Have Explicit Goals and a Time Horizon

Decision makers must have clear goals with explicit descriptions of the time horizon and the area over which IPM and IRM will be implemented and evaluated. A time horizon is a time period during which we observe, measure, and manage an ecological or agricultural system. A time horizon permits the specification of a clear point in time (at the end of the horizon) at which a final decision will be made. Thus, at the end of a given time horizon, the effectiveness of management is evaluated. For example, when you invest money in stocks or bonds, your goal is to earn a significant return on your investment over a time horizon. During the period, stock values will rise and fall, but the primary issue is whether you will have significantly more money at the end of the time horizon when you want to spend the money.

A time horizon also defines the set of observations and calculations that can be compared with those from another scenario for the same period. Time horizons and spatial scales are rarely specified, even in a vague way, except in model simulations. Modelers encourage stakeholders to be more rigorous in the specification of their goals, time horizons, and spatial scales.

Goals and management plans can take many forms. Is the goal to delay evolution of a gene over 20 generations in one county? Is the goal to control the pest over 40 generations in the Cotton Belt of the United States? Is the goal to minimize cost, including damage and control costs, on one farm until retirement of the farmer? Each stakeholder may have a goal: some similar and some different. These will influence the goals of the agencies, corporations, and individuals that make the actual IPM and IRM decisions. Thus, political and economic factors influence which goals are emphasized.

Account for the Issues Related to Coordination of Stakeholders

Coordination is critical for successful IRM (Chapter 1). At a minimum, coordination permits management of resistance across a large region and over a long period of time. The coordination may lead to actual cooperation among stakeholders and the sharing of resources and costs. Otherwise coordination can be implemented, possibly less effectively, by a centralized agency with legal authority to require certain IRM practices. Mosquito control agencies are an obvious example of the latter scenario. In the interest of public health, one or two agencies coordinate regional control of mosquito populations and manage the resistance to control tactics.

In agriculture, the Arizona case with transgenic insecticidal cotton and *P. gossypiella* clearly demonstrates the value of cooperation among public agencies and groups of growers. Data collection and education were facilitated by this cooperation (Chapter 9). Cox and Forrester (1992) present a fine summary of the characteristics of the cotton insecticide IRM program in Australia and highlight the important roles that coordination and cooperation played in its success. In fact, they defined IRM as "social technology." Caprio (1994) demonstrated how cooperation by farmers influences evolution of resistance in one of his models. His results suggest that some stakeholders will not always benefit directly from cooperation. Nevertheless, coordination may be able to take advantage of many resources and behaviors not typically included in models. For instance, the coordination of IRM and IPM across major cropping systems for major pests in Arizona, including *Bemisia*, is a lesson that should be learned by all working on future IRM problems.

Adjust the IRM Plan for Local Social and Environmental Conditions

Head and Savinelli (2008) focused our attention on this issue. They contrasted cases in which social and educational conditions permit optimal or complex IRM plans with situations in which constraints in these conditions limit the IRM strategies and focus consideration on more practical options. The example of *Bemisia* IPM and IRM in Arizona also demonstrates how management can be adjusted and optimized for local, social, and environmental conditions. The scientists devising strategies accounted for the needs of local farmers growing multiple crops. The work of Onstad *et al.* (2002) showed that IRM for transgenic insecticidal crops must be different under irrigated and nonirrigated conditions. In the United States, transgenic insecticidal corn is regulated differently in regions with transgenic insecticidal cotton than it is in regions without cotton. In Texas where cotton is grown, refuges must be 50% of the cornfields, whereas in northern states the refuge can be 20% of cornfields. Chapter 10 describes many other situations in which the landscape must be considered in the management of insect resistance. The bottom line is that one to three strategic options for IRM do not cover all the pests in all environments managed according to all possible goals of stakeholders. IRM strategies must be flexible and dynamic.

Evaluate the Economics of Each IRM Strategy

Management requires goals and a time horizon, and both of these must reflect the interests and values of the stakeholders and decision makers. Human values may not always be quantifiable and capable of being expressed in traditional economic terms, but economics can certainly be a starting point for discussion of these values, trade-offs, and differences among stakeholders. It is often difficult to place an economic value on natural resources and ecosystem services, but it is also difficult to justify an extremely narrow focus on only one resource, such as pest susceptibility, without considering the consequences. At a minimum, stakeholders should discuss how they value goods and services in the present versus those in the future and which aspects of the pest management system they value the most (Chapter 2). Some stakeholders will value ecosystem services more than they value crop yield, while others will value crop yield more.

Economics provides methods for allocating resources and time to solve multiple problems with efficiency and equity. This allocation of money occurs at various scales, which may or may not be apparent to the stakeholder. For example, Cox and Forrester (1992) evaluated the economics of cotton insecticide IRM in Australia. They concluded that there were substantial economic returns from IRM and that cotton IRM benefited from a low cost of implementation. Society's choices will always be of the form: Should we allocate scarce resources for (1) choices related to IRM for a single pest, (2) choices for IPM concerning a single pest on a single crop, (3) protection of multiple crops from multiple pests, or (4) crop protection and other important environmental problems within the overall economic system? Without the use of economic analyses, allocations will likely be less equitable and efficient. Will our grandchildren believe that we allocated too much or too little for IRM compared to managing global climate change?

Understanding economics and human behavior will also help scientists better predict cooperation and compliance with IRM strategies and regulations (Chapter 13). Farmers will likely consider the economics of IRM adoption and compliance in the context of their overall business and farming operation.

Predict the Risks and Implement Preventative IRM

Scientific IRM must provide support for strategies that delay the evolution of resistance or at least manage it within the goals of the stakeholders (Chapter 1). Scientific IRM uses general hypotheses and models to initiate plans and then obtains and analyzes data specific to the pest species and the management system to improve the strategies.

Roush (1997, 1999), Curtis et al. (1993), and Siegfried et al. (1998) advocate preventative IRM instead of curative or reactive strategies. Roush (1997) states; "The substitution of new toxicants to replace those that have failed is not resistance management, as it does nothing to preserve susceptibility" (p. 272). Curtis et al. (1993) conclude that reactive IRM and the switching of chemicals "can hardly be called a strategy for resistance management" (p. 3).

Prediction is the foundation for prevention. Although Gould (1998) and Carrière (2003) have demonstrated how the history of species in natural environments can be used to explain and predict the consequences of IRM (see also Chapter 3), they likely would agree that modeling is one of the best tools for predicting future consequences and scenarios (Chapter 14). For example, Caprio (1998) used a stochastic model to evaluate five IRM strategies: sequential introduction of two toxins, rotations and mosaics of toxins, and half-and full-rate mixtures of the toxins. When an economic threshold for applying an insecticide based on pest density was adopted along with refuges, Caprio (1998) demonstrated that full-rate mixtures have the potential to effectively delay resistance evolution. Simulations comparing monogenic and polygenic inheritance of resistance indicated that resistance took twice as long to evolve in the polygenic simulations. Caprio *et al.* (2006) used a similar model to investigate the history and assess the future risks of insecticide resistance by *Diabrotica virgifera virgifera*.

Tabashnik and Carrière used a combination of modeling and empirical work to explain the success of IRM for *P. gossypiella*. In 1997, extensive plantings of transgenic insecticidal cotton began to exert selection pressure on populations in Arizona. Carrière and Tabashnik (2001) used models to understand how the high-dose/refuge IRM strategy could be used to delay and even reverse resistance. (The high dose of toxin is expected to kill almost all heterozygotes, making the resistance gene functionally recessive.) Monitoring for resistance indicated that mean frequency of resistance by *P. gossypiella* did not increase over an eight-year period (Tabashnik *et al.*, 2005). Tabashnik *et al.* (2005) concluded that this delay in resistance can be explained by four factors. Arizona farmers complied with IRM regulations by planting nontoxic cotton refuges, which were sources of susceptible moths. With regard to population genetics of the pest, resistance is recessive but fitness costs are significant. Tabashnik *et al.* (2005) also considered incomplete resistance, or the limited survival of resistant homozygotes on transgenic cotton, to be another factor causing the delay in resistance evolution.

Understand and Possibly Alter Pest Behavior

Case studies throughout the book have demonstrated the importance of pest behavior in IRM. Waldstein *et al.* (2001) measured the rate at which larvae of *Choristoneura rosaceana* abandoned feeding sites on apple branches. Larvae frequently changed feeding sites, switching from older leaves to actively growing foliage with sublethal insecticide residues.

Waldstein *et al.* (2006) postulated that this behavior may increase larval survival and could slow evolution of resistance to insecticide by providing a refuge for susceptible insects. At a minimum, behavior must be understood to improve IRM or prevent disasters (Gould, 1991). In some cases, we can take advantage of pest behavior to cause the targeted pest to react to management in a certain way. However, we also should be aware that pest behavior can evolve. One of the oldest examples of behavioral resistance was avoidance of malathion baits in *Musca domestica* (Schmidt and LaBreoque, 1959). Gould (1984) was the first to model the evolution of behavioral resistance.

Onstad and Buschman (2006) used a model of *Ostrinia nubilalis* to evaluate the effectiveness of oviposition deterrence in transgenic insecticidal cornfields (*Zea mays*) for IRM. The population genetics of two genes was simulated: one for resistance to transgenic insecticidal corn and one for insensitivity to deterrence. They simulated two types of hypothetical deterrence: one has moths reducing their oviposition because of lost opportunities to lay eggs, and the other has the deterred moths moving to the refuge to lay the eggs. Oviposition deterrence was clearly effective in extending the time to resistance to transgenic insecticidal corn. The time to 50% frequency for the allele for resistance to transgenic corn was similar for the two types of simulated deterrence, but the pest densities were 100-fold higher when the deterred moths oviposited in the refuge.

Monitor Resistance Only When the Benefits are Worth the Costs

To some extent, monitoring programs for IRM are based on pessimism. The ability of arthropods to evolve and the chaotic and stochastic dynamics of natural systems should not be underestimated. Thus, scientists and stakeholders will always consider designing and implementing monitoring plans.

Stanley (Chapter 15) states that monitoring programs can be very expensive if rare resistant arthropods are sampled for. We can more cheaply monitor for high levels of resistance in a population. Thus, it is feasible to determine whether a control tactic has already failed or will fail relatively quickly. On the other hand, we will monitor what we value. Hence, the more valuable the damage caused by the pest, the more likely we can justify the cost of a monitoring program. For this reason, we should expect greater monitoring for resistance in vectors of human and livestock diseases. The World Health Organization and the Insecticide

Resistance Action Committee (2006) encourage the monitoring of vectors of human disease using bioassays designed or approved by the World Health Organization, but the feasibility and costs of surveys are not highlighted in most planning documents.

Nyrop et al. (1986) suggested that the cost of sampling and the uncertainty of the information provided by the sample be incorporated into cost-benefit analyses for decision making on pest control. Without such a procedure, sampling too often is perceived as free by many scientists and stakeholders.

Sampling can be more easily justified if it is a normal part of IPM. Stakeholders may already sample for insects or the damage caused by the pests. Additional steps could be added to the process, such as increasing the number of units sampled or sending specimens to a central agency for processing. Sampling that is valuable for IPM may be adjusted to contribute to IRM. One approach would be to use the thousands of farmers or ranchers as the monitors for the consequences of resistance. This would permit the feasible monitoring of a much larger region to determine when resistance becomes a problem. In this scenario, much greater coordination and training would be required than that needed for IPM monitoring.

Prepare for Evolution of Detoxification, Behavioral Modification, and Mechanisms of Maturation

Arthropods can evolve resistance in many ways. Every control tactic used against the pests can be overcome by modification of more than one mechanism. Chapters 3-4 present the wide variety of molecular, biochemical, and anatomical mechanisms of detoxification. Chapters 1, 7, and 8 provide examples of new behaviors that have evolved in arthropods. Resistance to mating disruption in male moths that changes their behavioral responses to pheromones is an interesting example of behavioral resistance (Chapter 1). Crowder et al. (2012) review the use of models to explore the evolution of resistance to juvenile hormone mimics, which some people predicted, many years ago, could not happen. Arthropods can also be selected for changes in maturation that either shorten or prolong one or more stages in their life cycles (Chapters 7 and 14). The case of Cydia pomonella presented in the first section of this chapter demonstrates the significance of this phenomenon. Changes in maturation, particularly diapause, permit the pest to avoid harm from a control tactic (Carrière et al., 1995).

Of course, two or more mechanisms can evolve simultaneously. Thus, scientists must be prepared to deal with a variety of mechanisms. In fact, every process in a population dynamics model can be a mechanism of resistance evolution and be represented in a population-genetics model with genetic variation that can be selected for or against. The more details the demographic and behavioral model contains, the more complicated the genetic model can be. The key is to determine which processes are under the greatest selection pressure and how much genetic variation exists for that mechanism. Perhaps some day, we will have a large database with information about genetic variation that will help us make predictions.

Databases with information about resistance mechanisms may also allow us to more easily take advantage of negative cross-resistance (Chapter 11). Genomic information concerning intoxication and detoxification will obviously be valuable knowledge (Chapter 3).

Do not Delay Implementing IRM

Once we accept the high probability of resistance to effective pest management, then we must consider how quickly we should implement IRM. Certainly, economics and local social and cultural conditions influence implementation (Guidelines D and E). Nevertheless, the work of several authors supports the argument that IRM should be implemented relatively soon after a control tactic becomes part of an IPM program. We should not wait until we understand the pest and its ecosystem perfectly.

Croft (1990) stated that many scientists believe that, although genetics, biochemistry, and physiology determine whether resistance evolves, these are usually not the primary determinants influencing the rate of resistance evolution. Instead it is the ecological genetics of resistance (including gene flow and population dynamics of selection) that largely determines the progression of resistance evolution.

McCafferty (1999) reviewed the status of resistance by Lepidopteran species in the genera *Heliothis* and *Helicoverpa*. Pest species in these genera are polyphagous, distributed worldwide, and have evolved resistance to a variety of insecticides. McCafferty (1999) concluded that, despite our rapidly increasing knowledge of the biochemical and molecular nature of resistance, the most effective IRM for the heliothine Lepidoptera remains a strict control of insecticide use. He cited the cotton program in Israel, which included a dramatic reduction in the number of insecticide applications and the use of several IPM tactics, as the most successful IRM program for these pests.

Roush (1989) clearly stated that stakeholders should not wait until genetic or molecular mechanisms of resistance or fitness costs are understood before implementing IRM. He argued that most IRM strategies should be effective against most if not all mechanisms and that the strategy can and should evolve as more knowledge is learned about the system. He even believed that species- and system-specific modeling would not be necessary to choose and implement the initial IRM strategy for an insecticide. His main concerns about implementation are social issues, such as stakeholder compliance with a rational strategy (Roush, 1994).

CONCLUSION

Insect resistance management requires the best and the brightest scientists and stakeholders to accept the ultimate challenge of applied biological sciences: predict how one or more species will evolve in a dynamic, heterogeneous environment influenced by arrogant and often irrational humans. Furthermore, once the prediction or range of predictions is made, design a strategy that counteracts the insect's inherent advantage and the human weaknesses. Under these circumstances, pessimism may be expected. Nevertheless, society and IRM practitioners can do better with science than without it. So I encourage the use of scientific approaches for preventative IRM.

Someday, many more scientists will realize the intellectual challenge of resistance management. They will realize that evolution is happening all around us in response to traditional approaches to pest management, such as biological control and crop rotation, and even traditional plant breeding and pheromone-based mating disruption. Brilliant scientists will realize that focusing on simple or abstract systems is insignificant compared to the excitement of studying real microevolutionary systems. I hope professors describe these exciting challenges to them sooner rather than later.

REFERENCES

Alyokhin, A., Chen, Y.H., Udalov, M., Benkovskaya, G., Lindstrom, L., 2013. Evolutionary considerations in potato pest management. In: Alyokhin, A., Vincent, C., Giordanengo, P. (Eds.), Insect Pests of Potato: Global Perspectives in Biology and Management.. Academic Press, Oxford, UK, Chapter 19.

Boivin, T., Bouvier, J.C., Beslay, D., Sauphanor, B., 2003. Phenological segregation of insecticide resistance alleles in the codling moth *Cydia pomonella* (Lepidoptera: Tortricidae): a case study of ecological divergences associated with adaptive changes in populations. Gen. Res. 81, 169–177.

Boivin, T., Bouvier, J.C., Beslay, D., Sauphanor, B., 2004. Variability in diapause propensity within populations of a temperate insect species: interactions between insecticide resistance genes and photoperiodism. Biol. J. Linn. Soc. 83, 341–351.

Boivin, T., Chadoeuf, J., Bouvier, J.C., Beslay, D., Sauphanor, B., 2005. Modelling the interactions between phenology and insecticide resistance genes in the codling moth *Cydia pomonella*. Pest Manag. Sci. 61, 53–67.

Caprio, M.A., 1994. *Bacillus thuringiensis* gene deployment and resistance management in single- and multi-tactic environments. Biocontrol. Sci. Tech. 4, 487–497.

Caprio, M.A., 1998. Evaluating resistance management strategies for multiple toxins in the presence of external refuges. J. Econ. Entomol. 91, 1021–1031.

Caprio, M.A., Nowatzki, T.M., Siegfried, B.D., Meinke, L.J., Wright, R.J., Chandler., L.D., 2006. Assessing risk of resistance to aerial applications of methyl-parathion in western corn rootworm (Coleoptera: Chrysomelidae). J. Econ. Entomol. 99, 483–493.

Carrière, Y., 2003. Haplodiploidy, sex, and the evolution of pesticide resistance. J. Econ. Entomol. 96, 1626–1640.

Carrière, Y., Tabashnik, B.E., 2001. Reversing insect adaptation to transgenic insecticidal plants. Proc. Roy. Soc. Lond. B. 268, 1475–1480.

Carrière, Y., Roff, D.A., Deland, J.P., 1995. The joint evolution of diapause and insecticide resistance: a test of an optimality model. Ecology. 76, 1497–1505.

Carrière, Y., Ellers-Kirk, C., Harthfield, K., Larocque, G., Degain, B., Dutilleul, P., et al., 2012. Large-scale, spatially-explicit test of the refuge strategy for delaying insecticide resistance. Proc. Natl. Acad. Sci. USA. 109, 775–780.

Cox, P.G., Forrester, N.W., 1992. Economics of insecticide resistance management in *Heliothis armigera* (Lepidoptera: Noctuidae) in Australia. J. Econ. Entomol. 85, 1539–1550.

Croft, B.A., 1990. Developing a philosophy and program of pesticide resistance management. In: Roush, R.T., Tabashnik, B.E. (Eds.), Pesticide Resistance in Arthropods. Chapman and Hall, New York, Chapter 11.

Crowder, D.W., Carrière, Y., Tabashnik, B.E., Ellsworth, P.C., Dennehy, T.J., 2006. Modeling evolution of resistance to pyriproxyfen by the sweetpotato whitefly (Homoptera: Aleyrodidae). J. Econ. Entomol. 99, 1396–1406.

Crowder, D., Ellsworth, P., Naranjo, S., Tabashnik, B., Carriere, Y., 2012. Modeling resistance to juvenile hormone analogs: linking evolution, ecology, and management. In: Devillers, J. (Ed.), Juvenile Hormones and Juvenoids: Modeling Biological Effects and Environmental Fate. CRC Press, Boca Raton, FL, pp. 99–126.

Crowder, D.W., Ellsworth, P.C., Tabashnik, B.E., Carrière, Y., 2008. Effects of operational and environmental factors on evolution of resistance to pyriproxyfen in the sweetpotato whitefly (Hemiptera: Aleyrodidae). Environ. Entomol. 37, 1514–1524.

Curtis, C.F., Hill, N., Kasim, S.H., 1993. Are there effective resistance management strategies for vectors of human disease? Biol. J. Linn. Soc. 48, 3–18.

Ellsworth, P.C., Martinez-Carrillo, J.L., 2001. IPM for *Bemisia tabaci*: a case study from North America. Crop Prot. 20, 853–869.

Flexner, J.L., Theiling, K.M., Croft, B.A., Westigard, P.H., 1989. Fitness and immigration: factors affecting reversion of organotin resistance in the twospotted spider mite (Acari: Tetranychidae). J. Econ. Entomol. 82, 996–1002.

Flexner, J.L., Westigard, P.H., Hilton, R., Croft, B.A., 1995. Experimental evaluation of resistance management for two spotted spider mite (Acari: Tetranychidae) on southern Oregon pear: 1987–1993. J. Econ. Entomol. 88, 1517–1524.

Gould, F., 1984. Role of behavior in the evolution of insect adaptation to insecticides and resistant host plants. Bull. Entomol. Soc. Amer. 30, 34–41.

Gould, F., 1991. Arthropod behavior and the efficacy of plant protectants. Annu. Rev. Entomol. 36, 305–330.

Gould, F., 1998. Sustainability of transgenic insecticidal cultivars: integrating pest genetics and ecology. Annu. Rev. Entomol. 43, 701−726.

Head, G., Savinelli, C., 2008. Adapting insect resistance management programs to local needs. In: Onstad, D.W. (Ed.), Insect Resistance Management: Biology, Economics and Prediction, first ed. Academic Press, Burlington, MA, Chapter 5.

Imai, C., 1987. Control of insecticide resistance in a field population of houseflies *Musca domestica* by releasing susceptible flies. Res. Pop. Ecol. 29, 129−146.

Insecticide Resistance Action Committee, 2006. Prevention and management of insecticide resistance in vectors and pests of public health importance. IRAC, p. 50.

McCafferty, A.R., 1999. Resistance to insecticides in heliothine Lepidoptera: a global view. In: Denholm, I., Pickett, J.A., Devonshire, A.L. (Eds.), Insecticide Resistance: From Mechanisms to Management. CABI Publishing, Wallingford, UK, pp. 59−74.

Naranjo, S.E., Ellsworth, P.C., 2009. Fifty years of the integrated control concept: moving the model and implementation forward in Arizona. Pest Manag. Sci. 65, 1267−1286.

Nyrop, J.P., Foster, R.E., Onstad, D.W., 1986. The value of sample information in pest control decision making. J. Econ. Entomol. 79, 1421−1429.

Onstad, D.W., Crowder, D.W., Mitchell, P.D., Guse, C.A., Spencer, J.L., Levine, E., et al., 2003. Economics versus Alleles: balancing IPM and IRM for rotation-resistant western corn rootworm (Coleoptera: Chrysomelidae). J. Econ. Entomol. 96, 1872−1885.

Onstad, D.W., Buschman, L.L., 2006. Evaluation of oviposition deterrence in the management of resistance to transgenic corn by European corn borer (Lepidoptera: Crambidae). J. Econ. Entomol. 99, 2100−2109.

Onstad, D.W., Guse, C.A., Porter, P., Buschman, L.L., Higgins, R.A., Sloderbeck, P.E., et al., 2002. Modeling the development of resistance by stalk-boring Lepidoptera (Crambidae) in areas with transgenic corn and frequent insecticide use. J. Econ. Entomol. 95, 1033−1043.

Palumbo, J.C., Ellsworth, P.C., Dennehy, T.J., Nichols, R.L., 2003. Cross-commodity Guidelines for Neonicotinoids in Arizona. The University of Arizona Cooperative Extension, IPM Series No. 17, AZ1319−5/2003. Online at: <http://www.cals.arizona.edu/pubs/insects/az1319.pdf>.

Roush, R.T., 1989. Designing resistance management programs: how can you choose? Pestic. Sci. 26, 423−441.

Roush, R.T., 1994. Managing pests and their resistance to *Bacillus thuringiensis*: can transgenic crops be better than sprays?. Biocontrol. Sci. Tech. 4, 501−516.

Roush, R.T., 1997. Managing resistance to transgenic crops. In: Carozzi, N., Koziel, M. (Eds.), Advances in Insect Control: The Role of Transgenic Plants. Taylor & Francis, London, Chapter 15.

Roush, R.T., 1999. Strategies for resistance management. In: Hall, F.R., Menn, J.J. (Eds.), Methods in Biotechnology Vol. 5 Biopesticides: Use and Delivery. Humana Press, Totowa, NJ, Chapter 30.

Roush, R.T., Tingey, W., 1992. Evolution and management of resistance in the Colorado potato beetle, *Leptinotarsa decemlineata*. In: Denholm, I., Devonshire, A.L., Hollomon, D.W. (Eds.), Resistance 91: Achievements and Developments in Combatting Pesticide Resistance. Elsevier Science Publishers, Ltd., Essex, England, pp. 61−74, p. 367.

Schmidt, C.H., LaBreoque, G.C., 1959. Acceptability and toxicity of poisoned baits to house flies resistant to organophosphorous insecticides. J. Econ. Entomol. 52, 345−346.

Siegfried, B.D., Meinke, L.J., Scharf, M.E., 1998. Resistance management concerns for areawide management programs. J. Agric. Entomol. 15, 359−369.

Tabashnik, B.E., Dennehy, T.J., Carrière, Y., 2005. Delayed resistance to transgenic cotton in pink bollworm. Proc. Nat. Acad. Sci. USA. 102, 15389−15393.

Waldstein, D.E., Reissig, W.H., Nyrop, J.P., 2001. Larval movement and its potential impact on the management of the obliquebanded leafroller (Lepidoptera:Tortricidae). Can. Ent. 133, 687−696.

Index

A

Printed and bound by CPI Group (UK) Ltd, Croydon, CR0 4YY

03/10/2024

01040416-0011